LIFE-CYCLE ANALYSIS AND ASSESSMENT IN CIVIL ENGINEERING: TOWARDS AN INTEGRATED VISION

PROCEEDINGS OF THE SIXTH INTERNATIONAL SYMPOSIUM ON LIFE-CYCLE CIVIL ENGINEERING (IALCCE 2018), 28–31 OCTOBER 2018, GHENT, BELGIUM

Life-Cycle Analysis and Assessment in Civil Engineering
Towards an Integrated Vision

Editors

Robby Caspeele & Luc Taerwe
Ghent University, Ghent, Belgium

Dan M. Frangopol
Lehigh University, Bethlehem, PA, USA

CRC Press is an imprint of the
Taylor & Francis Group, an **informa** business

A BALKEMA BOOK

CRC Press/Balkema is an imprint of the Taylor & Francis Group, an informa business

© 2019 Taylor & Francis Group, London, UK

Except for the following contributions, where copyright lies with the authors:

"Pocket-Monitoring" for the fatigue safety verification of a RC bridge deck slab
I. Bayane & E. Brühwiler

Fatigue safety verification of a steel railway bridge using short term monitoring data
B. Sawicki, E. Brühwiler & M. Nesterova

The effects of deterioration models on the value of damage detection information
L. Long, S. Thöns & M. Döhler

Fatigue reliability analysis of Cret De l'Anneau Viaduct: a case study
A. Mankar, S. Rastayesh & J.D. Sørensen

Typeset by MPS Limited, Chennai, India

All rights reserved. No part of this publication or the information contained herein may be reproduced, stored in a retrieval system, or transmitted in any form or by any means, electronic, mechanical, by photocopying, recording or otherwise, without written prior permission from the publishers.

Although all care is taken to ensure integrity and the quality of this publication and the information herein, no responsibility is assumed by the publishers nor the author for any damage to the property or persons as a result of operation or use of this publication and/or the information contained herein.

Published by: CRC Press/Balkema
Schipholweg 107C, 2316 XC Leiden, The Netherlands
e-mail: Pub.NL@taylorandfrancis.com
www.crcpress.com – www.taylorandfrancis.com

ISBN: 978-1-138-62633-1 (hardback + USB)
ISBN: 978-1-315-22891-4 (eBook)

Life-Cycle Analysis and Assessment in Civil Engineering: Towards an Integrated Vision – Caspeele, Taerwe & Frangopol (Eds)
© 2019 Taylor & Francis Group, London, ISBN 978-1-138-62633-1

Table of contents

Preface	XXIX
Sponsors, Supporting Associations and Academic Partners	XXXI
Symposium Organization	XXXV
Acknowledgements	XXXIX

FAZLUR R. KHAN LECTURE

Durability of bridges   3
M.C. Tang

KEYNOTE LECTURES

Optimal reliability-based aseismic design of high-rise buildings   11
A.H.-S. Ang, D. De Leon & W.L. Fan

Data standardization for life-cycle performance evaluation of a suspension bridge
with multi-pylons   16
A. Chen, D. Wang, Z. Pan, S. Xu & R. Ma

Promoting societal well-being by designing sustainable and resilient infrastructure:
Engineering tools and broader interdisciplinary considerations   23
P. Gardoni

Corrosion and its effects on deterioration and remaining safe life of civil infrastructure   29
C.-Q. Li

New architecture created from high performance structures   37
M.P. Sarkisian

Optimal planning of operation and maintenance for offshore wind turbines   38
J.D. Sørensen

Reliability assessment of deteriorating structures: Challenges and (some) solutions   45
D. Straub

Compliance testing for probabilistic durability design purposes   48
C. Thiel & C. Gehlen

MINI-SYMPOSIA

MS1: Load testing of new and existing structures
Organizers: A. de Boer, D.A. Hordijk, E.O.L. Lantsoght & Y. Yang

Critical proof load for proof load testing of concrete bridges based on
scripted FEM analysis   53
X. Chen, Y. Yang, P. Evangeliou & H. van der Ham

Proposed stop criteria for proof load testing of concrete bridges and verification   54
E.O.L. Lantsoght, C. van der Veen & D.A. Hordijk

Nonlinear finite element analysis of beam experiments for stop criteria   55
J.E. Paredes & E.O.L. Lantsoght

Load testing on a high railway bridge to determine the longitudinal stiffness
of the substructure   56
M. Wenner, F. Wedel, T. Meier & S. Marx

V

Bridge diagnostic load testing in Ecuador—case studies 57
J. Bonifaz, S. Zaruma, A. Robalino & T. Sánchez

Determination and assessment of the behavior of a semi-integral railway viaduct 58
M. Käding, M. Wenner, H. Liao & S. Marx

Development of a stop criterion for load tests based on the critical shear
displacement theory 59
K. Benitez, E.O.L. Lantsoght & Y. Yang

Verification of flexural stop criteria for proof load tests on concrete bridges based
on beam experiments 60
A. Rodriguez Burneo & E.O.L. Lantsoght

Load testing and rating of the KY 220 road bridge 61
A. Peiris, J. Hudson & I. Harik

Strength evaluation of prestressed concrete bridges by load testing 62
E.S. Hernandez & J.J. Myers

Static load deflection experiment on a beam for damage detection using the deformation
area difference method 63
D. Erdenebat, D. Waldmann & F.N. Teferle

Follow-up assessment on an old concrete road bridge based on operational
dynamic bridge behaviour—analysis of structural integrity and determination
of loading capacity 64
R. Veit-Egerer, J. Bursa & J. Synek

MS2: Vibration-based structural health monitoring, damage identification and residual lifetime estimation
Organizers: E.P.B. Reynders, G. Lombaert, E. Chatzi & C. Papadimitriou

Robust computation of the observability of large linear systems with
unknown parameters 67
X. Shi, M.N. Chatzis & M.S. Williams

Towards the use of UHPFRC in railway bridges: The rehabilitation of Buna Bridge 68
*H. Martín-Sanz, K. Tatsis, E. Chatzi, E. Brühwiler, I. Stipanovic, A. Mandic,
D. Damjanovic & A. Sanja*

Online tracking of inputs, states and parameters of structural dynamic systems 69
F. Karlsson, K. Maes & G. Lombaert

State estimation of geometrically non-linear systems using reduced-order models 70
K. Tatsis, L. Wu, P. Tiso & E. Chatzi

Modal strain-based damage identification on a prestressed concrete beam 71
D. Anastasopoulos, G. De Roeck & E.P.B. Reynders

Damage detection of steel bridge by numerical simulations and measurements 72
B.T. Svendsen, G.T. Frøseth & A. Rønnquist

Parametric identification and damage assessment of real buildings 73
G.S. Wang, Y.R. Li & F.K. Huang

Deep neural network for structural damage detection in bridges 74
D. Darsono & M. Torbol

Vibration-based damage indicators for corrosion detection in tubular structures 75
O.E. Esu, Y. Wang & M.K. Chryssanthopoulos

Pre-image reconstruction for compensation of environmental effects in structural health
monitoring by kernel PCA 76
C. Rainieri & E.P.B. Reynders

MS3: Probabilistic assessment of existing structures
Organizers: M. Sýkora, M. Holicky, R. Caspeele, D. Diamantidis & R.D.J.M. Steenbergen

Reducing semi-probabilistic methods to acceptable structural safety deficits in deterministic assessments of existing concrete structures
D. Zwicky — 79

The added value of full-probabilistic nonlinear finite element analysis for the assessment of reinforced concrete structural members
Á. Rózsás, A. Slobbe, D.L. Allaix, W.M.G. Courage, A. Bigaj van Vliet & H.G. Burggraaf — 80

Reliability assessment of infrastructure in Germany: Approaching a holistic concept
A. Panenka, F. Nyobeu, H. Schmidt-Bäumler, J. Sorgatz, R. Rabe & M. Reinhardt — 81

Probabilistic aspects relating to assessment of existing structures
M. Holicky — 82

Hidden safety in equilibrium verification of a steel bridge based on wind tunnel testing
J. Žitný, P. Ryjáček, J. Marková & M. Sýkora — 83

Analysis for repair effect in each layer of expressway pavements
S. Araki, T. Kazato, K. Kaito, K. Kobayashi & A. Tanaka — 84

Advancements in bridge specific assessment live loads for long span structures
N. Benham, C. Mundell & C.R. Hendy — 85

Probabilistic assessment of durability of hydroelectric power plants
J. Marková, K. Jung & M. Sýkora — 86

Reliability analysis of corroded reinforced concrete beam with regards to anchorage failure
M. Blomfors, D. Honfí, O. Larsson Ivanov, K. Zandi & K. Lundgren — 87

Historic roof structures: Life-cycle assessment and selective maintenance strategies
E. Garavaglia, N. Basso & L. Sgambi — 88

Development of an optimization-based and practice orientated assessment scheme for the evaluation of existing timber structures
M. Loebjinski, H. Pasternak, J. Köhler & W. Rug — 89

Reliability assessment of large hydraulic structures with spatially variable measurements
S. Geyer, I. Papaioannou, D. Straub & C. Kunz — 90

Uncertainties in the assessment of existing masonry structures
D. Müller & C.-A. Graubner — 91

MS4: Risk and reliability acceptance criteria
Organizers: D. Diamantidis, M. Faber, K. Fischer, M. Holicky, M. Sýkora,
J. Köhler & T. Vrouwenvelder

Obsolescence rate: Framework, analysis and influence on risk acceptance criteria
D. Diamantidis, M. Sýkora & E. Bertacca — 95

Reliability targets for semi-probabilistic design standards
J. Köhler — 96

Structural reliability analysis of wind turbines for wind and seismic hazards in Mexico
A. López López, L.E. Pérez Rocha, C.J. Muñoz Black, M.A. Fernández Torres & L.E. Pech Lugo — 97

Risk analysis of chemical industrial complex using parallel CUDA algorithms
M. Jeremiah & M. Torbol — 98

Acceptance criteria for structural design or assessment in accidental situations due to gas explosions
R. Hingorani, P. Tanner & C. Zanuy — 99

Safety requirements for the design of ancillary construction equipment
P. Tanner, R. Hingorani & J. Soriano — 100

Target reliability indices for existing quay walls derived on the basis of the LQI criterion 101
A.A. Roubos, D.L. Allaix, R.D.J.M. Steenbergen, K. Fischer & S.N. Jonkman

MS5: Early BIM for life-cycle performance
Organizers: P. Schneider & P. Geyer

Design-integrated environmental performance feedback based on early-BIM 105
A. Hollberg, I. Agustí-Juan, T. Lichtenheld & N. Klüber

Seamless integration of simulation and analysis in early design phases 106
A. Zahedi & F. Petzold

Early-design integration of environmental criteria for digital fabrication 107
I. Agustí-Juan, A. Hollberg & G. Habert

Efficient management of design options for early BIM 108
H. Mattern & M. König

The energy grey zone—uncertainty in embedded energy and greenhouse gas emissions
assessment of buildings in early design phases 109
H. Harter, P. Schneider-Marin & W. Lang

The early BIM adoption for a Contracting Authority: Standard and methods in the
ANAS approach 110
A. Osello, N. Rapetti & F. Semeraro

How to make decision-makers aware of sustainable construction? 111
H. Kreiner, M. Scherz & A. Passer

Information exchange scenarios between machine learning energy prediction model and
BIM at early stage of design 112
M.M. Singh, S. Singaravel & P. Geyer

A multi-LOD model representing fuzziness and uncertainty of building information
models in different design stages 113
J. Abualdenien & A. Borrmann

Intelligent substitution models for structural design in early BIM stages 114
D. Steiner & M. Schnellenbach-Held

MS6: Performance of concrete during life-cycle
Organizers: J. Li, X. Gao & X. Ren

Remaining life prediction of aged bridges under salt and carbonation damages by
concrete core tests 117
A. Miyamoto

Service life prediction of concrete under freeze-thaw deicing salt attack with
intermittent dry periods 118
C. Thiel, C. Gehlen & F. Foestl

Effect of choice of functional units on comparative life cycle assessment of concrete
mix designs 119
D. Kanraj, C.J. Churchill & D.K. Panesar

Evaluation of crack width in reinforced concrete beams subjected to variable load 120
T. Arangjelovski, G. Markovski & D. Nakov

Steel corrosion in ASR deteriorated concrete affected by de-icing salts 121
Y. Kubo, S.H. Ho, S. Kikuchi & Y. Ishikawa

Analysis of the influencing factors for shear capacity of reinforced concrete beam-column joints 122
X. Gao, D. Xiang, Y. He & J. Li

Study on physical property and durability of concrete with high content of blast furnace slag 123
K. Eguchi, Y. Kato & D.N. Katpady

Estimation of compressive strength at early age using electric conductivity 124
L.L. Chacha Costa, T. Shibuya & T. Iyoda

Investigating the effect of improving the low grade recycled aggregate by carbonation
A.A. Abdulkadeer, N. Matsuda & T. Iyoda
125

Influence of ASR expansion on concrete deterioration observed with neutron imaging of water
Y. Yoshimura, M. Mizuta, H. Sunaga, Y. Otake, Y. Kubo & N. Hayashizaki
126

Cement-based geotechnical elements exposed to chemical attack—provisions for durability in standards
F. Wagemann, F. Schmidt-Döhl & A. Rahimi
127

Shrinkage characteristics of ground granulated blast furnace slag high content cement
H. Mizuno & T. Iyoda
128

Performance evaluation on chlorine ion immobilizing ability of concrete using calcium aluminate aggregate and additive
Y. Nakanishi, S. Ito & T. Iyoda
129

Effect of chloride and corrosion of reinforcing steel on heat transfer of reinforced concrete
P. Sancharoen, D. Im, P. Julnipitawong & S. Tangtermsirikul
130

Long-term performance of cementitious materials used in nuclear waste disposal—a case study at SCK•CEN
Q.T. Phung, S. Seetharam, J. Perko, N. Maes, D. Jacques, S. Liu, E. Valcke, T. Seeman, G. Hoder, L. Lemmens, A. Varzina & R.A. Patel
131

Investigating the effect of surplus filler of asphalt plant on setting time and compressive strength of cement mortar
M. Kioumarsi & H. Vafaeinejad
132

Effect of initial curing temperature on mechanical strength of concrete
M. Kioumarsi & H. Vafaeinejad
133

A sustainable approach to the design of reinforced concrete beams
A.P. Fantilli, F. Tondolo, B. Chiaia & G. Habert
134

MS7: Life-cycle engineering for hydraulic structures, levees, and other water related infrastructure
Organizers: F. den Heijer, J. Wessels, H. Yokota & M. Hoffmann

Asset management maturity for flood protection infrastructure: A baseline across the North Sea region
B. Gersonius, R. Ashley, F. den Heijer, W.J. Klerk, P. Sayers & J. Rijke
137

Asset management of water and sewer networks, and levees: Recent approaches and current considerations
C. Curt, R. Tourment, Y. Le Gat & C. Werey
138

Working towards one and the same objective: How serious gaming makes storm surge barriers more reliable
M. Schelland, J.N. Huibregtse, M. Walraven & E.C.J. Bouwman
139

A framework for assessing information quality in asset management of flood defences
W.J. Klerk, R. Pot, J.M. van der Hammen & K. Wojciechowska
140

Fundamental study on effective utilization of caisson in breakwater
E. Kato, Y. Kawabata, K. Uno & H. Yokota
141

Efficient investments in the waterway Danube with vessel traffic analysis and fairway optimization
M. Hoffmann, A. Haberl, C. Konzel, T. Hartl, S. Simon & M. Simoner
142

Risk based inspection of flood defence dams: An application to grass revetments
W.J. Klerk, K.L. Roscoe, A. Tijssen, R.P. Nicolai, J. Sap & F. Schins
143

Comparative analysis of the reliability levels in hydraulic structures using partial safety factors and full probabilistic methods
A. Tahir & C. Kunz
144

A maintenance management plan for port mooring facilities based on cost-benefit
analysis—a case study 145
Y. Kawabata, E. Kato, Y. Tanaka & H. Yokota

Sustainable and future-proof port infrastructure 146
J.G. de Gijt, E.J. Broos, C. Bosschieter, P. Taneja & H.E. Pacejka

Robustness as a decision criterion for construction and evaluation of ship lock chambers 147
C. Kunz

Improvement of the planning process of flood protection assets by using experiences
with operation and maintenance—Hamburg case study 148
P. Fröhle, N. Manojlovic, S. Shaikh, P. Jordan, L. Seumenicht, M. Schaper, J.-C. Schmidt & M. Roth

Peer reviews as an asset management tool 149
B. Lassing, P. van Poorten, M. Walraven, S. van Herk, B. Vonk & R. Windsor

Expert interviews in long-term damage analysis for bottom and bank revetments along
German inland waterways 150
J. Sorgatz, J. Kayser & H. Schüttrumpf

MS8: IEA EBC Annex 72: Assessing life-cycle related environmental impacts caused by buildings
Organizers: R. Frischknecht, H. Birgitsdottir, T. Lützkendorf & A. Passer

Potential for interconnection of tools for cost estimation and life cycle assessment of partial
carbon footprint in the building sector in Czechia 153
A. Lupíšek, M. Nehasilová, J. Železná, P. Hájek, B. Pospíšilová & M. Hanák

Building life cycle assessment tools developed in France 154
B. Peuportier & P. Schalbart

LCA benchmarks for decision-makers adapted to the early design stages of new buildings 155
A. Hollberg, P. Vogel & G. Habert

Principles for the development and use of benchmarks for life-cycle related environmental
impacts of buildings 156
T. Lützkendorf & M. Balouktsi

Assessing life cycle related environmental impacts caused by buildings (IEA EBC Annex 72) 157
R. Frischknecht

Lessons learned from establishing environmental benchmarks for buildings in Switzerland 158
L. Tschümperlin & R. Frischknecht

Belgian approach to mainstream LCA in the construction sector 159
*D. Trigaux, K. Allacker, F. De Troyer, W. Debacker, W.C. Lam, L. Delem, L. Wastiels,
R. Servaes & E. Rossi*

Life cycle assessment benchmarks for Danish office buildings 160
F.N. Rasmussen & H. Birgisdóttir

Using BIM-based methods to obtain life cycle environmental benchmarks for buildings 161
A. García-Martínez, J.C. Gómez de Cózar & M. Ruiz Alfonsea

Development of a simplified methodology for creating embodied energy database of construction
materials and processes in India 162
S. Palaniappan & V. Bindu Inti

The coupling of BIM and LCA—challenges identified through case study implementation 163
M. Röck, A. Passer, D. Ramon & K. Allacker

Environmental impacts of future electricity production in Hungary with reflect on building
operational energy use 164
B. Kiss, Zs. Szalay & E. Kácsor

Modular methodology for building life cycle assessment for a building stock model 165
Zs. Szalay & B. Kiss

X

Effects of LCA impact categories and methodology on the interpretation of a
building's environmental performance 166
L. Delem, L. Wastiels & K. Allacker

Analyzing the life cycle environmental impacts in the Chinese building design process 167
W. Yang, Q.Y. Li, L. Yang, J. Ren & X.Q. Yang

Review of existing service lives' values for building elements and their sensitivity on
building LCA and LCC results 168
S. Lasvaux, M. Giorgi, D. Favre, A. Hollberg, V. John & G. Habert

Improving the reliability and specificity of an Input-Output-based Hybrid (IOH) method
for computing embodied energy of a building 169
M.K. Dixit & V. Venkatraj

Life cycle assessment of alternative masonry to concrete blocks 170
J. Dahmen, J. Kim & C.M. Ouellet-Plamondon

MS9: Multi-hazard resilience assessment in a life-cycle context
Organizers: J. Ghosh & J.E. Padgett

Resilience assessment of highway bridges under multiple natural hazards 173
Y. Dong & D.M. Frangopol

A discussion on the need for flexibility in infrastructure development 174
S. Torres & M. Sánchez-Silva

Modeling the resilience of aging concrete bridge columns subjected to
corrosion and extreme climate events 175
H. Almansour, A. Mohammed & Z. Lounis

Uncertainty analysis and impact on seismic life-cycle cost assessment of highway
bridge structures 176
S. Shekhar & J. Ghosh

Life-cycle based resilience assessment of bridges under seismic hazard 177
B. Sharanbaswa & S. Banerjee

Stochastic modeling of post-repair performance and integration into bridge life-cycle assessment 178
N. Vishnu & J.E. Padgett

Probabilistic resilience assessment of infrastructure—a review 179
C. de Paor, L. Connolly & A. O'Connor

MS10: Next generation asset management of civil infrastructure systems
Organizers: A. Michel, H. Stang, M.R. Geiker & M.D. Lepech

Comparison of optimization approaches for pavement maintenance and rehabilitation
policies on road section and network level 183
V. Donev & M. Hoffmann

Coupled mass transport, chemical, and mechanical modelling in cementitious materials:
A dual-lattice approach 184
A. Michel, V.M. Meson, H. Stang, M.R. Geiker & M. Lepech

Measuring critical chloride contents in structures and the influence on service life modeling 185
C. Boschmann Käthler, U.M. Angst & B. Elsener

A framework for modeling corrosion-related degradation in reinforced concrete 186
Z. Zhang, U.M. Angst & A. Michel

Design and maintenance of concrete structures requires both engineering and sustainability
limit states 187
M.R. Geiker, A. Michel, H. Stang, H. Vikan & M.D. Lepech

Long-term planning within complex and dynamic infrastructure systems 188
M. Havelaar, W. Jaspers, A.R.M. Wolfert, G.A. van Nederveen & W.L. Auping

Scheduling of waterways maintenance interventions applying queueing theory 189
F. Marsili, J. Bödefeld, H. Daduna & P. Croce

MS11: Life-cycle performance of structure and infrastructure under uncertainty
Organizers: M. Akiyama & D.M. Frangopol

Hazard analysis for bridge scour evaluation at watershed level considering climate change impact 193
D.Y. Yang & D.M. Frangopol

Sensitivity analysis of time dependent reliability of RC members in general climate environment 194
D.-G. Lu, W.-H. Zhang & W. Wang

Seismic fragility of corroding RC bridge substructures in marine environment 195
F. Cui & M. Ghosn

Time-dependent structural reliability analysis of shield tunnels in coastal regions 196
Z. He, M. Akiyama, C. He & D.M. Frangopol

Non linear structural analyses of prestressed concrete girders: Tools and safety formats 197
B. Belletti, F. Vecchi, M.P. Cosma & A. Strauss

Digitalization is the road to collaboration: A case study regarding the use of digital
support tools on sewage and water pumping systems at Schiphol and Haarlem 198
W. Jaspers, L.S. van Duffelen & J.-J. de Jong

Sensitivity analysis of fatigue lifetime predictions of pre-stressed concrete bridges
using Sobol'-indices 199
D. Sanio, M.A. Ahrens & P. Mark

A comparative study on fatigue assessment of orthotropic steel decks based on
long-term WIM data 200
B. Wang, A. Chen & H. De Backer

Reliability analysis for geotechnical structures using iterative particle filter 201
T. Shuku & I. Yoshida

Optimal inspection planning based on Value of Information for airport runway 202
Y. Tasaki & I. Yoshida

A sampling-based approach to identifying optimal inspection and repair strategies for
offshore jacket structures 203
R. Schneider, A. Rogge, S. Thöns, E. Bismut & D. Straub

Numerical durability simulation of reinforced concrete structures under consideration
of polymorphic uncertain data 204
K. Kremer, P. Edler, S. Freitag, M. Hofmann & G. Meschke

MS12: Life-cycle redundancy, robustness and resilience indicators for aging structural systems under multiple hazards
Organizers: F. Biondini & D.M. Frangopol

Influence of seismic hazard on RC buildings' resilience based on ANN 207
G. Bunea, F. Leon & G.M. Atanasiu

Evaluation of road network performance considering capacity degradation on numerous links 208
H. Nakajima & R. Honda

Probabilistic life-cycle resilience assessment of aging bridges and road networks under
seismic and environmental hazards 209
L. Capacci & F. Biondini

Innovative methodology of assessing the residual structural safety margin of reinforced
concrete structures—application to cooling towers 210
N.C. Tran, C. Toulemonde, F. Beaudouin & C. Mewisse

Life-cycle seismic performance prediction of deteriorating RC structures using
artificial neural networks 211
S. Bianchi & F. Biondini

Modelling of physical systems for resilience assessment 212
G. Tsionis, A. Caverzan, E. Krausmann, G. Giannopoulos, L. Galbusera & N. Kourti

Robustness analysis of 3D base-isolated systems 213
P. Castaldo, D. Gino & G. Mancini

MS13: Advances in Structural Health Monitoring for real-world applications
Organizers: C.W. Kim, P.J. McGetrick, A. Cunha & T. Kitahara

Identification of damaged regions in dynamically loaded dams 217
M. Alalade, T. Lahmer & F. Wuttke

Installation and results from the first 18 months of operation of the dynamic monitoring
system of Baixo Sabor arch dam 218
S. Pereira, F. Magalhães, A. Cunha, J. Gomes & J.V. Lemos

Monitoring of corrosive environment focusing on dew condensation in steel bridges 219
Z. Rasoli, K. Nagata & T. Kitahara

Optimising circumferential piezoelectric transducer arrays of pipelines through
linear superposition analysis 220
X. Niu, H.P. Chen & H.R. Marques

Particle Swarm Optimization for damage identification in beam-like structures 221
A. Barontini, M.-G. Masciotta, L.F. Ramos, P.B. Lourenço & P. Amado-Mendes

Experimental investigation on crack detection using imbedded smart aggregate 222
C. Du, Y. Yang & D.A. Hordijk

Development of a remote monitoring system with wireless power-saving sensors
for analyzing bridge conditions 223
E. Sasaki, P. Tuttipongsawat, N. Sinsamutpadung, H. Nishida & K. Takase

Investigation of Bayesian damage detection method for long-term bridge health monitoring 224
Y. Goi & C.W. Kim

A trial vibration measuring for evaluating performance of small bridge with small FWD system 225
H. Onishi, K. Ouchi, N. Kimura & D. Yaegashi

Structural health monitoring of a steel girder bridge utilizing reconstructed
sparse-like system matrix 226
T. Mimasu, C.W. Kim & Y. Goi

Damage identification of bridge structures using the Hilbert-Huang Transform 227
J.J. Moughty & J.R. Casas

Experimental studies on the feasibility of drive-by bridge inspection method using
an appropriate vehicle model 228
S. Nakajima, C.W. Kim, K.C. Chang & S. Hasegawa

MS14: Monitoring of structures for informed-decision making
Organizers: A. Strauss & D.M. Frangopol

RAMS evaluation for the steel-truss arch high-speed railway bridge based on SHM system 231
Y.L. Ding, H.W. Zhao & A.Q. Li

Crowd load prediction on pedestrian bridges using Fiber Bragg Grating Sensors 232
K. Hassoun, J. Karaki, S. Mustapha, A. Kassir, Z. Dawy & H. Abi-Rached

A case study of structural monitoring as a control in the restoration process of heritage structures:
The strengthening of the Vistabella Church's Tower 233
M. Llorens, R. Señís, S. Pavón & B. Moreno

XIII

Levels of assessment for chloride model parameters of existing concrete structures 234
A. Strauss, C. Matzenberger, K. Bergmeister, M. Somodikova & T. Zimmermann

MS15: Probability-based service life design of reinforced concrete structures exposed to reinforcement corrosion
Organizers: G. De Schutter & S. Keßler

Evaluation of the chloride migration and carbonation coefficients of Belgian ready mixed and precast concrete for a performance-based design 237
P. Minne, E. Gruyaert, L. De Winter, B. Craeye, R. Caspeele & G. De Schutter

Probabilistic evaluation of service-life of RC structures subjected to carbonation 238
R.A. Couto & S.M.C. Diniz

Durability design of concrete structures regarding chloride-induced corrosion by means of nomograms 239
A. Rahimi

Sensitivity analysis of a service life model linked to chloride induced corrosion focusing on the critical chloride content 240
G. Kapteina

The importance of the size effect in corrosion of steel in concrete for probabilistic service life modeling 241
U.M. Angst

Evaluation of half-cell potential measurement and its impact on the condition assessment 242
S. Keßler

Serviceability and residual working life assessment of existing bridges 243
J. Marková & V. Navarova

MS16: Life-cycle maintenance and management for urban infrastructures with big data
Organizers: A. Chen, Y. Yuan & X. Ruan

Performance comparison for pipe failure prediction using artificial neural networks 247
S. Kerwin, B. Garcia de Soto & B.T. Adey

Long-term mechanical reliability evaluation of the main cable of a suspension bridge 248
D. Wang, Y. Zhang, A. Chen, L. Li & H. Tian

Empirical Bayes-based Markov chain deterioration modelling for municipal sewer systems 249
P. Lin, X.X. Yuan & R. Rashedi

Incremental launching construction of Chajiaxia Yellow River Bridge with data feedback 250
L. Zhou, J. Peng, Y. Wu & Z. Yin

Bayesian formula based dynamic information updating of bridge traffic flow response probability model 251
X.J. Wang, X. Ruan & K.P. Zhou

Experimental study on the mechanical property evolution of main cable steel wires of suspension bridges under corrosion state 252
R. Ma, C. Cui, A. Chen, L. Li & H. Tian

A risk management system of Hainan interchange merge area in maintenance zone: Risk analysis, identification, assessment and countermeasure 253
B. Liu, H. Yan & W. Zhao

Buckling analysis of a long span steel cable-stayed bridge 254
Y. Wei, Y.T. He, X. Ruan & L. Ding

MS17: Life-cycle management as focus area within asset management
Organizers: J. Bakker, H. Roebers, M. Hertogh & J.F.M. Wessels

Quantifying the impact of variability in railway bridge asset management 257
P.C. Yianni, L.C. Neves, D. Rama, J.D. Andrews, N. Tedstone & R. Dean

Quantifying the Performance Age of highway bridges 258
Y. Xie, D. Schraven, J. Bakker & M. Hertogh

A simplified approach to address uncertainty in Life Cycle Costing (LCC) analysis 259
Y. Sun & D.G. Carmichael

Life cycle approach for sustainable pavement options for infrastructure projects 260
B. Czarnecki

Comparison of truck fuel consumption measurements with results of existing models and
implications for road pavement LCA 261
F. Perrotta, T. Parry, L.C. Neves, T. Buckland, E. Benbow & H. Viner

Technical management risks for transport infrastructures along whole life cycle:
Identification and analysis 262
D. García-Sánchez, J. Aurtenetxe, M. Zalbide, R. Socorro, A. Pérez-Hernando & D. Inaudi

Computational framework for a railway bridge maintenance strategies affected by
gradual deterioration 263
J. Fernandes, J.C. Matos, D.V. Oliveira & A.A. Henriques

A many-objective optimization model for sustainable pavement management considering
several sustainability metrics through a multi-dimensionality reduction approach 264
J. Santos, V. Cerezo, G. Flintsch & A. Ferreira

Expert-driven and data-driven risk-centred maintenance decision-making approaches
for railway transport assets 265
F. Dinmohammadi

Evaluation and application of AHP, MAUT and ELECTRE III for infrastructure management 266
Z. Allah Bukhsh, I. Stipanovic, A. Hartmann & G. Klanker

Assessing approximation errors caused by truncation of cash flows in public infrastructure
net present value calculations 267
R. Treiture, L. van der Meer, J. Bakker, M. Van den Boomen, R. Schoenmaker & R. Wolfert

A survey of health monitoring techniques for the Dutch transportation infrastructure 268
J.F.M. Wessels, P.J. van der Mark, K.E. Bektas, J. Bakker & M. van der Voort

Considerations on the use of data to better predict long term replacement and renovation
activities in transport infrastructure 269
J.F.M. Wessels, H. van Meerveld, J. Bakker & K.E. Bektas

Life cycle cost analysis for short span bridges in Indiana 270
S.L. Leiva & M.D. Bowman

Lifecycle management and replacement strategies: Two of a kind? 271
M. Zandvoort, M.J. van der Vlist, R. Haitsma & E. Oosterveld

A holistic approach to Life Cycle Management at Rijkswaterstaat 272
J. Bakker & H. Roebers

Lessons learned from data analytics, applied to the track maintenance of the
Dutch high speed line 273
R. Schalk, A. Zoeteman & A. Núñez

Asset information management using linked data for the life-cycle of roads 274
B. Luiten, M. Böhms, D. Alsem & A. O'Keeffe

Mechanisms for managing changes in construction projects 275
H.Ç. Demirel, L. Volker, W. Leendertse & M. Hertogh

MS18: Serviceability of underground structures
Organizers: Y. Yuan, E. Bilotta, H. Yu & Q. Ai

The effect of hydrogen embrittlement on durability of buried steel pipes 279
M. Wasim, M. Mahnoodian, D. Robert & C.-Q. Li

XV

Rapid detecting equipment for structural defects of metro tunnel 280
K. Wang & X. Yao

A state-oriented maintenance strategy of tunnel structure 281
Q. Ai, Y. Yuan & X. Jiang

Changes in rheology of printable concrete during pumping process 282
Y. Yuan, Y. Tao & X. Wang

Settlement control, monitoring and analysis of utility tunnel on soft soil foundation 283
J. Huang, H. Wang & J. Wang

Dynamic soil normal stresses on side wall of a subway station 284
Z.M. Zhang, Y. Yuan, E. Bilotta, H.T. Yu & H.L. Zhao

Shaking table model tests on tunnels at different depths 285
X. Zhao, R.H. Li, M. Zhao & L.J. Tao

Two-dimensional finite element analysis of the seismic performance in complex
underground structure 286
X.S. Cai, Z. Ye & Y. Yuan

Shaking table test on shaft ingate of shield tunnel 287
J. Zhang, X. Tu, X. Zhang, F. Li & Y. Yuan

MS19: Circular economy to improve sustainability of infrastructure
Organizers: S. de Vos-Effting & R. Hofman

Are recycled and low temperature asphalt mixtures more sustainable? 291
M. Hauck, E. Keijzer, H. van Meerveld, B. Jansen, S. de Vos-Effting & R. Hofman

Relevance of the information content in module D on circular economy of building materials 292
K. Krause & A. Hafner

PAPERCHAIN project: Establishment of new circular economy models between pulp
and paper industry and construction industry to create sustainable infrastructures 293
M.S. Martín-Castellote, J.J. Cepriá-Pamplona, M.M. Pintor Escobar & E. Guedella-Bustamante

Introducing the circular economy in road construction 294
W.L. Leendertse, M.E.M. Schäffner & S. Kerkhofs

CEO & CAMO ontologies: A circulation medium for materials in the construction industry 295
E.M. Sauter, R.L.G. Lemmens & P. Pauwels

SUP&R DST: SUstainable Pavement & Railways Decision Support Tool 296
J. Santos, S. Bressi, V. Cerezo & D. Lo Presti

SPECIAL SESSIONS

SS1: Structural Health Monitoring and decision making for infrastructures in multi-hazard environment
Organizers: M.P. Limongelli, J.R. Casas, M.G. Stewart & B. Imam

Stochastic differential equations for modeling deterioration of engineering systems and
calibration based on Structural Health Monitoring data 299
L. Iannacone & P. Gardoni

Information requirements for effective management of an ageing transport network 300
J.H. Paulissen, S.H.J. van Es, W.H.A. Peelen & H.E. Klatter

ROC-based performance analysis and interpretation of image-based damage diagnostic
tools for underwater inspections 301
M. O'Byrne, V. Pakrashi, F. Schoefs & B. Ghosh

Real-time monitoring as a non-structural risk mitigation strategy for river bridges 302
F. Ballio, G. Crotti & A. Cigada

Fiber optic sensing in an integrated Structural Health Monitoring system 303
R. Blin & D. Inaudi

Structural and climate performance indicators in service life prediction of concrete
bridges in multi-hazard environment 304
M. Kušter Marić, A. Mandić Ivanković & J. Ožbolt

SS2: Climate adaptation engineering
Organizers: E. Bastidas-Arteaga, M.G. Stewart & Y. Li

Balancing payoff and regret: A bi-objective formulation for the optimal adaptation
of riverine bridges under climate change 307
A. Mondoro & D.M. Frangopol

Modeling the climate change effects on storm surge with metamodels 308
A. Contento, H. Xu, P. Gardoni & S. Guerrier

Impact of climate change on optimal wood pole asset management 309
A.M. Salman, Y. Li & E. Bastidas-Arteaga

Climate change impact on safety and performance of existing and future bridges 310
A. Nasr, O. Larsson Ivanov, I. Björnsson, J. Johansson, D. Honfi & E. Kjellström

A tool to evaluate effectiveness of climate change adaptation measures for houses subjected
to coastal flood risks 311
A. Creach, M. Gonzva, E. Bastidas-Arteaga, S. Pardo & D. Mercier

Evaluating the effect of climate change on thermal actions on structures 312
P. Croce, P. Formichi, F. Landi & F. Marsili

SS3: Quality control procedures on the life-cycle management
of existing bridges
Organizers: J.C. Matos & J.R. Casas

COST Action TU1406 and main results on bridge lifecycle management 315
J.C. Matos & J.R. Casas

Performance based design and assessment—levels of indicators 316
A. Strauss, L. Mold, K. Bergmeister, A. Mandic, J.C. Matos & J.R. Casas

The indicator readiness level for the classification of research performance indicators
for road bridges 317
M.P. Limongelli, A. Orcesi & A. Vidovic

Effects of multivariate data reduction of condition assessment of bridge networks on the
Value of Information 318
L. Quirk, C.M. Hanley, J.C. Matos & V. Pakrashi

Regular bridge inspection data improvement using non-destructive testing 319
M. Kušar, N. Galvão & S. Sein

First results from a benchmarking of Quality Control Frameworks 320
A. Kedar & S. Sein

Standardizing the quality control of existing bridges 321
V. Pakrashi & H. Wenzel

The case study of Chile—how quality control could improve better lifecycle
management of bridges 322
M.A. Valenzuela

SS4: Modeling time-dependent behavior and deterioration of concrete
Organizers: R. Wan-Wendner, M. Alnagger, G. Di Luzio & G. Cusatis

Size and shape effect in shrinkage based on chemo-mechanical simulations 325
L. Czernuschka, I. Boumakis, J. Vorel & R. Wan-Wendner

Three types of errors in the international norms for the design of concrete and reinforced concrete 326
R.S. Sanjarovskiy, T.N. Ter-Emmanuilyan & M.M. Manchenko

XVII

Deflections of reinforced concrete beams made with recycled and waste materials under sustained
load: Experiment and *fib* Model Code 2010 predictions 327
N. Tošić, S. Marinković, I. Ignjatović & A. de la Fuente

An investigation into influential factors affecting the time to concrete cover cracking
in reinforced concrete structures 328
F. Chen, H. Baji & C.-Q. Li

Development of fatigue life prediction for RC slabs under traveling wheel-type loading 329
K. Takeda & Y. Sato

Early damage detection of fastening systems in concrete under dynamic loading—model
details and health monitoring framework 330
M. Hoepfner & P. Spyridis

Study on the time variant alteration of chloride profiles for prediction purpose 331
F. Binder, S.L. Burtscher & A. Limbeck

A rapid numerical simulation method of chloride ingress in concrete material 332
Y. Li, X. Ruan & Z.R. Jin

Concrete cover cracking under chloride-induced time-varying non-uniform steel corrosion 333
J. Zhang, P. Wang & Z. Guan

Analysis of coupled exposures considering the rapid chloride migration test and the
accelerated carbonation test 334
M. Vogel, S. Schmiedel & H.S. Müller

Modelling of corrosion induced cracking in reinforced concrete 335
I. Lau, G. Fu, C.-Q. Li & S. De Silva

Optimization of service life design of concrete infrastructures in corrosive environments
under a changing climate 336
Z. Lounis

Simulation of crack propagation owing to deformed bar corrosion 337
S. Okazaki, C. Okuma, H. Yoshida & M. Kurumatani

SS6: TRUSS ITN—reducing uncertainty in structural safety
Organizer: A. González

Mechanical characterisation of braided BFRP rebars for internal concrete reinforcement 341
S. Antonopoulou, C. McNally & G. Byrne

Use of post-installed screws in the compressive strength assessment of in-situ concrete 342
M.S.N.A. Sourav, S. Al-Sabah & C. McNally

Impact of input variables on the seismic response of free-standing spent fuel racks 343
A. Gonzalez Merino, L. Costas de la Peña & A. González

Surrogate infill criteria for operational fatigue reliability analysis 344
R. Teixeira, A. O'Connor & M. Nogal

Probabilistic decision basis and objectives for inspection planning and optimization 345
G. Zou, K. Banisoleiman & A. González

Characterization of hoisting operations on the dynamic response of the lifting boom
of a ship unloader 346
G. Milana, K. Banisoleiman & A. González

Monitoring crack movement on a masonry type abutment using optical camera
system—a case study 347
F. Huseynov, E. O'Brien, J. Brownjohn, K. Faulkner, Y. Xu & D. Hester

Outlier detection of point clouds generating from low-cost UAVs for bridge inspection 348
S. Chen, L.C. Truong-Hong, E. O'Keeffe, D.F. Laefer & E. Mangina

Using step-by-step Bayesian updating to better estimate the reinforcement loss due to
corrosion in reinforced concrete structures 349
F. Schoefs, B. Heitner, T. Yalamas, G. Causse & E.J. O'Brien

Structural health monitoring of bridges: A Bayesian network approach 350
M. Vagnoli, R. Remenyte-Prescott & J. Andrews

Noninvasive empirical methods of damage identification of bridge structures using vibration data 351
J.J. Moughty & J.R. Casas

Bridge condition evaluation using LDVs installed on a vehicle 352
A.D. Martínez Otero, A. Malekjafarian & E.J. O'Brien

On the bonding performance of Distributed Optical Fiber Sensors (DOFS) in structural concrete 353
A. Barrias, J.R. Casas & S. Villalba

A machine learning approach for the estimation of fuel consumption related to road
pavement rolling resistance for large fleets of trucks 354
F. Perrotta, T. Parry, L.C. Neves & M. Mesgarpour

Fuzzy-random approach to debris model for riverbed scour depth investigation at bridge piers 355
L. Sgambi, N. Basso & E. Garavaglia

SS7: Application of probabilistic methods in fire safety engineering
Organizers: D. Rush, L. Bisby & R. Van Coile

Effect of modelling on failure probabilities in structural fire design 359
M. Shrivastava, A.K. Abu, R.P. Dhakal & P.J. Moss

Numerical analysis on the fire behavior of a steel truss structure 360
L.M. Lu, G.L. Yuan, Q.J. Shu & Q.T. Li

The application of an LQI reliability based methodology to determine the fire resistance
requirements for two Mumbai residential towers 361
D. Hopkin, S. Lay & A. Henderson

Target safety levels for insulated steel beams exposed to fire, based on Lifetime
Cost Optimisation 362
R. Van Coile & D.J. Hopkin

SS8: Bespoke models for marine structural management
Organizer: M. Collette

Integrated Computational Materials Engineering (ICME) techniques to enable a
materials-informed digital twin prototype for marine structures 365
C.R. Fisher, K. Nahshon, M.F. Sinfield & D. Kihl

Adapting life-cycle management of ship structures under fatigue considering
uncertain operation conditions 366
Y. Liu & D.M. Frangopol

Probabilistic service life management of fatigue sensitive ship hull structures
considering various sea loads 367
S. Kim & D.M. Frangopol

Ship motion and fatigue damage estimation via a digital twin 368
M. Schirmann, M. Collette & J. Gose

SS9: Novel materials and systems for life-cycle structural health monitoring
Organizers: V. Pakrashi, P. Cahill & P. Michalis

Innovative soft-material sensor, wireless network and assessment software for
bridge life-cycle assessment 371
K. Loupos, Y. Damigos, A. Tsertou, A. Amditis, S. Lenas, C. Chatziandreoglou, C. Malliou,
V. Tsaoussidis, R. Gerhard, D. Rychkov, W. Wirges, B. Frankenstein, S. Camarinopoulos,
V. Kalidromitis, C. Sanna, S. Maier, A. Gordt & P. Panetsos

The potential of energy harvesting for monitoring corroding metal pipes
F. Okosun & V. Pakrashi
372

SS10: Value of structural health monitoring information for the life-cycle management of civil structures
Organizers: S. Thöns, G. Lombaert & M.P. Limongelli

Cost-based optimization of the performance of a damage detection system
A.C. Neves, J. Leander, R. Karoumi & I. González
375

The integration of bridge life cycle cost analysis and the value of structural health monitoring information
G. Du & J. Qin
376

The value of monitoring the service life prediction of a critical steel bridge
J. Leander & R. Karoumi
377

Structural monitoring and inspection modeling for structural system updating
A. Agusta & S. Thöns
378

The effects of deterioration models on the value of damage detection information
L. Long, S. Thöns & M. Döhler
379

The value of visual inspections for emergency management of bridges under seismic hazard
M.P. Limongelli, S. Miraglia & A. Fathi
380

A Bayesian network based approach for integration of condition-based maintenance in strategic offshore wind farm O&M simulation models
J.S. Nielsen, J.D. Sørensen, I.B. Sperstad & T.M. Welte
381

Structural integrity management with unmanned aerial vehicles: State-of-the-art review and outlook
M. Kapoor, E. Katsanos, S. Thöns, L. Nalpantidis & J. Winkler
382

Metamodeling strategies for value of information computation
M.S. Khan, S. Ghosh, J. Ghosh & C. Caprani
383

SS11: Design for robustness of steel and steel-concrete composite structures
Organizers: J.-F. Demonceau & J.-P. Jaspart

Robustness of steel structures subjected to a column loss scenario
J.-F. Demonceau, M. D'Antimo & J.-P. Jaspart
387

Investigation of the column loss scenario of one composite steel and concrete frame
G. Roverso, N. Baldassino & R. Zandonini
388

Design of steel and composite structures for robustness
N. Hoffmann, U. Kuhlmann & G. Skarmoutsos
389

Behaviour of an innovative joint solution under impulsive loading
M. D'Antimo, J.-F. Demonceau & J.-P. Jaspart
390

Influence analysis of group studs stiffness in accelerated construction steel-concrete composite small box girder bridges
Y. Xiang & S. Guo
391

Development of a design-oriented structural robustness index for progressive collapse
C. Praxedes & X.-X. Yuan
392

Performance metrics for seismic-resilient steel braced frame buildings
O. Serban & L. Tirca
393

Post-failure torsion capacity and robustness of encased tubular arch spring connections
Ph. Van Bogaert, K. Schotte & H. De Backer
394

SS12: Repair and self-repair of concrete
Organizers: N. De Belie, K. Van Tittelboom, D. Snoeck & E. Gruyaert

Optimizing nutrient content of microbial self-healing concrete   397
Y.Ç. Erşan & Y. Akın

Microencapsulated spores and growth media for self-healing mortars   398
K. Paine, I. Horne, L. Tan, T. Sharma, A. Heath, R. Cooper, J. Virgoe, D. Palmer & A. Kerr

Use of fiber-reinforced self-healing cementitious materials with superabsorbent polymers
to absorb impact energy   399
D. Snoeck, T. De Schryver, P. Criel & N. De Belie

The role of silicate salts in self-healing properties of cement pastes   400
M. Stefanidou, V. Kotrotsiou & F. Kesikidou

Life cycle assessment of Self-Healing Engineered Cementitious Composite (SH-ECC) used for
the rehabilitation of bridges   401
P. Van den Heede, N. De Belie, F. Pittau, G. Habert & A. Mignon

Self-healing concrete vs. conventional waterproofing systems in underground structures:
A cradle to gate LCA comparison with reference to a case study   402
S. Rigamonti, E. Cuenca, A. Arrigoni, G. Dotelli & L. Ferrara

Efficiency of manual and autonomous healing to mitigate chloride ingress in cracked concrete   403
K. Van Tittelboom, B. Van Belleghem, R. Callens, P. Van den Heede & N. De Belie

Establishment of spraying repair technology for concrete structures using drone   404
T. Iyoda, K. Nimura & T. Hasegawa

Polymer Flexible Joint as a structural repair method for reducing stress concentrations
in cracked concrete structures   405
Ł. Zdanowicz, M. Tekieli, B. Zając & A. Kwiecień

Structural column retrofitting of school building using Ferrocement Composites in Vigan,
Ilocos Sur, Philippines   406
J.M.C. Ongpeng, V. Pilien, A. Del Rosario, A.M. Dizon, K.B. Aviso & R.R. Tan

An optimum strategy for FRP-strengthening of corrosion-affected reinforced concrete columns   407
H. Baji, C.Q. Li, F. Chen & W. Yang

SS13: Life-cycle of slope and river bank protection system considering soil bioengineering as well as conventional structures
Organizers: G. Kalny, H.P. Rauch & A. Strauss

Degradation processes of wooden logs in soil bioengineering structures   411
G. Kalny, K. Rados, B. Berntatz, B. Winkler & H.P. Rauch

Development of a concept for a holistic LCA model for soil bioengineering structures   412
*M. von der Thannen, S. Hoerbinger, H.P. Rauch, R. Paratscha, R. Smutny,
A. Strauss & T. Lampalzer*

Soil bioengineering: Requirements, materials, applications   413
H.P. Rauch, M. von der Thannen & C. Weissteiner

Service life planning for Austrian river bank protection structures   414
R. Paratscha, A. Strauss, R. Smutny, M. von der Thannen, H.P. Rauch & T. Lampalzer

Specialisation for the ecoengineering sector in the Mediterranean environment ECOMED   415
P. Sangalli, G. Tardío & G. Zaimes

Development and challenges of soil bioengineering applications to vegetated riprap   416
P. Raymond, S. Tron & I. Larocque

The limits of mechanical resistance in bioengineering for riverbank protection   417
A. Evette, D. Jaymond, A. Recking, G. Piton, H.P. Rauch & P.-A. Frossard

SS14: Advanced NDT for visualization and quantification of concrete deterioration and repair effects
Organizers: T. Shiotani, E. Verstrynge, D.G. Aggelis & P. Pahlavan

Case study on determination of remaining bearing capacity of cantilevered balconies of high rise buildings
B. Craeye, W. Gijbels, L. De Winter, M. Maes, T. Soetens & D. Vanermen
421

Localization and characterization of damage modes in reinforced concrete by means of acoustic emission monitoring during accelerated corrosion and pull-out testing
C. Van Steen, E. Verstrynge & M. Wevers
422

Analysis of fatigue behaviour of single steel fibre pull-out in a concrete matrix with micro-CT and acoustic emission
M. De Smedt, C. Van Steen, K. De Wilder, L. Vandewalle & E. Verstrynge
423

Non-destructive inspection method for bending strength estimation of polymer concrete
C. Saito, M. Okutsu, M. Nakagawa, S. Yanagi & H. Takahashi
424

Study of the applicability of the polarization resistance method in analysis and experimental measurement of electric conductivity
C. Okuma, S. Okazaki & H. Yoshida
425

Damage mechanisms analysis of reinforced concrete beams in bending using non-destructive testing
S. Pirskawetz, G. Hüsken, K.-P. Gründer & D. Kadoke
426

A comparative study of acoustic emission tomography and digital image correlation measurement on a reinforced concrete beam
Y. Yang, F. Zhang, D.A. Hordijk, K. Hashimoto & T. Shiotani
427

Ultrasound pulse velocity to measure repair efficiency of concrete containing a self-healing vascular network
E. Tsangouri, J. Lelon, P. Minnenbo, D.G. Aggelis & D. Van Hemelrijck
428

Combining X-Ray imaging and acoustic emission to measure damage progression in ultra-high-performance-concrete
R. Kravchuk, D. Loshkov & E.N. Landis
429

SS15: PROGRESS—Provisions for Greater Reuse of Steel Structures
Organizers: P. Kamrath & P. Hradil

Environmental- and life cycle cost impact of reused steel structures: A case study
S. Vares, P. Hradil, S. Pulakka, V. Ungureanu & M. Sansom
433

Assessment of reusability of components from single-storey steel buildings
P. Hradil, L. Fülöp & V. Ungureanu
434

Calculating the climate impact of demolition
P. Kamrath
435

Deconstruction, recycling and reuse of lightweight metal constructions
P. Kamrath, M. Kuhnhenne, D. Pyschny & K. Janczyk
436

Modelling and experimental testing of interlocking steel connection behaviour
P. Matis, T. Martin, P.J. McGetrick & D. Robinson
437

SS16: Life-cycle asset management for railway-structures (LeCIE)
Organizers: T. Petraschek, A. Hüngsberg, N. Friedl, U. Staindl, G. Lener & A. Strauss

Life cycle assessment for civil engineering structures of railway bridges made of steel
G. Lener & J. Schmid
441

Approach on network-wide sustainable asset management focused on national funding
R. Liskounig
442

Decision-making framework and optimized remediation for railway concrete bridges deteriorated by carbonation and chloride attack
A. Vidovic, I. Zambon, A. Strauss & D.M. Frangopol
443

XXII

Short-, mid- and long-term LCM prognosis of heavy maintenance and replacement demand for
bridge structures at 3 selected railway routes analysing different maintenance strategies 444
R. Veit-Egerer, G.J. Rajasingam, T. Petraschek, L. Rossbacher, N. Friedl & U. Staindl

Execution time estimation of recovery actions for a disrupted railway track inspection schedule 445
M.H. Osman & S. Kaewunruen

Structural assessment and rehabilitation of old stone railway bridges 446
P.G. Malerba & D. Corti

SS17: INFRASTAR—fatigue reliability analysis of wind turbine and bridge structures
Organizers: E. Brühwiler, E. Niederleithinger & J.D. Sørensen

Fatigue reliability analysis of Cret De l'Anneau viaduct: A case study 449
A. Mankar, S. Rastayesh & J.D. Sørensen

"Pocket-monitoring" for fatigue safety verification of a RC bridge deck slab 450
I. Bayane & E. Brühwiler

Fatigue safety verification of a steel railway bridge using short term monitoring data 451
B. Sawicki, E. Brühwiler & M. Nesterova

SS18: The impact of BIM and web technologies in the life-cycle of our built environment
Organizers: P. Pauwels, K. McGlinn & V. Malvar

Investigation of the lifetime extension of bridges, using three-dimensional CIM data 455
T. Yamamoto, K. Konuma, T. Yaguchi, H. Furuta, H. Tsuruta & N. Ueda

A generic model for the digitalization of structural damage 456
A. Hamdan & R.J. Scherer

Modelling risk paths for BIM adoption in Singapore 457
X. Zhao

Using semantic technologies to improve FM asset information management processes 458
J.W.B. Kibe

GENERAL SESSIONS

GS1: Probability theory and applied structural reliability methods

Vulnerability of critical slopes by using continuous Bayesian networks 461
D. De León, D. Delgado, E. Solorio & L. Esteva

A fast and efficient approach to solution of structure system reliability 462
N. Xiao, Y. Chen & F.Y. Lu

Failure probability of a designed nonlinear structure taking into account the uncertainty
of Fourier phase 463
R. Huang, T. Sato, C. Wan, A. Ahamed & L. Zhao

Iterative point estimate method for probability moments of function 464
W. Fan, H. Guo, J. Wei, Z. Li & P. Deng

An assessment of the inherent reliability of SANS 10162-2 for cold-formed steel
columns using the Direct Strength Method 465
M.A. West-Russell, C. Viljoen & E. van der Klashorst

Failure probability estimation in high dimensional spaces 466
K. Breitung

Reliability based design of temporary structures 467
E. Vereecken, W. Botte & R. Caspeele

Reliability-based analysis of tensile surface structures designed using partial factors 468
E. De Smedt, M. Mollaert, L. Pyl & R. Caspeele

GS2: Durability

The resistance to salt penetration of the high-strength fly-ash concrete used to make
composite girder bridges 471
H. Ito, K. Kubota, H. Kuriyama & T. Izumiya

Influence of re-application of surface penetrant on progression of carbonation of
concrete with surface penetrant 472
Y. Sakoi, M. Aba & Y. Tsukinaga

Formulation of the conditions of the destruction of the passivation film of steel bar
by chloride ion in high pH environment 473
N. Hashimoto & Y. Kato

Simulating low-frequency and long-term fatigue loading for life-cycle structures 474
F. Li & J. Zhao

Study on the spatial distribution of the chloride ion supply in the superstructure of an
open-type wharf 475
Y. Tanaka, Y. Kawabata & E. Kato

GS3: Concrete structures

Application of 2D micro-scale image analysis on concrete surface for evaluating
concrete durability under various environments 479
T. Chlayon, M. Iwanami & N. Chijiwa

Robustness of flat slabs against progressive collapse due to column loss 480
T. Molkens

Prestressed concrete roof girders: Part I—deterministic and stochastic model 481
A. Strauss, B. Krug, O. Slowik & D. Novák

Prestressed concrete roof girders: Part II—surrogate modeling and sensitivity analysis 482
D. Lehký, D. Novák, L. Novák & M. Šomodíková

Prestressed concrete roof girders: Part III—semi-probabilistic design 483
D. Novák, L. Novák, O. Slowik & A. Strauss

Basalt fiber for strengthening of compressed structural elements in concrete and reinforced
concrete: Finite element modeling 484
T. Zhelyazov

GS4: Nonlinear analysis and structural optimization

Optimized design and life cycle cost analysis of a duplex welded girder bridge 487
B. Karabulut, B. Rossi, G. Lombaert & D. Debruyne

Decision criteria for life cycle based optimisation in early planning phases
of buildings 488
C. Dotzler, P. Schneider-Marin, C. Röger & W. Lang

Nonlinear reliability analysis of RC columns designed according to Chinese codes 489
D.-G. Lu, J.-S. Wang & Z.-M. Chang

Risk analysis for the impact on traffic sign bridges 490
T. Braml, M. Keuser & S. Petry

Dependency of punching shear resistance and membrane action on boundary conditions
of reinforced concrete continuous slabs 491
B. Belletti, S. Ravasini, F. Vecchi & A. Muttoni

Kriging-based heuristic optimization of a continuous concrete box-girder pedestrian bridge 492
V. Penadés-Plà, T. García-Segura, V. Yepes & J.V. Martí

GS5: Earthquake engineering

A simple estimation method of the probability distribution of residual displacement and
maximum bending moment for pile supported wharf by earthquake 495
T. Nagao & P. Lu

Machine learning implementation for a rapid earthquake early warning system 496
F. Sihombing & M. Torbol

Experimental investigation of seismic behaviour of corroded RC bridge piers 497
X. Ge, N.A. Alexander & M.M. Kashani

Seismic isolation design for Chaijiaxia Yellow River Bridge with steel triangular plate damper 498
L. Zhou, J. Peng, Y. Wu & Z. Yin

Assessing economic risk for businesses subject to seismic events 499
L. Hofer, M.A. Zanini, F. Faleschini & C. Pellegrino

Electricity supply reliability modelling for public sector facilities in view of seismic disaster risk 500
G. Shoji & I. Matsushima

Non-Gaussian stochastic features hidden in earthquake motion phase 502
T. Sato

Implications of performance-based seismic design of nonstructural building components
on life cycle cost of buildings 503
G. Karaki

Reliability of base-isolated structures with sliding hydromagnetic bearings considering
stochastic ground motions 504
L.C. Ding, R. Van Coile, R. Caspeele, Y.B. Peng & J.B. Chen

GS6: Traffic load modelling

Development and validation of a full probabilistic model for traffic load of bridges
based on Weigh-In-Motion (WIM) data 507
J. Kim & J. Song

The effect of traffic load model assumptions on the reliability of road bridges 508
M. Teichgräber, M. Nowak, J. Köhler & D. Straub

Extended extrapolation methods for robust estimates of extreme traffic load effects on bridges 509
M. Nowak, D. Straub & O. Fischer

GS7: Bridge engineering

Steel bridge structural retrofit: Innovative and light-weight solutions 513
A. Pipinato, R. Pavan, P. Collin, R. Hallmark, S. Ivanov, R. Geier & M. van der Burg

Assessment procedures for existing bridges: Towards a new era of codes and standards 514
R. Pavan & E. Siviero

Stress concentration factor in concrete-filled steel tubular K-joints under balanced axial loading 515
I.A. Musa & F.R. Mashiri

Influence of fly-ash mixture to give to life cycle cost and constructability of
composite girder bridge 516
K. Kubota, H. Ito, H. Kuriyama & T. Izumiya

Assessment of Barton High Level Bridge approach span superstructures 517
D.M. Day

Lifecycle performance of HSS bridges 518
M. Seyoum Lemma, C. Rigueiro, L. Simoes da Silva, H. Gervásio & J.O. Pedro

Refurbishment of Swanswell Viaduct
C.G. West
519

Resilience and economical sustainability of a FRP reinforced concrete bridge in Florida: LCC analysis at the design stage
T. Cadenazzi, M. Rossini, S. Nolan, G. Dotelli, A. Arrigoni & A. Nanni
520

GS8: Life-cycle assessment

ProLCA—treatment of uncertainty in infrastructure LCA
O. Larsson Ivanov, D. Honfi, F. Santandrea & H. Stripple
523

LCA of civil engineering infrastructures in composite materials—ACCIONA Construction's experience
M.M. Pintor-Escobar, E. Guedella-Bustamante & C. Paulotto
524

Method and assessment decisions in the evaluation of the LCA-results of timber construction components
S. Ebert & S. Ott
525

Comparative evaluation of the ecological properties of timber construction components of the dataholz.eu platform
S. Ott & S. Ebert
526

The impact of structural system composition on reduced embodied carbon
M. Sarkisian, D. Shook, C. Horiuchi & N. Wang
527

Benchmarking embodied carbon in structural materials
C. De Wolf & D. Davies
528

Sustainable model-based lifecycle cost analysis of real estate developments
M. Moesl & A. Tautschnig
529

EFIResources: A novel approach for resource efficiency in construction
H. Gervasio & S. Dimova
530

Sustainability rating of lightweight expanded clay aggregates using energy inputs and carbon dioxide emissions in life-cycle analysis
F.M. Tehrani, R. Farshidpour, M. Pouramini, M. Mousavi & A. Namadmalian Esfahani
531

Standardization of condition assessment methodologies for structures
J. Engelen, R. Kuijper, D. Bezemer & L. Leenders
532

Life cycle assessment of asphalt mixtures healed by induction heating
E. Lizasoain-Arteaga, I. Indacoechea-Vega & D. Castro-Fresno
533

Service life prediction of pitched roofs clad with ceramic tiles
R. Ramos, A. Silva, J. de Brito & P.L. Gaspar
534

Energy consumption evaluation of a passive house through numerical simulations and monitoring data
C. Tanasa, V. Stoian, D. Stoian & D. Dan
535

A review of retrofit strategies for Large Panel System buildings
E. Romano, O. Iuorio, N. Nikitas & P. Negro
536

Investigation, assessment and suggestion to the existing traditional house of low-income family in Cambodia based on the principles of passive house design
A. Vann & G.Q. He
537

Interchange of economic data throughout the life cycle of building facilities in public procurement environments
F. Salvado, N. Almeida & Á. Vale e Azevedo
538

Smart grid integration towards sustainable retrofitting of large prefabricated concrete panels collective housing built in the 1970s
D.M. Muntean & V. Ungureanu
539

GS9: Assessment of existing structures

Condition assessment based on results of qualitative risk analyses 543
A. Panenka & F. Nyobeu

Development and operation of non-destructive inspection device for stay cables 544
H. Sakai

Point-based POMDP risk based inspection of offshore wind substructures 545
P.G. Morato, Q.A. Mai, P. Rigo & J.S. Nielsen

Acoustic emission based fracture analysis in masonry under cyclic loading 546
N. Shetty, E. Verstrynge, M. Wevers, G. Livitsanos, D. Aggelis & D. Van Hemelrijck

A comparative study on load response of long-span bridges derived by the macro and
micro scale methods 547
Z.R. Jin, X. Ruan & Y. Li

Reschedule or not? Use of benefit-cost indicator for railway track inspection 548
M.H. Osman & S. Kaewunruen

Instrumentation, truck, track and bridge on operation railway 549
R. Montoya, L. Fernando Martha, J. Fernando Rodriguez, A. Merheb, A. Sisdelli & F. Masini

Adaptive direct policy search for inspection and maintenance planning in structural systems 550
E. Bismut & D. Straub

Author index 551

Life-Cycle Analysis and Assessment in Civil Engineering: Towards an Integrated Vision – Caspeele, Taerwe & Frangopol (Eds)
© 2019 Taylor & Francis Group, London, ISBN 978-1-138-62633-1

Preface

Civil engineering structures are expected to be designed to meet long-term requirements related to structural safety, robustness, serviceability, durability, resilience, sustainability and cost. Life-cycle civil engineering provides the knowledge and tools to accommodate for these requirements and relates to the design, inspection, monitoring, assessment, maintenance and rehabilitation of structures and infrastructure systems taking into account their lifetime perspective and duly considering uncertainties involved.

During past two decades there is a growing awareness of the importance of a life-cycle perspective in civil engineering and challenges exist to tackle the contemporary needs, for example related to the long-term prediction of material behaviour and structural response, the assessment of existing structures, inspection and maintenance strategies, life-cycle optimization, among others. Considering these needs, the objective of the International Association for Life-Cycle Civil Engineering (IALCCE) is to promote international cooperation in this field of expertise.

IALCCE is a young Association founded in October 2006. Its activities encompass all aspects of life-cycle assessment, design, maintenance, rehabilitation, and monitoring of civil engineering systems. To this intent, the Association organizes biennial Symposia, bringing together the top experts in this field and providing a unique international platform for the advance of research and practice. These events have been held worldwide since 2008, with previous symposia at Varenna, Lake Como (Italy) (IALCCE'08), Taipei (Taiwan) (IALCCE2010), Vienna (Austria) (IALCCE2012), Tokyo (Japan) (IALCCE2014) and Delft (The Netherlands) (IALCCE2016).

The sixth International Symposium on Life-Cycle Civil Engineering, IALCCE2018, is held in Ghent (Belgium) from October 28 to 31, 2018, under the auspices of Ghent University and in particular the Department of Structural Engineering. The mission of IALCCE2018 is to bring together all cutting-edge research in the field of Life-Cycle Civil Engineering and so to advance both the state-of-the-art and state-of-practice in the field. The more particular IALCCE2018 objective is summarized in its theme "Towards an Integrated Vision for Life-Cycle Civil Engineering", in which a focus is put on new interdisciplinary developments that enable bringing life-cycle assessment in civil engineering to a higher level. During this symposium a particular focus is put on the cross-fertilization between different sub-areas of expertise and the development of an overall vision for life-cycle analyses.

This volume contains the papers presented at IALCCE2018. It consists of a book of extended abstracts and a USB device with 399 full papers from all over the world, including the Fazlur R. Khan lecture and 8 keynote lectures. Contributions relate to design, inspection, assessment, maintenance or optimization in the framework of life-cycle analysis of civil engineering structures and infrastructure systems. Life-cycle aspects that are developed and discussed range from structural safety and durability to sustainability, serviceability, robustness and resilience. Applications relate to buildings, bridges and viaducts, highways and runways, tunnels and underground structures, off-shore and marine structures, dams and hydraulic structures, prefabricated design, and infrastructure systems, among others.

The aim of the Editors is to provide a valuable source of cutting edge information for anyone interested in life-cycle analysis and assessment in civil engineering, including researchers, practicing engineers, consultants, contractors, decision makers and representatives from local authorities. Besides authors promoting their own research and professional work, participants learn about and discuss the latest accomplishments, innovations and potential future directions in Life-Cycle Civil Engineering. They connect with each other in order to build innovative and lasting networks and collaborations, benefitting from profound professionalism and stimulating cross-fertilization that enables new developments.

Being held in Ghent, with its rich history and superbly preserved building patrimony which is sometimes referred to as the Manhattan of the Middle Ages, the Editors trust to have provided a unique setting to reflect on the challenges in life-cycle civil engineering and hope that this may yield new fruitful ideas and stimulate new initiatives for the future.

October 2018

Robby Caspeele	*Luc Taerwe*	*Dan M. Frangopol*
Ghent University	Ghent University	Lehigh University
Ghent, Belgium	Ghent Belgium	Bethlehem, Pennsylvania, USA
Chair, IALCCE2018	Co-Chair, IALCCE2018	Co-Chair, IALCCE2018

Sponsors, Supporting Associations and Academic Partners

ORGANIZING ASSOCIATION

IALCCE
International Association for Life-Cycle Civil Engineering

ORGANIZING INSTITUTION

Ghent University, Ghent, Belgium

SPONSORS

Gold sponsors

BESIX, Brussels, Belgium

Jan De Nul Group, Hofstade-Aalst, Belgium

Silver sponsors

TUC RAIL, Brussels, Belgium

INFRABEL, Brussels, Belgium

Bronze sponsors

FRANKI, Kontich, Belgium

 DENYS, Wondelgem, Belgium

 ArcelorMittal, Ghent, Belgium

 SBE, Sint-Niklaas, Belgium

Sponsoring research foundation

FWO
Research Foundation – Flanders

SUPPORTING ASSOCIATIONS

IALCCE
International Association for Life-Cycle Civil Engineering

IABMAS
International Association for Bridge Maintenance and Safety

fib
The International Federation for Structural Concrete

RILEM
International Union of Laboratories and Experts
in Construction Materials, Systems and Structures

IABSE
International Association for Bridge and Structural Engineering

InfraQuest – A collaboration of Rijkswaterstaat, TNO, Delft University of Technology

ACADEMIC PARTNERS

Ghent University, Ghent, Belgium

KU Leuven, Leuven, Belgium

University of Liège, Liège, Belgium

ATLSS Engineering Research Center, Bethlehem, PA, USA

Lehigh University, Bethlehem, PA, USA

BOKU – University of Natural Resources and Life Sciences, Vienna, Austria

Life-Cycle Analysis and Assessment in Civil Engineering: Towards an Integrated Vision – Caspeele, Taerwe & Frangopol (Eds)
© 2019 Taylor & Francis Group, London, ISBN 978-1-138-62633-1

Symposium Organization

ORGANIZING ASSOCIATION

IALCCE
International Association for Life-Cycle Civil Engineering (http://www.ialcce.org).
The objective of IALCCE is to promote international cooperation in the field of life-cycle civil engineering for the purpose of enhancing the welfare of society.

IALCCE EXECUTIVE BOARD

President
Dan M. Frangopol — *Lehigh University, Bethlehem, PA, USA*

Honorary President
Alfredo H-S. Ang — *University of California, Irvine, CA, USA*

Vice-Presidents
Harald Budelmann — *Technical University of Braunschweig, Braunschweig, Germany*
Hitoshi Furuta — *Kansai University, Osaka, Japan*

Members
Mitsuyoshi Akiyama — *Waseda University, Tokyo, Japan*
Jaap Bakker — *Dutch Ministry of Infrastructure and the Environment, Utrecht, The Netherlands*
Fabio Biondini — *Politecnico di Milano, Milan, Italy*
Shi-Shuenn Chen — *National Taiwan University of Science and Technology, Taipei, Taiwan*
Luis Esteva — *National University of Mexico, Mexico City, Mexico*
Leo Klatter — *Dutch Ministry of Infrastructure and the Environment, Utrecht, The Netherlands*
Mark P. Sarkisian — *Skidmore, Owings & Merrill LLP, San Francisco, CA, USA*

Secretary General
Fabio Biondini — *Politecnico di Milano, Milan, Italy*

SYMPOSIUM CHAIRS IALCCE2018

Robby Caspeele — *Ghent University, Ghent, Belgium*
Luc Taerwe — *Ghent University, Ghent, Belgium*
Dan M. Frangopol — *Lehigh University, Bethlehem, PA, USA*

STEERING COMMITTEE

Dan M. Frangopol (chair) — *Lehigh University, Bethlehem, PA, USA*
Mitsuyoshi Akiyama — *Waseda University, Tokyo, Japan*
Alfredo H-S. Ang — *University of California, Irvine, CA, USA*
Gabriela M. Atanasiu — *Technical University "Gheorghe Asachi" of Iasi, Iasi, Romania*
Jaap Bakker — *Dutch Ministry of Infrastructure and the Environment, Utrecht, The Netherlands*
Konrad Bergmeister — *BOKU – University of Natural Resources and Applied Life Sciences, Vienna, Austria*
Fabio Biondini — *Politecnico di Milano, Milan, Italy*
Robby Caspeele — *Ghent University, Ghent, Belgium*
Airong Chen — *Tongji University, Shanghai, China*
David De Leon — *Autonomous University of Mexico State, Toluca, Mexico*
Bruce Ellingwood — *Colorado State University, Fort Collins, CO, USA*
Luis Esteva — *National University of Mexico, Mexico City, Mexico*
Hitoshi Furuta — *Kansai University, Osaka, Japan*

Michel Ghosn	*The City College of New York, New York, NY, USA*
Ichiro Iwaki	*Nihon University, Sendai, Japan*
Hyun-Moo Koh	*Seoul National University, Seoul, South Korea*
Pier Giorgio Malerba	*Politecnico di Milano, Milan, Italy*
Robert Melchers	*The University of Newcastle, Callaghan, Australia*
Torgeir Moan	*Norwegian University of Science and Technology, Trondheim, Norway*
Terry Neimeyer	*KCI, Sparks, MD, USA*
Mark Sarkisian	*Skidmore, Owings & Merrill LLP, San Francisco, CA, USA*
Luc Taerwe	*Ghent University, Ghent, Belgium*
Man-Chung Tang	*T.Y. Lin International, San Francisco, CA, USA*

INTERNATIONAL SCIENTIFIC COMMITTEE

Luc Taerwe (chair)	*Ghent University, Ghent, Belgium*
Fabio Biondini (co-chair)	*Politecnico di Milano, Milan, Italy*
Airong Chen (co-chair)	*Tongji University, Shanghai, China*
Sreenivas Alampalli	*New York State Department of Transportation, Albany, NY, USA*
André Beck	*University of Sao Paulo, Sao Paulo, Brazil*
Túlio N. Bittencourt	*University of Sao Paulo, Sao Paulo, Brazil*
Eugen Brühwiler	*Ecole Polytechnique Fédérale De Lausanne, Lausanne, Switzerland*
Joan R. Casas	*Technical University of Catalonia – BARCELONATECH, Barcelona, Spain*
Robby Caspeele	*Ghent University, Ghent, Belgium*
Jan Cervenka	*Cervenka Consulting, Prague, Czech Republic*
Eleni Chatzi	*ETH Zurich, Zurich, Switzerland*
Jianbing Chen	*Tongji University, Shanghai, China*
Matthew Collette	*University of Michigan, Ann Arbor, MI, USA*
Paulo Cruz	*University of Minho, Guimaraes, Portugal*
Alvara Cunha	*University of Porto, Porto, Portugal*
Donald W. Davies	*Magnusson Klemencic Associates – MKA, Seattle, WA, USA*
Nele De Belie	*Ghent University, Ghent, Belgium*
Ane De Boer	*Dutch Ministry of Infrastructure and environment, Utrecht, The Netherlands*
Geert De Schutter	*Ghent University, Ghent, Belgium*
Jean-François Demonceau	*Université de Liège, Liège, Belgium*
Sofia Diniz	*Federal University of Minas Gerais, Minas Gerais, Brazil*
Panos Diplas	*Lehigh University, Bethlehem, PA, USA*
Christoph Gehlen	*Technical University of Munich (TUM), Munich, Germany*
Thomas Gernay	*Université de Liège, Liege, Belgium*
Petr Hajek	*Czech Technical University, Prague, Czech Republic*
Marcel Hertogh	*Delft University of Technology, Delft, The Netherlands*
Dick Hordijk	*Delft University of Technology, Delft, The Netherlands*
Chul-Woo Kim	*Kyoto University, Kyoto, Japan*
Sunyong Kim	*Wonkwang University, Iksan, South Korea*
Takeshi Kitahara	*Kanto Gakuin University, Yokohama, Japan*
Leo Klatter	*Dutch Ministry of Infrastructure and environment, Utrecht, The Netherlands*
Chung-Qing Li	*RMIT University, Melbourne, Australia*
Geert Lombaert	*KU Leuven, Leuven, Belgium*
Zoubir Lounis	*National Research Council Canada, Ottawa, Canada*
Giuseppe Mancini	*Politecnico di Torino, Turin, Italy*
José Matos	*University of Minho, Guimaraes, Portugal*
Stuart Matthews	*Building Research Establishment, Garston, United Kingdom*
Ayaho Miyamoto	*Yamaguchi University, Ube, Japan*
Drahomir Novak	*Brno University of Technology, Brno, Czech Republic*
André D. Orcesi	*French Institute of Science and Technology for Transport - IFSTTAR, Marne-la-Vallée, France*
Jamie Padgett	*Rice University, Houston, TX, USA*
Mahesh Pandey	*University of Waterloo, Waterloo, ON, Canada*
Kok Kwang Phoon	*National University of Singapore, Singapore, Singapore*
Dirk Proske	*Axpo Power AG, Zurich, Switzerland*

Han Roebers	*Dutch Ministry of Infrastructure and environment, Utrecht, The Netherlands*
Xin Ruan	*Tongji University, Shanghai, China*
Mauricio Sanchez-Silva	*Los Andes University, Bogota, Colombia*
Mohamed Soliman	*Oklahoma State University, Stillwater, OK, USA*
John Dalsgaard Sørensen	*Aalborg University, Aalborg, Denmark*
Bill F. Spencer	*University of Illinois, Champaign, IL, USA*
Raphaël Steenbergen	*TNO, Delft, The Netherlands*
Marc G. Stewart	*The University of Newcastle, Callaghan, Australia*
Daniel Straub	*Technical University of Munich (TUM), Munich, Germany*
Alfred Strauss	*BOKU – University of Natural Resources and Life Sciences, Vienna, Austria*
Miroslav Sykora	*Czech Technical University, Prague, Czech Republic*
Sebastian Thöns	*DTU – Technical University of Denmark, Kgs. Lyngby, Denmark*
Yiannis Tsompanakis	*Technical University of Crete, Crete, Greece*
Thomas Ummenhofer	*Karlsruhe Institute of Technology, Karlsruhe, Germany*
Klaas van Breugel	*Delft University of Technology, Delft, The Netherlands*
Kim Van Tittelboom	*Ghent University, Ghent, Belgium*
Celeste Barnardo Viljoen	*University of Stellenbosch, Stellenbosch, South Africa*
Naiyu Wang	*University of Oklahoma, Norman, OK, USA*
Roman Wendner	*BOKU – University of Natural Resources and Life Sciences, Vienna, Austria*
Yiqiang Xiang	*Zhejiang University, Hangzhou, China*
Victor Yepes	*Polytechnic University of Valencia, Valencia, Spain*
Yong Yuan	*Tongji University, Shanghai, China*

LOCAL ORGANIZING COMMITTEE

Robby Caspeele (chair)	*Ghent University, Ghent, Belgium*
Didier Droogné (technical secretary)	*Ghent University, Ghent, Belgium*
Wouter Botte	*Ghent University, Ghent, Belgium*
Pieterjan Criel	*Ghent University, Ghent, Belgium*
Hans De Backer	*Ghent University, Ghent, Belgium*
Wouter De Corte	*Ghent University, Ghent, Belgium*
Bart De Waele	*Ghent University, Ghent, Belgium*
Jean-François Demonceau	*Université de Liège, Liege, Belgium*
Elke Gruyaert	*KU Leuven, Leuven, Belgium*
Geert Lombaert	*KU Leuven, Leuven, Belgium*
Christel Malfait	*Ghent University, Ghent, Belgium*
Boyan Mihaylov	*Université de Liège, Liège, Belgium*
Jens Mortier	*Ghent University, Ghent, Belgium*
Pieter Pauwels	*Ghent University, Ghent, Belgium*
Marijke Reunes	*Ghent University, Ghent, Belgium*
Nicky Reybrouck	*Ghent University, Ghent, Belgium*
Didier Snoeck	*Ghent University, Ghent, Belgium*
Luc Taerwe	*Ghent University, Ghent, Belgium*
Ruben Van Coile	*Ghent University, Ghent, Belgium*
Philip Van den Heede	*Ghent University, Ghent, Belgium*
Jolien Van Der Putten	*Ghent University, Ghent, Belgium*
Kizzy Van Meirvenne	*Ghent University, Ghent, Belgium*
Tim Van Mullem	*Ghent University, Ghent, Belgium*
Kim Van Tittelboom	*Ghent University, Ghent, Belgium*
Els Verstrynge	*KU Leuven, Leuven, Belgium*

IALCCE2018 WEBSITE

http://www.ialcce2018.org

IALCCE WEBSITE

http://www.ialcce.org

Life-Cycle Analysis and Assessment in Civil Engineering: Towards an Integrated Vision – Caspeele, Taerwe & Frangopol (Eds)
© 2019 Taylor & Francis Group, London, ISBN 978-1-138-62633-1

Acknowledgements

The Editors would like to take this opportunity to express their sincere thanks to the authors, the reviewers, the organizers of Mini-Symposia and Special Sessions, and the members of the Steering Committee, the International Scientific Committee and the Local Organizing Committee for the time and efforts they have devoted to make IALCCE2018 a successful event.

The Editors wish to wholeheartedly thank all sponsors, the supporting associations and the academic partners, whose support contributed to the success of this Symposium.

Ghent University is acknowledged for hosting this conference. Especially the support from the Department of Structural Engineering and all its members is highly appreciated. Without their support, efforts and teamwork, the organization of this Symposium would not have been possible.

The support of the International Association for Life-Cycle Civil Engineering (IALCCE) is gratefully acknowledged, especially through the support of Dan Frangopol and Fabio Biondini. Furthermore, the help, advice and promotion from the previous main organizers is strongly appreciated, i.e. through Alfred Strauss (IALCCE2012), Mitsuyoshi Akiyama (IALCCE2014) and Jaap Bakker (IALCCE2016).

The Editors would like to especially thank Didier Droogné, member of the Local Organizing Committee, for his intensive, very dedicated and unconditional support of the main chair for the management of all aspects of the Symposium. Finally, the dedicated support of Marijke Reunes and Christel Malfait is gratefully acknowledged and proved indispensable for the logistical organization of the Symposium.

FAZLUR R. KHAN LECTURE

Durability of bridges

M.C. Tang
T.Y. Lin International, San Francisco, CA, USA

ABSTRACT Life cycle cost is defined as the sum of initial costs, construction costs, maintenance costs, rehabilitation costs, and demolition costs. This is the total direct cost of building a bridge from beginning to end until it has been removed from the site. It is a direct cost in that it does not include costs that incurred indirectly by the bridge. The costs included in the above definition are all monetary costs. Missing in the definition are two major, non-monetary items: social costs and environmental costs. These may be considered two major costs as compared to other costs. Thus, a project with the least life cycle costs as defined above could actually be more expensive than a project that has less of an impact on society and the environment. We will discuss this in this article.

1 INTRODUCTION

Life cycle cost is defined as the sum of initial costs, construction costs, maintenance costs, rehabilitation costs, and demolition costs. This is the total direct cost of building a bridge from beginning to end until it has been removed from the site. It is a direct cost in that it does not include costs that incurred indirectly by the bridge. The costs included in the above definition are all monetary costs. Missing in the definition are two major, non-monetary items: social costs and environmental costs. These may be considered two major costs as compared to other costs. Thus, a project with the least life cycle costs as defined above could actually be more expensive than a project that has less of an impact on society and the environment. We will discuss this in this article.

Figure 1 shows the timeline of these costs: the initial costs all happen before the bridge is opened to traffic, from time point A to B; the cost of operation, maintenance, and rehabilitation all of which occur during the time that the bridge is in operation, from time point B to point C; and the cost of disposal, which occurs after the bridge has been closed to traffic, from time point C to point D.

A bridge offers a value only between time point B and time point C, that is, within its service life. The value of a bridge consists of functional value, aesthetic value, and in some cases monumental value. Functional value is the advantage received by using the bridge such as the time we save because time has value. In most cases, functional value is the only value that we consider when we assess the feasibility of a bridge project. Aesthetic value refers to the beauty of the bridge. A beautiful bridge is a tourist attraction of a city. In addition, the surroundings enhanced by the beauty of a bridge offers citizens a much more enjoyable environment. A bridge has monumental value if it is built to celebrate certain important events, such as the Alexander III Bridge in Paris, which was built to commemorate the friendship between France and Russia.

2 DURABILITY

Service life is the length of time a bridge can be used safely accordingly to the design criteria of the bridge. At the time when we design a bridge, we do not know how long it will last. In present design specifications, we are required to "consider" a design life. The design life of a bridge is the assumed service life of the bridge based on all known and assumed conditions. Most specifications specify a design life of 100 years for major bridges and 75 years for regular bridges and viaducts. Longer design life may be required for specific bridges for various reasons. For example, the Eastern Replacement Spans of the San Francisco-Oakland Bay Bridge (SFOBB) have been designed to a 150-year design life and the Hong Kong Zhuhai Macao Bridge has been designed for a 120-year design life. These two projects are very large projects and are very important for the local economy. The SFOBB, for example, is a lifeline structure that

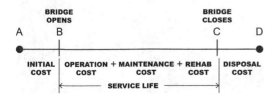

Figure 1. Time line of the service life of a bridge.

must be kept operable even after a major earthquake in the area. Its critical importance justifies a longer design life.

However, the actual service life of a bridge is not necessarily equal to its design life. A bridge in good condition at the end of its design life will not be demolished if it is still healthy and capable of carrying the design traffic. On the other hand, a bridge will probably be replaced or dismantled when it becomes unsafe or cannot serve its purposes any longer, even if it has not reached its design life.

The basic reason we build a bridge is to carry traffic, both vehicular and pedestrians across a discontinuity in a road. But, a bridge must be safe, functional, economical, and look good. When we say that a bridge cannot serve its purpose any longer, we do not refer to only the structural safety of the bridge. If a bridge is functionally obsolete, or if it becomes too expensive to maintain, or if it becomes so ugly that it is an eye sore of the area, it will likely be dismantled. When it is dismantled, it is at the end of its service life.

A bridge is durable if its service life is sufficiently long. Thus, durability can be expressed by the length of the actual service life of the bridge. Since a bridge must be safe, functional, economical, and good looking, it implies that a bridge can be durable only if it is durable in safety, durable in functionality, durable in economy, and durable in aesthetics.

3 DURABILITY IN SAFETY

The required safety of a bridge is usually defined in the codes and specifications by an equation in the form of:

$$\phi C \geq \sum \gamma_i L_i \qquad (1)$$

or

$$C \geq \sum \gamma_i L_i / \phi \qquad (1A)$$

where C is the structural capacity of the structure, ϕ is a strength factor, usually less than 1.0, L_i is the demand caused by load i and γ_i is the corresponding load factor to be applied to load i. γ_i are usually larger than 1.0. For example, for a bridge under permanent load and live load, AASHTO (2002) specifies that for checking the flexural capacity of a cross section, the load factors shall be 1.37 for permanent load (DL), 2.17 for live load (LL), and the strength reduction factor $\phi = 0.9$, so that the flexural capacity, M, of a member shall be

$$M \geq (1.37 \, M_{DL} + 2.17 \, M_{LL}) / 0.90$$

For simplicity's sake, we may rewrite Eq. (1A) as

$$\phi C / \sum \gamma_i L_i \geq 1.0 \qquad (2)$$

If we define a safety index

$$\alpha = \phi C / \sum \gamma_i L_i \qquad (3)$$

Thus,

$$\alpha \geq 1.0 \qquad (4)$$

is the criteria for the structural safety of the bridge. We further define the corresponding safety index at the time the bridge is opened to traffic as α_0. In most cases, the value of α is higher than 1.0 at the time the bridge is opened to traffic because engineers are relatively conservative. This means, α_0 is usually larger than 1.0.

If we assume that the codes and specifications will not be revised throughout the life of the bridge, then the value of ϕ and $\gamma_i L_i$ in this equation will remain unchanged within its service life. However, the value of C, which is the capacity of the bridge, may change with time due to corrosion, erosion, fatigue, or other reasons, as the bridge is subjected to traffic and environmental attacks. Thus, the value of α will vary with time, or more correctly, α will decrease with time if no intervention takes place.

However, all terms in $\sum \gamma_i L_i$ are "desirable values" used in a design. They contain certain safety factors and $\alpha = 1.0$ does not mean the bridge is at the brink of imminent failure. As a matter of fact, when we rate an existing bridge, we usually allow it to have lower loads and/or lower load factors. Thus, when we determine if a bridge structure is still sufficiently safe to carry the design traffic, we require its safety index to be at least α_{min}, which is usually less than 1.0. In other words, we allow the value of α to vary between α_0 and α_{min}. See Figure 2.

There are two ways to prolong the structural service life of a bridge: proper maintenance and rehabilitation. Proper maintenance includes ongoing activities such as painting, cleaning, minor repairs, and routine partial replacements. For example, it is reported that the Golden Gate Bridge is being painted from one end to the other continuously to protect the steel members from corrosion that may be caused by the salty air of the area. Routine replacement of stay cables on a cable-stayed bridge or routine replacement of bearings also belong to this category of maintenance. Currently, the consensus is that today's cables will not last a 100-year

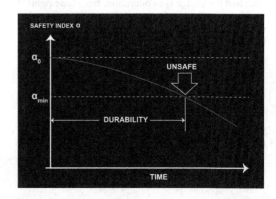

Figure 2. Safety index.

design life. So, they are to be replaced at a certain interval as specified in the original design. Rehabilitation is a major operation to be performed on a bridge that is not routine. One example is the replacement of cables that may have corroded before the end of its design life, or the repair of cracks in a steel bridge deck.

The effect of maintenance and rehabilitation on the service life of a bridge can be expressed as seen in Figures 4 and 5. Maintenance prolongs the service life of a bridge, while rehabilitation increases or restores its service life.

Design specifications may change while a bridge is in service. Seismic design is a good example. For example, when the Golden Gate Bridge was designed, engineers believed that a horizontal force equal to 5% of the gravity load would be sufficient for seismic design. Therefore, all bridges designed at that time would not have had sufficient capacity to accommodate major earthquake as required by present-day design specifications. These bridges must be strengthened to meet the new requirement. Some of them, such as the eastern spans of the SFOBB and the Carquinez Bridge, both in California, USA, must be replaced.

4 DURABILITY IN FUNCTIONALITY

The main function of a bridge is to carry traffic from one end to the other end over a discontinuity in a road or highway. Traffic can be vehicles, both motorized and non-motorized, people, water and/or pipeline, and any other facilities that must pass over the discontinuity. The traffic may be on the bridge or under the bridge. Keep in mind that a bridge must also provide the required navigation clearance when crossing a river.

Sight distance is another important design requirement. Sight distance is related to the vertical and horizontal curvatures as well as the design speed of the roadway. If the design speed of vehicles on a bridge is increased, the existing sight distance may not be sufficient.

These functional requirements may change with time. Traffic volume often increases with time, especially in developing countries. When a bridge is no longer capable of carrying the actual traffic volume, it may be widened or demolished and replaced. This happens rather often.

Insufficient navigation clearance is another possible cause for a bridge to be demolished or replaced. For example, both the Sidney Lanier Bridge and the Talmadge Bridge in the State of Georgia, USA, were replaced due to insufficient navigation clearance. The old Talmadge Bridge's vertical clearance was insufficient for today's ocean-going ships, so it suffered a few collisions by tall ships and so the Georgia DOT replaced it with a new bridge that has a higher clearance. The original Sidney Lanier Bridge was a lift bridge with a horizontal clearance of about 76 meters. It was also hit a few times by relatively wide ships. It was replaced with a 430 m-span cable-stayed bridge, see Figure 6. Ships are getting bigger and taller with time. Many older bridges have similar problems.

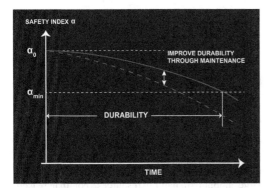

Figure 3. Safety increased by maintenance.

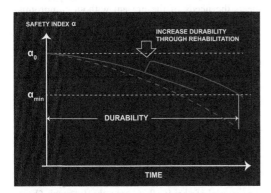

Figure 4. Durability increased by rehab.

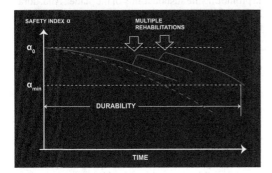

Figure 5. Multiple rehabs.

Figure 6. Original Sidney Lanier Bridge and its replacement.

Durability in functionality must fully consider the possible future changes in traffic patterns and volume during the design stage.

5 DURABILITY IN ECONOMY

During the lifetime of a bridge, there are ongoing operational and maintenance costs. These costs often depend on the quality of the design and construction. If the quality is good, these costs will be less. If the quality is not good, these costs may be very high. When the operational and maintenance costs exceed the value of the bridge, it may be demolished and/or replaced.

6 DURABILITY IN AESTHETICS

Bridges are typically very visible structures in a city. Therefore, aesthetics is one of the most important design requirements. Bridges should look nice and they should be in harmony with the cityscape where they are located. Even though it is seldom that a bridge is torn down just because it looks bad, it is a fact that a nice-looking bridge will get more attention and that people would be more willing to pay for higher maintenance expenses, or just to keep it as a monument of the area even if it becomes functionally obsolete. Therefore, it is important that a bridge must not deteriorate aesthetically with time. High-quality construction, both in material and in workmanship, are essential to achieving durability in aesthetics. Certainly, proper maintenance is equally important.

7 SERVICE LIFE

The service life of a bridge begins when it is opened to traffic and ends when it is closed to traffic for demolition. See Figure 1. If the service life of a bridge is equal to or longer than its design life, we may label it as durable. If a bridge's service life is less than its design life, then it is non-durable. Thus, the durability of a bridge is synonymous to the length of its service life. Again, a bridge must be durable in safety, functionality, economy, and aesthetics.

In the old days, for example, during the Roman Empire, engineers tried to build the best bridge they could build. They did not consider its design life. Quite a number of Roman bridges are still standing, almost 2,000 years after their construction. The Zhaozhou Bridge (see Fig. 7) along with a few other old bridges in China is still carrying traffic. How long will our bridges last?

8 SUSTAINABILITY

By anyone's estimate, human civilization, which we can characterize as the period during which any human building activities occurred, is less than 100,000 years old. The planet Earth's age is said to be about 3 billion light years. So, the span of human activities on earth is only a very short period compared to the age of the planet. However, human activities have significantly changed the planet's environment. We have built roads, buildings and bridges to further our civilization and to improve our way of life. But, at the same time, we have exploited and exhausted a large amount of natural resources. For that reason, the United Nations in 1987 recommended sustainable development, which it defined as: "Development that meets the needs of the present without compromising the ability of future generations to meet their own needs." There are two major concerns to consider—the depletion of our natural resources and the negative effects of human building activity on the environment. Natural resources include the direct use of construction materials and non-renewable energies. Harmful gases produced as a by-product of construction and the destruction and denuding of our planet's forests and wetlands have caused grave impacts on the environment. Both are inversely proportional to the length of service life of structures. The longer the service life, the less material we consume per unit over time. A bridge that lasts 300 years consumes only one-third the material of a bridge that lasts only 100 years. Therefore, increasing the service life of bridges is important in promoting the sustainability of the planet. And, durability is the key to increasing the service life of a bridge!

9 DESIGN LIFE OF A BRIDGE

The design life of a bridge is the assumed service life of the bridge. As mentioned above, most major bridges are designed for a design life of 100 years. There are certain items in a bridge that cannot last long. These can be divided into two categories: replaceable items and non-replaceable items. Examples of replaceable items are cables in a cable-stayed bridge and bearings and expansion joints in general. Their replacement is

Figure 7. Zhaozhou Bridge, Hebei, China.

part of a design. So, the method of replacement should be included in the original design. Replaceable items are items that can be replaced repetitively for as long as we wish. They do not limit the service life of a bridge.

In many cases, non-replaceable items will limit the service life of a bridge. Because if a non-replaceable part of a bridge is essential for the safety of the bridge, its service life in and of itself will determine the service life of the bridge as a whole. For example, if the piles of a bridge were corroded to a degree that they could no longer support the bridge, then that would be the end of the bridge's service life. Another example of non-replaceable items are steel details subject to fatigue. Fatigue of steel details can be minimized by reduction of the cyclic stress range. Based on the current code, the allowable number of stress cycles is usually inversely proportional to the cube of the stress range. The number of load cycles is usually directly proportional to the length of time the bridge is under the assumed loading. Thus, if we could reduce the stress range by 30%, for example, $1/(1-0.3)^3 = 2.83$, we could increase the fatigue life of a steel detail to about 283%, so a bridge designed for a 100-year life could be increased to almost 300 years by reducing the stress range 30%.

In order to help the sustainability of the planet, the writer has been promoting the idea of increasing the design life of bridges to at least 300 years. The longer, the better! Based on my personal observation, increasing the design life to 300 years will not increase cost much. Most materials can last for 300 years if they are maintained properly. If some Roman bridges can stand up more than 2,000 years, we should be ashamed even to mention that our bridges can only last for 300 years, after 2,000 years of progress in civilization and with all the advances in science and technology.

10 TOTAL LIFE CYCLE COST

The present definition of "Total Life Cycle Cost" is the sum of initial costs, construction costs, maintenance costs, rehabilitation costs, and demolition costs. However, a bridge's impact on society and the environment must also be properly considered. Bridges range widely in type and they can have different alignments so that each alternative scheme may have different impacts on the local community and surrounding environment. These costs are usually difficult to estimate, but they are real.

11 CONCLUSION

Sustainability is a very important topic that deserves our attention. There are two common ways to achieve sustainability: by finding better ways to build bridges that reduce the consumption of construction materials and by increasing the service life of bridges. As engineers, the former is not under our control; but the latter can be achieved by increasing the durability of structures. It is in the writer's opinion that we should increase the design life of our bridges to at least 300 years. It is doable with relatively little effort!

REFERENCE

American Association of State Highway and Transportation Officials. (2002) *Standard Specifications for Highway Bridges: 2002.* 17th edition. AASHTO, Washington, DC

KEYNOTE LECTURES

Life-Cycle Analysis and Assessment in Civil Engineering: Towards an Integrated Vision – Caspeele, Taerwe & Frangopol (Eds)
© 2019 Taylor & Francis Group, London, ISBN 978-1-138-62633-1

Optimal reliability-based aseismic design of high-rise buildings

A.H.-S. Ang
University of California, Irvine, California, USA

D. De Leon
Autonomous University of Mexico State, Toluca, Mexico

W.L. Fan
Chongqing University, Chongqing, China

ABSTRACT   The reliability-based design of structural components has been well established based on calibration. However, the reliability-based design of structures as a system remains an issue that has not been addressed adequately. This issue requires the assessment of the system reliability of a complete structure, considering the effects of both the aleatory and epistemic types of uncertainty. This can be addressed effectively using the PDEM (probability density evolution method) for the effects of the aleatory uncertainty, resulting in the PDF of the critical state of a system. The effects of the epistemic uncertainty may be included as the error of the mean-value of the critical state of the system. For illustration, the method is applied to the minimum life-cycle cost aseismic design of a high-rise building in Mexico City.

1 INTRODUCTION

Thus far, standards for the reliability-based design of structural components, such as beams and columns, are well-known, that were developed on the basis of calibration. No such standards, however, are available for the design of a complete structure as a system. For that matter, no reliability-based procedure exists for the system design of complex structures.

As every system is unique, there cannot be a uniform standard for its design. A systematic procedure is required for the design of a complete system. Proposed here is a practical probabilistic procedure for the reliability-based optimal design of a structure as a complete system. In other words, the proposed procedure will determine the reliability-based safety index for the optimal design of a complete structure as a system.

1.1 On uncertainty in structural designs

In engineering, uncertainties are unavoidable. In fact, the main objective of the reliability approach is to handle uncertainties in a proper and rational basis. For practical purposes, engineering uncertainties may be classified into two broad types – namely the *aleatory* and the *epistemic* types. These two types of uncertainty have been referred to, respectively, as "data based" and "knowledge based" uncertainties.

Being an inherent part of nature, the aleatory type cannot be reduced; whereas, the epistemic type may be reduced with improved knowledge of the true state of nature. In practice, it is seldom practical to reduce the epistemic uncertainty; however, minimizing its effects is practically feasible and important in the development of an engineering design.

2 DETERMINING OPTIMAL DESIGN

To determine the structural design with the minimum expected life-cycle cost, E(*LCC*), proceed to design the structure with varying design safety indices, β, and estimate the corresponding E(*LCC*). Plot the resulting designs as shown in Fig. 1 showing the various specific designs. From Fig. 1, the optimal design with the minimum E(*LCC*) and corresponding β can be identified.

2.1 Expected life-cycle cost, E(LCC), design of structures

In minimizing the cost of a structure, it is the expected whole life cost, or expected life-cycle cost, E(*LCC*), that is pertinent. This should include all the cost items over the life (normally >40–50 years) of the structure. Specifically, the total E(*LCC*) would consist of the following expected costs

$$C_T = C_I + C_M + C_S \tag{1}$$

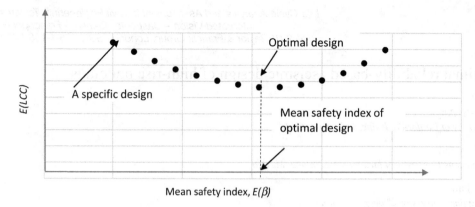

Figure 1. Plot of various designs with varying mean safety indices E(β) and associated E(LCC).

where,
C_I = the initial cost, including the design and construction costs;
C_M = the cost of maintenance, including the costs of inspection and repair, potential damage and failure;
C_S = social cost, including possible loss of lives, injuries, loss of revenues, etc.

For consistency, all cost items must be transformed or expressed in "present value" taking into account the discount rate of the region.

2.2 Reliability of optimal design

For the optimal design identified from Fig. 1, its underlying reliability is strictly due to the effects of the aleatory uncertainty or variability of information. For design, the effects of the epistemic uncertainty must also be included.

For a complex system, the PDEM or *probability density evolution method* developed recently by Li and Chen (2009) is highly effective for determining the PDF of the ultimate system performance Z_{max}, i.e., $f_{Z\,max}(z)$, from which the mean reliability, R, can be assessed through the one-dimensional integration,

$$R = \int_\Omega f_{Z\,max}(z)\,dz \qquad (2)$$

where, Ω is the safe domain of the system.

For design, the effects of the epistemic uncertainty must also be included. The question is "how to do this systematically in the context of the reliability approach?"

For this purpose, the results obtained above by the PDEM is also useful and significant. In particular, the PDF of Z_{max} provides a rational and practical basis for including the effects of the epistemic uncertainty in the reliability-based design of a complex system.

2.3 System reliability including effects of epistemic uncertainty

Equation 2 gives the mean reliability (a single value) by the PDEM due to the aleatory uncertainty. To include the effects of the epistemic uncertainty, it is reasonable to limit this uncertainty as the error in the estimation of the mean-value of Z_{max}. That is, the range of possible mean-values of Z_{max} will represent the effects of the epistemic uncertainty.

In other words, the mean-value of Z_{max}, μ_Z, becomes a random variable; its PDF may be assumed (for convenience) to be a lognormal distribution with a mean of 1.0 and a specified c.o.v. (or equivalent range of values). Therefore, with the PDF of Z_{max}, i.e., $f_{Z\,max}$ and the PDF μ_Z, or $f_{\mu_z}(\mu)$ the convolution integration of these two PDFs will then yield the PDF of the overall system reliability,

$$R = \int_\Omega f_{Z|\mu}(z) \int_0^\infty f_{\mu_z}(\mu)\,dz\,d\mu \qquad (3)$$

where Ω is the safe domain of the system.

Simple Monte Carlo simulation should yield the histogram of the reliability of the system, or its safety index. This histogram contains the effects of both types of uncertainty in the calculated system reliability.

From this histogram, values of the calculated safety index, will be associated with respective levels of statistical "confidence". On this basis, a high confidence level of the safety index would be appropriate and may be specified for the safe design of a complete structural system.

It is well to emphasize also that a high confidence level in design serves to minimize the effects of the epistemic uncertainty.

3 EXAMPLE OF A COMPLEX SYSTEM

As an example of a complex system, consider the 15-story reinforced concrete building shown in Fig. 2

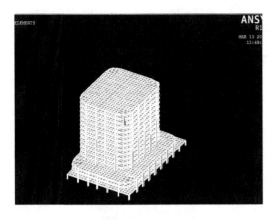

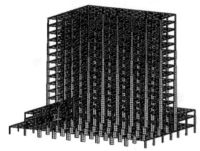

Figure 2. A 15-Story R/C building in Mexico City and its 3D FE model.

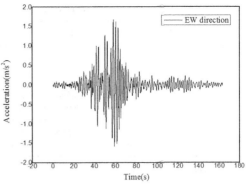

(a) *N-S Ground Motions*

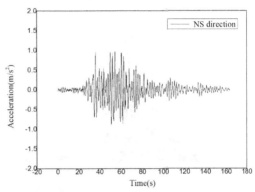

(b) *E-W Ground Motions*

Figure 3. Ground Motions of the 1985 Mexico Earthquake.

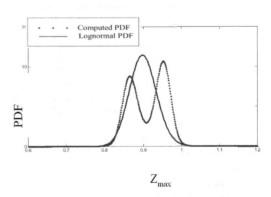

Figure 4. Computed PDF of Z_{max} of the building and corresponding fitted lognormal.

together with its three dimensional finite element model.

The building is located in Mexico City that was designed and built of reinforced concrete. It was subjected to the 1985 earthquake; the ground motions are described in Fig. 3.

Structural properties of the building are described in Ang, De Leon, and Fan (2017).

3.1 Response analyses and optimal design of 15-story building

For this building, the calculations of its response for the 3D model of the building is quite involved and complicated. The number of finite elements is over 205,000, and the total number of nodes and dofs are over 140,000 and 800,000, respectively. The total number of nodes and dofs are over 63,000 and 200,000, respectively, for the model used in the PDEM analysis.

With the PDEM (Li and Chen, 2009), about 135 representative points or samples of deterministic building responses, with their respective probabilities, were necessary to obtain the PDF of the ultimate performance function, Z_{max} of the building as shown in Fig. 4 (dotted line with twin modes); and the safety index $\beta = 2.73$. This PDF represents only the effect of the underlying aleatory uncertainty. Introducing the distribution fitting procedure, the same aleatory uncertainty can be modeled approximately with the fitted lognormal PDF with a mean value of 0.9 and standard deviation of 0.035 (solid line also shown in Fig. 4); the corresponding mean safety index of the fitted lognormal PDF would be $\beta = 2.77$.

For this example, assume that the epistemic uncertainty (representing the inaccuracy of the mean-value of Z_{max}) can be modeled also with a lognormal PDF with a mean-value of 1.0 and a c.o.v. of 0.10. With Eq. 3, convolution integration of this lognormal PDF and

Table 1. Safety Index β and E(*LCC*) for all case.

Case (% of original)*	80%	90%	95%	100%	105%	110%	120%	130%
Mean β	0.023	0.465	1.71	2.72	2.85	2.99	3.31	3.74
E(*LCC*)**	876	576	82.86	11.24	9.59	8.46	7.34	7.21
90% conf. β	0.025	0.524	1.91	3.05	3.23	3.39	3.71	4.22
E(*LCC*)**	875	539	55.14	7.46	6.80	6.59	6.69	7.07
95% conf. β	0.026	0.545	2	3.18	3.37	3.55	3.87	4.39
E(*LCC*)**	874	526	45.67	6.73	6.36	6.30	6.60	7.05

*in percentage of original as-built structure
**in million U.S. dollars

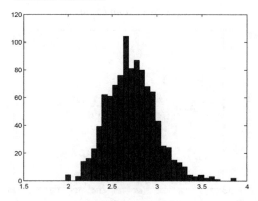

Figure 5. Histogram of β of the original 15-story building based on lognormal PDF of Z_{max}.

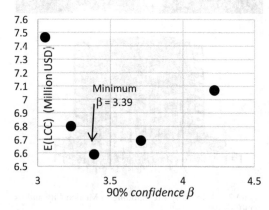

Figure 6. Plots of alternative designs with 90% confidence β versus respective E(*LCC*).

the lognormal PDF of the aleatory uncertainty with a mean of 0.9 and standard deviation of 0.035, then yields the histogram of all possible values of the safety index for the 15-story building as shown in Fig. 5.

Similar complicated PDEM calculations were performed for the same building with different percentages of the original building design; namely, 80%, 90%, 95%, and 105%, 110%, 120%, 130% of the 100% original as-built structure. The results of all these cases, together with the corresponding expected life-cycle costs, E(*LCC*) in million U.S. dollars, are summarized in Table 1.

The statistics of the histogram in Fig. 5 can be summarized as follows:

mean $\beta = 2.7$
90% confidence $\beta = 3.0$
95% confidence $\beta = 3.18$

Plots of the safety index β (with 90% and 95% confidence) versus the corresponding E(*LCC*), are shown below in Figs. 6 and 7, respectively.

Figs. 6 and 7 clearly show that with 90–95% confidence the optimal design of the 15-story building would have required a safety index of $\beta = 3.39$–3.55 which is 110% of the original design; i.e., to obtain the minimum E(*LCC*) design would require slightly stronger, 110%, of the original as-built building. Observe from Table 2, that the original as-built structure was designed with a safety index of $\beta = 3.05$–3.18 with the same 90–95% confidence.

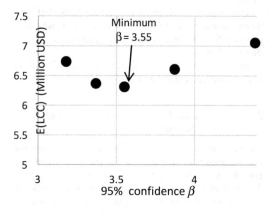

Figure 7. Plots of alternative designs with 95% confidence β versus respective E(*LCC*).

4 CONCLUSIONS

The main conclusions of this study may be summarized as follows:

- The required safety in the design of a structural system cannot be prescribed as that of a structural component. Each structural system is unique; thus, the required safety for its design must be determined independently.

- A practical procedure for determining the safety index for the design of a structural system is proposed; the procedure is based on achieving the minimum expected life-cycle cost design of the structure.
- As illustrated in the example, the proposed procedure shows that the 15-story building in Mexico City for earthquake resistance could have been designed with a higher safety level and a lower life-cycle cost.

ACKNOWLEDGEMENTS

The extensive and complicated PDEM calculations for all the design cases of the 3D FEM model of the 15-story building were performed at the Chongqing University in China with the assistance of Mr. Runyu Liu under the direction of Prof. Wenliang Fan. Whereas, all the information on costs, including the associated estimations of the damage costs and expected life-cycle costs for all the cases of the 15-story building designs were provided by Prof. David De Leon of the Autonomous Metropolitan University in Tuluca, Mexico.

REFERENCES

Ang, A.H-S., De Leon, D. & Fan, W.L. (2017) Optimal reliability-based design of high-rise buildings in Mexico. In: *Proceedings of the Mexican National Congress in Earthquake Engineering*

De Leon, D. & Ang, A.H-S. (2008) Confidence bounds on structural reliability estimations for offshore platforms, *Journal of Marine Science and Technology*, 13 (3), 308–315.

Han, S.H. & Ang, A.H-S. (2008) Optimal design of cable-stayed bridges based on minimum life-cycle cost. In: *Proceedings. IABMAS'08*, Seoul, Korea.

Li, J. & Chen, J.B. (2009) *Stochastic Dynamics of Structures.* Singapore, John Wiley & Sons, 103(GT11), 1227–1246.

Life-Cycle Analysis and Assessment in Civil Engineering: Towards an Integrated Vision – Caspeele, Taerwe & Frangopol (Eds)
© 2019 Taylor & Francis Group, London, ISBN 978-1-138-62633-1

Data standardization for life-cycle performance evaluation of a suspension bridge with multi-pylons

A. Chen, D. Wang, Z. Pan, S. Xu & R. Ma
College of Civil Engineering, Tongji University, Shanghai, China

ABSTRACT   Because of the environmental and loading effects, complex deterioration processes of structural performances will be encountered in the management and maintenance of long-span cable-stayed and suspension bridges during the service life. Hence, the methods to evaluate structural performance need to be proposed and improved. The Taizhou bridge is a suspension bridge with three pylons and two main spans (1080 m + 1080 m). The special components such as steel middle pylon and elastic cable result in significant differences between the Taizhou bridge and normal suspension bridges, and more complex deterioration mechanisms of structural performances as well. The relationship between structural performance and performance evaluation is unclear, which cannot provide objective and reasonable guidelines to the management, maintenance and repair activities of the bridge during the service life. Furthermore, the collection, storage, maintenance and application of the data from in-situ testing and monitoring do not have a unified standard. This paper mainly studies the standard of data for performance evaluation of the Taizhou bridge in the following aspects: 1) the brief framework of the data-driven assessment method is schematically established and the entire framework is divided into four levels; 2) the parameters to represent structural performance in aspects of safety, service and durability are proposed according to the available research achievements, specifications and structural analysis; 3) the standards for data collection are established based on the available experiment, testing and monitoring techniques. The final objective of this research is to set up the data standard for performance evaluation of the Taizhou Bridge, which can guide the data collection and storage and serve for the daily management and maintenance activities of large-span bridges.

1 INTRODUCTION

Large-span bridges play an important role in maintaining the sustained and steady development of the national economy. With the rapid development of structural analysis theory, material technology and construction technique, bridges are striving toward greater span. In China, the span of the cable-stayed bridge of SuTong Yangtze River Bridge has reached 1088 m. The span of the suspension bridge of XiHouMen Bridge has reached 1650 m. The main span of the cable-stayed bridge of Hong Kong-Zhuhai-Macau Bridge has reached 1150 m, and the total length has reached 55 km, becoming the longest sea-crossing bridge in the world. As the span of bridge gradually increases, the structure becomes more and more complex, which puts forward higher requirements for the design theory of bridge structure, but also greatly promotes the development of advanced construction technology. However, how to evaluate the performance of the bridge during operation stage and take appropriate maintenance management measures have become an increasingly prominent problem. During the long-term operation of bridges, it is inevitable to experience various kinds of damage, leading to the deterioration

of the bridge performance. There are many kinds of factors, such as improper maintenance of structure, vehicle impact damage, earthquake, strong wind, environmental erosion and so on. The occurrence of bridge safety accidents at home and abroad has prompted the engineering community to pay more attention to effectively evaluate the performance of bridge structure. As early as half a century ago, bridge engineers and scholars had realized the importance of bridge structural performance evaluation. However, due to the backward acquisition method of bridge performance evaluation data, the result was not ideal in practical application. In 1980s, with the rapid development of monitoring and sensing technology, some countries in Europe and the United States formally put forward the concept of bridge health monitoring and successively established health monitoring and evaluation system on many important large bridges.

At the same time, the structural disease identification analysis based on computer vision has gradually shifted from qualitative description to quantitative description. Computer vision is a subject that studies how to use digital images and videos to obtain information. It is considered as an important part of artificial intelligence. The research of computer vision originated in the 70s of last century, and the early research focused on the reconstruction of the objective three-dimensional world from the two-dimensional world of images as well as the modeling of non-polygonal objects. Early computer vision studies often underestimate the complexity of visual tasks, and the description of visual problems is mostly limited to qualitative descriptions. Today, computer vision technology has penetrated into all aspects of people's daily life. In the civil field, computer vision technology is used for optical character recognition and face recognition. In the manufacturing industry, high resolution images are used to realize rapid detection of products to improve the production efficiency and improve the production process. In the medical field, using computer vision technology to help doctors quickly identify the patient's disease is already a mature application. Computer vision has a wide application prospect and market space, and the image data used by computer vision will become an indispensable data type in bridge performance evaluation.

The research of structural performance evaluation in China began in 1990s. With the development of large-scale infrastructure construction in China, the number of large bridges has increased, and the bridge structural performance evaluation theory can be carried out and verified in these structures. However, there are few systematic studies on the evaluation index of bridge structural performance. Moreover, most of the previous studies have developed a monitoring program for a single bridge, and the indexes for structural performance evaluation have great arbitrariness. There is little research about whether the data obtained by the monitoring system is reliable, whether it is critical, how to save it and how to evaluate it. The non-standardization of the performance evaluation results in the common problems of the current bridge performance evaluation data: 1) The monitoring system is idle or the monitoring data is numerous and jumbled, so the analysis and processing can't start and the performance of the bridge can't be judged by the data; 2) The correlation between different bridge types is unknown, so that the performance of the target bridge structure can't be evaluated through the comparison of the different bridge state data; 3) The data content, format, storage and transmission mode obtained by different bridges are different, so it is difficult to collect and process big data in a unified way, which is not conducive to the development of bridge monitoring, management and maintenance toward the direction of big data informatization.

In the background of big data era, modern operation and management of bridge engineering can no longer like the traditional era, relying solely on human resources, relying on empirical formula to manage the project, as well as only relying on experience to draw up the methods of data acquisition, storage and evaluation of bridge performance evaluation. Although in the short run, relying on experience to develop a set of monitoring and supervision program for a single bridge project has very high efficiency, but considering

Figure 1. Taizhou Bridge.

the vast territory of China, the large number of bridges, different characteristics of bridge structures and the big difference of bridge operation environment in the status quo, combining with the background that big data technology is currently gaining popularity, it is the only way for China's bridge maintenance and management to enter into a new era by formulating the standardization model of bridge state data and unifying the data structure of each independent project. In view of this, it is urgent to carry out a standardized study of the various parts of bridge data, including data acquisition, preservation, transmission and analysis and processing indicators. Taking the Taizhou Bridge as the research object, according to the characteristics of the three-tower suspension bridge of Taizhou Bridge, combining with the historical data of Taizhou Bridge and the multi-scale finite element model analysis, this paper studies the standardization of the behavior data of the Taizhou Bridge.

Taizhou Bridge is located between GaoGang district of Taizhou city and YangZhong city in Jiangsu Province (As shown in Figure 1), which is the world's first kilometer-level three tower suspension bridge and one of the world's few multi-tower suspension bridge. Taizhou Bridge has both features of the traditional two tower suspension bridge and the structure characteristics as well as the structure components of multi-tower suspension bridge itself: 1) The longitudinal displacement of the main girder is restrained by the longitudinal elastic cable to prevent the longitudinal floating of the deck system and reduce the longitudinal displacement at the end of the beam under the live load and the bridge wind load; 2) The anti-slip problem between the main cable and the saddle of the middle tower should be better solved and the friction between the main cable steel wire and the saddle should be taken as the guarantee for the stability against sliding; 3) The strength and safety of the middle tower must be fully guaranteed, and the stability should meet the requirements. These structural characteristics and special components determine that the long-term performance evolution law of Taizhou Bridge must be different from the ordinary long span bridge, and its research methods and research emphasis are not in complete accord.

The main bridge across the river is the control project of this project. The 390 m + 2 × 1080 m + 390 m three-tower two-span steel box suspension bridge scheme is the pioneer in the world. As for the cable system, each main cable consists of 169 strands, and each strand is about 3100 m long, which is composed of 91 high-strength galvanized steel wire with a diameter of 5.2 mm. The main beam adopts flat streamlined steel box girder with a height of 3.5 m and a width of 39.1 m. There are three towers. The north and south edge towers are the reinforced concrete portal frame construction which has a height of 180 m. Considering the middle tower's stiffness balance, finally the vertical herringbone and horizontal portal frame structure form steel tower is adopted, which has a height of 200 m. Each tower is divided into 21 segments, connected by high strength bolts. Floating system is used between tower and beam. To limit the longitudinal drift of the girder, Taizhou Bridge is equipped with a special girder-middle tower elastic cable.

As a representative large-span suspension bridge, Taizhou Bridge is the world's first three-tower suspension bridge with a main span of more than 1000 m. The completion of Taizhou Bridge marks the modern multi-tower suspension bridge construction technology entered a new stage. The multi-pylon structure also provides a new bridge type for the broad coastal waters and islands link project. Research on data standardization of based on Taizhou Bridge can not only obtain the universal standard of large-span suspension bridge, but also set the corresponding standard for the particularity of the three-tower two-span suspension bridge.

2 FRAMEWORK STUDY

2.1 Philosophy

The bridge condition assessment is one of the most important tasks during the service life. Only with the accurate condition assessment can the proper decisions on efficiency maintenance strategy can be made. The traditional assessment of the bridge condition is mainly based on the engineering experiences, which is sometimes over-subjective. To improve the objectivity of the bridge condition assessment, the concept of "Big Data" from the fields of finance and Internet Technology (IT) can be introduced into the bridge engineering, which can eventually lead to the data-driven method for the bridge condition assessment. The brief framework of the data-driven assessment method is schematically shown in Fig. 2. The entire framework can be divided into four levels. The bridge condition is the ultimate output, which is controlled by the performances of different members such as the girder, tower and cable. The member performances as the second level are evaluated by three methods which can categorized into two types: direct and indirect methods. Thus, the evaluation method is the third level in this

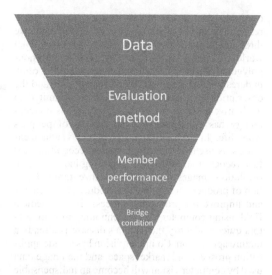

Figure 2. Framework of data-driven method for bridge condition assessment.

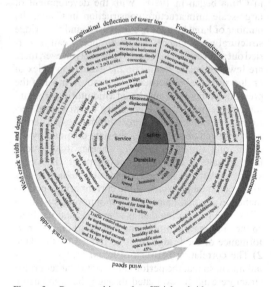

Figure 3. Parameter hierarchy of Taizhou bridge steel tower.

framework. To conduct the evaluation, different types of data as the last level need to be collected and analyzed.

With the framework in Fig. 2, for each bridge member, a pie chart with different levels can be finally obtained, as shown in Fig. 3. In the chart, three types of bridge member parameters are distinguished by three colors, the innermost fan-type annotates the performance type, and outer rings denote: parameter, data source, evaluation method and corresponding countermeasures. The outermost arrow indicates that some parameters may affect different performances at the same time. For example, the crack width can decrease the service performance of the concrete member by causing an uneasy feeling of the structural safety, and affect the durability performance at the same time, as the crack will facilitate the penetration of harmful agents such as chloride ions and carbon dioxide. The pie chart in Fig. 3 can provide a preliminary suggestion to the decision making on the maintenance strategy.

2.2 Performance

The performances of different bridge members can be categorized into three types: safety performance, service performance and durability performance. These performances may have different proportions for different members. For example, the performance of the deck pavement is dominated by its service performance as it is closely related to the driving comfort. But for the suspender cable, the safety is absolutely the first concern of its performance evaluation.

As a common approach, the bridge condition is usually assessed based on the performance evaluation of each member. Thus, to facilitate the condition assessment, the bridge members need to be classified. In the current engineering practice, the members of large-scale suspension bridges are usually classified into structural member and auxiliary facilities based on their importance levels. But this classification is over-simplified and does not fully consider the above different proportions of performances for different members. Thus, a proper and refined classification of bridge members is needed in the framework of the data-driven bridge condition assessment.

2.3 Evaluation method

The evaluation performance of each bridge member can be conducted by the direct or indirect methods. The direct evaluation means that the member performance can be assessed only based on the data collected from the health monitoring system or periodic inspections. For example, based on the longitudinal and transversal deflections at the top of the tower which can be obtained from GPS monitoring, the safety performance of the tower can be evaluated according to the permitted deflections in standards and specifications.

In the indirect method, collected data need to be further analyzed to generate any valuable results. The indirect method can be further divided into numerical analysis and machine learning. The former is often used when the relationship between input parameters and output results is clear. For example, the carbonation depth is an important parameter to represent the risk of rebar corrosion in concrete members. But the in-situ test of the carbonation depth will cause a local damage on the surface of concrete members. To avoid this damage, a non-destructive test method may be adopted, but has not been reported yet. An alternative approach is to use the environmental parameters such as temperature, relative humidity and concentration of carbon dioxide which are easily obtained and the available numerical models of concrete carbonation to calculate the carbonation depth, then use the calculated depth to evaluate the risk of rebar corrosion. Obviously, this method requires a fully understanding of the concrete carbonation mechanism.

If the relationship between input parameters and output results is not clear, the technique of "machine learning" can be used to find the implicit relationship between the input and output. The term "implicit" here means that, with machine learning, we know that typical inputs will lead to corresponding outputs, but we do not know how. In other words, we do not know the exact mechanism hidden behind the data. However, in most cases, this implicit relationship is already enough for the decision making on the maintenance strategy.

2.4 Data

Compared with the data commonly used in structural design, the term "data" in Fig. 4 has two special characteristics. The first one is "big", which means that the data has a huge size which requires a large database, as well as a special approach to storage and process. For example, the health monitoring system of the Sutong bridge will generate about 10GB data per day, which is about 3.65 TB per year. Thus, an efficiency algorithm is obviously needed to search for the target data from a huge number of records, and sort, analyze, compress the data if necessary.

The other characteristic of the data is "multi-source", which means that the data used in the bridge condition assessment nowadays is not limited to the classical time-domain signals usually obtained from sensors in the health monitoring system. Images, records of sound and even texts can be also used. Because of the diversity of the data source, another concept of "data fusion" or "information fusion" may be introduced into the bridge engineering. The data fusion means that the decisions are not made based on only one data source, e.g., data from a typical sensor, but with the help of different data sources, e.g., data from a series of sensor, images, etc. The application of

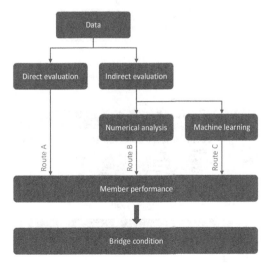

Figure 4. Evaluation method flow chart.

the data fusion can enable a more comprehensive and convincing assessment of the bridge condition.

The data obtained from multiple sources will be dealt with by following the flowchart in Fig. 5. Firstly, a pre-analysis is conducted to judge which kind of and how much data is needed. The pre-analysis can also provide suggestions to the settlement of sensors in the health monitoring system. The sensors should be installed at key positions of the bridge member which are identified from a structural analysis.

After the pre-analysis, the required data is acquired from in-situ monitoring or inspection. The collected data is then transferred to the control center, where the raw data is preprocessed, e.g., cleansing, repairing and compression. This step is the core of the data standardization and of great importance to the further data processing.

The preprocessed data is further uploaded and stored in the cloud. The techniques of cloud computing and cloud storage can be adopted to increase the efficiency of data transfer, storage and processing. With the cloud computing, the heavy task of data mining can be divided and distributed to various kinds of terminal devices such as the computing server, workstation and personal computer (PC). With the cloud storage, the qualified end-users can access the data everywhere in the world for further structural analysis, performance evaluation, decision making, etc.

3 CASE STUDY: INDIRECT EVALUATION OF SAFETY PERFORMANCE OF STEEL MIDDLE TOWER

The change of the status of a single member in the bridge will not only change its own performance, but also affect the whole structure. Based on existing research results and specifications, the parameters linked to the performance of different members in the bridge can be figured out. For example, the mid-span deflection is obviously an important parameter to evaluate the usability of a suspension bridge. But only one parameter cannot represent the condition of the entire bridge. To get a full evaluation, the indirect method such as the finite element analysis should be used to further quantitative evaluate the effect of status change of a single member on the performances of its own and the entire bridge structure.

Based on the perspective and design details, a finite element model of Taizhou Bridge can be established, which is shown in Fig. 6. Through the finite element model, the influence on bridge of each parameter can be analyzed one by one. All the parameters obtained in the investigation process which are mentioned in other bridge programs are included in the research scope. The first batch of parameters include the tower top offset, the base settlement, the system temperature change, the proportion of the broken wire of the main cable, the proportion of the broken wire of the sling, etc.

Taking the tower top offset as an example, the method of finite element analysis is used to figure out the correlation between structure reaction and the offset. First, different tower top offset conditions were simulated. Then, by observing the stress and displacement of the key sections under each condition, the relationship between the offset and the reaction of bridge can be find. Finally, the influence of tower top offset to the whole structure can be comprehensive evaluated in consider of the ultimate limit state and the serviceability limit state.

Figure 6. Finite element model of Taizhou Bridge.

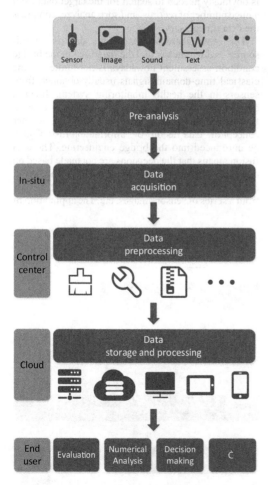

Figure 5. Data standardization analysis process.

Figure 7. Tower top offset condition.

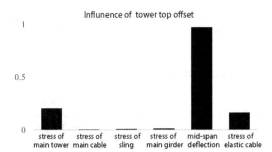

Figure 8. Results of sensitive analysis on influence of tower top offset.

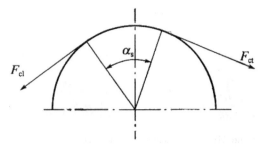

Figure 9. The schematic of pulling forces and included angle.

Fig. 8 shows the sensitivity of every control index of Taizhou Bridge to the tower top offset. According to the results, the tower top offset almost has no influence on the stress of main cable, sling and main girder. But it will affect the main tower and elastic cable. What more, it has greatly influence on the mid-span deflection. So, tower top offset can be considered as an important state parameter.

With the finite element numerical model analysis, the parameters of safety performance, durability performance and usability performance of each component can be determined. What's more, the positions where each parameter data is obtained, and the number of sensor arrangements also can be determined. The location of data acquisition and the number of sensor arrangements can be used as important parts of data standardization for multi-pylon suspension bridges life cycle performance evaluation.

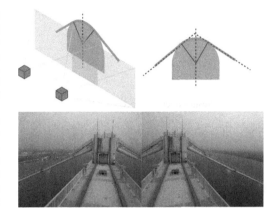

Figure 10. Camera deployment layout and captured images.

4 CASE STUDY: COMPUTER-VISION BASED METHOD TO EVALUATE SAFETY PERFORMANCE OF MAIN CABLE

For multi-pylon suspension bridges, the angles between the main cable and steel saddles is a significant parameter to represent the safety performance of the bridge. However, the exact value of the angle is very different to obtain based on the classical data from sensors in health monitoring systems and in-situ tests. In this case study, a new data source, i.e., image, is used to calculate the angle based on computer vision and evaluate the safety performance of the bridge.

The anti-sliding safety of the main cable is one of the core performances related to the safety of the whole suspension bridge. As a classical approach, the anti-sliding safety of the main cable is usually guaranteed through the theoretical analysis and model test. So far, studies mainly focus on design and construction measures to ensure anti-sliding safety of the main cable, and rare researches can be found on monitoring the anti-sliding safety during the operation stage of the bridge.

According to Chinese specifications for design of highway suspension Bridge (JTG/T D65-05-2015), the checking formula for the anti-sliding safety of main cable is expressed as:

$$K = \frac{\mu \alpha_s}{\ln\left(\dfrac{F_{ct}}{F_{cl}}\right)} \quad (1)$$

where K is the safety factor of anti-sliding which is required to be larger than 2.0, μ is the friction factor between the main cable and saddle, α_s is the included angle between the cable sides, F_{ct} and F_{cl} are the pulling forces in the tight side and loose side. The meanings of the above symbols are schematically shown in Fig. 9.

Based on the model test in design stage and available research data, the friction factor μ is usually assumed as 0.2. The pulling forces can be obtained through sensors in the main cable. The included angle needs to be precisely calibrated.

A fast measurement method based on digital image processing is proposed to obtain the included angle, which is achieved by deploying cameras outside the saddle room. Firstly, two cameras are used to capture the orthographic projection of the main cable, as shown in Fig. 10. Secondly, the image stitching technique (Brown and Lowe, 2003, Brown and Lowe, 2007) is

Figure 11. Panorama of main cable after image stitch.

Figure 12. Results of edge detection.

Figure 13. Results of line detection.

used to get the panorama of the main cable at the top of the tower, as shown in Fig. 11.

The main cable can be considered as line elements in the image. Thus, the line detection algorithms in image processing are applied to measure the included angle. Three assumptions are proposed during the image processing: 1) the imaging plane is parallel to the saddle room, which ensures that the captured image is an orthographic projection. 2) the main cable is considered as straight line (the curvature is neglected). 3) The upper and lower edges of the main cable (including handrails) are parallel to the cable axis. The Hough line transform is applied to measure the included angle. Before that, the Canny edge detection algorithm (Canny, 1986) is operated to get edge image. Then Hough line transform (Hough, 1962, Duda and Hart, 1972) is operated to detect lines in the image. As shown in Fig. 12, totally 34 lines are detected in the captured panorama image.

Based on the detected lines, the included angle is finally calibrated as 0.8324143 rad (47.693°). Therefore, the anti-sliding safety factor can be verified together with the pulling forces and friction factor to evaluate the safety performance of the main cable.

5 CONCLUSION

This paper expounds the standardized structure of bridge structural performance evaluation data for the three-tower suspension bridge of Taizhou Bridge. The main conclusions are as follows:

1) The brief framework, with four levels, of the data-driven assessment method is schematically established. The entire framework can be divided into four levels. The bridge condition is the ultimate output, which is controlled by the performances of different members such as the girder, tower and cable. The member performances as the second level are evaluated by three methods which can categorized into two types: direct and indirect methods. Thus, the evaluation method is the third level in this framework. To conduct the evaluation, different types of data as the last level are collected and analyzed.
2) With the finite element analysis, the parameters of safety performance, durability performance and usability performance of each component can be determined. What's more, the positions where each parameter data is obtained, and the number of sensor arrangements can be also determined. The location of data acquisition and the number of sensor arrangements can be used as important parts of data standardization for multi-pylon suspension bridges life cycle performance evaluation.
3) The included angle between the main cable and saddle can be effectively calibrated by the computer-vision technique.

ACKNOWLEDGEMENTS

The research in this paper is partially supported by the Natural Science Foundation of China with the grant No. 51608377, which is appreciated.

REFERENCES

Brown, M. & Lowe, D.G. (2003) Recognising panoramas. In: *IEEE International Conference on Computer Vision.*

Brown, M. & Lowe, D.G. (2007) Automatic panoramic image stitching using invariant features. *International Journal of Computer Vision,* 74, 59–73.

Canny, J. (1986) *A Computational Approach to Edge Detection.* Washington, DC, IEEE Computer Society.

Duda, R.O. & Hart, P.E. (1972) Use of the Hough transformation to detect lines and curves in pictures. *CACM,* 15, 11–15.

Hough, P.V.C. (1962) *Method and Means for Recognizing Complex Patterns.* US Patent 3 069 654.

Life-Cycle Analysis and Assessment in Civil Engineering: Towards an
Integrated Vision – Caspeele, Taerwe & Frangopol (Eds)
© 2019 Taylor & Francis Group, London, ISBN 978-1-138-62633-1

Promoting societal well-being by designing sustainable and resilient infrastructure: Engineering tools and broader interdisciplinary considerations

P. Gardoni
University of Illinois at Urbana-Champaign, Urbana, IL, USA

ABSTRACT   Modern societies rely on large-scale interdependent infrastructure, including transportation, water and wastewater, electric power, communication and information networks, that are critical for economic growth and societal well-being. Such infrastructure is vulnerable to natural hazards, such as earthquakes and tsunamis, hurricanes, tornadoes, floods, and wildfires; as well as anthropogenic hazards from industrial accidents, diseases and malevolence. Past disasters have shown that the societal consequences of the damage and failure of infrastructure often significantly exceed the physical damage to such systems, are typically not limited to the immediate aftermath of a damaging event but can be long term, and are unequal among individuals. Inequalities in the impact might be spatial (between and within communities) and temporal (across different generations). Aging and deterioration of physical systems, population growth, economic development in regions particularly vulnerable to natural hazards such as coastal regions, and climate change can exacerbate risks. This paper presents some of the engineering tools for the modeling of the impacts on infrastructure components, systems and communities and the modeling of their recoveries. The paper also introduces sustainability and resilience as two of the most important elements in risk evaluation and the development infrastructure able to address societal needs.

1   INTRODUCTION

Decision-making is often based on information about risks, which capture the possible consequences of future events and the associated probabilities (Gardoni 2017, Gardoni & Murphy 2014, 2018, Gardoni et al. 2016). Risk analysis requires the use of accurate models to quantify the probabilities, as well as definitions of consequences that capture the relevant aspects in the decision process.

In the past decades there has been significant progress in assessing the performance and safety of different physical systems including structures (e.g. bridges, buildings, and electrical sub-stations – Gardoni et al. 2002, Ramamoorthy et al. 2008, Xu & Gardoni 2016) and infrastructure (e.g. transportation, potable water, and electric power networks (Guidotti et al. 2017a, Kang et al. 2008). In general, accurate predictions of the physical damage due to a hazard need to consider the effects of aging and deterioration of the physical systems, as well as of climate change. In addition, there are dependencies/interdependencies among physical systems as well as among physical and non-physical systems (like socio-economic systems) (Ellingwood et al. 2016). As a result, the loss of functionality of a supporting system can have significant effects on the supported (or dependent) systems. To model such cascading effect, we need to convert the

damage of a physical system into a reduction or loss of functionality of such system and then assess the effects on the supported systems. Finally, given the change in the functionality of the physical systems, we need to assess and evaluate the societal impacts. Such assessment and evaluation requires defining relevant measures of impact. Murphy & Gardoni (2006) noted that common measures of societal impact (such as direct physical damage, dollar loss and duration of downtime – Gardoni & LaFave 2016) are generally incomplete. There is a need to define more informative measures that capture the spatial and temporal variability of the impact. One aspect often overlooked in regional risk analysis is the possible inequality of the impacts among individuals. Inequalities might be spatial (between and within communities) and temporal (across different generations). Inequalities might be present before the occurrence of a hazard, and then be exacerbated by the hazard occurrence (Gardoni et al. 2016). The evaluation of a risk (e.g. deciding whether a risk is acceptable, tolerable or not tolerable – Murphy & Gardoni 2008) should consider the duration of the impact (which calls for consideration of resilience), as well as the spatio-temporal variability of the impacts (i.e. inequalities) (which calls for consideration of sustainability and social justice). Once the measures of societal impact are defined, we need to develop multidisciplinary models that convert the impact of a hazard on the physical systems into the selected measures of societal impact. Such models need to be probabilistic and capture the relevant uncertainties to be able to predict the probabilities of the selected measures of impact (Gardoni et al. 2002, Murphy et al. 2011, Gardoni 2017).

The opportunities that individuals have define their level of well-being (Sen 1999). Opportunities are functions of what an individual has (e.g. personal resources, skills, and knowledge) as well as what an individual can do with what he/she has given legal, economic, and social constraints and the characteristics of the physical systems. For example, structures and infrastructure support the opportunities of being mobile, being nourished, having shelter and being educated. Similarly, the recovery of the physical systems, at least in part, defines the recovery of individuals by re-establishing their lost opportunities.

So Murphy & Gardoni (2006, 2007, 2008, 2010, 2011) Gardoni & Murphy (2008, 2009, 2010, 2014, 2018), and Tabandeh et al. (2018) proposed to use opportunities that individuals have as the relevant measures of societal well-being before and after the occurrence of a hazard. Specifically, they proposed to use changes to such opportunities as measures of impact and recovery. By looking at the spatial and temporal variability of the opportunities, we can also capture the spatial and temporal variability of the impact.

This paper presents a holistic approach to regional risk analysis that considers the impact of hazards on physical and non-physical systems, and ultimately on the well-being of individuals. The paper discusses the impact of aging and deterioration on physical systems, the impact of climate change, the modelling of dependencies/interdependencies among systems, and the modeling of the cascading effects of the loss of functionality of the physical systems. The paper discusses some of the relevant issues related to the spatio-temporal variability of the impact and presents a normative framework to evaluate risks based on a consideration of resilience and sustainability.

2 REGIONAL RISK ASSESSMENT

2.1 Modeling the impact on physical systems

There are three main steps in assessing the impact of hazards on physical systems. First, we need to define the relevant intensity measures (IMs) of the hazard over the region of interest. For example, in the case of seismic hazards, this can be done using attenuation relationships. Then, we need to define the characteristics of the physical systems that influence their vulnerability and functionality. For example, for buildings, we need to define the construction type and material, and the design specifications (which could be inferred from the year of construction). For spatially distributed networks, we need to define the different constitutive elements of the network as well as the

network topology (i.e. how the difference elements are connected) (Guidotti et al. 2016). Finally, we need to estimate the level of physical damage of each structure and infrastructure component for the given IMs at their site. The damage assessment can be done using fragility curves for punctual elements located at a specific site (e.g. Gardoni et al. 2002), and repair rate curves for linear elements that are spatially distributed (Gardoni 2017).

2.2 Modeling the impact of aging and deterioration of physical systems

Multiple mechanisms can deteriorate physical systems over time. Such deterioration can result in a reduced functionality, reliability and overall service life of a system (Choe et al. 2010, Gardoni & Rosowsky 2011, Kumar et al. 2015). In most cases deterioration mechanism can interact with each other resulting in a faster deterioration than when simply superimposing each mechanism. Jia & Gardoni (2018a, b) developed a state-dependent formulation to model the impact of multiple and possibly interacting deterioration mechanisms on the state and reliability of physical systems. These models can be used in a Renewal Theory-based Life-cycle Analysis (Kumar & Gardoni 2014, Kumar et al. 2015, Jia et al. 2017) to model the service life of a physical system.

2.3 Modeling the impact of climate change

Climate change is impacting regional risk analysis in four fundamental ways (Gardoni et al. 2016, Murphy et al. 2018). First, climate change is impacting the likelihood of occurrence of extreme natural events like heat waves and droughts, severe precipitations, and hurricanes. Second, climate change is impacting the deterioration processes. Sudden (shock) deteriorations due to the occurrence of severe natural events might become more frequent and more significant in magnitude due to the increased likelihood of extreme events. The rate of gradual deterioration might change due to changes in the environmental conditions that govern such mechanisms. Third, climate change is likely to exacerbate social differences and inequalities since individuals that are worst off often live in areas that are most likely to experience natural hazards impacted by climate change (e.g. flood). Finally, climate change is bringing additional uncertainties in the prediction of the physical damage and the societal impact.

2.4 Modelling dependencies/interdependencies

The state of a network can be characterized using measures of connectivity like the network diameter and efficiency (Crucitti et al. 2003, Guidotti et al. 2016, 2017a). Such measures can be used in the definition of the failure of a network (Kang et al. 2008). However, such measures consider only individual networks without accounting for their dependencies/interdependences. To overcome this limitation, Guidotti et al. (2017b) proposed a Multi-layer Heterogeneous Network modelling approach that can assess the reliability of dependent/interdependent systems. With a paradigm shift with respect to current formulations, the Multi-layer Heterogeneous Network modelling defines the different elements in the dependent/interdependent systems as different component types, and the different dependencies/interdependencies as different layers.

2.5 Modeling the reduction or loss of functionality of physical systems

After assessing the physical damage to the physical systems, we need to assess the loss or reduction in their functionality. Such change in the functionality might be due to the physical damage of the system under consideration as well as to the loss or reduction of functionality of any supporting system (Ellingwood et al. 2016, Guidotti et al. 2016). For example, building functionality is based on the building damage but also on the service provided by the supporting water and power infrastructure. Similarly, the functionality of the water infrastructure depends on the power at the pumping stations and control systems provided by the supporting power infrastructure. Guidotti et al. (2016) developed a probabilistic procedure for modelling the reduction or loss of functionality of physical systems accounting for their physical damage and the possible reduction or loss of functionality of the supporting systems.

2.6 Modeling the societal impact using a capability approach

For a structure or structural system, the likelihood that a given hazard causes significant damage or collapse is based on the vulnerability of the structure or structural system. Vulnerability captures the propensity of the structure or system to be impacted. In the same way, for an individual or household the likelihood that a hazard turns into a disaster is based on the vulnerability of the individual or household (World Commission on Environment and Development 1987, Ribot 1995, Adger 2006, Gardoni et al. 2016). Vulnerability now captures the propensity of the individual or household to be impacted. Like structural characteristics define the structural vulnerability, socio-economic characteristics defined the societal vulnerability (Mileti 1999, Peacock & Girard 1997, Kajitani et al. 2005). As a result, after assessing the reduction or loss of functionality of the physical systems we need to consider the social vulnerability to estimate the impact of a hazard on the well-being of individuals. The characteristics that define the social vulnerability are also relevant in predicting the recovery time and in developing mitigation/recovery strategies that promote social justice (as discussed in more detail later in the paper).

To combine the modelling of the reduction or loss of functionality of the physical systems with the social vulnerability and obtain the societal impact of hazards, Murphy & Gardoni (2006, 2007, 2008, 2010, 2011) and Gardoni &Murphy (2008, 2009, 2010, 2014, 2018) proposed a Capability Approach. In this approach, capabilities are the opportunities that individuals have to do or become things of value like being mobile, being nourished, having a shelter and being educated (Sen 1999). The capabilities that individuals have define their level of well-being. Capabilities are functions of what an individual has (e.g. personal resources, skills, and knowledge) as well as what an individual can do with what he/she has given legal, economic and social constraints, and the characteristics of the physical systems. For example, structures and infrastructure support the opportunities of being mobile, being nourished, having shelter and being educated. At the same time, the recovery of the physical systems, at least in part, defines the recovery of individuals by helping re-establish their lost opportunities.

So Murphy & Gardoni (2006, 2007, 2008, 2010, 2011), Gardoni & Murphy (2008, 2009, 2010, 2014, 2018), and Tabandeh et al. (2018) proposed to use opportunities that individuals have as the relevant measures of societal well-being before and after the occurrence of a hazard. Specifically, they proposed to use changes to such opportunities as measures of impact and recovery. By looking at the spatial and temporal variability of the opportunities, we can also capture the spatial and temporal variability of the impact.

A Capability Approach to risk assessment and evaluation allow us to

– focus on what is most crucial, namely the societal well-being;
– account for the socio-economic and institutional factors that define the vulnerability of individuals;
– integrate the impact of hazards on physical systems with the vulnerability of individuals; and
– evaluate a risk as acceptable or tolerable (as discussed later).

Details on how a capability approach can be operationalized can be found in Gardoni & Murphy (2009) and Tabandeh et al. (2018).

3 RESILIENCE

Resilience is generally defined as "the ability of a system to withstand external perturbation(s), adapt, and rapidly recover to the original or a new level of functionality" (Gardoni & Murphy 2018). Examples of systems include structures and infrastructure (i.e. physical systems) and also individuals, communities, or economic systems (i.e. the society).

3.1 Recovery and resilience of physical systems

Sharma et al. (2017) developed a stochastic formulation to model mathematically the recovery of physical systems. The formulation models: 1) the completion time of each group of recovery activities that improve the functionality or reliability of the system, 2) the possible occurrence of disrupting shocks during the recovery process, and 3) the system state as a function of time. In addition, Guidotti et al. (2016) considered the impact of dependencies/interdependencies on the resilience of physical systems, and found that dependencies among physical systems can significantly slow down the recovery while dependencies among physical and non-physical systems could be beneficial.

3.2 Recovery and resilience of society

The functionality of physical systems plays a critical role on the well-being of individuals. As a result, the recovery of the physical systems plays a critical role on the recovery of communities impacted by a hazard. The Capability Approach presented earlier in this paper to gauge the societal impact of hazards can also be used to model the societal recovery as described in Gardoni & Murphy (2018). In this case, the estimates of functionality obtained from the models of the recovery of the physical systems can be used as inputs in the Capability Approach to obtain the well-being of individuals as a function of time. Tabandeh et al. (2017) proposed a Dynamic Bayesian Network (DBN) to mathematically implement such prediction process.

4 SUSTAINABILITY AND SOCIAL JUSTICE

There are three normative considerations related to sustainability that should influence the design of physical systems and the evaluation of risks (Murphy & Gardoni 2008, and Gardoni & Murphy 2018). These normative considerations are a) environmental justice, b) global (or distributive) justice, and c) intergenerational justice.

Environmental justice is about the state of the natural ecosystem. Having a flourishing natural ecosystem might be good in itself as well as instrumentally since an ecosystem might support the well-being of individuals. Environmental justice calls for the design of physical systems and recovery strategies that protect and ideally promote the flourishing of natural ecosystems (Anderson & Woodrow 1989). A Capability Approach allow us to incorporate environmental justice considerations by selecting a capability that captures explicitly the impact on individuals of the flourishing of ecosystems and/or by defining a functional relation between the flourishing of ecosystems and the opportunities individuals have (Martins 2011, Ballet et al. 2011, 2013, Gardoni & Murphy 2018).

Global justice is about the fairness of the distribution of the impact and recovery opportunities. Past disasters have shown that certain population groups are often more impacted than other ones and also recover more slowly. Global justice calls for the design of physical systems and the distribution of mitigation and recovery resources that promote fairness in the impact and recovery of communities (World Commission on Environment and Development 1987, Alexander 2002). A Capability Approach allows us to look at the fairness in the genuine opportunities that different individuals have over time. Capturing the spatial heterogeneity of the impact and recovery requires a sufficient level of granularity in the models.

Finally, intergenerational justice is about giving a fair consideration to future generations (Gardoni & Murphy 2008, Gardoni et al. 2016). Intergenerational justice calls for the design of physical systems, and the development of mitigation and recovery strategies that respect and ideally promote intergenerational equity (Mileti 1999, World Commission on Environment and Development 1987). Climate change, population growth (in particular in coastal regions) and movement (for example from rural to urban areas), the likely advent of new technologies, and the additional uncertainties associate with looking at far future scenarios are example of factors that need to be considered when discussing intergenerational justice (Goodin 1999, Murphy et al. 2011).

These three normative considerations might be competing with each other and pull in opposite directions (MacLean 2016). As a result, addressing them can be challenging and might require different trade-offs.

5 RISK EVALUATION

Murphy & Gardoni (2008) proposed to evaluate a risk by comparing a measure of the capabilities of individuals with an acceptable threshold and a tolerable threshold. The acceptable threshold sets the minimum level of a capability that should be permissible over any period of time to any individual. The definition of the acceptable threshold captures a demand of justice, "necessary condition of justice for a public political arrangement is that it delivers to citizens a certain basic level of capability" (Nussbaum 2000).

In rare occasions (like in the case of the occurrence of an extreme natural event), Murphy & Gardoni (2008) argued that a lower level of the capabilities should be permitted. However, being below the acceptable threshold should be temporary, and still above a tolerability threshold. The tolerability threshold is defined as the "absolute minimum level of capabilities any individual should have at any time." (Murphy & Gardoni 2008)

Using the acceptable and tolerable thresholds, a risk can be classified as acceptable, not acceptable and not tolerable based on the capabilities immediately after the occurrence of a hazard (see Figure 1 from Gardoni & Murphy (2018) for a schematic representation). When a risk is classified as not acceptable, we need to consider how quickly the capabilities improve to assess whether the risk is tolerable or not. This can

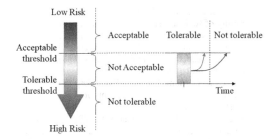

Figure 1. Schematic representation of individual risk evaluation (Adapted from Gardoni & Murphy 2018).

be done by modelling the recovery of the physical systems and society as described in Section 3. As a result, the resilience of physical systems and society informs the evaluation of a risk. If the physical systems and society are more resilient, the recovery time tends to be shorter, and risks tend to be more tolerable.

As for the evaluation of risk at the individual level, we can account for global and intergenerational justice in the risk evaluation by defining a permissible threshold of inequality (Gardoni & Murphy 2018). If the inequalities are likely to be exacerbated beyond the inequality threshold, the risk might not be acceptable because of its social injustice. At the same time, the risk might be tolerable as long as the exacerbation is temporal and reversed in a sufficiently short time. The definitions of the acceptable, tolerable and inequality thresholds can be based on normative obligations informed by human rights as well as incorporate information solicited by democratic deliberations.

6 CONCLUSIONS

Natural and anthropogenic hazards can significantly impact society. This paper focused on the question of how we can predict the societal impact of hazards. Since physical systems (like structures and infrastructure) play a critical role on the well-being of individuals, the paper started by describing how we can predict the reduction or loss of functionality of physical systems. Then, the paper presented a Capability Approach to convert the functionality of physical systems into predictions of the genuine opportunities that individuals have. Such opportunities (also called capabilities) are measures of well-being. A Capability Approach is used to convert the functionality of the physical systems into opportunities of individuals considering social vulnerability factors. A Capability Approach also helps in the process of evaluating risks giving proper consideration about resilience, sustainability and social justice. The results of a regional risk and resilience analysis conducted within the conceptual framework of the Capability Approach can guide in the design of physical systems and the development of mitigation and recovery strategies that promote well-being and social justice.

ACKNOWLEDGEMENTS

The insightful comments from Prof. Colleen Murphy, Armin Tabandeh, Roberto Guidotti, and Jessica Boakye at the University of Illinois at Urbana-Champaign are acknowledged. The research presented in this paper was supported in part by the Center for Risk-Based Community Resilience Planning funded by the U.S. National Institute of Standards and Technology (NIST Financial Assistance Award Number: 70NANB15H044) and the MAE Center: Creating a Multi-hazard Approach to Engineering at the University of Illinois at Urbana-Champaign. The views expressed are those of the author, and may not represent the official position of the sponsors.

REFERENCES

Adger, W.N. (2006) Vulnerability. *Global Environmental Change*, 16 (3), 268–281.

Choe, D., Gardoni, P. & Rosowsky, D. (2010) Fragility increment functions for deteriorating reinforced concrete bridge columns. *ASCE Journal of Engineering Mechanics*, 136 (8), 969–978.

Crucitti, P., Latora, V., Marchiori, M. & Rapisarda, A. (2003) Efficiency of scale-free networks: Error and attack tolerance. *Physica A: Stat. Mech. and its Applications*, 320, 622–42.

Ellingwood, B.R., Cutler, H., Gardoni, P., Peacock, W.G., van de Lindt, J.W. & Wang, N. (2016) The Centerville virtual community: A fully integrated decision model of interacting physical and social infrastructure systems. *Sustainable and Resilient Infrastructure*, 1 (3–4), 95–107.

Gardoni, P. (ed.) (2017) *Risk and Reliability Analysis: Theory and Applications*. Cham, Springer.

Gardoni, P., Der Kiureghian A. & Mosalam K.M. (2002) Probabilistic capacity models and fragility estimates for RC columns based on experimental observations. *ASCE Journal of Engineering Mechanics*, 128 (10), 1024–1038.

Gardoni, P. & LaFave, J. (eds.) (2016) *Multi-hazard Approaches to Civil Infrastructure Engineering*. Cham, Springer.

Gardoni, P. & Murphy, C. (2008) Recovery from natural and man-made disasters as capabilities restoration and enhancement. *International Journal of Sustainable Development and Planning*, 3 (4), 1–17.

Gardoni, P. & Murphy, C. (2009) Capabilities-based Approach to measuring the societal impacts of natural and man-made hazards in risk analysis. *ASCE Natural Hazards Review*, 10 (2), 29–37.

Gardoni, P. & Murphy, C. (2010) Gauging the societal impacts of natural disasters using a capabilities-based approach. *Disasters*, 34 (3), 619–636.

Gardoni, P. & Murphy, C. (2014) A scale of risk. *Risk Analysis*, 34 (7), 1208–1227.

Gardoni, P. & Murphy, C. (2018) Society-based design: Developing sustainable and resilient communities. *Sustainable and Resilient Infrastructure*. doi: 10.1080/23789689.2018.1448667.

Gardoni, P., Murphy, C. & Rowell, A. (eds.) (2016) *Societal Risk Management of Natural Hazards*. Cham, Springer.

Gardoni, P. & Rosowsky, D. (2011) Seismic fragility increment functions for deteriorating reinforced concrete bridges. *Structure and Infrastructure Engineering*, 7 (11), 869–879.

Goodin, R. (1999) The sustainability ethic: Political, not just moral. *Journal of Applied Philosophy*. doi: 16(3), 247–254.10.1111/japp.1999.16.issue-3.

Guidotti, R., Chmielewski, H., Unnikrishnan, V., Gardoni, P., McAllister, T. & van de Lindt, J. (2016) Modeling the resilience of critical infrastructure: The role of network dependencies. *Sustainable and Resilient Infrastructure*, 1 (3–4), 153–168.

Guidotti, R., Gardoni, P. & Chen, Y. (2017a) Network reliability analysis with link and nodal weights and auxiliary nodes. *Structural Safety*, 65, 12–26.

Guidotti, R., Gardoni, P. & Chen, Y. (2017b) Multi-layer heterogeneous network model for interdependent infrastructure systems. In: *Proceedings of the 12th International Conf. on Structural Safety & Reliability* (ICOSSAR 2017), August 6–10, 2017, Vienna, Austria.

Jia, G. & Gardoni, P. (2018) State-dependent stochastic models: A general stochastic framework for modeling deteriorating engineering systems considering multiple deterioration processes and their interactions. *Structural Safety*, 72, 99–110.

Jia, G. & Gardoni, P. (2018) Simulation-based approach for estimation of stochastic performances of deteriorating engineering systems. *Probabilistic Engineering Mechanics*, 52, 28–39.

Jia, G., Tabandeh, A. & Gardoni, P. (2017) Life-cycle analysis of engineering systems: Modeling deterioration, instantaneous reliability, and resilience. In: Paolo Gardoni (ed.) *Risk and Reliability Analysis: Theory and Applications*. Cham, Springer.

Kajitani, Y., Okada, N. & Tatano, H. (2005) Measuring quality of human community life by spatial-temporal age group distributions-case study of recovery process in a disaster-affected region. *Natural Hazards Review*, 6 (1), 41–47.

Kang, W.H., Song, J. & Gardoni, P. (2008) Matrix-based system reliability method and applications to bridge networks. *Reliability Engineering and System Safety*, 93, 1584–93.

Kumar, R. & Gardoni, P. (2014) Renewal theory-based life-cycle analysis of deteriorating engineering systems. *Structural Safety*, 50, 94–102.

Kumar, R., Cline, D. & Gardoni, P. (2015) A stochastic framework to model deterioration in engineering systems. *Structural Safety*, 53, 36–43.

Mileti, D.S. (1999) *Disasters by Design: A Reassessment of Natural Hazards in the United States*. Washington, DC, Joseph Henry Press.

Murphy, C. & Gardoni, P. (2006) The role of society in engineering risk analysis: A capabilities-based approach. *Risk Analysis*, 26 (4), 1073–1083.

Murphy, C. & Gardoni, P. (2007) Determining public policy and resource allocation priorities for mitigating natural hazards: A capabilities-based approach. *Science and Engineering Ethics*, 13 (4), 489–504.

Murphy, C. & Gardoni, P. (2008) The acceptability and the tolerability of societal risks: A capabilities-based approach. *Science and Engineering Ethics*, 14 (1), 77–92.

Murphy, C. & Gardoni, P. (2010) Assessing capability instead of achieved functionings in risk analysis. *Journal of Risk Research*, 13 (2), 137–147.

Murphy, C. & Gardoni, P. (2011) Evaluating the source of the risks associated with natural events. *Research Publica*, 17 (2), 125–140.

Murphy, C., Gardoni, P. & Harris, C.E. (2011) Classification and moral evaluation of uncertainties in engineering modeling. *Science and Engineering Ethics*, 17 (3), 553–570.

Murphy, C., Gardoni, P. & McKim, R. (eds.) (2018) *Climate Change and Its Impact: Risks and Inequalities*. Cham, Springer.

Nussbaum, M. (2000) Aristotle, politics, and human capabilities. *Ethics*, 111 (1), 102–140.

Peacock, W.G. & Girard, C. (1997) Ethnic and racial inequalities in hurricane damage and insurance settlements. In: W.G. Peacock, B.H. Morrow & H. Gladwin (eds.) *Hurricane Andrew: Ethnicity, Gender and the Sociology of Disasters*. Routledge, London.

Ramamoorthy, K.S., Gardoni, P. & Bracci, J.M. (2008) Seismic fragility and confidence bounds for gravity load designed reinforced concrete frames of varying height. *ASCE Journal of Structural Engineering*, 134 (4), 639–650.

Ribot, J. (1995) The causal structure of vulnerability: Its application to climate impact analysis. *GeoJournal*, 35 (2), 119–22.

Sen, A. (1999) *Development as Freedom*. New York, Anchor Books.

Sharma, N., Tabandeh, A. & Gardoni, P. (2017) Resilience analysis: A mathematical formulation to model resilience of engineering systems. *Sustainable and Resilient Infrastructure*. doi:10.1080/23789689.2017.1345257.

Tabandeh, A., Gardoni, P. & Murphy, C. (2018) Reliability-based capability approach: A system reliability formulation for the capability approach. *Risk Analysis*, 38 (2), 41–424.

Tabandeh, A., Gardoni, P., Murphy, C. & Myers, N. (2017) A dynamic bayesian network for capability-based risk and resilience analysis. In: *Proceedings of the 12th International Conf. on Structural Safety & Reliability* (ICOSSAR 2017), August 6–10, 2017, Vienna, Austria.

World Commission on Environment and Development (1987) *Our Common Future*. New York, Oxford University Press.

Xu, H. & Gardoni, P. (2016) Probabilistic capacity and seismic demand models and fragility estimates for reinforced concrete buildings based on three-dimensional analyses. *Engineering Structures*, 112, 200–214.

Life-Cycle Analysis and Assessment in Civil Engineering: Towards an Integrated Vision – Caspeele, Taerwe & Frangopol (Eds)
© 2019 Taylor & Francis Group, London, ISBN 978-1-138-62633-1

Corrosion and its effects on deterioration and remaining safe life of civil infrastructure

C.-Q. Li
RMIT University, Melbourne, Australia

ABSTRACT   This paper summarises results produced from a comprehensive research program on investigation of ferrous metal corrosion and its effect on remaining safe life of ferrous metal structures, including both above ground structures, e.g., steel bridges and underground structures, e.g., cost iron pipelines. The experimental results include corrosion effects on mechanical property of metal at both macro level, e.g., the fundamental stress and strain, and micro level, e.g., element content changes, thereby presenting a cause and effect relationship for corrosion induced material degradation. Numerical results include corrosion effects on fracture failure for both single mode and mixed mode fractures. Analytical results include new solutions to first passage probability which is the most rational method in time-dependent reliability employed for service life prediction of structures. A new solution to upcrossing rate for non-stationary and non-Gaussian processes is presented. Examples are provided to elaborate on new findings presented in the paper.

1 INTRODUCTION

Corrosion of ferrous metal structures, both steel and cast iron and both above ground and underground structures, has been a global problem and poses a great challenge to research community of civil engineering. The cost related to corrosion induced maintenance, repairs and unexpected failures is beyond comprehension which can be the motivation for a large number of researchers working in the subject area.

Corrosion of steel may also lead to degradation of its mechanical properties by changing three microstructural features: (1) grain size, (2) phase composition, and (3) formation of corrosion pits. Zhou and Yan (2016) pointed out that corrosion reduces the grain size through intergranular corrosion. The bonding force among grains can be weakened during intergranular corrosion, which degrades the mechanical properties of steel. Shanmugam et al. (2007) and Olasolo et al. (2011) indicated that steel contains two main phases judging from its crystal structure, namely ferrite and pearlite. Ferrite is known as α-iron (α-Fe) which provides steel with ductility and yield strength, and pearlite is composed of ferrite (α-Fe) and cementite (Fe_3C) that makes steel brittle. Sun et al. (2014) revealed ferrite is corrosion prone and cementite is corrosion resistant. Furthermore, Revie (2008) and Turnbull (2014) stated that the mechanical properties declined due to the existence of pitting corrosion, as stress concentration happens at corrosion pits.

Cast iron pipes have had a historic use for water, oil and gas transmission since the early 19th century. Substantial literature is reported on the corrosion of buried metals, with most extensive investigations in field (Moore and Hallmark 1987, Norin and Vinka 2003, Petersen et al. 2013, Romanoff 1957, 1964), or laboratory tests using simulated solutions by accelerated means (Mohebbi and Li 2011, Liu et al. 2010, Wu et al. 2010). A limited number of laboratory investigations using real soils (Li et al. 2017) were performed in the past. It is known that corrosion of cast iron pipes experiences different rates within service life, imposing difficulties in quantifying corrosion rate and predicting its behavior.

Corrosion induced surface cracks have long been recognized as a major cause of potential failures for cylindrical structures, e.g., pipes. This is because, under external loads and exacerbated with material deterioration, these cracks can quickly grow and the stresses around the crack region intensify, which may result in sudden failure of the structure. Literature review suggests that most of the previous research focused on strength of metal water pipes for their failure assessment. For cast iron pipes, Rajani et al. (2000) developed a methodology to estimate the remaining service life of grey cast iron water mains. Based on this study, Sadiq et al. (2004) used Monte Carlo simulation (MCS) to conduct reliability analysis of cast iron water mains, with both axial and hoop stresses considered in the limit state function.

In contrast to the above research on the strength failure of pipes, only a few reliability based studies are carried out on metal pipes, applying fracture as the failure criterion. Camarinopoulos et al. (1999) developed a multi-dimensional time dependent failure surface method for brittle cast iron water mains. Li and Mahmoodian (2013) and Mahmoodian and Li (2016) employed the up-crossing method and the stochastic gamma process respectively for the service life prediction of cast iron pipes. All the above three studies only considered Mode I stress intensity factors in the limit state function.

It has been known that most of the parameters associated with pipe failures, such as pipe geometry, material properties and corrosion process, exhibit various degrees of variations. They also change with time. To take into account the uncertainty and time-variance in these parameters, a stochastic approach is essential for the assessment of pipe failures and prediction of remaining service life.

2 CORROSION OF STEEL

2.1 Experimental program

Design of test specimens. G250 mild steel was used as the test material. Two series of steel specimens were designed in this study: specimens for tensile test (Figure 1) following ASTM E8/E8M-16a (2016), and specimens for hydrogen concentration test (Figure 2) following ASTM F1113-87 (2017). In addition, the tensile test specimens were wrapped with acid resistant tape at both ends, so that corrosion only takes place in the tested area (i.e., 50 mm gauge length in the middle). After hydrogen concentration measurement, specimens were further cut into $40 \times 14 \times 6$ mm samples (Sample 1) for element composition analysis, and $14 \times 6 \times 4$ mm samples (Sample 2, 3, 4, and 5) for microstructure study, according to ASTM E1621-13 (2013) and ASTM F1113-87 (2017) (Figure 2).

Mechanical tests. Tensile test was carried out on the specimens after they were cleaned and weighed, according to ASTM E8/E8M-16a (2016). The crosshead speed was set as 1.1 mm/s for entire specimens. Yield strength, ultimate strength, and failure strain were measured by tensile test and determined by the true stress-strain curve. True stress-strain curve can be computed based on Garbatov et al. (2014). Fatigue tests were also carried out to obtained S-N curve for corroded steel.

Microstructural tests. At the end of each immersion period, steel samples (i.e., sample 1 in Figure 2) were used for X-ray fluorescence (XRF) test. XRF test was conducted by Bruker Axs S4 Pioneer XRF equipment to detect the element compositions of each sample. Before grain size analysis, the prepared samples in Figure 4 were etched with 2% Nital for 30 seconds to make the grain boundaries appear (ASTM E340-15 2015). Grain size was characterized for etched samples at the end of each immersion period. Measuring location is same as that for steel phase, shown as Figure 4. By using ImageJ (a software designed to analyse and edit images), average grain size can be calculated using Linear Intercept Procedure (ASTM E112-13 2013).

2.2 Results and analysis

Corrosion progress. Corrosion is measured physically by mass loss of steel specimens as a function of time is shown in Figure 3. The results show that mass loss increases gradually with time. In details, mass loss is 1.33% in 0.00001 M HCl (pH = 5), 2.55% in 0.003 M HCl (pH = 2.5), 17.17% in 1M HCl (pH = 0), and 50.86% in 3 M HCl (pH = −0.5) after 28 days. In addition, the result indicates that a higher concentration of acid favours corrosion process. These findings are also recorded by previous literatures (Li et al. 2017).

Corrosion pits. The largest corrosion pits observed close to the boundaries of Sample 2, 3, and 4 after 28 days immersion are shown in Figure 4 as representatives. The width and depth of these corrosion pits were measured through ImageJ. In details, the width of the corrosion pits is 0.07 mm, 0.54 mm, 0.47 mm, and 1.21 mm for 00001 M HCl (pH = 0), 0.003 M HCl (pH = 2.5), 1 M HCl (pH = 0) and 3 M HCl (pH = −0.5) solutions respectively. The depth of corrosion pits is 0.11 mm, 1.17 mm, 1.73 mm, and 2.45 mm respectively.

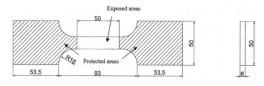

Figure 1. Specimen for tensile test (Unit: mm).

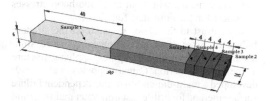

Figure 2. Specimen for hydrogen concentration, element composition and microstructure test (Unit: mm).

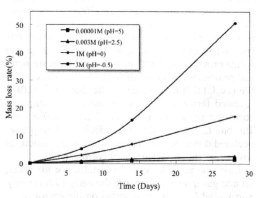

Figure 3. Mass loss with time.

Figure 4. SEM analysis of corrosion pits before and after 28 days' immersion.

Figure 5. Corrosion induced delamination.

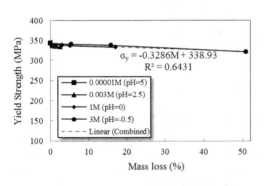

(a) Yield strength

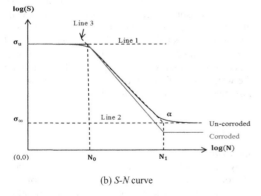

(b) S-N curve

Figure 6. Reduction of mechanical properties of corroded steel.

Corrosion induced delamination. Figure 5 shows the photo and analysis of corrosion induced delamination of specimens after 28 days immersion in pH solutions. As it is seen, visible splits of layers (delamination) occur in the middle cross the thickness. This is because the middle of the thickness has larger grain size; higher ferrite composition, and more impurities than the edge (Marcus 2011, Syugaev et al. 2008, Ralston et al. 2010).

Effect of corrosion on mechanical properties. The effect of corrosion on mechanical properties was investigated by tensile test and fatigue test, the results of which are shown in Figure 6. The graphs show that yield strength and fatigue decrease with mass loss increment. To be specific, after 28 days' corrosion, yield strength drops from 342.57 MPa to 334.56 MPa in 0.00001 M HCl (pH = 5) when mass loss reaches 1.33%, and reduces to 333.58 Mpa in 0.003 M HCl (pH = 2.5) when mass loss reaches 2.55%. In addition, yield strength falls to 332.79 MPa in 1 M HCl and 321.42 MPa in 3 M HCl, when mass loss rises to 17.7% and 50.86% respectively.

Effect of corrosion on microstructure. Changes of element contents of two key elements are shown in Figure 7. In Figure 7a, iron (Fe), which makes steel soft and ductile, decreases from 93.01% to 76.74% in 0.00001 M HCl (pH = 5), 74.78% in 0.003 M HCl (pH = 2.5), 58.60% in 1 M HCl (pH = 0), and 51.61% in 3 M HCl (pH = −0.5) after 28 days' immersion. This reduction also explains the degradation of steel ductility after corrosion. Simultaneously, oxygen (O) element increases from 5.92% to 20.8%, 22.78%, 38.27%, and 46.08% in 0.00001 M (pH = 5), 0.003 M (pH = 2.5), 1 M (pH = 0), and 3 M HCl (pH = −0.5) respectively (Figure 9b). This is due to the formation of iron oxide (Magentite, Lepidocrocite, etc) during corrosion.

From Figure 8, the average grain size of steel is 12.18 μm before corrosion. The grain size then reduces by 29.9% in 0.00001 M HCl (pH = 5), 40.1% in 0.003 M HCl (pH = 2.5), 42.9% in 1 M HCl, and 46.5% in 3 M HCl after 28 days' immersion. Intergranular corrosion is one of crucial mechanisms of the corrosion-induced degradation of mechanical properties. It is important to estimate the level of intergranular corrosion during corrosion by monitoring the grain size changes. More details of the test results and their analysis are available in Li et al. (2017).

3 CORROSION OF CAST IRON

3.1 *Experimental program*

Selection of materials. The material used in this study is gray cast iron due to its widespread application in water, oil and gas transmission sector. To manufacture pipe specimens with realistic material composition representing those of in-service pipelines,

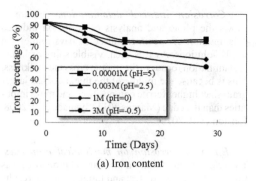

(a) Iron content

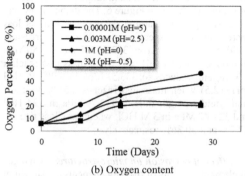

(b) Oxygen content

Figure 7. Changes of element contents due to corrosion.

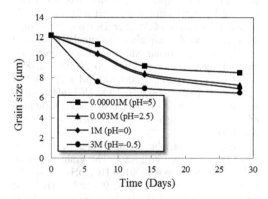

Figure 8. Quantification of grain size variation in various solutions after 28 days' immersion.

the element composition tests were conducted on two exhumed cast iron pipes obtained from local water utilities. The comparison of the element composition with current Australian specification of cast iron material shows that T220 Gray cast iron has a most similar composition with the exhumed pipe material compositions. Therefore, T220 gray cast iron was selected in this study to make pipe specimens.

Specimens for fracture toughness. In the current study, the fracture toughness of corroded pipes was determined using single-edge bend specimens (SENB)

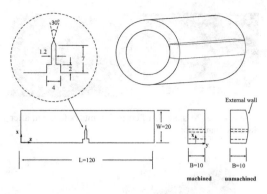

Figure 9. Fracture toughness specimens cut from pipe setion.

which were cut from pipe wall as illustrated in Figure 9. Two types of specimens were used in current fracture toughness tests. One is a standard specimen with square cross section recommended by ASTM E399/E1820 (2012). Since the external surface of pipe wall was flatten in making this type of specimens, they are called machined specimens in this paper. The machined specimens were prepared from both corroded and non-corroded pipes, and 4 duplicates were made from each pipe section to allow for the effects of material and test system variability. To consider the effect of external corrosion on fracture toughness of pipe, another type of specimen was made with the external surface of pipe unmachined (called unmachined specimens), and 4 duplicates were cut from each pipe specimen.

Test setup and procedure. In each soil container, the pipe specimens were buried end to end with 60 mm distance between each to eliminate the influence of galvanic action of one specimen on another. Both the bedding height and the distance between pipe surface and container wall were designed to be the same as burial depth, i.e., 300 mm. To have uniform soil density, clay soil was compacted with each layer of 50 mm in height. The uniform compaction was ensured by wet tamping using a medium scale drop hammer (weight 3.5 kg, 180 blows per layer). In each container, two moisture sensors, three thermocouples, and two pH electrodes were buried at the pipe burial level. All sensors were connected to a data logger which recorded the measured data every hour. One soil container was placed on the high capacity digital scale to measure the weight loss of container weekly. From the weight loss, the water was compensated.

3.2 Results and analysis

Corrosion progress. Corrosion is measured physically by mass loss of specimens as shown in Figure 10 for specimens exposed to soil with pH of 2.5, 3.5 and 5.0 for exposure of 365 days (12 months). As expected, it is found that in general with longer exposure times, more metal loss was caused and the

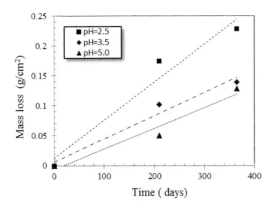

Figure 10. Mass loss of pipe sections in soil with various pH.

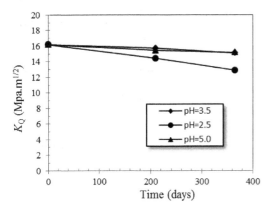

Figure 11. Reduction of fracture toughness in soil with various pH.

corrosion rate attains a relatively constant value in different soil conditions. For example, in specimens buried in soil with pH = 2.5, the corrosion induced mass loss is 0.17 g/cm² after 7 months' exposure and 0.23 g/cm² at the end of 12 months. In comparison, the pipe specimen buried in soil with pH 3.5 and pH 5.0 experienced less corrosion. The slopes of mass loss change in both pH 2.5 and 3.5 indicates a reduction of corrosion rates over exposure time. A clear variance of mass loss results can also be seen due to the stochastic nature of corrosion and measurement errors. Overall, this result is in a reasonable agreement with the preceding corrosion current measurement and those in published reports (e.g., Murray and Moran 1989 and Romoff 1957).

Fracture toughness reduction. The results of fracture toughness test of pipe specimens are summarized in Figure 11 which indicates a decreasing trend of fracture toughness with longer exposure time for specimens buried in soil with various pH levels. To be specific, the reduction of fracture toughness after 365 days corrosion exposure for pH 5.0, 3.5 and 2.5 are 6.22%, 6.70% and 20.29%, respectively. As anticipated, the reduction of fracture toughness of specimens exposed to soil with pH = 2.5 is larger than that of specimens in the soil of pH 3.5 and 5.0. Also, the specimens in soil with pH 2.5 had larger variation rate of corrosion current density (i.e., the slope of curve) than specimens in soil with pH 3.5 and 5.0.

More details of test results and their analysis are available in Wang et al. (2017).

4 CORROSION INDUCED FRACTURE FAILURE

4.1 Stress intensity factors for mixed mode fracture

There are three modes of fracture, namely Modes I, II and III (Anderson 2005). In Mode I, the applied load is normal to the crack plane and tends to open the crack. Mode II refers to in-plane shear mode and tends to slide one crack surface with respect to the other, while Mode III corresponds to out-of-plane shear. A cracked pipe can be subjected to a load that results in any of these three modes of fracture or a combination of two or all three modes. For cracked pipes, the formula for Mode I stress intensity factor at any location along an semi-elliptical crack in Raju and Newman (1982) can be generalized for the stress intensity factors for mixed mode fracture as follows

$$\boldsymbol{K} = \sigma\sqrt{\pi a/Q}\,\boldsymbol{F}(a/d, a/c, d/R, \theta, \xi) \quad (1)$$

where $\boldsymbol{K} = \{K_I\ K_{II}\ K_{III}\}^T$, K_I, K_{II} and K_{III} are stress intensity factors for Modes I, II and III respectively, σ is the applied stress, a is crack depth, Q is the shape factor for an ellipse and is given by the square of the complete elliptical integral of the second kind (Green and Sneddon 1950), d is the thickness of the pipe, c is half of crack length, R is the internal radius of the pipe, θ is the angle between the crack and the axial direction (Fig. 12), and $\xi = x'_P/x'_E$ is used to define the position of an arbitrary point P along the semi-elliptical crack. $\boldsymbol{F}(a/d, a/c, d/R, \theta, \xi) = \{F_I\ F_{II}\ F_{III}\}^T$, where F_I, F_{II}, F_{III} are the influence coefficient functions for Modes I, II and III respectively, which are a function of the above geometrical parameters.

4.2 Three-dimensional finite element modelling

Consider a pipe with an inclined crack as shown in Fig. 12. The origin of the Cartesian coordinate system is set at the centre of the pipe. The internal and external radii of the pipe are R_i and R_o respectively, the thickness of the pipe wall is d, and the whole pipe length is l. The crack depth and length are denoted as a and $2c$ respectively (Fig. 19). The Poisson's ratio is taken to be 0.3 for all pipe analyses. Two types of loads are applied to the pipes: far-field uniform tension and far-field bending. The far-field uniform tension stress and maximum bending stress are denoted by α_a and α_b

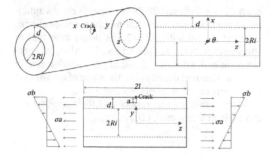

Figure 12. A pipe with an inclined crack subjected to axial tension and bending.

respectively as shown in Fig. 18. The general-purpose code ABAQUS (2011) is employed for all numerical analyses.

4.3 Results and analysis

The influence coefficients of stress intensity factors for different inclination angles are presented in Fig. 13 for bending ($d/R = 0.25$, $a/c = 1.5$, $a/d = 0.5$). From all analysed cases, it can be observed that mixed modes exist along the crack front when an inclined crack ($\theta \neq 90$) is subjected to bending. At the deepest points of the cracks, only Mode I and Mode III exist while at the other points all three modes take place. More details of corrosion induced fracture with mixed mode are available in Fu et al. (2017).

5 METHODOLOGY FOR SERVICE LIFE PREDICTION

5.1 Problem formulation

Service life of a pipe is in general defined as the time period at the end of which the pipe stops performing its intended functions (Li and Mahmoodian, 2013). In assessing the risk of failures for a pipe, a performance criterion should be established. In the theory of structural reliability, this criterion is expressed in the form of a limit state function as follows

$$G(R, L, t) = R(t) - L(t) \quad (2)$$

where $L(t)$ is the load or its effect at time t, $R(t)$ is the resistance. With the limit state function of Equation (8), the probability of pipe failure p_f can be determined by

$$p_f(t) = P[G(R, L, t) \leq 0] = P[L(t) \geq R(t)] \quad (3)$$

where P denotes the probability of an event. Equation (9) represents a typical up-crossing problem, which can be dealt with using time-dependent reliability methods (Melchers 1999). This is known as "first passage probability" and under the assumption of Poisson

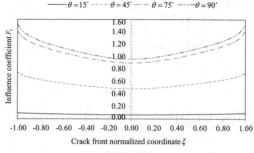

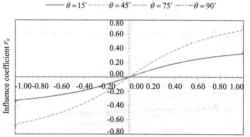

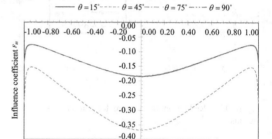

Figure 13. Influence coefficients of stress intensity factors for pipes under bending ($d/R = 0.25$, $a/c = 1.5$, $a/d = 0.5$).

processes it can be expressed as follows (Melchers 1999)

$$p_f(t) = 1 - [1 - p_f(0)]e^{-\int_0^t v d\tau} \quad (4)$$

where $p_f(0)$ is the probability of pipe failure at time $t = 0$ and v is the mean rate for the response process $L(t)$ to upcross the threshold $R(t)$.

At a time that $p_f(t)$ is greater than a maximum acceptable risk in terms of the probability of failure, p_a, it is the time that the pipe fails. This can be determined as follows

$$p_f(t_L) \geq p_a \quad (5)$$

where t_L denotes the service life based on the criterion employed. In principle, p_a can be determined from risk-cost optimization of the pipe during its whole service life.

5.2 New solution for upcrossing rate for lognormal process

In general, the mean upcrossing (or outcrossing) rate of Equation (4) can be calculated from the Rice formula as follows

$$v(t) = v_a^+(t) = \int_a^\infty (\dot{z} - \dot{a}) f_{Z\dot{Z}}(a, \dot{z}) d\dot{z} \qquad (6)$$

where v_a^+ is the mean upcrossing rate of the scalar process $Z(t)$ relative to $a(t)$, the deterministic barrier level to be upcrossed; $\dot{a}(t)$ is the slope of $a(t)$ with respect to time; $\dot{Z}(t)$ is the time derivative process of stochastic process $Z(t)$; and $f_{Z\dot{Z}}$ is the joint probability function for $Z(t)$ and $\dot{Z}(t)$.

To derive a solution to Eq. (14), the joint probability density function of Z and $\dot{Z}$, i.e., $f_{Z\dot{Z}}$, is expressed as $f_{z\dot{z}}(a,\dot{z}) = f_Z(a) \cdot f_{\dot{z}|z}(\dot{z}|a)$ where $f_{\dot{z}|z}$ is the conditional probability density function of $\dot{Z}$ given $Z = a$. The closed from solution to upcrossing rate can be derived as

$$v_a^+(t) = \frac{e^{\varepsilon_{\dot{z}|z}(t)}}{a(t)\varepsilon(t)} \varphi\left(\frac{\ln[a(t)] - \lambda(t)}{\varepsilon(t)}\right) \left\{\sqrt{e}\left[1 - \Phi\left(1 - \frac{\ln[\dot{a}(t)] - \lambda_{\dot{z}|z}(t)}{\varepsilon_{\dot{z}|z}(t)}\right)\right]\right.$$
$$\left. - \left(\frac{\ln[\dot{a}(t)] - e^{\lambda_{\dot{z}|z}(t)}}{e^{\varepsilon_{\dot{z}|z}(t)}}\right) \Phi\left(-\frac{\ln[\dot{a}(t)] - \lambda_{\dot{z}|z}(t)}{\varepsilon_{\dot{z}|z}(t)}\right)\right\}$$

More details of the derivation of the solution are available in Li et al. (2017).

6 WORKED EXAMPLES

6.1 Corrosion affected steel bridge

The example bridge is a railway steel viaduct in Melbourne, Australia built around 1889. It has 45 spans and each span has 3 steel girders. The entire length of the viaduct is around 800 m with individual span ranging from 10 m to 20 m. The girder is composed of riveted mild steel plates, with typical cross section shown in Fig. 14a. Connections between girders are considered as pin joints based on BS 5400.5 (2005), with details shown in Fig. 14b.

The results are shown in Fig. 15. The acceptance probability of failure (P_a) is 0.05% based on BS 5400.1 (1988). Based on fatigue damage determined at mid span, the fatigue life (T_c) for the girder is 113 years when auto-correlation coefficient $\rho = 0.9$, 136 years when $\rho = 0.5$, and 190 years when $\rho = 0.1$. Fig. 24 also indicates that probability of failure is sensitive to the auto-correlation between two points in time (ρ value). According to Miner's law, fatigue damage at any time point depends on the accumulate damage caused by load cycles in previous times. Therefore, it is reasonable to assume that there is high auto-correlation between different time ($\rho = 0.9$), based on which the fatigue life for the selected bridge girder (T_c) is 113 years.

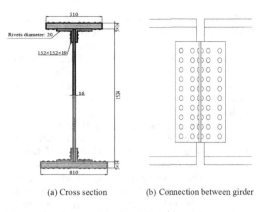

(a) Cross section   (b) Connection between girder

Figure 14. Structural details of the bridge girder.

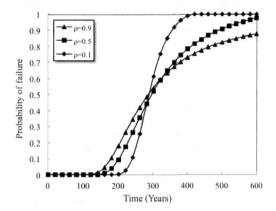

Figure 15. Probability of failure due to fatigue and corrosion for different coefficients of correlation (ρ).

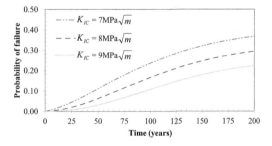

Figure 16. Probability of pipe failure due to corrosion induced crack for different fracture toughness.

6.2 Corrosion affected cast iron pipes

A pipe with an inclined crack made of cast iron is considered for illustration (Figure 12). The aspect ratio a/c of the crack is assumed to be 0.4. As a result, only the deepest point along the crack front is considered, as the stress intensity factor of which is the largest. The values of Mode I fracture toughness for cast iron and steel are taken to be 8 and 50 MPa$\sqrt{m}$ respectively.

At the deepest point along the crack front, only mixed modes I and III exist (Table 1). Therefore, only the fracture toughness for mixed mode I and III is required. Assume that K_C/K_{IC} for mixed mode I and III follows the trend of that in Raghavachary et al. (1990). The statistical information of the random variables involved in the analyses for both pipes is shown in Table 5. More details are available in Yang et al. (2017).

The results are presented in Figure 16 which indicates that different values of material fracture resistance would lead to different probability of failure. The greater the fracture toughness K_{IC} is, the smaller the probability of failure. This is again self-evident.

With the probability of failure known over time, the time for the pipe to be unsafe due to corrosion induced cracking failure, i.e., the service life t_L can be obtained for a given acceptable risk. For example, from Figure 16 and for acceptable risk of 0.1, it can be determined that $t_L = 35$ years, 70 and 100 years for a cast iron pipe respectively. More details of the example are available in Li et al. (2017).

7 CONCLUSION

Results produced from a comprehensive research program on investigation of ferrous metal corrosion and its effect on remaining safe life of ferrous metal structures have been presented in the paper. From the results and their analysis it has been found that (1) Corrosion of steel reduces the mechanical properties of steel, including the fatigue resistance; (2) corrosion of cast iron reduces fracture toughness of the cast iron; (3) corrosion pits induced cracks cause mixed mode fracture failure of cast iron pipes; (4) a new solution to upcrossing rate for lognormal process accommodates practical

application where distribution of design parameters cannot be negative. It can be concluded that the results presented in the paper can contribute to the body of knowledge of ferrous metal corrosion and its effects on remaining safe life of corrosion affected infrastructure.

REFERENCES

Fu, G.Y., Yang, W. & Li, C.Q. (2017) Stress intensity factors for mixed mode fracture induced by inclined cracks in pipes under axial tension and bending. *Theoretical and Applied Fracture Mechanics*, 89, 100–109.

Li, C.Q., Firouzi, A. & Yang, W. (2016) Closed Form Solution to First Passage Probability for Non-Stationary Lognormal Processes. *ASCE Journal of Engineering Mechanics*, 142 (12), doi: 10.1061/(ASCE)EM.1943-7889.0001160.

Li, L., Li, C.Q., Robert, D. & Mahmoodian, M. (2017) Effect of corrosion and hydrogen embrittlement on microstructure and mechanical properties of mild steel. *Construction and Building Materials*, revised.

Marcus, P. (ed.) (2011) *Corrosion Mechanisms in Theory and Practice*. Boca Raton, FL, CRC Press.

Rajani, B. (2000) *Investigation of Grey Cast Iron Water Mains to Develop a Methodology for Estimating Service Life*. Denver, CO, American Water Works Association.

Revie, R.W. (2008) *Corrosion and Corrosion Control*. Hoboken, NJ, John Wiley & Sons.

Romanoff, M. (1957) *Underground Corrosion*. Washington, DC, US Government Printing.

Wang, W.G., Li, C.Q., Robert, D. & Mahmoodian, M. (2017) Experimental investigation on corrosion effect on fracture toughness of buried cast iron pipes. *ASCE Materials in Civil Engineering* (in press).

Yang, W., Fu, G.Y. & Li, C.Q. (2017) Reliability based service life prediction of corrosion affected metal pipes with mixed mode fracture. *ASCE Journal of Engineering Mechanics*, 144 (2), 9.

Life-Cycle Analysis and Assessment in Civil Engineering: Towards an Integrated Vision – Caspeele, Taerwe & Frangopol (Eds)
© 2019 Taylor & Francis Group, London, ISBN 978-1-138-62633-1

New architecture created from high performance structures

M.P. Sarkisian
Skidmore, Owings & Merrill LLP, San Francisco, CA, USA

ABSTRACT

Invention is essential to address urgent needs in the built environment. As population growth continues and demands on our cities increase, new ideas are constantly needed to address urban planning, architecture, and engineering through creative use of materials, construction techniques, and building operations.

Inspirations from natural growth and human behavior are key to developing ideas that adapt to changing climate and depletion of resources. Awareness is key to solving complex issues and the collaboration between academic study, research, and practice is critical.

New ideas for considering structures for higher performance are being developed. One of the platforms for considering this performance is structural optimization where density analyses positions material within structures when subjected to various loading and support conditions. This technique has become particularly important not only in high-rise building design but in long-span structures with most recent developments apply to conventional framing structures. Resulting designs are structurally efficient, increase the structures life-cycle due to superior performance, and help to reduce the impact of emitted carbon into the environment.

Design is usually to achieve maximum stiffness with least material. For a building, a maximum tip displacement of the building due to lateral loads is typically considered. For a bridge, a maximum vertical displacement is considered do to imposed gravity and/or vehicular loads.

The optimal geometry and placement of material is usually derived based on elastic material behavior that is appropriate for wind and frequent earthquakes. With more significant seismicity, the structure is typically designed to exceed the elastic limit of material or perform in a pseudo plastic behavior (i.e. friction joints). In these cases, ductile or plastic elements should be incorporated into the stiffness of the lateral system to ensure energy dissipation and the consequential reduction of internal loads to the structure.

Over the last several years, optimization techniques have been used to inspire creative approaches to tall building design. The design of the structure for the super-tall Shenzhen Citic Financial was optimized to achieve the maximum stiffness of the building with least material, by considering tip displacement of the building when subjected to load. The resulting design is the stiffest structure that can be obtained for the volume fraction of material considered and satisfies the performance requirements under gravity and lateral loads (wind and seismic).

The application of this technique to post-tensioning tendon layouts provides opportunities for reductions in materials and increased performance. The proposed innovative design methodology enhances the structural designers understanding of gravity framing through the employment of topology optimization. This optimization method iteratively searches a continuous design space for the stiffest configuration of material given a set of support conditions and static loadings.

The use of topology optimization analysis to obtain the PT layout is opening a new window of opportunity for flat plate slab design. What is perhaps equally as important is the potential for this approach to apply to shear wall core lateral systems. Post-tensioning in these core wall systems increases resiliency, controls damage, and reduces occupancy downtime following a major seismic event.

The ultimate goal would be to reach essentially elastic systems when subjected to extreme loadings. This could be accomplished by combining optimized layouts of post-tensioned tendons with mechanical devices similar to Pin-Fuse Seismic Systems. These systems would focus hinged areas on areas that result in best performance.

Life-Cycle Analysis and Assessment in Civil Engineering: Towards an Integrated Vision – Caspeele, Taerwe & Frangopol (Eds)
© 2019 Taylor & Francis Group, London, ISBN 978-1-138-62633-1

Optimal planning of operation and maintenance for offshore wind turbines

J.D. Sørensen
Aalborg University, Aalborg, Denmark

ABSTRACT   Wind energy is an important renewable energy source. Reduction of the levelized cost of energy (LCOE) for wind turbines are very important in order to make wind energy competitive compared to other energy sources, especially for offshore wind turbines. All costs in the life-cycle as well as the energy produced contribute to the calculation of LCOE. Therefore the turbine components should be designed to have sufficient reliability but also not be too costly, and it is important to include in the reliability assessment the information obtained during operation (from condition monitoring and inspections) and the maintenance performed (incl. repairs and replacements). The paper presents models for reliability- and risk-optimal planning of Operation & maintenance incl. inspections.

1 INTRODUCTION

Wind energy is an important renewable energy source. Reduction of the levelized cost of energy (LCOE) for wind turbines are very important in order to make wind energy competitive compared to other energy sources, especially for offshore wind turbines. All costs in the life-cycle as well as the energy produced contribute to the calculation of LCOE. Therefore the turbine components should be designed to have sufficient reliability but also not be too costly, and it is important to include in the reliability assessment the information obtained during operation (from condition monitoring and inspections) and the maintenance performed (incl. repairs and replacements).

Wind turbines are exposed to dynamic loads and the response is very dependent on the wind turbine control system. Fatigue and extreme load effects are subject to significant uncertainties. Models are presented for uncertainty modeling and reliability assessment of especially the structural components such as tower, blades, substructure and foundation. Since the function of a wind turbine is highly dependent on many electrical and mechanical components as well as the control system also reliability aspects of these components are discussed and it is described how their reliability influences the reliability of the structural components. Probabilistic assessment of wind turbines are considered e.g. in Veldkamp (2006), Toft (2010), Agarwal (2008) and Sørensen & Toft (2010).

Further, an approach for risk-based cost optimal planning of operation & maintenance (OM) for offshore wind turbines is presented. Illustrative examples are presented incl. uncertainty modeling, and reliability assessment for substructure steel components

and for blades exposed to extreme loads and fatigue, respectively.

The risk-based approach for planning of OM is exemplified by reference to examples with offshore wind farm where especially planning of inspections and maintenance for blades are considered, and where inspections are planned for fatigue critical offshore wind turbine support structures.

2 RELIABILITY ASSESSMENT

Structural components in wind turbines are designed considering a number of load combinations, see (IEC 61400-1, 2005):

- Failure during normal operation in extreme load or by fatigue (DLC 1)
- Failure under fault conditions (e.g. failure of electrical/mechanical components or loss of grid connection) due to extreme loads or by fatigue (DLC 2)
- Failure during start up, normal shut down or emergency shut down (DLC 3, 4 and 5)
- Failure when the wind turbine is idling/parked and does not produce electricity. Failure can be by extreme loads or by fatigue (DLC 6)
- Failure during transportation and installation (DLC 7)
- Failure during transport, assembly, maintenance and repair (DLC 8)

Further, a reliability assessment requires that a system model is established for each wind turbine, and in some cases also for all wind turbines in a wind farm. This can especially be important for reliability- and risk-based planning of OM. The components can generally be divided in two groups:

1) Electrical and mechanical components modelled using classical reliability models, with the main descriptor being the failure rate, λ. Further, the bath-tub model is often used to describe the time dependent behaviour of the failure rate/hazard rate. The reliability is typically modelled by a Weibull models for the time to failure. Using e.g. FMEA (Failure Mode and Effect Analysis) or FTA (Failure Tree Analysis), system models can be established and the systems reliability can be estimated, see e.g. Tavner et al. (2007) and Tavner (2012). Reliability of drivetrain components (gear-box) has been considered in e.g. (Dong et al. 2013).

2) Structural elements such as tower, main frame, blades and the support structure/foundation with failure modes that can be described by limit state equations, $g_i(\mathbf{X})$. Failure of the tower can e.g. be buckling or fatigue. The parameters in the limit state equation $g(\mathbf{X})$ are assumed to be modelled by n stochastic variables $\mathbf{X} = (X_1, \ldots, X_n)$. The probability of failure, P_f can be estimated using Structural Reliability Methods, e.g. FORM/SORM/ simulation methods, see e.g. Madsen et al. (1986), Ronold et al. (1999) and JCSS (2002).

In many cases fault of an electrical component or the control system results in an increase of the fatigue damage level of the structural components or cause large extreme load effects. As an example, for the load case (DLC 6.2) where the wind turbine is parked, loss of the grid occurs and structural failure by an ultimate limit state. The annual failure rate for failure mode j for this DLC can be estimated from, see also Sørensen & Toft (2014):

$$\lambda_{F_j} = \left\{ \sum_i P\left(F_{i,j} \middle| \text{extreme wind} \cap \text{grid loss} \right) \cdot P_i \right\} \cdot v_{\text{grid loss}} \quad (1)$$

where $\lambda_{\text{grid loss}}$ is the annual failure rate for grid loss/loss of electrical network connection which e.g. can be estimated directly based on observed data. $P(F_{i,j}|\text{extreme wind} \cap \text{grid loss})$ is the probability of failure for a specific failure mode, j with extreme wind speed and 'misalignment position' i at grid loss. P_i is the probability of at 'misalignment position' i at grid loss.

$P(F_{i,j}|\text{extreme wind} \cap \text{grid loss})$ can be estimated by structural reliability methods for a specific failure mode modelling the maximum mean wind speed related to the time period for the grid loss.

For wind turbines the risk of loss of human lives in case of failure of a structural element is generally very small. Further, it can be assumed that wind turbines are systematically reconstructed in case of collapse or end of lifetime. Therefore, an appropriate target reliability level corresponding to a minimum annual probability of failure is considered be $5 \cdot 10^{-4}$ (annual reliability index equal to 3.3), see IEC 61400-1 (2018) and Sørensen & Toft (2014).

3 OPERATION AND MAINTENANCE

Planning of Operation & Maintenance for wind turbines and wind farms is a complex task involving a high degree of uncertainty due to diversity of assets and their corresponding damage mechanisms and failure modes, weather-dependent transport conditions, unpredictable spare parts demand, insufficient space or poor accessibility for maintenance and repair, limited availability of resources in terms of equipment and skilled manpower, etc. In (Shafiee & Sørensen 2017) a conceptual classification framework is presented for maintenance policy optimization and inspection planning of wind energy systems and structures (turbines, foundations, power cables and electrical substations). The framework addresses a wide range of theoretical and practical issues, including the models, methods, and the strategies employed to optimise maintenance decisions and inspection procedures in wind farms.

Wind turbine components are exposed to deterioration processes such as fatigue, wear and corrosion which may result in failures, for example in welded details, blades, bearings and gearboxes. In a number of cases it may be possible to detect these damages before actual failure, and thereby perform preventive maintenance instead of expensive corrective repair/maintenance. This requires information from condition monitoring or structural health monitoring systems to obtain information on the condition of the components. The information can either be online monitoring of e.g. vibrations, temperature, oil particles, etc. or manual inspections of blades, towers, gearboxes, etc. Use of preventive maintenance can reduce the costs, as repairs can be cheaper to perform before actual failure, and because the downtime due to limited weather windows (for offshore wind turbines) will be shorter compared to corrective maintenance. On the other hand, preventive maintenance may lead to more repairs in total, and optimally the maintenance effort should be optimized to minimize the total expected costs applying a combination of corrective and preventive maintenance.

In order to minimize the costs, various maintenance strategies can be considered, and the one resulting in the lowest costs should be chosen. To model the relationship between maintenance and reliability, and to take information from condition monitoring and inspections into account in a consistent way, methods can be developed using Bayesian decision theory, see Raiffa & Schlaifer (1961).

In the oil and gas industry, methods based on the Bayesian pre-posterior decision analysis have been used named risk-based inspection (RBI) see e.g. Faber et al. (2005). Here, the required reliability level for (manned) offshore platforms is generally higher compared to offshore wind turbines, and less information is available for the decision maker. A reliability- and risk-based approach for planning of operation & maintenance is described in e.g. Dong et al. (2012), Sørensen (2009) and Nielsen & Sørensen (2014).

A decision tree related to the life cycle of a wind turbine or wind farm can be constructed, see Sørensen (2009). The decisions are taken by the decision makers (designer/owner/…) and the observations of uncertain parameters (unknown at the time of the decision) are typically:

- At the design stage a decision on the optimal design parameters $\mathbf{z} = (z_1, ..., z_N)$ is made which in principle should maximize the total expected benefits minus costs during the whole lifetime such that safety requirements are fulfilled at any time. In practice requirements from standards and actual costs of materials are used to determine the optimal design.
- During the lifetime continuous monitoring of the wind turbines and inspections of critical components/details are performed. These consists of:

  - a decision on times and types of inspection/ monitoring for the rest of the lifetime
  - observations from inspection/monitoring
  - decision on eventual maintenance/repair based on the inspection/monitoring results

- Realisation of uncertain parameters such as wind and wave climate, strengths, degradation, model uncertainties will take place during the lifetime. It is noted that these uncertainties can be divided in aleatory and epistemic uncertainties. Aleatory uncertainty is inherent variation associated with the physical system or the environment – it can be characterized as irreducible uncertainty or random uncertainty. Epistemic uncertainty is uncertainty due to lack of knowledge of the system or the environment – it can be characterized as subjective uncertainty, reducible uncertainty.
- The total cost is the sum of all costs in the remaining part of the lifetime after the decision time.

The approach can be used for operation and maintenance planning related to different failure & error types in Gearbox, Generator, Rotor blades, Blade pitch mechanism, Yaw mechanism, Main shaft, Tower/ support structure (fatigue cracks, corrosion), etc.

Further, decisions related to operation and maintenance are related to different time scales:

- short (minutes) for decision related to e.g. parking the wind turbine,
- medium (days) for e.g. decisions on when to start offshore maintenance/repair actions depending on e.g. weather forecasts, and
- long (months/years) for e.g. preventive maintenance and inspection/monitoring planning for gear boxes.

An important step in risk-based inspection & maintenance planning is collection of data/information and probabilistic modelling of this information. Information can come from Condition Monitoring Systems (CMS), Structural Health Monitoring (SHM) or inspections. Typically information from CMS and SHM are indicators of the deterioration. Based on this information deterioration reliabilities and failure rates can be updated using Bayesian methods.

4 DAMAGE MODELLING

If a corrective maintenance strategy is used only information on failure rates is basically needed in order to estimate the expected total life-cycle costs. However, if a preventive maintenance strategy is used then also models for damage accumulation are needed. Damage in terms of e.g. corrosion, wear and cracks can be modelled in different ways. In the following defects are considered which can be modelled by crack growth models. Two approaches are followed: 1) application of fracture mechanics models allowing detailed planning of e.g. inspections of fatigue cracks in welded details, see e.g. Sørensen (2012); and 2) Markov type of models which can be used in cases where less detailed information on damage growth is available, see e.g. Florian & Sørensen (2017).

Preventive maintenance is often implemented by performing regular inspections either at fixed time intervals, or depending on the result of the last inspection. Based on the inspection result, different maintenance/repair actions can be performed, often based on a classification of the size of damage:

0 no damage
1 cosmetic
2 minor damage
3 major damage
4 serious damage
5 critical damage
6 collapse

It is noted that the degradation categories can represent quite different forms depending on the type of component. For composite materials used for wind turbine blades the categories can be related to defects/cracks in the shell, erosion of the outer skin, delamination of the carbon fiber sheets or de-bonding in various joints.

A Markov discrete state model can be used to represent the degradation process, see e.g. (Florian & Sørensen 2017). A state i is defined by a corresponding damage level and the amount of time T_i that the defect/crack spends in it. A stochastic modelling of T_i is modelled using a transition probability matrix $\mathbf{P}$. The initial conditions are assumed to be represented by a vector P_0.

It is noted that the probabilistic modelling of category 0–5 typically follows the Markov model, whereas the probabilistic modelling of collapse follows the structural reliability approach in section 2.

5 EXAMPLE – OPERATION AND MAINTENANCE PLANNING

In this section an example is presented where a reliability and risk based maintenance approach is used considering wind turbine blades with the NORCOWE reference wind farm, see Florian & Sørensen (2017) for details. The example also presents the important aspects related to weather constraints and logistics.

Deterioration of wind turbine blades is often detectable by cracks on the outer skin of the blade, or delamination and erosion of the carbon fiber layers. When these deterioration indicators become large enough to be detected at inspections, the remaining life of the blade is typically still sufficiently long enough to allow for detection and planning of an appropriate action.

In this example a Markov model is used for probabilistic modelling of defect growth and as basis for maintenance planning.

Longitudinal cracks in blade trailing edges are a major problem for some wind turbine blades. Installation of so-called D-strings may reduce the problem of longitudinal cracking, see Florian & Sørensen (2017). A probabilistic model is established based on data in the Guide-to-defect (G2D) database.

The maintenance strategy applied in the following is partly based on corrective and partly based on condition based maintenance planning. Condition based maintenance is applied for the blade, implying an inspection scheduling and a decision model for repair/maintenance, while corrective maintenance is used for all other wind turbine components.

It is noted that the base failure rates applied are prior to installation of the D-string, which in this example is considered to mitigate 20% of the damages. Further, representative blade maintenance cost and down times are applied

The overall maintenance policy is to perform the repairs as soon as possible, given that the weather conditions are favorable and the appropriate vessel and technicians are available. Further, in cases with insufficient number of vessels to carry out necessary actions, a priority system is used. Finally, for each vessel maximum wave and wind limits are applied.

A general maintenance strategy and cost mode is used for all other wind turbine components. This model is partly based on Dinwoodie et al. (2015). The failure rates are given in number of failures per turbine per year, and cover failures for all components other than blades. Weather limitations are only dependent on vessels, and similar to the blade maintenance model. Repairs are made as soon as possible, with respect to weather conditions, work force availability and priority order.

Finally, the reliability of the applied inspection technique is accounted for by a probability for not detecting a damage assumed to be proportional to the level of degradation.

For optimal decision making, a simulation based approach is used where lifetime (25 years) simulations are performed for a considered offshore wind farm; using the maintenance model presented above. For this example the NORCOWE reference wind farm layout (Bak et al. 2016) is used, with the modification that 3 MW turbines are considered, since the input provided for the blade maintenance model is for this size of turbines.

A time step of 1 month is used for the Markov model, corresponding to the transition probabilities.

The time to failures are assumed to follow an exponential distribution.

The vessel logistics and the time required for a repair activity is accounted for using the distance between the turbine and the base harbor and the weather conditions. Weather conditions are given as 25 year wind and wave time series at the reference wind farm location, comprising of 3 hour mean values.

When the O&M cost and the availability are computed, the levelized cost of energy (LCOE) is estimated based on the example data in Bak et al. (2017). For this case study, two scenarios are simulated and compared. A base case scenario, where there is no installation of the D-strings and one where the strings are scheduled to be installed in year 5 on all turbines in the wind farm, see Florian & Sørensen (2017) for details.

The main output from the simulations shows that LCOE is decreased by 5% if D-strings are installed after 5 years. This reduction should be compared to the cost of the installation of the D-strings on the existing offshore wind turbine. The uncertainty (COV) of the LCOE estimate is approximately 2.2% and is reduced slightly to 1.9% with D-strings installed.

Another important parameter is the availability which is increased slightly from approximately 92.7% to 92.3%. The uncertainty of the availability estimate is approximately 0.1%, i.e. quite low.

The main contributors to the blade OM are with D-string installed in ( ): Heavy-Lift Vessel: 47% (41%), Replacement of blade: 26% (22%), work boat: 10% (13%), category 4 maintenance: 6% (7%) and category 3 maintenance: 7% (9%). A major difference which has a large effect on the OM costs is the relative decrease in Heavy-Lift Vessels.

Similar results are obtained for downtime for maintenance with the major contributors: category 3 maintenance: 33% (35%), category 4 maintenance: 29% (27%), inspections: 13% (16%) and category 5 maintenance: 10% (9%).

6 EXAMPLE – INSPECTION PLANNING

This example describes how Fatigue design Factors, *FDF* values can be calibrated taking into account inspections as part of O&M planning for offshore wind turbines, see Sørensen (2012) and Sørensen & Toft (2014) for details. The theoretical basis for reliability-based planning of inspection and maintenance for fatigue critical details in offshore steel substructures is described in e.g. Madsen & Sørensen (1990), Faber et al. (2005), Moan (2005), Straub (2004) and Sørensen (2009). Risk- and reliability-based inspection planning is widely used for inspection planning for oil & gas steel jacket structures. Fatigue reliability analysis of jacket-type offshore wind turbine considering inspection and repair is also considered in Dong et al. (2010) and Rangel-Ramírez & Sørensen (2010). In this section examples on how much *FDF* values can be reduced if inspections are performed.

For the fatigue sensitive details/joints to be considered in an inspection plan, the acceptance criteria for the annual probability of fatigue failure may be assessed using a measure for the decrease in ultimate load bearing capacity given failure of each of the individual joints to be considered together with the annual probability of joint fatigue failure. For offshore structures the RSR (Reserve Strength Ratio) is often used as a measure of the ultimate load bearing capacity.

Inspection planning as described above requires information on costs of failure, inspections and repairs. Often these are not available, and the inspection planning is based on the requirement that the annual probability of failure $\Delta P_F(t)$ in all years has to satisfy the reliability constraint

$$\Delta P_F(t) \le \Delta P_{F,\max} \qquad (2)$$

where $\Delta P_{F,\max}$ is the maximum acceptable annual probability of failure taking into account the consequence of fatigue failure. $\Delta P_{F,\max}$ is equal to $5 \cdot 10^{-5}$ following section 2.

This implies that the annual probabilities of fatigue failure have to fulfill (1). Further, in risk-based inspection planning the planning is often made with the assumption that no cracks are found at the inspections. If a crack is found, then a new inspection plan has to be made based on the observation.

A Fracture Mechanics modeling of the crack growth is applied assuming that the crack can be modelled by a 2-dimensional semi-elliptical crack. It is assumed that the fatigue life may be represented by a fatigue initiation life and a fatigue propagation life. It is therefore:

$$N = N_I + N_P \qquad (3)$$

where N is the number of stress cycles to failure, N_I is the number of stress cycles to crack propagation and N_P is the number of stress cycles from initiation to crack through.

The number of stress cycles from initiation to crack through is determined on the basis of a two-dimensional crack growth model. The crack is assumed to be semi-elliptical. The crack growth can be described by two coupled differential equations, see (Faber et al. 2005). The stress range $\Delta\sigma$ is obtained from

$$\Delta\sigma = X_{Wind} X_{SCF} \cdot \Delta\sigma^e \qquad (4)$$

where X_{Wave}, X_{SCF} are model uncertainties and $\Delta\sigma^e$ is the equivalent stress range.

The limit state equation is written

$$g(\mathbf{X}) = N - n\, t \qquad (5)$$

where n is the number of fatigue cycles per year and t is time in years.

The following stochastic model is applied, see below and Sørensen & Toft (2014) for details.

Table 1. Uncertainty modelling used in the fracture mechanical reliability analysis. D: Deterministic, N: Normal, LN: LogNormal, W: Weibull.

Variable	Dist.	Expected value	Standard deviation
N_I	W	μ_0 (reliability based fit to SN approach)	$0.35\,\mu_0$
a_0	D	0.1 mm (high material control)/ 0.5 mm (low material control)	
$\ln C$	N	$\mu_{\ln C_C}$ (reliability based fit to SN approach)	0.77
m	D	m-value (reliability based fit to SN approach)	
X_{SCF}	LN	1	0.05
X_{Wind}	LN	1	0.20
a_c	D	T (thickness)	

$\ln C$ and N_I are correlated with correlation coefficient $\rho_{\ln C, N_I} = -0.5$

To model the effect of different weld qualities, different values of the crack depth at initiation a_0 is used. The corresponding assumed length is 5 times the crack depth. The critical crack depth a_c is taken as the thickness of the tubular member.

The parameters m, $\mu_{\ln C}$ and μ_0 are fitted such that difference between the probability distribution functions for the fatigue live determined using the SN-approach and the fracture mechanical approach is minimized.

The reliability of inspections can be modeled in many different ways. Often POD (Probability Of Detection) curves are used to model the reliability of the inspections, e.g. an exponential model:

$$POD(x) = 1 - \exp\left(-\frac{x}{\lambda}\right) \qquad (6)$$

where λ is the expected value of the smallest detectable crack size.

The crack width $2c$ is obtained from the following model for $a/2c$ as a function of the relative crack depth a/B, where B is the thickness:

$$\frac{a}{2c} = 0.06 - 0.03\ln\left(\frac{a}{B}\right) \qquad (7)$$

If an inspection has been performed at time T_I and no cracks are detected then the probability of failure can be updated by

$$P_F^U\left(t \,|\, \text{no-detection at time } T_I\right) = $$
$$P\left(g(t) \le 0 \,|\, h(T_I) > 0\right) \qquad , t > T_I \qquad (8)$$

where $h(t)$ is a limit state modeling the crack detection. If the inspection technique is related to the crack length then $h(t)$ is written:

$$h(t) = c_d - c(t) \qquad (9)$$

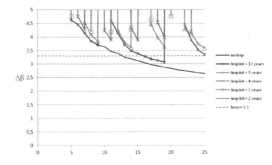

Figure 1. Annual reliability index without and with inspections. Inspection time intervals 2, 3, 4, 5 and 10 years and $\lambda = 10$ mm, partial safety factor $\gamma_m = 1.0$.

where $c(t)$ is the crack length at time t and c_d is smallest detectable crack length. c_d is modelled by a stochastic variable with distribution function equal to the POD-curve. Similarly if the inspection technique is related to the crack depth then $h(t)$ is written $h(t) = a_d - a(t)$ where $a(t)$ is the crack length at time t and a_d is smallest detectable crack length.

It is emphasized that the inspection planning is based on a no-find assumption. This way of inspection planning is the one which if most often used. Often this approach results in increasing time intervals between inspections.

Results show among others that

- if the fatigue partial safety factor is chosen to 1.0 then one inspection is needed at year 13 with at least a reliability which corresponds to an expected value of the smallest detectable crack equal to 2 mm, see Figure 1.
- if the fatigue partial safety factor is chosen to 1.0 then inspection intervals of maximum 5 years should be performed with at least a reliability which corresponds to an expected value of the smallest detectable crack equal to 10 mm.
- if the fatigue partial safety factor is chosen to 1.1 then inspection intervals of maximum 10 years should be performed with at least a reliability which corresponds to an expected value of the smallest detectable crack equal to 10 mm.

These results can be used as basis for planning of inspections of offshore wind turbines, and also as basis for modifying partial safety factors to be used for design of wind turbines.

7 SUMMARY

This paper describes the basic considerations for reliability analysis of wind turbines with special focus on structural components, and thus also provides part of the basis for probabilistic design of wind turbines. Based on the main design load cases to be considered in design of wind turbine components the corresponding reliability modelling is described including the effects of the control system and possible faults due to failure of electrical/mechanical components. Further, the target reliability level to be used for probabilistic design of wind turbine structural components is discussed.

Operation & Maintenance planning is an important part of lowering the LCOE of wind energy. Basic principles of cost optimal planning of OM is discussed with focus on use of a risk-based life-cycle approach where the total expected costs during the whole lifetime. This requires a probabilistic modelling of the damage accumulation for deteriorating components. Two examples are described considering planning of OM and inspections for offshore wind turbines.

REFERENCES

Agarwal, P. (2008) *Structural Reliability of Offshore Wind Turbines*. PhD Thesis, University of Texas at Austin.

Bak, B., Graham, A., Florian, M., Sørensen, J.D., Knudsen, T., Hou, P. & Chen, Z. (2017) Baseline layout and design of a 0.8 GW reference wind farm in the North Sea. *Wind Energy*, 2017. doi:10.1002/we.2116.

Dinwoodie, I.A., Endrerud, O.-E., Hofmann, M., Martin, R., & Sperstad, I.B. (2015) Reference cases for verification of operation and maintenance simulation models for offshore wind farms. *Wind Engineering*, 39 (1), 1–14.

Dong, W.B., Gao, Z. & Moan, T. (2010) Fatigue reliability analysis of jacket-type offshore wind turbine considering inspection and repair. In: *Proceedings of the EWEC 2010, EWEA Brussels*.

Dong, W., Moan, T. & Gao, Z. (2012) Fatigue reliability analysis of the jacket support structure for offshore wind turbine considering the effect of corrosion and inspection. *Reliability Engineering & System Safety*, 106, 11–27.

Dong, W.B., Moan, T. & Gao, Z. (2013) Reliability-based gear contact fatigue analysis for wind turbines under stochastic dynamic conditions. In: *Proceedings of the ICOSSAR 2013*, New York.

Faber, M.H., Sørensen, J.D., Tychsen, J. & Straub, D. (2005) Field implementation of RBI for jacket structures. *Journal of Offshore Mechanics and Arctic Engineering*, 127, 220–226.

Florian, M. & Sørensen, J.D. (2017) Case study for impact of D-string® on levelized cost of energy for offshore wind turbine blades. *International Journal of Offshore and Polar Engineering*, 27 (1), 63–69.

IEC 61400-1 (2005) *Wind Turbine Generator Systems – Part 1: Safety Requirements*. 3rd edition. International Electrotechnical Commission.

IEC 61400-1 (2018) *Wind Turbine Generator Systems – Part 1: Safety Requirements*. FDIS draft of 4th edition. International Electrotechnical Commission.

JCSS (2002) (Joint Committee on Structural Safety): Probabilistic Model Code. http://www.jcss.byg.dtu.dk/

Madsen, H.O. & Sørensen, J.D. (1990) Probability-based optimization of fatigue design inspection and maintenance. In: *Proceedings of the International Symposium on Offshore Structures*, July 1990, Glasgow.

Madsen, H.O., Krenk, S. & Lind, N.C. (1986) *Methods of Structural Safety*. Mineola, NY, Dover Publications, Inc.

Moan, T. (2005) Reliability-based management of inspection, maintenance and repair of offshore structures. *Structure and Infrastructure Engineering*, 1 (1), 33–62.

Nielsen, J.S. & Sørensen, J.D. (2014) Methods for risk-based planning of operation and maintenance. *Energies*, 7, 6645–6664.

Raiffa, H. & Schlaifer, R. (1961) *Applied Statistical Decision Theory*. Cambridge, Harvard University Press/Cambridge University Press.

Ronold, K.O., Wedel-Heinen, J. & Christensen, C.J. (1999) Reliability-based fatigue design of wind-turbine rotor blades. *Engineering Structures*, 21, 1101–1114.

Shafiee, M. & Sørensen, J.D. (2017) Maintenance optimization and inspection planning of wind energy assets: models, methods and strategies. *Reliability Engineering and System Safety*. doi:10.1016/j.ress.2017.10.025.

Straub, D. (2004) *Generic Approaches to Risk Based Inspection Planning for Steel Structures*. PhD Thesis, Swiss Federal Institute of Technology, Zurich.

Sørensen, J.D. (2009) Framework for risk-based planning of operation and maintenance for offshore wind turbines. *Wind Energy*, 12, 493–506.

Sørensen, J.D. & Toft, H.S. (2010) Probabilistic design of wind turbines. *Energies*, 3, 241–257.

Sørensen, J.D. (2012) Reliability-based calibration of fatigue safety factors for offshore wind turbines. *International Journal of Offshore and Polar Engineering*, 22 (3), 234–241.

Sørensen, J.D. & Toft, H.S. (2014) *Safety Factors – IEC 61400-1 ed. 4 – Background Document*. DTU Wind Energy-E-Report-0066 (EN).

Tavner, P.J., Xiang, J. & Spinato, F. (2007) Reliability analysis for wind turbines. *Wind Energy*, 10, 1–18.

Tavner, P. (2012) *Offshore Wind Turbines: Reliability, Availability and Maintenance*. London, Institution of Engineering and Technology.

Toft, H.S. (2010) *Probabilistic Design of Wind Turbines*. PhD Thesis, Aalborg University.

Veldkamp, D. (2006) *Chances in Wind Energy – A Probabilistic Approach to Wind Turbine Fatigue Design*. PhD Thesis, DUWIND Delft University, Wind Energy Research Institute, Delft.

Reliability assessment of deteriorating structures: Challenges and (some) solutions

D. Straub
Engineering Risk Analysis Group, Technische Universität München, Germany

ABSTRACT

The assessment of deteriorating structural systems has been one of the major applications of probabilistic analysis and structural reliability theory, for multiple reasons. Firstly, with the exception of fatigue, the quantitative assessment of deterioration is not considered in most current structural codes based on the partial safety factor format. Hence, it is challenging to demonstrate the reliability of deteriorating structural systems with standard (semi-probabilistic) safety concepts alone. Secondly, deterioration is often assessed for existing structures in which damages have been observed. For such structures, data is typically available from past inspections, monitoring or tests performed during the construction and operation of the structure. These data can be used to update the parameters of structural models, deterioration models or the reliability itself. Such an integration is best performed through a probabilistic analysis. Thirdly, in addition to deterioration, existing structures are often subject to changes in demand or loss of capacity, which may cause non-compliance of the structure with current code requirements. In some cases, a probabilistic assessment can demonstrate that a structure is nevertheless safe for future usage.

All these applications of structural reliability require a proper probabilistic description of the structural model and its parameters. A major challenge thereby is that the data available from existing structures cannot be explained with the simple models assumed in classical structural design and simple reliability analysis. Instead, the data reflect the real behavior of the structure and the spatial variability of the parameters. For this reason, the assessment of existing structures often necessitates more sophisticated structural and probabilistic models than the design of new structures. Thereby, the proper modeling choices are crucial, as overly simple models can lead to entirely wrong results, whereas overly complicated models lead to an unnecessary effort for analysis and – because of the complexity of the reliability analysis – are also more error-prone.

The modeling challenges in reliability analysis of deteriorating structures are related to:

- Deterioration modeling
- Structural system modeling

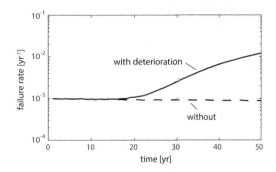

Figure 1. Hazard function (failure rate) of the example structure.

- Probabilistic modeling
- Dependence modeling
- Modeling of inspection and monitoring

In this contribution, I review these challenges and outline strategies for approaching the modeling tasks. The aim is to discuss when simple models are sufficient, and when more advanced models of the structural system and spatially variable parameters are required. The selected modeling approach has direct implications on the computational aspect of the reliability analysis. I will highlight recent computational developments that facilitate the application of advanced models in engineering practice. To illustrate the points made in the discussion, I introduce a simple example structure (Figure 1 shows the hazard function of the system).

The modeling of deterioration processes in structures has been the subject of a large amount of research efforts. Nevertheless, the success in accurately predicting deterioration, both deterministically and stochastically, is limited (e.g., Figure 2). For this reason, assessing deterioration in structures is mostly based on inspections and monitoring. An additional challenge is the need for understanding the dependence among deterioration processes at different elements or locations of the structure. Understanding such dependence is of limited importance in the design phase of the structure, but can be central in assessing existing deteriorating structures. In particular, it affects the system reliability, typically in a negative manner, but also facilitates a more efficient integrity management, because information obtained for one part of

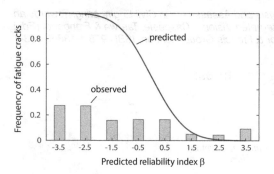

Figure 2. Observed frequency of fatigue cracks in structural details of offshore steel structures, in function of the calculated fatigue reliability of these details. The predicted frequency is the one according to theory, i.e. $\Phi(-\beta)$. The discordance between the prediction and the observation indicates that the uncertainty in the fatigue performance is much larger than implied by the stochastic model. Figure adapted from Aker Offshore Partner (1999).

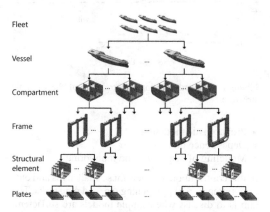

Figure 3. Hierarchical model for corrosion in ship structures, from Luque et al. (2017). At each level of the hierarchy, common factors are introduced to represent the dependence among elements belonging to the same instance. At the lowest level (e.g. plates in a compartment), the model allows for a random field. The model is learned from thickness measurements in Luque et al. (2017).

the structure can be utilized to learn about other parts of the structure. I briefly review the use of hierarchical models (Figure 3) and random field models for representing such dependence and evaluating its effects.

I also review the challenges in implementing accurate structural system reliability models in practice. As with the deterioration models, accurate system models are typically more relevant for assessing deteriorating structures than for the design of new-built structures. In particular, the code-based formats for ensuring the reliability of structures can be strongly conservative, which makes them unsuited for assessing existing structures (Figure 4).

A special focus is put on the effect of inspections and monitoring, which represent the most common strategy for dealing with deterioration in structures.

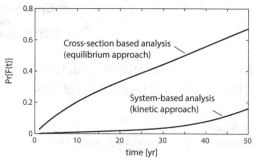

Figure 4. Estimates of the probability of failure of the example structure, computed with a cross-section-based model in accordance with the classical design methodology in Eurocode, compared to the probability of failure evaluated with a system-based analysis.

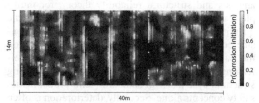

Figure 5. Probability of corrosion initiation (depassivation) in the reinforcement of a RC parking deck. Probabilities are evaluated by Bayesian updating of the corrosion model with spatially distributed inspection data from half-cell potential measurements, cover depth measurements and chloride profiles. Figure from Straub, Fischer et al. (in prep), based on data from Gehlen & Von Greve-Dierfeld (2010).

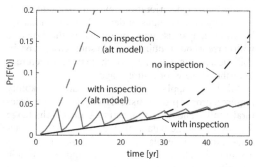

Figure 6. Probability of failure of the example structure, with and without inspections. Results are computed for the original model and an alternative model with increased uncertainty in the deterioration model. This alternative model mimics a situation with a poor deterioration model. With increasing amount of inspection results, the relevance of the prior model assumptions decreases and the predictions with the two models almost coincide.

As illustrated in Figure 6, inspections can ensure a certain degree of reliability even when the modeling of deterioration is subject to large uncertainty. Increasingly, inspection and monitoring data is available in a spatially distributed manner. This requires algorithms that enable the spatially-explicit inclusion

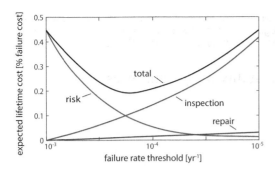

Figure 7. Expected lifetime cost in function of the threshold on the failure rate. Inspections are planned just before the failure rate estimate exceeds the threshold.

of inspection and measurement results into the predictive models, ideally based on Bayesian analysis, as illustrated in Figure 5.

Because inspection and monitoring can contribute significantly to the total life-cycle cost, there is an interest in optimizing inspection efforts. Ideally, owners and operators of structures try to identify the optimal trade-off between the cost of inspections and the risk of failure (Figure 7). Such risk- or reliability-based inspection planning has been one of the successful applications of structural reliability methods in practice. However, multiple challenges as well as opportunities remain also in this area, some of which I discuss in the paper.

Finally, main challenges in the assessment of deteriorating structures are associated with the management and the organization of owners and operators. These include data and model availability, development of codes and standards as well as general organizational aspects. I address these throughout the manuscript.

In conclusion, structural reliability assessment provides a solid foundation for optimal management of deteriorating structures. However, looking back on over 40 years of research and development in the area of structural reliability, the outcome has been mixed. Structural reliability methods still play a minor role in the management of existing deteriorating structures in practice. While a lot of progress has been made, there is still a long way ahead until reliability-informed planning and assessment becomes the norm rather than the exception. Research efforts are needed in particular on an improved modeling of system reliability that is compatible with standard structural assessment approaches, and on the understanding of the real reliability associated with current conservative modeling assumptions. Opportunities arise from improved IT, sensor and communication technology, which should be embraced to enhance our models. Most of all, our community needs to work on changing the current prescriptive approach to management and assessment of structures, in order to provide incentives to the structural engineering community for more realistic and optimal predictions in lieu of conservative assumptions.

REFERENCES

Aker Offshore Partner. (1999) Validation of Inspection Planning Methods, Volume 1999/59 of Offshore Technology Report. Health & Safety Executive, UK.

Gehlen, C. & von Greve-Dierfeld, S. (2010). Optimierte Zustandsprognose durch kombinierte Verfahren. *Beton-und Stahlbetonbau*, 105 (5), 274–283.

Luque, J., Hamann, R. & Straub, D. (2017). Spatial probabilistic modeling of corrosion in ship structures. *ASCE-ASME Journal of Risk and Uncertainty in Engineering Systems, Part B: Mechanical Engineering*, 3 (3), 031001.

Straub, D., Fischer, J., Geyer, S. & Papaioannou, I. Spatial probabilistic prediction of corrosion initiation in RC based on combined measurements. Manuscript in preparation.

Compliance testing for probabilistic durability design purposes

C. Thiel & C. Gehlen
Technical University of Munich, Munich, Germany

ABSTRACT

Against the background of resource-saving construction, material performance is gaining in importance and higher durability (>100 a) becomes more desirable. For this to succeed, knowledge of the progress of possible damage is decisive. In design situations, input values representing material resistances and expected field exposure conditions are needed in order to compute the probability of reaching a given structural state as a function of service time. Usually, the material resistances are obtained from laboratory compliance tests. The laboratory tests (direct or indirect tests, tests under accelerated or under natural conditions) should be reliable, reproducible, cheap, easy to perform and provide results in short time. Unfortunately, the tests rarely fulfil all these requirements. Accelerated methods must always be questioned critically since the correlation with field behaviour and the precision of the test only applies to the concrete compositions actually considered. The present contribution illustrates this problem and develops a solution strategy based on the example of compliance tests in the field of reinforcement corrosion induced by the carbonation of concrete.

Figure 1 shows the basic procedure for durability design based on performance criteria. The carbonation resistance can be determined with validated compliance tests. However, the carbonation tests presently available are criticized as follows.

– Poor or no relation to real carbonation resistance (indirect test methods)
– Long time (i.e. natural carbonation: 365 d)
– Differently changed microstructure for different cementitious materials so that a valid correlation with field conditions cannot be established (accelerated tests)

Indirect compliance testing is difficult since carbonation depends on both carbonatable mass and microstructure (porosity, microcracks and internal moisture conditions). When carbonation proceeds, both these factors significantly change and affect one another (Papadakis et al. 1991). Ongoing research deals with the development of new models taking into account the effects of diffusion, hydration degree, moisture conditions and microstructure. Simplified tests for the determination of diffusion coefficients and the binding capacity fulfilling the requirements of a compliance test are also being developed.

Another approach is direct testing under natural carbonation conditions, 20°C and 65% relative humidity. However, this procedure is time-consuming and has the drawback that small changes in the boundary conditions (i.e. slight increase in temperature or CO_2 concentration) significantly affect the outcome. Rapid tests are therefore favoured.

Due to the limits in accelerating carbonation by increasing temperature and CO_2 concentration, a new test method has been developed (Thiel 2018). First of all, the practical behaviour was studied in long-term tests. Based on this, the effect of penetrating CO_2 with different gas pressures up to 10 bar (1 MPa) and different concentrations (0.04 to 10 vol.%) was studied with special regard to moisture conditions for a variety of concrete compositions. The combination of moderately increasing gas pressure by 2 to 3 bar and using an average relative humidity of 50–70% combined with a CO_2 concentration of 2 vol.% was found to best fulfil all requirements of a compliance test. The test results were compared with samples stored under natural conditions for up to five years and found to correlate well.

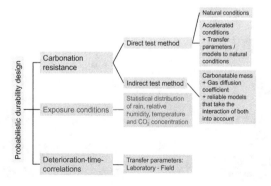

Figure 1. Input for probabilistic durability design.

Further current research now deals with the automation of the test setup to reduce costs and the variation of the boundary conditions.

However, it has to be kept in mind that standardized compliance tests do generally exclude the effect of specific construction measures (i.e. curing, compaction, geometry). Therefore, probabilistic service life models are necessary to combine material resistance, determined under laboratory conditions, with time-dependent deterioration laws while considering specific construction and exposure conditions. This enables accurate service life prediction.

REFERENCES

Papadakis, V.G., Vayenas, C.G. & Fardis, M.N. (1991) Fundamental modeling and experimental investigation of concrete carbonation. *ACI Material Journal*, 1, 363–373.

Thiel, C. (2018) Einfluss von CO_2-Druck und Betonfeuchtegehalt auf das Porengefüge des Betons während der Carbonatisierung. Doctoral thesis, Technical University of Munich.

MINI-SYMPOSIA

MS1: Load testing of new and existing structures

Organizers: A. de Boer, D.A. Hordijk, E.O.L. Lantsoght & Y. Yang

Critical proof load for proof load testing of concrete bridges based on scripted FEM analysis

X. Chen & Y. Yang
Faculty of Civil Engineering and Geosciences, Delft University of Technology, Delft, The Netherlands

P. Evangeliou
DIANA FEA BV, Delft, The Netherlands

H. van der Ham
Rijkswaterstaat, Utrecht, The Netherlands

ABSTRACT

As the bridge stock in The Netherlands and Europe is ageing, various methods to analyse the capacity of existing bridges are being studied. Proof load testing is one of the method to test the capacity of bridges by applying loads on the existing concrete bridges with small spans. Because of the fact that neither the actual traffic load nor the design traffic load required by Eurocode can be directly applied on the target bridge in real-life proof load testing, an equivalent wheel load has to be applied instead. The magnitude and the location of the equivalent wheel load is determined in such a way that it generates the same magnitude of inner forces in the cross section. Such calculation is usually done by linear finite element analyses (FEA). Whereas, different bridges have different geometry such as length, width, thickness, skewness, number of spans and lanes etc. For each configuration, FEA has to be done first to determine the loading position. The main aim of this paper is to study the relation between bridge geometry and unfavourable loading positions. Based on that, a guidance tool is developed for the determination of the critical proof load testing locations for the practice. To achieve this goal, a Python script has been developed using the general purpose FEM platform DIANA FEA. The script enables the automatic generation and analysis of a bridge model with different geometries and loading conditions. By applying the Eurocode Load Model 1 at variable locations, the most unfavourable loading positions for the proof load are obtained at the corresponding boundary conditions.

Skewness, span length, span width, averaging thickness and thickness ratio, their influence on the critical proof load are analysed. Using the data provided by Rijkswaterstaat (Ministry of Infrastructure and Water Management). The ranges of the parameters are selected in a way that most of the existing bridges and viaducts configurations in Netherlands are covered.

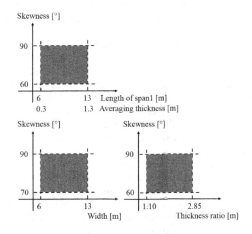

Figure 1. Scope and coupling of different parameters.

Figure 1 shows the scope of the parameters that analysed in this paper and the corresponding coupling.

In total, 166 sets of data are obtained by performing thousands of linear finite element analyses. Formulas for predicting the critical position factor k_p and the equivalent loading factor k_q are obtained, which can be used to calculate the location of the critical proof load and its corresponding magnitude.

$$k_p = 0.002lr^2 - 0.0123lr - 0.000038l\alpha - 0.0043\alpha + 0.9684$$
$$k_q = 0.0055lr - 0.0012r\alpha + 0.000043\alpha + 1.1518$$

where,
l is the length of tested span. Unit: meter [m].
α is the angle of skewness. Unit: degree [°].
r is the thickness ratio.

Recommendations are given for a better application of the script that developed in this paper. Most of them can be achieved by directly modifying the existing python script. The output of the study provides a convenient tool for future proof load testing.

Life-Cycle Analysis and Assessment in Civil Engineering: Towards an Integrated Vision – Caspeele, Taerwe & Frangopol (Eds)
© 2019 Taylor & Francis Group, London, ISBN 978-1-138-62633-1

Proposed stop criteria for proof load testing of concrete bridges and verification

E.O.L. Lantsoght
Concrete Structures, Delft University of Technology, Delft, The Netherlands
Politecnico, Universidad San Francisco de Quito, Quito, Ecuador

C. van der Veen & D.A. Hordijk
Concrete Structures, Delft University of Technology, Delft, The Netherlands

ABSTRACT

Proof load testing can be used to directly evaluate if a given structure fulfils the code requirements. When analytical methods cannot address additional sources of capacity, or when the effect of degradation on the capacity is unknown, proof load testing can be used. In a proof load test, a large load representing the factored live load is applied. This large load implies a risk to the structure and the employed personnel. Therefore, the measurements become critical. Stop criteria are used to interpret the measurements, and to know when further loading can result in irreversible damage or even collapse.

The existing codes and guidelines that provide stop and acceptance criteria are the German guidelines (Deutscher Ausschuss für Stahlbeton, 2000), ACI 437.2M-13 (ACI Committee 437, 2013), and the Czech and Slovak guidelines (Frýba and Pirner, 2001). The limitation to these stop criteria is that they only consider the failure mode of flexure, and that the German guidelines and ACI 437.2M-13 are developed for buildings. These existing criteria formed the starting point for the development of stop criteria for proof load testing of concrete bridges, and in particular reinforced concrete slab bridges.

To develop a proposal of stop criteria, as given in Table 1, the following actions were taken: analysis of existing stop criteria, a series of pilot proof load tests, additional testing on beams in the laboratory, and the development of stop criteria based on theoretical considerations. The proposed stop criteria are then verified with the data from the proof load tests and the tests to the ultimate. The results from the proof load tests show that the proposed stop criteria are not overly conservative, since they were not exceeded during the pilot proof load tests, or were only exceeded during the last load step. The analysis of the pilot proof load tests does not give an insight in the margin of safety of the proposed stop criteria. For this purpose, tests to collapse need to be studied. Thus, in a second analysis, the beam tests from the laboratory and the results of testing the Ruytenschildt Bridge were used to study the margin of

Table 1. Currently proposed stop criteria.

Failure mode	Cracked in bending or not?	
	Not cracked in bending	*Cracked in bending*
Bending moment	Concrete strain ε_{stop}	Concrete strain ε_{stop}
	$w_{max} \leq w_{stop}$	$w_{max} \leq w_{stop}$
	$w < 0.05$ mm	$w < 0.05$ mm
	$=> w \approx 0$ mm	$=> w \approx 0$ mm
	$w_{res} \leq 0.3\, w_{max}$	$w_{res} \leq 0.2\, w_{max}$
	25% reduction of stiffness	25% reduction of stiffness
	Deformation profiles	Deformation profiles
	Load-deflection diagram	Load-deflection diagram
Shear	Concrete strain ε_{DAfstB}	Concrete strain ε_{DAfstB}
	$w_{max} \leq 0.4\, w_{ai}$	$w_{max} \leq 0.75\, w_{ai}$
	$w < 0.05$ mm	$w < 0.05$ mm
	$=> w \approx 0$ mm	$=> w \approx 0$ mm
	25% reduction of stiffness	25% reduction of stiffness
	Deformation profiles	Deformation profiles
	Load-deflection diagram	Load-deflection diagram

safety of the proposed stop criteria. It was found that the stop criteria are exceeded between 40–65% of the maximum load, and thus provide sufficient safety in a field test. However, the available experiments that can be used to come to this conclusion are limited, and only in three cases a shear failure was obtained. Therefore, further validation with experiments on slabs subjected to a cyclic loading protocol and failing in shear and flexure is necessary.

REFERENCES

ACI Committee 437 (2013) *Code Requirements for Load Testing of Existing Concrete Structures (ACI 437.2M-13).* Commentary Farmington Hills, MA.

Deutscher Ausschuss Für Stahlbeton (2000) *DAfStb-Guideline: Load Tests on Concrete Structures (in German).* Deutscher Ausschuss fur Stahlbeton.

Frýba, L. & Pirner, M. (2001) Load tests and modal analysis of bridges. *Engineering Structures*, 23, 102–109.

Nonlinear finite element analysis of beam experiments for stop criteria

J.E. Paredes
Universidad San Francisco de Quito, Quito, Pichincha, Ecuador

E.O.L. Lantsoght
Delft University of Technology, Delft, The Netherlands
Universidad San Francisco de Quito, Quito, Pichincha, Ecuador

ABSTRACT

Existing bridges may not fulfill the safety requirements of the governing codes. However, replacing such structures or performing rehabilitation works may not be economically viable. Proof load testing has proven to be a satisfactory method to determine the capacity and guarantee adequate performance of the structure (Koekkoek et al. 2016). Proof load testing implies contradictory requirements as the maximum load should be as high as possible in order to prove sufficient structural capacity and to gather information about the behavior of the structural system, nevertheless the maximum load is also limited so that irreversible damage or reduction of the capacity due to excessive load levels is prevented. Stop criteria, based on measurements taken during proof load tests, determine if a test should be stopped before reaching the target proof load in order to maintain structural integrity. The German guideline (Deutscher Ausschuss für Stahlbeton 2000), and ACI 437.2M-13 (ACI Committee 437 2013) provide a detailed test method developed for reinforced concrete buildings and contain a thorough description of stop criteria or acceptance criteria.

A nonlinear finite element model is proposed to investigate stop criteria. A reinforced concrete beam with plain reinforcement is modeled. The goal is to develop a reliable finite element model with adequate material constitutive models to analyze available stop criteria from existing codes. Constitutive models and material properties are selected according to Guidelines for Nonlinear Finite Element Analysis of Concrete Structures from Rijkswaterstaat (RWS) (Rijkswaterstaat 2017) and Model Code 2010 (fib 2012). The finite element model is verified in terms of strains by comparing the strain caused by the externally applied loads from the model and the strain measurements from the experimental study (Lantsoght et al. 2017), as shown in Figure 1. Stop criteria from ACI 437.2M-13 and the German guideline are analyzed for the beam model. The presented analysis shows that nonlinear finite element models can be used for the

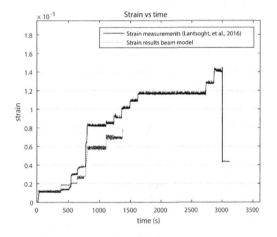

Figure 1. Strains measured on beam experiment P804B and strains resulting from beam model.

evaluation of stop criteria for proof load testing to limit the required number of laboratory tests.

REFERENCES

ACI Committee 437 (2013) *Code Requirements for Load Testing of Existing Concrete Structures (ACI 437.2M-13)*. Commentary Farmington Hills, MA.

Deutscher Ausschuss Für Stahlbeton (2000) *DAfStb-Guideline: Load Tests on Concrete Structures (in German)*. Deutscher Ausschuss für Stahlbeton.

FIB (2012) *Model Code 2010: Final Draft*. Lausanne, International Federation for Structural Concrete.

Koekkoek, R., Lantsoght, E.O.L., Yang, Y., De Boer, A. & Hordijk, D. (2016) Defining loading criteria for proof loading of existing reinforced concrete bridges. In: Beushausen, H. (ed.) *Performance-Based Approached for Concrete Structures*. Cape Town, South Africa.

Lantsoght, E.O.L., Yang, Y., Van Der Veen, C., De Boer, A. & Hordijk, D.A. (2017) Beam experiments on acceptance criteria for bridge load tests. *ACI Structural Journal*, 114, 1031–1041.

Rijkswaterstaat (2017) *Guidelines for Nonlinear Finite Element Analysis of Concrete Structures*.

Load testing on a high railway bridge to determine the longitudinal stiffness of the substructure

M. Wenner & F. Wedel
Marx Krontal GmbH Beratende Ingenieure, Hannover, Germany

T. Meier
Baugrund Dresden Ingenieurgesellschaft mbH, Dresden, Germany

S. Marx
Leibniz Universität Hannover, Institute of Concrete Construction, Hannover, Germany

ABSTRACT

The Itz valley bridge is a 868 m long railway viaduct on the new high speed line between Nürnberg and Erfurt in Germany, see Figure 1.

Due to the length of the bridge and its static system, the track-bridge-interaction takes a major role in the design of the bridge structure and of the track. The longitudinal stiffness of the substructure is one of the main parameters in this calculation. The higher the stiffness, the lower the displacements of the deck and the lower the additional rail stresses. By performing the calculation of the stiffness using a simple pile grillage model with the assumptions of the geotechnical report (see also method 1), the stiffness is often underestimated. In order to make a realistic prediction of the longitudinal stiffness, a large-scale test has been performed.

For this test high forces had to be introduced in the pier top and the corresponding displacements had to be measured. The forces have been introduced with hydraulic jacks in the bridge joints.

Before the test, geotechnical calculations predicted the longitudinal stiffness of the substructures with two different methods:

Method 1: pile grillage in the form of beam structures for which the ground is considered in a simplified way by means of subgrade reaction approaches based on empirical values.
Method 2: 3D FEA continuum model of the foundation and the soil with appropriate contact formulations for the interfaces between subsoil and structural elements.

Figure 2 shows the results of both the large-scale test and the two calculation methods. It shows that the real stiffness of the system is much higher than the calculation results according to method 1. For the numerical investigations according method 2, the results correspond very well with the experimental results.

Figure 1. Photograph of the Itz valley railway viaduct, parallel to it in the background the motorway viaduct of the A 73.

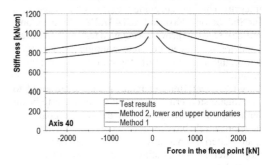

Figure 2. Comparison between measured and calculated stiffness of the substructure using the example of axis 40.

With the real stiffness, the calculation of the track-bridge interaction could be performed successfully.

Especially for long railway bridges, the stiffness of the substructure has a major influence on the design and calculation of the bridge system. The performed experimental and numerical investigations shows, that for these cases a simple calculation considering the pile grillage model is not sufficient and a detailed calculation of this stiffness using a 3D-FE-continuum model can be necessary.

Life-Cycle Analysis and Assessment in Civil Engineering: Towards an Integrated Vision – Caspeele, Taerwe & Frangopol (Eds)
© 2019 Taylor & Francis Group, London, ISBN 978-1-138-62633-1

Bridge diagnostic load testing in Ecuador—case studies

J. Bonifaz, S. Zaruma, A. Robalino & T. Sánchez
ADSTREN Cía. Ltda., Quito, Ecuador

ABSTRACT

Load testing has been performed throughout the history of bridges and different procedures have been developed for new and existing bridge structures. Diagnostic load tests are used in new bridges to verify that the structure behaves as predicted by analytical models used during design. These tests reduce uncertainties related to different parameters that can affect the behavior of the structure, such as material properties, boundary conditions and cross-section contributions. As a result, greater confidence in the analytical model is achieved and knowledge about the real load capacity of the bridge is obtained. To determine the response in all critical members, the test load is placed at different positions on the bridge. Futhermore, the magnitude of the test load should be high enough to obtain measurable responses but within the elastic range (NCHRP 1998).

Diagnostic load testing was conducted for two new steel bridge projects located in the suburbs of Quito, Ecuador: "La Armenia 1" interchange and "Los Pajaros" Bridge. Structural analyses were performed in LARSA-4D to establish the magnitude and location of the test load so that the induced deflections were measurable. The test load consisted of a number of dump trucks that was determined through the analyses and was located over the bridges to produce the maximum moments. Deflections from the load test were compared with deflections from the analytical models, and the results are discussed in this paper.

In some cases, the measured deflections were not close to those predicted by analytical models in magnitude. Nevertheless, in those cases, the measured values were lower which showed that the bridges might have a higher load capacity than expected. Furthermore, the deformed shape was similar enough to conclude that the bridges behave as designed and could be safely opened to traffic.

This work proved the usefulness of diagnostic load testing in new bridges to better understand the behavior of the structure and obtain relevant information that could be utilized latter during the life of the bridge. For example, for the cases presented in this paper, further research can be conducted to determine the source of the difference in magnitude of deflections and update the analytical models for future analyses of the bridges. Likewise, this information can be utilized to update the bridge load rating.

REFERENCE

NCHRP (1998) *Manual for Bridge Rating Through Load Testing.*

Determination and assessment of the behavior of a semi-integral railway viaduct

M. Käding, M. Wenner & H. Liao
Marx Krontal GmbH, Hannover, Germany

S. Marx
Institute of Concrete Construction, Leibniz Universität Hannover, Hannover, Germany

ABSTRACT

For the first time in history of the DB AG new concepts for supporting structures for valley viaducts have been realized on the newly built line Ebensfeld-Erfurt-Halle/Leipzig by using semi-integral structures (Schenkel et al., 2009). Because of the variance to the previous design principles, the authorities require the investigation of the real behavior of the viaducts in order to verify the assumptions from the calculation and to gain experience with semi-integral structures. A concept for the determination of the long-term behavior (monitoring) and the behavior due to traffic load (load-testing) was developed and implemented.

One of these viaducts is the 576,5 m long Scherkonde valley viaduct (Marx et al. 2010). Almost all piers and one abutment are monolithically connected to the superstructure. The installation and commissioning of the sensors started during the stage of erection in 2009. Today almost eight years of continuous data are available. Through the analysis of the longitudinal displacements in the joint and the metrological data, the value of the thermal expansion coefficient was determined, a separation between the thermal and the permanent displacements was archived and the value for the final stage of creep and shrinkage was prognosticated. Shortly after the completion of the ballastless track the viaduct was loaded with two heavy weight freight trains. Slow rides and precise load positioning were performed to obtain influence lines of the real structure response to traffic loads.

The measurement results were compared with the calculated behavior of a FE model and with a model for creep and shrinkage according to the Eurocode considering the real construction process. Especially for the creep and shrinkage an astonishing good concordance was found. Regarding the general bending behavior of the superstructure and piers, the participation of the ballastless track and other coupled equipment to the

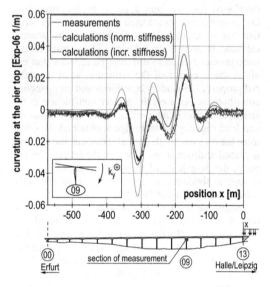

Figure 1. Influence line of the curvature of the pier top in axis 09 due to traffic load of two heavy weight freight train, comparison of the measurements and calculations.

bending stiffness of the deck and the favorable effect of the soil conditions on the stress in the substructure could be analyzed (Fig. 1).

REFERENCES

Marx, S., Krontal, L., Bätz, S. & Vehlow, A. (2010) Die Scherkondetalbrücke, die erste semi-integrale Talbrücke der DB AG auf der Neubaustrecke Erfurt – Leipzig/Halle VDE 8.2. *Beton-und Stahlbetonbau*, 105 (3), 134–141.

Schenkel, M., Marx, S. & Krontal, L. (2009) Innovative Großbrücken im Eisenbahn-Hochgeschwindigkeitsverkehr am Beispiel der Neubaustrecke Erfurt-Leipzig/Halle. *Beton-und Stahlbetonbau*, 104 (11), 782–789.

Development of a stop criterion for load tests based on the critical shear displacement theory

K. Benitez
Universidad San Francisco de Quito, Ecuador

E.O.L. Lantsoght
Delft University of Technology, Delft, The Netherlands
Universidad San Francisco de Quito, Ecuador

Y. Yang
Delft University of Technology, Delft, The Netherlands

ABSTRACT

Bridges that were built decades ago were not designed for the actual traffic loads. Checking the integrity and performance of these bridges under actual conditions is required. Proof load testing is one of the available ways to evaluate if the provisions of the current codes are satisfied. Bridges assessed through proof loading are usually evaluated for flexure since the stop criteria for proof load tests provided by the guidelines are suitable for flexure only. Shear assessment of existing reinforced concrete slab bridges in the Netherlands (Lantsoght et al. 2013) showed that a large number do not fulfill the code requirements for shear. Therefore there's a need to develop stop criteria that could be used for tests carried out on shear-critical bridges.

The Critical Shear Displacement Theory (Yang et al. 2016) is used as the basis for a newly proposed stop criterion. For the application to a practical stop criterion suitable for field testing, the analysis of the theory is converted to a measurable response. For this purpose, strains are used. The aim of the theoretical derivation is to develop a limiting strain.

The shear capacity of the section (V_{CSDT}) is obtained using the Critical Shear Displacement Theory, then, the corresponding bending moment (M_{CDST}) is calculated. The moment-curvature relation of the section is obtained using the parabolic stress-strain relation proposed by Thorenfeldt. Then the moment M_{CSDT} is located in the moment-curvature diagram. For a shear-critical beam, this bending moment typically lies between the points of cracking and yielding. The associated curvature and internal strain state is then obtained. For the sectional analysis, a parabolic stress block is assumed for the concrete. The strain at the bottom of the cross-section can then be considered as the limiting strain, see Figure 1. For a proof load test, the strain caused by permanent loads should be subtracted.

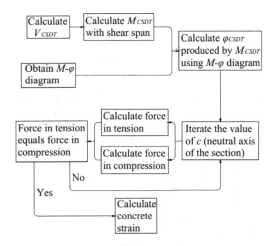

Figure 1. Flow chart for calculating limiting strain for stop criterion.

The proposed strain-based stop criterion has a theoretical basis and is a significant improvement as compared to a stop criterion that uses an arbitrary limiting strain. From the comparison to experimental results, it seems to have a sufficient margin of safety, and can thus be used for the application to proof load testing of reinforced concrete slab bridges.

REFERENCES

Lantsoght, E.O.L., Van Der Veen, C., De Boer, A. & Walraven, J.C. (2013) Recommendations for the shear assessment of reinforced concrete slab bridges from experiments. *Structural Engineering International*, 23, 418–426.

Yang, Y., Den Uijl, J.A. & Walraven, J. (2016) The Critical Shear Displacement theory: On the way to extending the scope of shear design and assessment for members without shear reinforcement. *Structural Concrete*, 17, 790–798.

Verification of flexural stop criteria for proof load tests on concrete bridges based on beam experiments

A. Rodriguez Burneo
Universidad San Francisco de Quito, Quito, Pichincha, Ecuador

E.O.L. Lantsoght
Delft University of Technology, Delft, The Netherlands
Universidad San Francisco de Quito, Quito, Pichincha, Ecuador

ABSTRACT

Civil structures subjected to continuous loading and environmental factors deteriorate with time. Therefore, they must be evaluated to determine if they are still suitable for use. As a structure deteriorates, mechanical properties of its materials may vary. This presents problems when using computational models to evaluate such structures, since the input parameters of the materials might be inaccurate. Proof loading is another way of evaluating existing structures with a high level of assessment, given that data from the structure is obtained in real time while the proof load test is being carried out. Nonetheless, there are some risks when performing a proof load test. If applied loads are minimal, no relevant data can be obtained, and if loads are too high, structures may suffer irreparable damage or even collapse. Consequently, building codes have established protocols to perform proof load tests, which include stop criteria. Stop criteria often refer to the evaluation of data that is taken as the test is being carried out to assure that the structure does not suffer irreparable damage.

Other proposals of stop criteria have been submitted to improve safety of proof load tests. This paper analyses the results obtained from 4 experiments on 2 cast-in-laboratory beams, and compares them to the values obtained with the stop criteria established by the ACI 437.2M-13 and the German guidelines of the DAfStB. In addition, a new proposal for stop criteria by Werner Vos from TU Delft in the Netherlands is also compared to the experimental results. Beam layout and characteristics of each experiment are shown on Figure 1, where the position of the load a the span length l and the beam's height h varied between experiments.

This research aims to analyze under which circumstances it is better to apply a specific stop criterion, which are the limitations on the criteria from the codes

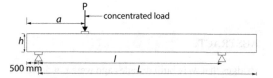

Figure 1. Beam experiment layout.

and the new proposal, and how reliable they are to be applied not only to buildings but to concrete bridges.

After comparing the results to the threshold values established by each stop criterion the following results were found. For the ACI stop criteria, the loading protocol is imperative to evaluate stop criteria. However a force-controlled loading protocol required by the code is not as safe for bridges as a displacement-controlled loading.

As for the other proposals, depending on the margin of safety considered to avoid permanent damage, reliable results were found for the concrete strain criterion established by the DAfStB, and deflection criterion established by Werner Vos. Further investigation in flexural stop criteria would help to develop better ways to calculate the maximum allowable values, which will lead to a better and safer application of proof load tests.

REFERENCES

ACI Committee 437 (2013) *Code Requirements for Load Testing of Existing Concrete Structures (ACI 437.2M-13)*. Commentary Farmington Hills, MA.

Deutscher Ausschuss Für Stahlbeton (2000) *DAfStb-Guideline: Load Tests on Concrete Structures (in German)*. Deutscher Ausschuss fur Stahlbeton.

Vos, W. (2016) *Stop Criteria for Proof Loading – The Use of Stop Criteria for a Safe use of 'Smart Proof Loading'*. MSc Thesis, Delft University of Technology.

Life-Cycle Analysis and Assessment in Civil Engineering: Towards an Integrated Vision – Caspeele, Taerwe & Frangopol (Eds)
© 2019 Taylor & Francis Group, London, ISBN 978-1-138-62633-1

Load testing and rating of the KY 220 road bridge

A. Peiris, J. Hudson & I. Harik
University of Kentucky, Lexington, Kentucky, USA

ABSTRACT

The National Bridge Inspection Standards (NBIS) regulations in the United States require that all bridges on public roads be assigned a load rating. The load rating is carried out in accordance with the guidelines provided in the American Association of State Highway and Transportation Officials (AASHTO) Manual for Bridge Evaluation (MBE) (AASHTO, 2011). The objective of a load rating is to evaluate the safe live load carrying capacity of a bridge, based on as-built construction plans and material properties while accounting for any structural damage that may exist. The theoretical load rating, expressed as a Rating Factor (RF) or in terms of a particular truck weight in tons, tends to be conservative due to many of the assumptions made in the calculation process. Numerous studies have shown, field testing bridges can provide more accurate and reliable information regarding their current condition. This paper discusses a steel girder bridge with a non-composite concrete deck that was field load tested in Kentucky.

The bridge subjected to load testing traverses Martins Branch Creek on KY 220 in Hardin County, K.Y. It was built in 1935 and no original design or construction plans were available. A single span bridge, it is 7.47 m (24'–8") long and composed of six W18 × 50 girders spaced at 1.22 m (4'–0") intervals with a 178 mm (7") non-composite concrete deck on top. This paper evaluates the maximum load carrying capacity of the KY 220 Bridge under different truck types.

A Load Factor Rating (LFR), based on the bridge plans and estimated material properties, was carried out and compared with the load rating derived from field load test results. The field load test was carried out using a loaded dump truck. Eight load cases were evaluated to obtain a range of strain and deflection data. The results from the load case that yielded the

Table 1. AASHTO theoretical load rating vs. field load testing load rating for KY 220 bridge.

AASHTO & KY trucks	LFR inventory rating			
	AASHTO		Field load test	
	RF	Load posting	RF	Load posting
HS20	0.92	29.0 t (32 US t)	1.83	59.0 t (65 US t)
KY Type 1	0.92	16.3 t (18 US t)	1.83	32.7 t (36 US t)
KY Type 2	0.71	18.1 t (20 US t)	1.42	36.3 t (40 US t)
KY Type 3	0.63	20.9 t (23 US t)	1.25	41.7 t (46 US t)
KY Type 4	0.98	35.4 t (39 US t)	2.21	79.8 t (88 US t)

greatest distribution factor and the maximum moment load case are included in this paper. The bridge had a 15.4 t (17 US t) load posting. Field load tests revealed that the load rating factor for strength was adequate and that the load posting can be removed.

Table 1 summarizes the load rating results for Inventory level ratings for an AASHTO HS20 Truck and Kentucky Legal Truck Types (Types 1-4). The field load testing resulted in an increase to the load rating of the bridge for all truck types.

REFERENCE

American Association of State Highway and Transportation Officials (2011) *The Manual for Bridge Evaluation.* 2nd edition. Washington, DC, AASHTO.

Strength evaluation of prestressed concrete bridges by load testing

E.S. Hernandez & J.J. Myers
*Department of Civil, Architectural and Environmental Engineering,
Missouri University of Science and Technology, Rolla, Missouri, USA*

ABSTRACT

Field tests confirm additional strength capacity in existing bridges, particularly in the case of precast-prestressed concrete bridges, despite their visual condition and age. Sources that explain the differences are multiple and may be related to different site-specific parameters that are not considered during the strength evaluation of a bridge structure. This study aimed at presenting a load rating protocol using experimental data collected with robust and reliable measurement devices to conduct an experimental strength evaluation of prestressed concrete bridges in LRFD format. The case study presented herein demonstrates how site-specific parameters, that are measured during a

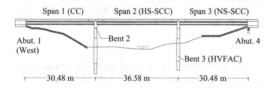

Figure 1. Bridge A7957. (a) elevation.

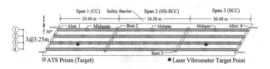

Figure 2. Bridge A7957 plan view and instrumentation layout.

Figure 3. Load test of Bridge A7957.

Table 1. Bridge A7957 strength evaluation results.

LR	Span 1	Span 2	Span 3
Analytical load rating			
LR_{int}	1.26	1.43	1.26
LR_{ext}	1.06	1.15	1.06
Experimental load rating			
LR_{int}	1.75	1.91	1.75
LR_{ext}	2.06	2.25	2.06

diagnostic load test, can be incorporated into the strength evaluation of a bridge to improve its load rating capacity. Bridge A7957's experimental data and evaluation methodology are expected to encourage more discussion among bridge designers and evaluators to better understand and improve current bridge analysis and evaluation practices of precast-prestressed concrete bridges.

By incorporating site-specific data measured during a diagnostic load test, a more precise load rating was obtained in comparison to the theoretical load rating values obtained following the AASHTO LRFD procedure. The diagnostic load test presented herein proved that Bridge A7957 superstructure possesses a larger strength capacity than the predicted by the theoretical approach (see Table 1). This difference can be explained by the fact that a diagnostic load test incorporates in-situ parameters that are beneficial to the bridge's service response. These parameters are not considered by the current AASHTO LRFD design (AASHTO 2012) and evaluation specifications (AASHTO 2010).

REFERENCES

American Association of State Highway and Transportation Officials (2010) *The Manual for Bridge Evaluation (2nd Edition) with 2011, 2013, 2014 and 2015 Interim Revisions*. Washington, DC.

American Association of State Highway and Transportation Officials (2012) *LRFD Bridge Design Specifications*. 6th edition. Washington, DC.

Life-Cycle Analysis and Assessment in Civil Engineering: Towards an
Integrated Vision – Caspeele, Taerwe & Frangopol (Eds)
© 2019 Taylor & Francis Group, London, ISBN 978-1-138-62633-1

Static load deflection experiment on a beam for damage detection using the deformation area difference method

D. Erdenebat, D. Waldmann & F.N. Teferle
Institute for Civil Engineering and Environment (INCEEN), University of Luxembourg, Luxembourg

ABSTRACT

A reliable and safe infrastructure for both transport and traffic is becoming increasingly important today. The condition assessment of bridges remains difficult and new methods must be found to provide reliable information. A meaningful in-situ assessment of bridges requires very detailed investigations which cannot be guaranteed by commonly used methods.

It is known that the structural response to external loading is influenced by local damages. However, the detection of local damage depends on many factors such as environmental effects (e.g. temperature), construction layer (e.g. asphalt) and accuracy of the structural response measurement. Within the paper, a new so-called Deformation Area Difference (DAD) Method is presented. The DAD method is based on a load deflection experiment and does not require a reference measurement of initial condition. Therefore, the DAD method can be applied on existing bridges. Moreover, the DAD method uses the most modern technologies such as high precision measurement techniques and attempts to combine digital photogrammetry with drone application.

The DAD method uses information given in the curvature distribution from a theoretical model of the structure and compares it to real measurements. The paper shows results from a laboratory load-deflection experiment with a steel beam with a span of 5.60 m (Figure 1).

The laboratory test includes several steps of damage scenarios which are created by slitting the bottom flange stepwise (Figure 2 and Figure 3). The loading is carried out path-controlled whereby the maximum deflection did not exceed the limit of the serviceability limit state. The successful application of the DAD method contributes to the state-of-the-art with a new non-destructive method for damage localization and assessment.

The paper includes the presentation of the method based on theoretical numerical and experimental

Figure 1. Test specimen: HEA180 S235 steel beam.

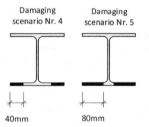

Figure 2. Damaging scenario exemplary for Nr. 4 and Nr. 5.

Figure 3. Damaging of the cross-section at the damage position 1 damage level 4 due to slitting the bottom flange.

results. Numerical results showed the capacity of the method for detection of damage without influence of measurement noise and environmental influences. The method allows a clear and reliable identification of damages within the serviceability limit state based on theoretical values. The photogrammetry enabled the highly precise measurements of the deflection line with minimum effect of noise. However, the multiple derivation of the deflection line for calculation of the inclination angle and curvature leads to increased noise effects. The distribution of the DAD values resulting from the curvature showed peaks and discontinuities in the area of manually generated damages.

Life-Cycle Analysis and Assessment in Civil Engineering: Towards an Integrated Vision – Caspeele, Taerwe & Frangopol (Eds)
© 2019 Taylor & Francis Group, London, ISBN 978-1-138-62633-1

Follow-up assessment on an old concrete road bridge based on operational dynamic bridge behaviour—analysis of structural integrity and determination of loading capacity

R. Veit-Egerer
VCE Vienna Consulting Engineers, Vienna, Austria

J. Bursa
MDS Projekt, Vysoké Mýto, Czech Republic

J. Synek
Road construction and maintenance department of Pardubice region, Pardubice, Czech Republic

ABSTRACT

The aim of the investigation was the analysis of the integrity of a concrete road bridge structure crossing the Elbe river and having been operated since 1924. In accordance to a baseline investigation from 2015 in depth dynamic measurements were conducted in 2017 with regard to the remaining loading capacity based on a dense grid of accelerometer data and additional displacement measurements.

The bridge assessment focused on:

- Analysis of operational condition, determination of effective structural resistance and remaining loading capacity with regard to ensuring the required bridge safety issues.
- In addition to characteristic operational loading conditions a tailored loading campaign was performed using a set of bridge trucks crossing the bridge at defined weight and velocity levels.
- Analysis of structural degradation, evaluation of structural integrity (condition and mechanical behavior of the load bearing concrete structure and the bridge bearings).
- With regard to the Czech standard ČSN 73 6209 an analytical finite element bridge model was used already in the course of the baseline investigation in 2015 to provide a reference for structural and modal analysis in order to compare the current bridge behavior with the expected, calculated one.
- The investigation was concluded by means of comparing the results (effective structural resistance and remaining loading capacity) from the follow-up campaign 2017 with those from the baseline assessment 2015.

- Subsequently the findings from the performed bridge analysis were transferred into recommendations for short- and midterm measures in terms of bridge maintenance, retrofit or reconstruction.

REFERENCES

Beards, C.F. (1996) *Structural Vibration-Analysis and Damping*. London, GB, Arnold (A Member of the Hodder Headline Group).

ČSN 73 6222 (2013) Zatížitelnost mostů pozemních komunikací (Load bearing capacity of road bridges), Praha, Úřad pro technickou normalizaci, metrologii a státní zkušebnictví. Czech Republic.

ČSN 736209 (2005) Zatěžovací zkoušky mostů (Loading tests of bridges). Praha, Český normalizační institut. Czech Republic.

Ředitelství silnic a dálnic ČR (1998) Hlavní Mostní Prohlídka. Pardubice, Czech Republic.

VCE (2017) Most ev. č. 3227 – 3 (most přes Labe). Následná nedestruktivní diagnostika mostu 2017 založená na průzkumném měření dynamického chování metodou BRIMOS®. Analýza provozního stavu, stanovení realní zatížitelnosti a porovnání s výsledky diagnostiky 2015. Project Report. Vienna Austria.

Veit-Egerer, R. & Hubka, M. (2008) Silniční most přes trat' ČD za obcí Komořany – Dynamické chování vzhledem k provoznímu stavu a zatížitelnosti metodou BRIMOS® – Determination of Structural Integrity and Explicit Load Capacity of a Prestressed Concrete Bridge by Means of Dynamic Monitoring (written & presented in Czech language). In: *Proceedings of the 13th International Symposium Bridges*. Sekurkon, Brno, Czech Republic.

Wenzel, H. (2009) Health Monitoring of Bridges. Chichester, England, J. Wiley and Sons Ltd.

Wenzel, H. & Pichler, D. (2005) Ambient Vibration Monitoring. Chichester, England, J. Wiley and Sons Ltd.

*MS2: Vibration-based structural health monitoring, damage
identification and residual lifetime estimation*
Organizers: E.P.B. Reynders, G. Lombaert, E. Chatzi & C. Papadimitriou

Life-Cycle Analysis and Assessment in Civil Engineering: Towards an Integrated Vision – Caspeele, Taerwe & Frangopol (Eds)
© 2019 Taylor & Francis Group, London, ISBN 978-1-138-62633-1

Robust computation of the observability of large linear systems with unknown parameters

X. Shi, M.N. Chatzis & M.S. Williams
University of Oxford, Oxford, UK

ABSTRACT

Observability of a dynamical system is used to account for whether time-invariant parameters and time-variant state of the system can be estimated and tracked based on knowledge of mathematical model of the system, nominal input and output measurements. If a state/parameter is unobservable/identifiable then no system identification methods will be successful in evaluating it properly. In structural health monitoring (SHM), the concept of observability plays an important role in determining suitable sensor placement strategies that will lead to successful estimation and tracking of structural properties and response from measured data. Avoiding unobservable setups before conducting measurement campaigns can result in savings of both time and money.

The Observability Rank Condition (ORC) is an efficient tool used for analyzing the observability of nonlinear dynamical systems. However, there is a practical constraint in implementing the ORC for large systems (e.g. long-span bridges or high-rise buildings) or systems with complicated equations (e.g. damaged structures), because the physical memory requirements to calculate the symbolic expressions of the ORC matrix often exceeds the computer's capacity.

The objective of this work is to explore a computationally efficient way to calculate the observability matrix. The method presented is based on the idea of expanding Lie derivatives of any order to the products between easier to calculate tensor gradients. This work specifically focuses on the nonlinear augmented systems occurring from identifying the structural properties of a linear system. The form of the resulting observability matrix is provided and an efficient recursion algorithm to compute the observability matrix numerically is developed. The new algorithm implemented on a standard desktop computer is capable of handling a linear civil engineering system with at least hundreds of degrees of freedom, while the symbolic algorithm can only deal with a similar system with a few dozen degrees of freedom.

Towards the use of UHPFRC in railway bridges: The rehabilitation of Buna Bridge

H. Martín-Sanz, K. Tatsis & E. Chatzi
Department of Civil, Environmental and Geomatic Engineering (IBK), ETH Zürich, Zürich, Switzerland

E. Brühwiler
Structural Maintenance and Safety Laboratory (MCS), EPFL, Laussane, Switzerland

I. Stipanovic
University of Twente, Enschede, The Netherlands

A. Mandic & D. Damjanovic
Faculty of Civil Engineering, University of Zagreb, Zagreb, Croatia

A. Sanja
Department of Materials and Laboratory of Concrete, Institut ZAG, Ljubljana, Slovenia

ABSTRACT

In the last decade, Ultra High Performance Fibre Reinforced cement-based Composites (UHPFRC) have been increasingly implemented for rehabilitation and strengthening purposes, rendering outstanding results. The ease of application, along with their superior mechanical and durability properties against other cementitious materials, constitute the drivers for their successful application. Despite this field being thoroughly explored and extensive literature already being available with respect to concrete and UHPFRC solutions, with particular focus on bridges or maritime environments, research on UHPFRC combined with steel in structures such as steel decks or railway bridges has only recently surfaced. This paper provides an example of the latter: the Buna Bridge in Croatia is a 9m non-ballasted steel bridge built in 1893, and repaired in 1953 albeit no longer in operation. The structure was transported in a laboratory setting for testing, envisioning a later strengthening by a prefabricated UHPFRC slab, connected to the original structure by means of steel studs. Dynamic and static test will be performed prior and after rehabilitation in order to compare the efficiency of the solution and in particular the bond between the two materials. A detailed analysis on fatigue will be developed, based on the updated Finite Element model obtained form the results of the test, helping to deliver an appropriate design for the future strengthening. The results summarize the effective capacity of the girder and estimate the extension in the residual life of the beam on the basis of prediction of fatigue accumulation under regular operational conditions.

Figure 1. Buna Bridge after transportation to the laboratory.

REFERENCES

Dzajic, I., S. A. a. I. (2014) Rehabilitation of steel railway bridges by implementation of uhpfrc deck. In: *3rd Internationl Conference on Road and Rail Infrastructure*, Split, Croatia.

Martín-Sanz, H., Chatzi, E. & Brühwiler, E. (2016) The use of ultra high performance fibre reinforced cement based composites in rehabilitation projects: A review. In: *Proceedings of the 9th International Conference on Fracture Mechanics of Concrete and Concrete Structures FraMCoS-9*, Berkeley, CA, USA.

Miner, M.A. (1945) Cumulative damage in fatigue. *ASME Applied Mechanics Transactions*, 12, 159–164.

Online tracking of inputs, states and parameters of structural dynamic systems

F. Karlsson
Department of Civil Engineering, KU Leuven, Leuven, Belgium
KTH Royal Institute of Technology, Stockholm, Sweden

K. Maes & G. Lombaert
Department of Civil Engineering, KU Leuven, Leuven, Belgium

ABSTRACT

Condition monitoring of civil engineering structures often involves the tracking of unknown states, inputs, and parameters. Many methods that allow for the identification of one or more of these quantities of interest have been developed. Most of these methods, however, require some prior knowledge on the inputs or parameters, in order to be able to estimate the remaining quantities. In cases where no prior knowledge is available, the system states, inputs, and parameters should be jointly estimated from the dynamic response of the system.

This paper presents a novel algorithm for joint input-state-parameter estimation. The algorithm is an extension of an existing joint input-state estimation algorithm, where the state-space equations are linearized around the current state to deal with the non-linearities introduced by the parameter dependence. The distinction between this algorithm and existing algorithms in the literature is that the estimated input does not require the choice of any regularization parameters. For applications in linear structural dynamics, a parametrization scheme for the structural matrices of the equations of motion is proposed, resulting in analytical expressions for the Jacobian matrices that occur in the presented algorithm.

The proposed methodology is verified by numerical simulations for a four story shear building (Figure 1).

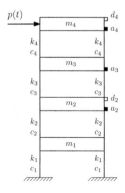

Figure 1. Four story shear building model.

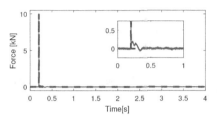

Figure 2. Time history of the estimated force (solid line) and the actual force (dashed line) acting at the fourth story.

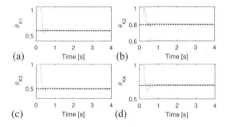

Figure 3. Time history of the estimated stiffness ratios θ_i (solid line) and the actual values of θ_i (dotted line) for each story of the shear building.

It is shown that the choice of the uncertainty assigned to the parameters is very important for the application of the algorithm, since it affects both the convergence rate of the parameter estimates and the final input and state estimates.

REFERENCES

Maes, K., Smyth, A., De Roeck, G. & Lombaert, G. (2016) Joint input-state estimation in structural dynamics. *Mechanical Systems and Signal Processing*, 70–71, 445–466.

Naets, F., Croes, J. & Desmet, W. (2015) An online coupled state/input/parameter estimation approach for structural dynamics. *Computer Methods in Applied Mechanics and Engineering*, 283, 1167–1188.

Simoen, E., De Roeck, G. & Lombaert, G. (2015) Dealing with uncertainty in model updating for damage assessment: A review. *Mechanical Systems and Signal Processing*, 56, 123–149.

Life-Cycle Analysis and Assessment in Civil Engineering: Towards an Integrated Vision – Caspeele, Taerwe & Frangopol (Eds)
© 2019 Taylor & Francis Group, London, ISBN 978-1-138-62633-1

State estimation of geometrically non-linear systems using reduced-order models

K. Tatsis
Department of Civil, Environmental and Geomatic Engineering, Institute of Structural Engineering, ETH Zürich, Switzerland

L. Wu
Faculty of Mechanical, Maritime and Materials Engineering, Delft University of Technology, Delft, The Netherlands

P. Tiso
Department of Mechanical and Process Engineering, Institute for Mechanical Systems, ETH Zürich, Switzerland

E. Chatzi
Department of Civil, Environmental and Geomatic Engineering, Institute of Structural Engineering, ETH Zürich, Switzerland

ABSTRACT

Slender structures with high stiffness-to-weight ratio form the main bearing element of modern engineering. This renders geometrical non-linear effects a key feature to be considered throughout the whole life-time of diverse structural components, such as Wind Turbine (WT) blades which inevitably follow the design trend of rapidly growing in size WTs. Although the Finite Element (FE) method constitutes a well-established tool for modeling such systems, the resulting models are often prohibitively expensive in terms of computational resources and thus cannot be implemented in online fashion.

The problem of state estimation for condition diagnostics and control applications is therefore rendered a challenging and intricate task when it comes to systems experiencing geometrical non-linearities. This is firstly due to the computationally demanding FE models associated with such systems and, secondly, to the requirement that estimation methods must consider non-linear phenomena. The problem is further pronounced in online applications, where real-time performance is required, as is commonly the case in structural health monitoring (SHM). Within this context, the focus is on computationally efficient models that operate on subspaces of significantly smaller size as compared to the full-order problem and which can be tailored to the framework of non-linear state estimation.

This study proposes the implementation of physics-based reduced-order models (ROMs) for response prediction of systems featuring geometrically non-linear effects. In so doing, the concept of modal derivatives (Weeger et al., 2016) is adopted and combined with a flexible multibody approach in order to capture second-order effects, e.g. twist-bend coupling, that arise as the system departs from the linear regime and exert a strong impact on the aeroelastic modal properties and stability of blades (Lobitz & Veers, 1999). In identifying the vibration response of such structures, the ROMs are compounded with the unscented Kalman filter (UKF) (Julier & Uhlmann, 2004) for the non-linear state estimation. The outlined approach is tested on the real-time response prediction of the NREL 5-MW reference WT blade, assuming that a limited number of artificial vibration measurements is available. The effectiveness of the scheme is assessed as a tool for online SHM and vibration control.

REFERENCES

Julier, S. & Uhlmann, J.K. (2004) Unscented filtering and nonlinear estimation. In: *Proceedings of the IEEE*, 92 (3), 401–422.
Lobitz, D.W. & Veers, P.S. (1999) Aeroelastic behavior of twist-coupled HAWT blades. In: *ASME/AIAA Wind Energy Symposium*.
Weeger, O., Wever, U. & Simeon, B. (2016) On the use of modal derivatives for nonlinear model order reduction. *International Journal for Numerical Methods in Engineering*, 108, 1579–1602.

Modal strain-based damage identification on a prestressed concrete beam

D. Anastasopoulos, G. De Roeck & E.P.B. Reynders
Department of Civil Engineering, KU Leuven (University of Leuven), Leuven, Belgium

ABSTRACT

The development of reliable structural health monitoring (SHM) techniques that can provide valuable information about a structure's condition is necessary, not only for maintenance purposes but also in a post-earthquake situation. This information is important for making inspections more timely and minimizing the uncertainty associated with post-earthquake damage assessments. Vibration-based Structural Health Monitoring is widely used for monitoring civil infrastructure as it constitutes a non-destructive condition assessment method, dependent on the identification of changes in the dynamic characteristics of a structure that are related to damage. However, the commonly employed dynamic characteristics, i.e. natural frequencies and displacement mode shapes, are relatively insensitive to local damage of moderate severity (Deraemaker et al., 2008).

Modal strains and curvatures are more sensitive to local damage (Unger et al., 2005; Anastasopoulos et al., 2018), but the direct monitoring of these quantities is challenging due to the very small (sub-microstrain) strain levels occurring during ambient, or operational excitation. The introduction of fiber-optic sensing systems, which can accurately measure dynamic strains while also offering ease of installation, resistance in harsh environment and long-term stability, contributed to an increased interest in adopting these systems for VBSHM applications. Fiber-optic Bragg grating (FBG) strain sensors share these advantages with other fiber-optic sensors, but additionally they are easy to multiplex (i.e. many different sensors with different operating wavelengths can be inscribed into the same glass fiber).

In the present work, the identification of the modal strains of a pre-stressed concrete beam (Figure 1), subjected to a four-point bending progressive damage test, is performed. Dynamic measurements are conducted on the beam at the beginning of each loading cycle. The response of the beam is recorded with uniaxial accelerometers and four chains of multiplexed FBG strain sensors. The evolution of the dynamic characteristics of the beam after each loading cycle is investigated. Changes of the natural frequency values, of the amplitude and the curvature of the strain mode shapes are observed even from an early

Figure 1. The experimental setup of the modal test and the PDT.

damaged state, indicating the presence of damage even when the cracks where nearly closed due to the prestressing force. No changes in the displacement mode shapes are observed, which confirms that accurately identified modal strains are much more sensitive to structural damage in prestressed concrete structures than conventional modal displacements.

REFERENCES

Anastasopoulos, D., De Smedt, M., Vandewalle, L., De Roeck, G. & Reynders, E. (2018) Damage identification using modal strains identified from operational fiber-optic bragg grating data. *Structural Health Monitoring*. doi:10.1177/1475921717744480.

Deraemaeker, A., Reynders, E., De Roeck, G. & Kullaa, J. (2008) Vibration based Structural Health Monitoring using output-only measurements under changing environment. *Mechanical Systems and Signal Processing*, 22 (1), 34–56.

Unger, J.F., Teughels, A. & De Roeck, G. (2005) Damage detection of a prestressed concrete beam using modal strains. *Structural Engineering*, 131 (9), 1456–1463.

Damage detection of steel bridge by numerical simulations and measurements

B.T. Svendsen, G.T. Frøseth & A. Rønnquist
The Norwegian University of Science and Technology, Trondheim, Norway

ABSTRACT

The Hell Bridge Test Arena is a riveted steel bridge taken out of service and moved to new foundations on land to serve as a full-scale laboratory for research within structural health monitoring, inspection, service life estimation and damage detection. The secondary load carrying system of the bridge consists of longitudinal girders connected to cross girders and lies below the bridge deck. Alternative ways of performing inspection and damage detection is explored due to low accessibility of this system.

Structural health monitoring (SHM) is defined as the process of implementing a strategy for identification of damage in structural systems (Farrar & Worden, 2007). Reducing the cost of maintenance and inspection and subsequently operational downtime is one of the most important motivations for performing SHM today.

This paper investigates the use of imposed vibration as a method to assess the condition of structural joints below the bridge deck. Results obtained from field measurements and numerical simulations are compared using analysis techniques in both time and frequency domain. Damage is identified by establishing differences in results from beam members connected to damaged and undamaged joints. The objective is to verify that 1) the results obtained from field measurements is caused by a structural damage and 2) the method considered is feasible to identify severe faults in the connections considered in the study.

In this paper, it is seen that results obtained from simulations and field measurements corresponds, verifying that a damage is present in the structural system. Figure 1 shows the comparison of cross correlation obtained from field measurements for sensors A01 and A06 located on an undamaged and damaged structural component, respectively.

It is concluded that both the existence and location of damage is determined. The existence of damage is determined based on the analysis methods in time and frequency domain, and the location of damage is determined based on the instrumentation setup. Using imposed vibration from a modal hammer as a method for performing inspection of the secondary load carrying system from the bridge deck, is further strengthened towards being considered as a valid approach.

REFERENCE

Farrar, C.R. & Worden, K. (2007) An introduction to structural health monitoring. *Philosophical Transactions of the Royal Society A: Mathematical, Physical and Engineering Sciences*, 365 (1851), 303–315. doi: 10.1098/rsta.2006.1928.

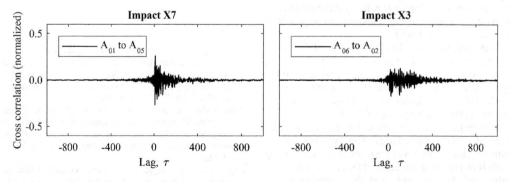

Figure 1. Normalized cross correlation for sensors A01 and A06, field measurements. A05 and A02 are reference sensors.

Parametric identification and damage assessment of real buildings

G.S. Wang & Y.R. Li
Department of Construction Engineering, Chaoyang University of Technology, Taichung, Taiwan

F.K. Huang
Department of Water Resources and Environmental Engineering, Tamkang University, New Taipei City, Taiwan

ABSTRACT

Nowadays, many civil structures in Taiwan area, including buildings and bridges, are installed with Strong-Motion Systems. Consequently, identification of either linear or nonlinear structural behaviors of these buildings, using the data collected from the accelerographs installed on them, has become important discipline due to the increasing need to estimate the behavior of a system with partially known dynamics. In addition to updating the structural parameters for better response prediction, system identification techniques also made possible to monitor the current state or damage state of the structures.

The parameters of some real buildings may vary with the change of the amplitude of vibration, and the identified parameters may not reflect the real state of the structure if we only use a model of a linear system to represent the structure. In this regard, a new system identification method called recursive hybrid genetic algorithm, which can identify the change of parameters, was developed for the purpose of identifying the system parameters of the buildings with accelerographs installed (Wang, Tsai & Huang, 2013). The time history of the measurement is divided into a series of time intervals, and then the model of equivalent linear system is employed to identify the modal parameters of the system.

In the implementation of the hybrid GA in the time domain, numerical integration is essential for solving the differential equation in the time domain. This integration procedure may result in a huge amount of computational time. In order to accelerate the identification process, a recursive hybrid GA in the frequency domain is developed. The differential equation can be transformed into the frequency domain by finite Fourier transform. The numerical time for solving the algebraic equations can be reduced tremendously compared to the time for solving the differential equation. The proposed algorithm has been explored by comparing the results of the predicted response with the measured response for both the simulated SDOF linear system and the simulated MDOF linear system with or without noise contamination (Wang & Huang, 2015).

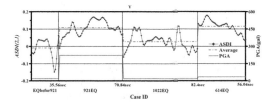

Figure 1. Variation of ASDI index of the top floor in the transverse direction for NCREE building.

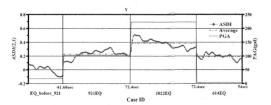

Figure 2. Variation of ASDI index of the top floor in the transverse direction for SHES building.

In this paper, this new identification strategy will first be applied to the identification of the office building of National Center for Research on Earthquake Engineering (NCREE) located in Taipei subjected to 4 sets of earthquake including the Chi-Chi earthquake. Then, the algorithim will also be applied to the identification of building of the Shih Hsien Elementary School located in Chai-Yi City. By employing the Maximum Softening Index and Approximate Story Damage Index, we can determine the damage states of these buildings.

REFERENCES

Wang, G.S., Tsai, Y.C. & Huang, F.K. (2013) Hybrid genetic algorithm in frequency domain to structural dynamic parameter identification. In: *11th International Conference on Structural Safety & Reliability*, New York, USA.

Wang, G.S. & Huang, F.K. (2015) Frequency-domain recusive hybrid GA to the identification of a real building. In: *15th Structural Engineers Congress*, Singapore, Singapore.

Life-Cycle Analysis and Assessment in Civil Engineering: Towards an Integrated Vision – Caspeele, Taerwe & Frangopol (Eds)
© 2019 Taylor & Francis Group, London, ISBN 978-1-138-62633-1

Deep neural network for structural damage detection in bridges

D. Darsono & M. Torbol
School of Urban & Environmental Engineering, Ulsan National Institute of Science and Technology, Ulsan, South Korea

ABSTRACT

A number of artificial intelligence techniques have been implemented to detect the damage in civil structure to partially replace onsite inspections. These artificial intelligence methods are primarily used to manipulate dataset and extract defect features, such as: the damage that occurs in concrete and steel bridges. This paper proposes a vibration-based method using a deep architecture of back propagation neural networks (BPNNs) for detecting damage. BPNNs can learn the features of images and signals. The robustness and adaptability of the proposed approach are tested on two different structures. The results show that the proposed method has better performances than existing ones.

The aging civil infrastructure has been challenging many researchers to study the condition civil structures to better face. Maintaining safe and reliable civil infrastructures for daily use is very important to human activities. Knowing the condition and integrity of the structure in terms of its age and usage, and its level of safety from chatastropic event is important and necessary. The process evaluation of determining and tracking structural integrity and assessing the nature of damage in a structure is often referred to as health monitoring.

Generaly, the observation of a structural system over time using periodically sampled response measurements from an array of sensors, the extraction of damage-sensitive features from these measurements, and the statistical analysis of these features to discriminate the actual structural condition for short or long-time periods. Then, once the normal condition has been successfully learned, the model can be used for rapid condition assessment to provide, in nearly real time, reliable information regarding the integrity of the structure.

It has been known that many researchers are working on various techniques for detection of structural damage. The recent trends in damage detection in structures are generally done by using optimization and computational method such as fuzzy logic, neural network, artificial intelligence, etc.

This study proposes a new method that uses machine learning in the form of neural network for the damage identification of bridges. Neural networks are capable of auto association, self learning, non-linear modeling and are excellent in pattern recognition (Jang 2011, Park 2009, Yeung and Smith 2005, Ko et al 2002). This study couples time series autoregressive with exogenous input, an artificial neural network (ANN), a finite element (FE) model of the bridge of interest, and signals from installed sensors to identify and to locate the damages in the bridge.

REFERENCES

Allemang, R.J. (2003) The modal assurance criterion – Twenty years of use and abuse. *Sound and Vibration*, 37 (8), 14–23.

Bandara, R.P., Chan, H.T. & Thambiratnam, D.P. (2014) Structural damage detection method using frequency response function. *Structural Health Monitoring*, 13 (4), 418–429.

Brincker, R., Zhang, L. M. & Andersen, P. (2000) Modal identification from ambient responses using frequency domain decomposition. In: *IMAC-XVIII: A Conference on Structural Dynamics, Vols 1 and 2, Proceedings*, 4062, 625–630.

Yun, C.B. & Bahng, E.Y. (2000) Sub structural identification using neural networks. *Computers & Structures*, 77 (1), 41–52.

Vibration-based damage indicators for corrosion detection in tubular structures

O.E. Esu, Y. Wang & M.K. Chryssanthopoulos
University of Surrey, Guildford, Surrey, UK

ABSTRACT

An increasing number of tubular steel structures such as pipelines, offshore jacket legs, wind turbine towers etc. are approaching or have exceeded their design service lives. Hence, monitoring of these structures is crucial in preventing any unforeseen failure and corresponding catastrophic consequences – safety or economic. As is well known, vibration-based structural health monitoring (SHM) presents non-destructive methods for damage identification in structural elements, though their application in corrosion problems appears somewhat limited.

In this paper, numerical models of a pipe in its intact and three distinct corroded patterns (Fig. 1) are built and analysed using ABAQUS®. Modal parameters from the first five vibration modes extracted from analyses results are employed in natural frequency and modeshape-based methods to identify corrosion at three levels: detection (level-1), localisation (level-2) and quantification (level-3).

The employed natural frequency method involved a comparison of the eigen-frequencies of the intact and damaged pipes. Results show that the frequency method is more capable of detecting local corrosion than global corrosion. It was also revealed that detection of corrosion by frequency comparison is more difficult to achieve if the damage is located towards the end of a pipe. The existing Modeshape Curvature (MSC) and Modal Strain Energy (MSE) methods were also evaluated and it was demonstrated that they are both capable of level-2 identification of the investigated corrosion patterns.

Finally, a potential damage indicator: Normalised Displacement Modeshape (NDM) is introduced. NDM operates on the relative difference between eigenvector co-ordinates of the intact and defective structure to identify damage according to Equation 1.

$$NDM(j) = \frac{\sum_{i=1}^{I}\left[1 - \frac{|(\psi_{ji}^*) - (\psi_{ji})|}{MAX_{j=1}^{J}|\psi_{ji}^* - \psi_{ji}|}\right]}{J} \quad (1)$$

where, ψ_{ji} and ψ_{ji}^* are the intact and damaged eigenvectors respectively; i is the vibration mode number; I is the total number of vibration modes investigated while j is the eigen-vector co-ordinate parameter and J represents the total number of eigen-vector coordinates (DOFs). The values obtainable are between 0.0 (most damaged region) and 1.0 (least damaged region). A severity factor, F_{NDM} was also developed to extend the applicability of NDM to level-3 identification. Figure 2 presents one of the cases (Case-3c) where a 100 mm corrosion pit (4.26% of the pipe's volume) is modelled with mid-point set at 750 mm from one end of the 2 m long test pipe.

The plot exhibits the efficiency of NDM at level-2 identification. For the same case, an F_{NDM} factor of 20.6 was obtained, compared to 2.5 and 5.9 when corrosion pits were 1.07% and 2.13% of pipe volume respectively; implying NDM's strong potential for level-3 identification. With these assuring results, future research will seek to combine the advantageous components of different methods to produce a universal indicator for high-level vibration-based corrosion identification.

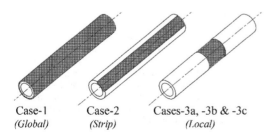

Figure 1. Corrosion patterns investigated.

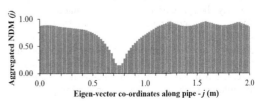

Figure 2. NDM Plot for Case-3c.

Pre-image reconstruction for compensation of environmental effects in structural health monitoring by kernel PCA

C. Rainieri
University of Molise, Campobasso, Italy

E.P.B. Reynders
Katholieke Universiteit Leuven, Leuven, Belgium

ABSTRACT

Popular damage detection methods are based on the analysis of the variations of modal parameters. While the principle of modal based damage detection is very simple, since it assumes that damage is reflected by changes in the mass or stiffness properties of the monitored structure, modal parameter estimates are sensitive not only to damage but also to environmental and operational variables (EOVs). Thus, approaches to compensate the environmental and operational variability of modal parameter estimates, possibly without measuring the EOVs, are highly desirable. A number of approaches have been proposed in the literature. The most popular is based on Principal Component Analysis (PCA) (Yan et al. 2005). However, linear relationships among the unknown EOVs are assumed, so the method shows limited performance when those are non-linear. Taking into account that the influence of EOVs on modal characteristics is in general nonlinear and it can affect the monitored features in different ways, the use of nonlinear system identification procedures, such as kernel PCA, for Structural Health Monitoring (SHM) in changing environmental conditions has been proposed (Reynders et al. 2014). Kernel PCA is a robust and computationally efficient procedure and the type of nonlinearity has not to be explicitly defined. However, it requires the setting of two parameters, which influence the quality of the results. A procedure for their optimal setting has been proposed in (Reynders et al. 2014). However, that method yields principal components as well as residues expressed in the feature space, that is to say, in the space defined by the application of a nonlinear function Φ, and not in terms of the physical quantities of the input space (namely, the space of the original – unmapped – data). The present paper presents a procedure for the reconstruction of the influence of EOVs on modal parameter estimates and the computation of the corresponding residues in the input space. As an additional advantage, the two user-defined parameters in kernel PCA are automatically selected in an optimal way.

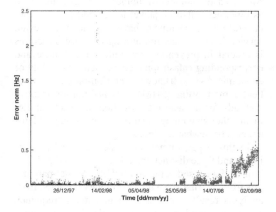

Figure 1. Misfit corresponding to sub-optimal parameter choice.

Application of the proposed methodology to selected case studies provided encouraging results. In particular, the advantage of the proposed approach to the automated selection of the optimal parameter set has been confirmed by the analysis of the error norms. In fact, when sub-optimal parameters are selected, the misfit evaluated either in the feature space or in the input space (Figure 1), might still be affected by environmental/operational conditions making the setting of appropriate thresholds for anomaly detection more difficult and affecting the reliability of modal based SHM.

REFERENCES

Reynders, E., Wursten, G. & De Roeck, G. (2014) Output-only structural health monitoring in changing environmental conditions by means of nonlinear system identification. *Structural Health Monitoring*, 13, 82–93.
Yan, A.-M., Kerschen, G., De Boe, P. & Golinval, J.-C. (2005) Structural damage diagnosis under varying environmental conditions—Part I: A linear analysis. *Mechanical Systems and Signal Processing*, 19, 847–864.

MS3: Probabilistic assessment of existing structures
Organizers: M. Sýkora, M. Holicky, R. Caspeele, D. Diamantidis & R.D.J.M. Steenbergen

Reducing semi-probabilistic methods to acceptable structural safety deficits in deterministic assessments of existing concrete structures

D. Zwicky
Head of Institute of Construction and Environmental Technologies iTEC, School of Engineering and Architecture of Fribourg HEIA-FR, University of Applied Sciences and Arts of Western Switzerland HES-SO, Switzerland

ABSTRACT

Probabilistic assessment of existing structures can be a powerful tool for efficiently prioritizing maintenance interventions within usually limited budgets for this activity of ever increasing importance. However, structural engineers in practice usually have nor sufficient know-how nor enough time at hand to apply full- or semi-probabilistic procedures. They need an easily applicable deterministic evaluation tool to judge what level of structural safety deficit is still acceptable if recommending disproportionate structural interventions shall be avoided.

Based on semi-probabilistic updating for normally distributed variables, considering generally accepted target reliability indexes β_0 as well as proportionality of maintenance interventions, a simplified proposal for determining acceptable structural safety deficits in deterministic assessments of concrete elements is derived and discussed in the full paper.

Varying material qualities, as often encountered in existing structures, and different types of failure modes are considered in linking a deterministic degree of compliance n to the reliability index β, where $n = R_{d,act}/E_{d,act}$, with $R_{d,act} =$ (updated) examination value of structural resistance and $E_{d,act} =$ (updated) examination value of action effect. It corresponds to the inverse of the "unity check" used elsewhere, and quantifies to what extent an existing structure meets the requirements of a comparable new structure. The basic requirement is $n \geq 1$ or $R_{d,act} \geq E_{d,act}$, respectively. The reliability index, in turn, can be related to cost efficiency of intervention measures.

Such reflections are applied in combination with Swiss code directives. The Figure below shows examples of correlations between n and β, allowing to identify acceptable structural deficits if disproportionate interventions shall be avoided (i.e. overly expensive measures, in relation to the attained risk reduction).

If n is relatively low or structural safety deficits are relatively high, respectively, urgent safety measures may be required, requiring execution within hours (or, possibly, days). The full paper also proposes a simplified approach how to address this question, referring to consequence classes for different use categories and taking advantage of practical and well-accepted structural engineering knowledge.

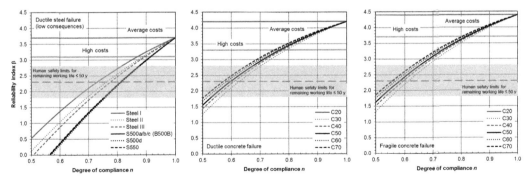

Figure 1. Correlations between n and β for different material qualities and failure types, considering cost efficiency of intervention measures.

Life-Cycle Analysis and Assessment in Civil Engineering: Towards an Integrated Vision – Caspeele, Taerwe & Frangopol (Eds)
© 2019 Taylor & Francis Group, London, ISBN 978-1-138-62633-1

The added value of full-probabilistic nonlinear finite element analysis for the assessment of reinforced concrete structural members

Á. Rózsás, A. Slobbe, D.L. Allaix, W.M.G. Courage, A. Bigaj van Vliet & H.G. Burggraaf
Department of Structural Reliability, Netherlands Organization for Applied Scientific Research (TNO), Delft, The Netherlands

ABSTRACT

The objective of this paper is to explore the added value of full-probabilistic nonlinear finite element analysis (NLFEA), which is illustrated for two reinforced concrete (RC) structural members: a RC deep beam and a RC continuous girder. This exploration is presented gradually: step by step advancing the approximation level of mechanical and the probabilistic models. The most advanced combination uses NLFEA coupled with full-probabilistic analysis (reliability analysis) where uncertain model parameters are represented by random variables. To discuss the added value of various methods, a comparison is made between the semi-probabilistic Eurocode method, the semi-probabilistic NLFEA methods from Eurocode and fib Model Code 2010, and the full-probabilistic NLFEA. A new quantitative measure of added value is introduced and used for this comparison. It expresses the largest value of relative deficit (non-compliance) that can be compensated by a more advanced method.

For the RC deep beam example, the measure of added value of the full-probabilistic NLFEA over the semi-probabilistic Eurocode method is found to be 0.56. This means that even if the semi-probabilistic Eurocode method shows that the design action is 56% higher than the design resistance, the compliance can be demonstrated (reserves uncovered) by full-probabilistic NLFEA. For the RC continuous girder, this measure of added value is 0.48. While in case of the deep beam the gain is largely attributed to the higher approximation level of NLFEA, for the continuous girder almost solely the probabilistic models contribute to the gain.

Though the added values are case dependent, the results indicate that a more detailed physical representation of the problem and an explicit treatment of the uncertainties can compensate a substantial deficit of simplified methods. Hence, these advanced methods may uncover reserves and offer a promising alternative in the assessment of existing structures, enabling to avoid expensive measures that might be needed based on simplified methods.

Reliability assessment of infrastructure in Germany: Approaching a holistic concept

A. Panenka, F. Nyobeu, H. Schmidt-Bäumler & J. Sorgatz
Federal Waterways Engineering and Research Institute, Karlsruhe, Germany

R. Rabe
Federal Highways Research Institute, Bergisch-Gladbach, Germany

M. Reinhardt
Federal Railway Authority, Bonn, Germany

ABSTRACT

This paper gives insight about the development process of a common reliability assessment concept, which meets the requirements of all relevant modes of transport in Germany. Given the current context of aging infrastructure and increasing environmental stressors, the concept aims at establishing a maintenance management strategy involving all modes of transport. Thus, a cross-cutting research programme, namely the BMVI Network of Experts (NoE) was initiated 2016 by the German Federal Ministry of Transport and Digital Infrastructure (BMVI), as owner of the infrastructure (Federal Ministry of Transport and Digital Infrastructure 2017).

In this context, Research Area 3 investigates methods and concepts, which could help in increasing the reliability of the transportation infrastructure again. State-of-the-art field tests methods to assess the condition of the infrastructure and modern evaluation tools in combination with the knowledge about the availability and vulnerability under extreme weather events result in the development of effective rehabilitation measures under operation (see Fig. 1).

In order to evaluate the reliability in an appropriate way, different methods and techniques of reliability assessment are tested and evaluated. This includes quantitative approaches considering statistical models, probabilistic approximation methods, and advanced structural analyses as well as qualitative approaches, such as the failure mode and effect analysis.

The development of dynamic load models for railway bridges provides valuable insights into the analytic processes, which evolves from theoretical assumptions about physical mechanisms to their practical validation by means of in-situ measurements and monitoring, and thereafter to a standardised load model ready for actual design applications.

The definition of structural reliability implies that relevant limit states of the assessed structures are generally known. If this information is not available, it has to be elicited systematically using appropriate methods, like guided expert interviews to collect

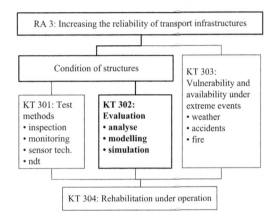

Figure 1. Key Topics of NoE Research Area 3.

systematically (long-term) observations based on which model tests could be designed.

Apart from the technical aspects, a future-oriented maintenance management system has also to consider consequences of possible failure mechanisms. Consequently, the capabilities of using multiple criteria decision analyses using key figures characterising risks of structural failure are also investigated.

The projects support the up-to-date discussion on a reliable and highly available transportation infrastructure by providing modern analytical methods, modelling techniques and software solutions. While the first results of each project are related to a specific mode of transport, the context, fundamental considerations and applied methods for producing these results are relevant for all involved parties.

REFERENCE

Federal Ministry of Transport and Digital Infrastructure. (2017) *BMVI Network of Experts. Knowledge – Ability – Action.* Berlin: Printed In-house. Available online under http://www.bmvi-expertennetzwerk.de/DE/Publikationen/Medien/Brochure-Expertennetzwerk.pdf

Life-Cycle Analysis and Assessment in Civil Engineering: Towards an Integrated Vision – Caspeele, Taerwe & Frangopol (Eds)
© 2019 Taylor & Francis Group, London, ISBN 978-1-138-62633-1

Probabilistic aspects relating to assessment of existing structures

M. Holicky
Czech Technical University in Prague, Prague, Czech Republic

1 INTRODUCTION

The final draft of CEN Technical Specification (TS) on assessment of existing structures is linked to the probabilistic concepts and fundamental requirements of the EN Eurocodes. The document concerns all types of buildings, bridges and construction works, including geotechnical structures, exposed to all kinds of actions. The final draft of the Technical specification contains requirements, general framework of assessment, data updating, structural analysis (linear, nonlinear, dynamic), verifications (partial factors, probabilistic methods, risk assessment), past performance, interventions, annexes (flowchart, time-dependent reliability, assessment of heritage structures). The submitted contribution provides background information on the principles accepted in the TS. Particular attention is given to verification of the target reliability levels and operational procedures in engineering practice including partial factors method, design value method, probabilistic approach and methodology of risk assessment.

2 CONCLUDING REMARKS

The main principles of the upcoming European document on assessment of existing structures can be summarized as follows:

- Currently valid codes for verification of structural reliability should be applied, historic codes valid in the period when the structure was designed, should be used as guidance documents only;
- Actual characteristics of structural material, action, geometric data and structural behaviour should be considered.

The most important step of the whole assessment procedure is evaluation of available data and updating of prior information concerning actions, strength and structural reliability. It appears that a Bayesian approach can provide an effective tool.

Typically, assessment of existing structures is a cyclic process in which the first preliminary assessment is often supplemented by subsequent detailed investigations, data evaluation and updating. A report on structural assessment prepared by an engineer assessing the structure should include a recommendation on possible structural and operational interventions.

The following conclusions are related to the target reliability level:

(1) The target reliability levels recommended in various national and international documents are inconsistent in terms of the values and the criteria according to which the appropriate values are to be selected.
(2) In the latest draft or revision of prEN 1990 (1997) the target reliability level is indicated only by annual levels related to one year.
(3) Transformation formulae for adjustment of reliability level to different reference periods taking into account mutual dependence of failure probabilities in subsequent years are missing.
(3) Proposed transformation formula for reliability index β_{nk} depends on the reference period n and independence interval k.
(4) Reliability index β_{nk} decreases with the reference period n and increases with the independence interval k.
(5) When determining the target reliability index the assumption of annual independency of failures ($k = 1$) may be unsafe.

ACKNOWLEDGEMENT

This contribution has been developed as a part of the research project GAČR 16-11378S, "Risk based decision making in construction" supported by the Czech Grant Agency.

Hidden safety in equilibrium verification of a steel bridge based on wind tunnel testing

J. Žitný, P. Ryjáček, J. Marková & M. Sýkora
Czech Technical University in Prague, Prague, Czech Republic

ABSTRACT

The Cervena Bridge is a long-span steel truss bridge built in the Czech Republic in 1886. The assessment by the partial factor method in Eurocodes reveals its insufficient reliability. In particular the bridge fails to satisfy the equilibrium limit state when strong wind occurs simultaneously with an unloaded train on the bridge. To avoid structural interventions or expensive

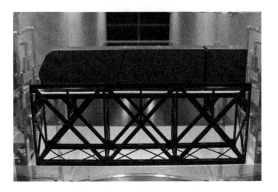

Figure 1. Model of the bridge with a suburban train fixed in the force balance.

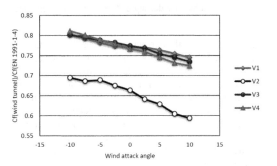

Figure 2. Ratio of force coefficients – the wind tunnel to EC1-4, for four types of vehicles and a selected EC1-4 approach.

traffic restrictions, a model of the bridge was tested in a wind tunnel to obtain force coefficients of wind pressure for the specific shape of the bridge and selected types of light-weight trains (Figure 1). The values of force coefficients based on wind tunnel tests correspond to 70–80% of those provided in EN 1991-1-4 (EC1-4, Figure 2).

Traffic flow records for the railway line under consideration are utilised to obtain distributions of weights and heights of light-weight trains. Detailed probabilistic analysis reveals that bridge reliability is close to the target level when the wind tunnel force coefficients are considered along with free-field wind and railway traffic records.

Many existing light-weight iron and historic steel railway bridges fail to comply with the criteria on equilibrium in Eurocodes while no problems have been experienced over decades. The case study based on measurements in the wind tunnel, free-field wind velocity and traffic flow records indicates that:

1. EC1-4 gives ambiguous guidance on how to determine wind loads on truss girder bridges. Several alternatives provided therein lead to distinctly different wind pressures. The force coefficients updated by wind tunnel tests correspond to 70–80% of those given in EC1-4; the range 60–95% is indicative for other types of steel bridges.
 - The partial factor method in Eurocodes includes "hidden safeties" that may be removed by measurements. They may be related to permanent actions, geometry, wind pressure (shape, orography, terrain roughness, wind directionality, global dynamic effect on a large structure, basic wind velocity for local conditions), traffic load (realistic weights of unloaded trains), and the combination of strong wind and railway traffic.

ACKNOWLEDGEMENTS

The study was partly supported by Grants TE01020168 and FV 20585.

Analysis for repair effect in each layer of expressway pavements

S. Araki
Graduate School of Engineering, Osaka University, Osaka, Japan

T. Kazato
Nippon Expressway Research Institute Company Limited, Tokyo, Japan

K. Kaito
Graduate School of Engineering, Osaka University, Osaka, Japan

K. Kobayashi
Graduate School of Management, Kyoto University, Kyoto, Japan

A. Tanaka
Graduate School of Engineering, Osaka University, Osaka, Japan

ABSTRACT

The asphalt pavements of the expressway in Japan managed by NEXCO have layered structures, and consist of paving, bedding, upper subbase, and under subbase from the surface. NEXCO RI conducts FWD survey nationwide in order to inspect pavements' deterioration, and it utilizes the results of the survey when they make decisions for repair projects. Due to recent restrictions on budget for repair projects and many deteriorated infrastructures in Japan, they must implement efficient repair projects.

Deterioration of pavement progresses with the passage of time. The larger the value of FWD which represents load bearing capacity, the more the pavement is deflected and the more the pavement is deteriorated. Further, the value of FWD gets small after pavement repair. Figure 1 shows the deterioration process of spots once repaired. Mizutani et al. (2016) revealed statistically that the increasing process of FWD values is subject to exponential function.

We represent the FWD measured value (see Figure 1) x_i as following:

$$x_i = \exp(-B_i)\, g_T - \xi_i \quad (1a)$$

$$B_i = \mathbf{z}_i \boldsymbol{\theta}' + \sigma w_i \quad (1b)$$

Here, B_i is an index reflecting the heterogeneity of the deterioration characteristics of spot i and can be expressed by the sum of characteristic variable term $\mathbf{z}_i \boldsymbol{\theta}'$ and error term σw_i as shown in Equation (1b). In Equation (1b), $\mathbf{z}_i = (z_i^1, \ldots, z_i^M)$ is a characteristic variable vector affecting deterioration of the spot i, $\boldsymbol{\theta} = (\theta^1, \ldots, \theta^M)$ is a parameter vector, w_i is a probability variation term representing the deterioration factor peculiar to the spot i, and the σ is a scale parameter. Also, g_T is linear combination of exponential functions. ξ_i represents the amount of recovery by repair in "Recovery process" in Figure 1. Assume that ξ_i is

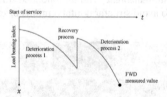

Figure 1. The transition process of the load bearing capacity on the spot where once repaired.

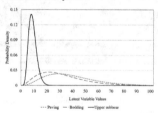

Figure 2. Probability density distributions that latent variables follow.

subject to Weibull distribution, and using the parameters of the distribution we estimate the difference in the repair effects of target layers. Also, assume that the probability variation term w_i is subject to Gumbel distribution, we define the likelihood function and estimate the value of the other parameters.

Figure 2 shows the results of estimated distributions of repair effects. The horizontal axis means reduction of the values of FWD, and the vertical axis means the probability density.

REFERENCE

Mizutani, D., Kobayashi, K., Kazato, T., Kaito, K. & Matsushima, K. (2016) A continuous deterioration hazard model: Application to pavement structure. *JSCE*, D3, 72 (2), 191–210.

Life-Cycle Analysis and Assessment in Civil Engineering: Towards an Integrated Vision – Caspeele, Taerwe & Frangopol (Eds)
© 2019 Taylor & Francis Group, London, ISBN 978-1-138-62633-1

Advancements in bridge specific assessment live loads for long span structures

N. Benham, C. Mundell & C.R. Hendy
Atkins, Epsom, UK

ABSTRACT

In the UK there are a number of long span suspension bridges, with many of these now requiring routine inspection, assessment and maintenance to ensure their continued durability. The UK Design Manual for Roads and Bridges (DMRB) has explicit guidance on the traffic loading for assessment lengths up to 50 m, however beyond this the assumptions become conservative. In these instances, the assessment of these structures requires a Bridge Specific Assessment Live Load (BSALL) to be derived. Although a number of methodologies exist to derive BSALLs, there are several parameters that may significantly affect their results and there is little published guidance on the subject.

Atkins have now calculated BSALLs for three of the UK's most iconic suspension bridges, being the Severn Bridge, the Forth Road Bridge and the Humber Bridge. Through this work, Atkins have done many parametric studies, considering different distribution methods and the relative importance of the various parameters involved, including the mix of traffic of the bridge and the impact of tolling and changes in tolling.

This paper discusses the above themes and outlines the advancements made by Atkins in this field, highlighting the critical parameters to consider, the advantages and limitations of the various approaches, and a recommended approach based on our findings to date.

Probabilistic assessment of durability of hydroelectric power plants

J. Marková, K. Jung & M. Sýkora
Klokner Institute, Czech Technical University in Prague, Prague, Czech Republic

ABSTRACT

Application of partial factor method for the assessment of existing water construction works including hydroelectric power plants according to current codes for structural design might lead to conservative estimations. The probabilistic assessment of existing structures makes it possible to effectively estimate remaining working life of structures and to plan their maintenance.

Presently Eurocodes do not give guidance for the design and verification of the water construction works. That is why supplementary national provisions were developed in the revised Czech standards focused on water construction works and structures in hydroelectric power plants which are presently harmonised with Eurocodes. Additional reliability elements were recommended for the static and hydrodynamic water pressures. Water construction works were classified into reliability classes RC1 to RC3. The coefficient K_{FI} recommended in EN 1990, Annex B is also applied in the revised Czech national standard, however with an increased value 1.1 for the reliability class RCII.

Selected results of reliability analyses of a steel structural member of water construction works based on four sets of different partial factors for actions and material properties are analysed. It appears that the reliability elements given in CSN 73 0038 with a set of National Annexes to ISO 13822 is well calibrated and might be applied for the national verification of existing structures.

The probabilistic reliability assessment of existing water construction works is illustrated by the case study of an industrial bridge where the new technology of cleaning machinery is planned. Inspection revealed considerable rusting of steel members and their joints, deformations due to overloading of some supports and lack of regular maintenance. It was decided on structural strengthening and also partial replacement of structural members. Concrete piers will be also rehabilitated.

Probabilistic methods represent suitable tool for decision about alternative technological procedures

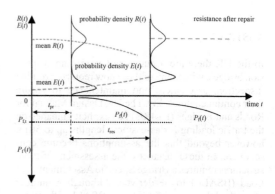

Figure 1. Probabilistic assessment of the remaining working life of an existing structure.

and evaluation of actual conditions or remaining working life of structures as illustrated in Fig.1.

Probabilistic assessment may help to authorise further use of partly deteriorated structures for shorter working lives without the need of immediate structural interventions or limitations of their use.

REFERENCES

CSN 75 6303 (2012) *Basis of Structural Design and Actions on Hydro-technical Structures*. ÚNMZ (in Czech).

CSN 73 0038 (2014) *Assessment and Verification of Existing Structures – Supplementary Provisions*. ÚNMZ (in Czech).

ISO 13822 (2010) *Bases for Design of Structures – Assessment of Existing Structures*. ISO, p. 35.

Holický, M. & Marková, J. (2007) Probabilistic design of structures for durability. In: *ESREL 07*, pp. 1757–1762.

Holický, M., Tanner, P., Steenbergen, R., Nuti, C., Shave, J. & Schnell, J. (2017) *Assessment of Existing Structures (final draft of the technical specification, October 2017)*. CEN TC250/ WG2.T1, p. 37.

Markova, J. & Holicky, M. (2014) Reliability of structures in national codes to Eurocodes. In: *Safety and Reliability, Methodology and Applications*. Wroclaw, 2015, pp. 2207–2212.

Reliability analysis of corroded reinforced concrete beam with regards to anchorage failure

M. Blomfors
RISE CBI, Gothenburg, Sweden
Chalmers University of Technology, Gothenburg, Sweden

D. Honfi
RISE Research Institutes of Sweden, Gothenburg, Sweden

O. Larsson Ivanov
Lund University, Lund, Sweden

K. Zandi & K. Lundgren
Chalmers University of Technology, Gothenburg, Sweden

ABSTRACT

Reinforcement corrosion is a common problem in reinforced concrete infrastructure today (Bell 2004) and it is expected to increase in the future (Wang et al., 2010). To simply replace the corroded structures with new ones requires large resources, both in financial and environmental terms. Therefore it is important that existing structures are used to their full potential, also after the onset of corrosion. Here, the reliability of the anchorage capacity of a reinforced concrete beam (presented in Figure 1) subjected to reinforcement corrosion is studied.

As expected, the results show that the reliability is reduced with corrosion; the magnitude depends to a large extent on the modelling uncertainty used for the bond model for corroded reinforcement. The influence of the corrosion level on the reliability without considering the modelling uncertainty is presented in Figure 2.

The sensitivity of the input parameters on the model response was also studied and the corrosion level was shown to be influential also in this respect. This was expected based on the properties of the underlying bond model. In Figure 3 the sensitivity of the anchorage capacity with respect to the basic variables is presented.

This paper demonstrates that probabilistic evaluations can give valuable insight regarding the reliability of corroded reinforced concrete structures. The knowledge can be used to prolong the service-life of existing infrastructure and save both money and the environment.

Figure 1. Studied beam with uniformly distributed loads; g is permanent load and q is variable load.

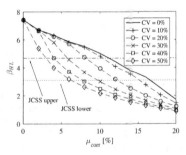

Figure 2. β_{HL} versus corrosion level for several CVs for the corrosion, without considering resistance model uncertainty.

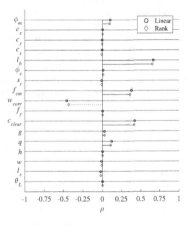

Figure 3. Sensitivity of the anchorage capacity with respect to the basic variables when considering corrosion.

REFERENCES

Bell, B. (2004) *Sustainable bridges. D1.3 European Railway Bridge Problems.* sustainablebridges.net.

Wang, X., et al. (2010) *Analysis of Climate Change Impacts on the Deterioration of Concrete Infrastructure – Synthesis Report.* Canberra: CSIR.

Historic roof structures: Life-cycle assessment and selective maintenance strategies

E. Garavaglia
Department of Civil and Environmental Engineering, Politecnico di Milano, Milan, Italy

N. Basso
Department of Architecture, School of Creative Science and Engineering, Waseda University, Tokyo, Japan

L. Sgambi
Faculty of Architecture, Architectural Engineering and Urbanism, Université Catholique de Louvain, Louvain-la-Neuve, Belgium

ABSTRACT

In the last two decades there has been growing interest in the conservation and adaptive-reuse of disused military and industrial buildings (Neaverson & Palmer 1998).

Especially in Europe, the historic heritage includes several good examples of rehabilitation and reuse of industrial archaeology buildings. When it comes to transform old constructions for a new or different use, social and economic factors as well as preservation and structural reliability levels must be deeply investigated. Whenever those aspects are not well considered and analysed, there is no possibility to define proper, successful intervention scenarios.

Many industrial and military constructions built between late 19th and early 20th century show long span roof structures with slender steel truss members. This characteristic makes the structures almost invisible and the inner space appears larger than it is. Despite disuse and negligence, after 100 years or more, these structures are still in good conditions. Because of their architectural and historic relevance, these buildings are usually protected by Cultural Heritage regulations. In this regard, three examples are here introduced and analysed in order to evaluate their residual reliability and robustness.

A Monte Carlo simulation implemented with a damage law has been applied to investigate, member by member, the current structural reliability of three roof truss systems of military buildings in Pavia, Northern Italy (Garavaglia et al. 2016). Once the members most damaged were identified, a sudden collapse simulation has been run to estimate the structural response associated with each of those members' failure, and evaluate the residual load-bearing capacity in terms of structural robustness. This analysis provides an initial overview on where and when maintenance actions must be performed in order to extend the structural lifespan and ensure its possible reuse responding to an acceptable safety level.

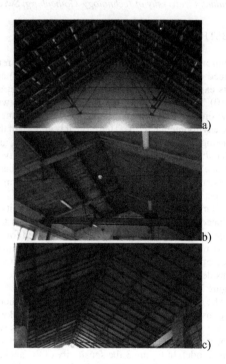

Figure 1. Arsenal and Rossani barrack complex in Pavia (North Italy): a) Pavilion in Rossani barracks timber-and-iron roof system, b) Arsenal Building 65, timber-and-iron roof system, c) Arsenal Building 15, iron roof system.

REFERENCES

Garavaglia, E., Sgambi, L. & Basso, N. (2016) Markovian simulation-based life-cycle assessment method for an old roof steel truss. In: *Proc. of IALCCE16*, Delft, The Netherland, October 16–19, 2016. Bakker, Frangopol & van Breugel (eds.) Taylor & Francis Group, London, 2017, 1677–1682.

Neaverson, P. & Palmer, M. (1998) *Industrial Archaeology: Principles and Practice*. New York, Routledge, p. 181.

Life-Cycle Analysis and Assessment in Civil Engineering: Towards an Integrated Vision – Caspeele, Taerwe & Frangopol (Eds)
© 2019 Taylor & Francis Group, London, ISBN 978-1-138-62633-1

Development of an optimization-based and practice orientated assessment scheme for the evaluation of existing timber structures

M. Loebjinski & H. Pasternak
Brandenburg University of Technology, Cottbus, Germany

J. Köhler
Norwegian University of Science and Technology, Trondheim, Norway

W. Rug
Eberswalde University for Sustainable Development, Eberswalde, Germany

ABSTRACT

Improving the modelling of load-bearing capacities of existing structure is an important component of a sustainable economy and contributes to the reduction of waste and the saving of resources and energy. Besides, the preservation of existing structures and the protection of cultural heritage is of societal interest. Current Eurocodes are orientated on the construction of new structures. If no national recommendations exist, the principles of EN 1990:2012 together with load and material specific codes have to be applied for the verification of load-bearing capacities of existing structures. In this contribution, an assessment and evaluation scheme that allows a stepwise evaluation of existing structures taking into account data gained in situ is presented. The concept has been developed for timber structures but is applicable on further construction materials.

The procedure starts with a defined initial point, i.e. *doubts concerning load-bearing capacities* or a *defined inspection time*. If it is decided to carry out an assessment, the first step is the *definition of requirements and goals* for the next steps. If within a *preliminary evaluation of conditions* safety can be ensured it can be decided to do nothing. In general, more data will be required. Hence a *preliminary investigation of possible damages and causes* is being done. For timber structures these causes for damages can be due to *biological* (insects, fungi), *chemical* (maceration) or *mechanical* (overloading, structural alterations) reasons. If significant damage is indicated, immediate safety measures have to be carried out.

Further investigation is divided into *Knowledge Levels*. The term is based on Luechinger et al. (2015), but the concept is widened into four levels. Table 1 summarizes the evaluation steps.

The application of the concept is illustrated by a structural example. It is shown that by a more detailed

Table 1. Summary of evaluation formats for Knowledge Levels.

Knowledge level	Evaluation format
KL 0	Semi-probabilistic No determination of strength grade → PSF to be calibrated
KL 1	Semi-probabilistic → PSF from current Eurocodes
KL 2	Semi-probabilistic → Updated PSF using reference property Probabilistic → Updated distribution function using reference property
KL 3	Probabilistic → Updated distribution function using directly updated property

evaluation taking into account data gained in situ load-bearing capacities can be modelled more accurately. The concept is compared to the current praxis in CEN member countries without national specific rules and to SIA 269:2011. The procedure is a contribution to the development of a consistent concept for the evaluation of existing structures within the concept of the Eurocodes.

REFERENCE

Luechinger, P., Fischer, J., Chrysostomou, C., Dieteren, G., Landon, F., Leivestad, S., Malakatas, N., Mancini, G., Markova, J., Matthews, S., Nolan, T., Nutti, C., Osmani, E., Rønnow, G., Schnell, J., Tanner, P., Dimova, S., Pinto, A. & Denton, S. (2015) *New European Technical Rules for the Assessment and Retrofitting of Existing Structures.* Luxembourg, Publications Office of the European Union.

Reliability assessment of large hydraulic structures with spatially variable measurements

S. Geyer, I. Papaioannou & D. Straub
Engineering Risk Anaylsis Group, Technische Universität München, Munich, Germany

C. Kunz
Bundesanstalt für Wasserbau, Karlsruhe, Germany

ABSTRACT

A large portion of existing hydraulic structures in Germany have an age of more than 70 years and about 25% of them are already past their intended service life of 100 years. An efficient life-cycle management is necessary to manage these large concrete structures safely and economically. Inspections and material testing form an important part of this process, as they enable an improved assessment of the condition and properties of the structure and its material. Traditionally, the limited data available from these tests is used to fit probability distributions of material parameters; characteristic values for the design are then obtained from the fitted distributions. Spatial correlation between the measurement locations or different material layers are typically neglected. In this contribution, we propose to model the spatial variability of the material parameters with random fields. We use the available data from measurements on three core samples to update the distribution of the random fields with Bayesian analysis. Here, we present a special case where an analytical solution is available for the Bayesian analysis. The posterior mean and standard deviation of the lognormal random fields are illustrated in Figure 1. We calculate the structural reliability by application of subset simulation, an adaptive sampling approach. We show that through the employed random field modeling approach, a more detailed statement about the condition of a structure can be made by using not only the measurements of the material properties, but also the information about their spatial location. The results are summarized in Table 1. They show that consistent modeling of the spatial variability of the concrete materials can potentially increase the reliability estimate of large hydraulic structures when measurement information is included.

Table 1. Estimates of the probability of failure $\hat{P}_F$ and its coefficient of variation $\hat{CV}_{\hat{P}_F}$ for the different modeling approaches obtained with subset simulation.

Modeling approach	$\hat{P}_F$	$\hat{CV}_{\hat{P}_F}$
Random variables	4.57×10^{-3}	0.19
Random fields	2.15×10^{-3}	0.20
Random fields with update	8.2×10^{-5}	0.28

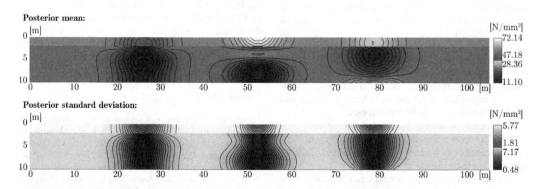

Figure 1. Posterior mean and standard deviation of the lognormal random fields representing the basic compression strength of two concrete layers of different type (Prior moments of the top layer: $\mu = 47.35$ N/mm^2, $\sigma = 5.87$ N/mm^2; Prior moments of the bottom layer: $\mu = 17.89$ N/mm^2, $\sigma = 7.67$ N/mm^2).

Uncertainties in the assessment of existing masonry structures

D. Müller & C.-A. Graubner
Technische Universität Darmstadt, Germany

ABSTRACT

For the assessment of existing structures, reliability analyses or the adjustment of partial safety factors can be very helpful if a conventional verification of structural safety according to design codes is not possible anymore. However, if probabilistic methods are applied, it is crucial to quantify all uncertainties realistically. Compared to the uncertainties in the design of new structures, additional types of uncertainty arise in the assessment of existing structures. This includes statistical and testing uncertainty, which can both be classified as epistemic. In contrast to that, aleatory uncertainties only consist in the spatial variability of material properties. This paper aims at categorizing and describing the uncertainties involved in the assessment of existing structures as well as their effect on structural reliability. It addresses masonry structures in particular but mostly gives general thoughts that are also valid for other types of structures, e.g. steel or concrete structures.

It is shown that the influence of spatial variability on the reliability of masonry walls depends on the failure mode of the wall, the material ductility of masonry as well as the length of the wall, see also Müller et al. (2017). In usual cases, it is conservative to neglect spatial variability and use homogenous probabilistic models. Only if the structure almost behaves like a series system, a negative effect of spatial variability must be taken into account.

The importance of considering statistical uncertainties in the determination of required design values for material properties or corresponding partial factors based upon test results is emphasised. Procedures for modifying partial factors for existing structures like the adjusted partial factor method (fib 2016) can be extended by a consideration of the statistical uncertainty, which is introduced by a small number of tests n. This can be done by the Bayesian method with vague prior distributions as described in ISO 2394 (1998). Figure 1 shows the influence of the number of tests

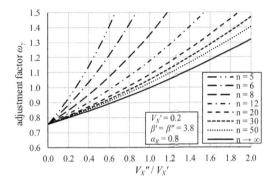

Figure 1. Influence of the ratio V_X''/V_X' and the number of tests n on the factor ω_γ for adjusting a material partial factor γ_M.

n and the sample coefficient of variation of the tested material property V_X'' on the required adjustment factor ω_γ, which is the ratio of the required partial safety factor for the assessed existing structure γ_M'' to the partial safety factor for new structures γ_M'. It can be seen that, based upon test results and if no prior information is taken into account, a reduction of the partial safety factor is only possible for a high number of tests in combination with a sample coefficient of variation of those tests V_X'' much smaller than that for new structures V_X'.

REFERENCES

fib (2016) *Bulletin 80 – Partial Factor Methods for Existing Concrete Structures: Recommendation.* Lausanne, Switzerland, Fédération internationale du béton.

ISO 2394 (1998) *General Principles on Reliability for Structures.* ISO Standard.

Müller, D., Förster, V. & Graubner, C.-A. (2017) Influence of material spatial variability on required safety factors for masonry walls in compression. *Mauerwerk – European Journal of Masonry*, 21 (4), 209–222.

MS4: Risk and reliability acceptance criteria
Organizers: D. Diamantidis, M. Faber, K. Fischer,
M. Holicky, M. Sýkora, J. Köhler & T. Vrouwenvelder

Life-Cycle Analysis and Assessment in Civil Engineering: Towards an Integrated Vision – Caspeele, Taerwe & Frangopol (Eds)
© 2019 Taylor & Francis Group, London, ISBN 978-1-138-62633-1

Obsolescence rate: Framework, analysis and influence on risk acceptance criteria

D. Diamantidis
Ostbayerische Technische Hochschule, Regensburg, Germany

M. Sýkora
Czech Technical University, Prague, Czech Republic

E. Bertacca
University of Pisa, Italy
(Guest Researcher, OTH Regensburg, Germany)

ABSTRACT

Obsolescence of buildings and infrastructure is a major issue and a threat for built property. Buildings and infrastructure are major long lasting assets of the society and business enterprises and form a significant part of infrastructure management. They require preventive or corrective maintenance and in many cases they can become inappropriate for their original purpose due to obsolescence and are demolished. Obsolescence is the process of declining performance due to changes in the functional, economic, ecological, social, legal, political or physical environment. It reflects the development of the society and environment around the still standing structure and it mostly depends on human activities and demands, while physical and chemical processes normally play a minor role.

The present contribution explores the obsolescence of buildings and in particular the related influencing factors, the resulting rate and the associated uncertainties. Useful definitions of the obsolescence rate are presented based on the physical and useful (service) life. The experienced useful life of buildings in Finland, Germany and North America is statistically analysed, utilizing available databases and comparisons with the design working life specified in standards are made. Probability of not exceeding a design working life is discussed.

Whereas many factors affecting the obsolescence rate are beyond the control of civil engineers, particular attention needs to be paid to cases where a useful life of the structure is determined by physical aspects related to deterioration. These cases are identified and recommendations to improve structural design are offered. The useful life and related renewal rate is influencing the risk acceptance criteria given in ISO 2394:2015. A statistical model for the obsolescence rate is derived and its influence on the target reliability is investigated for a representative failure limit state.

The main conclusions of the presented study are summarised as follows:

- Obsolescence rate relates the theoretical physical life of a building to its experienced useful life.
- The 'obsolescence' rate used in optimization studies should be rather termed as a renewal rate and defined as a reciprocal value of the useful lifetime.
- The experienced useful life of buildings is in most cases higher than the design working life; the latter thus represents a lower fractile and should not in principle be used in optimisation studies. The numerical error resulting from this misinterpretation is, however, small.
- Better insight into the obsolescence and its control can be gained by analysing the probability that the structure will exceed its design lifetime.
- The evaluation of databases in Finland, Germany and North America shows that the design lifetime of buildings is exceeded with region-dependent probability – an indicative value is 70%.
- It appears that risk acceptance criteria are normally insignificantly affected when actual useful life (influenced by obsolescence) is incorrectly replaced by a design working life.
- It is more important to make distinction between these two when considering time-dependent phenomena such as corrosion, fatigue or severe changes in the environment. The analysis based on the design life – lower fractile of the expected useful lifetime – can lead to misleading results of optimisation studies.

ACKNOWLEDGEMENTS

This work has been partly supported by Grants 16-11378S and LTT18003, by the Regensburg Center of Energy and Resources of OTH Regensburg and by the Technology- and Science Network Oberpfalz.

Life-Cycle Analysis and Assessment in Civil Engineering: Towards an
Integrated Vision – Caspeele, Taerwe & Frangopol (Eds)
© 2019 Taylor & Francis Group, London, ISBN 978-1-138-62633-1

Reliability targets for semi-probabilistic design standards

J. Köhler

Department of Structural Engineering, Norwegian University of Science and Technology (NTNU),
Trondheim, Norway

ABSTRACT

The codes and standards constitute the primary decision support for practicing structural engineers; the vast majority of structures is designed and controlled according to these legal regulations. Since a large proportion of a societys wealth is invested in the continuous development and maintenance of the built infrastructure, it is essential that structural design codes represent a rationale that facilitates design solutions that balance expected adverse consequences (e.g. in case of failure or deterioration) with investments into more safety (e.g. larger cross sections). Structural design codes should therefore be calibrated on the basis of associated risks or, in the same vein, on the basis of associated failure probability. In the present paper, reliability targets for reliability based code calibration will be discussed. Special reference to the concurrent revision of the EUROCODES is made.

The design of a structure is in principle a decision problem with the objective to identify a structural design with a performance that maximizes the expected utility for the corresponding decision maker. A common attribute of these decisions is that they have to be performed subject to uncertainties. In general, different levels of detail for the assessment of the structural performance are distinguished; risk-informed decision making, reliability based design, and semi-probabilistic design.

In risk-informed decision making the maximization of the utility of the decision maker/owner/society is considered explicitly and the corresponding decision/design alternative (that also complies with superior regulations e.g. related to acceptability of risk to human lives) is identified. It is interesting to note that formally, the risk-informed decision making also includes the decision for a level of detail in system representation (mechanical modelling, uncertainty representation, consideration of data and other available information all with strong implications on the kind of engineering expertise that is hired in to inform the decision making). i.e. maximization of utility is only possible if the possible options of system representation are treated explicitly as a part of the decision problem. Reliability requirements for risk-informed decision making are not necessary in principle. However, legal requirements often require the introduction of regulative risk acceptance criteria as constrains for the risk assessment.

In reliability based design a structure or a structural component is designed in order to fulfill a set of reliability requirements, generally communicated as maximum admissible failure probabilities or minimum values for the FORM reliability index. Obviously is the reliability-based design approach less detailed than the risk-informed one as the maximization of utility as e.g. expressed as minimization of expected costs is not addressed explicitly. However, with a discrete differentiation of reliability requirements dependent on the consequences of failure and the cost of a safety measure it is possible to account for these risk aspects at least implicitly.

In semi-probabilistic design a structural design is identified such that it complies to a set of design inequalities that contain the most important variables representing the corresponding design situation. The variables are represented as deterministic numbers that are called design values. However, the number are quantified in a way that the fulfillment of the design inequalities corresponds to a desired safety-level.

Reliability requirements for structural design and assessment depend always on the context to which they are applied. This is rather obvious for different anticipated consequences and therefore reliability requirements are often differentiated for different consequence classes. The marginal cost for increasing the reliability is also a characteristic of the design or assessment situation. Especially for cases with large uncertainty in regard to the assessment of existing structures this is of relevance. A very important aspect, especially for a regulative context, is related to the level of modeling detail as discussed in this article. As codes and standards in general include reliability assessment methods on rather different levels of modeling sophistication, it is very problematic to prescribe a single set of reliability requirements and treat them as absolute values for all different contexts that they are applied to in the code. Alongside the Example of the EUROCODES this is illustrated, and recommendations on the further treatment of reliability targets are made.

Structural reliability analysis of wind turbines for wind and seismic hazards in Mexico

A. López López, L.E. Pérez Rocha, C.J. Muñoz Black, M.A. Fernández Torres & L.E. Pech Lugo
Instituto Nacional de Electricidad y Energías Limpias, Cuernavaca, Morelos, Mexico

ABSTRACT

Nowadays the use of wind farms to generate electricity has become an important issue around the world. In particular, due to the fact that Mexico has been committed (COP21, 2015) to increase the generation of electricity in 30% for 2024 by using clean energies, wind farms projects are growing for sustainable production of energy. In order to achieve safety designs of these structures, a reliability analysis requires special considerations to take into account both wind and seismic hazards.

During the last three years, the wind and seismic hazards for Mexico has been updated for the Wind and Seismic Design Chapters of the Manual of Civil Works Design (MDOC-DS 2015, MDOC-DV in prep., for its acronym in Spanish) and used by the Federal Electricity Commission of Mexico (CFE, for its acronym in Spanish.

A methodology to perform structural reliability analysis of wind turbines is presented. First, an updated wind and seismic hazards for Mexico are presented and optimal bending moments for combined hazards are determined (Figure 1). Then, by convolving the hazards with vulnerability functions of a wind turbine, the annual failure probability rates of the structure are estimated.

Likewise, reliability indexes for wind, seismic and combined hazards are compared in terms of the base bending moments. Afterwards, selected results using contour maps for the whole country are shown (Figure 2) and allow to point out that this type of structures can be strongly affected by both hazards, mainly depending on their frequency of occurrence and the design thresholds.

From the results obtained, reliability indexes lower than 3 were found for the majority of the Pacific coast, so this is the more exposed region. For wind turbines design practice, it is recommended reliability indexes values varying from 3 to 5 (DNV/Riso, 2002). In the future, corrosion effects and other design situations will need to be addressed as they contribute the overall wind turbine safety.

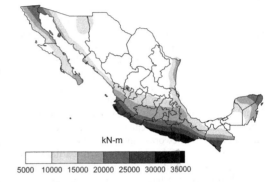

Figure 1. Optimal base bending moment for wind and seismic actions on wind turbine studied, for a structure factor of $S_L = 50$.

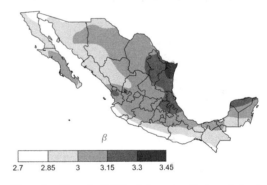

Figure 2. Map of reliability indexes considering wind and seismic actions on wind turbine studied for $S_L = 50$.

REFERENCES

Energy Mexico Forum (2018) https://www.gob.mx/sener/prensa/mexico-es-la-sexta-economia-mundial-mas-atractiva-para-las-energias-renovables-pjc

Pérez, L.E., et al. (2011) Wind optimal design for wind turbines in Mexico. In: *Proceedings of the 13th International Conference on Wind Engineering*, The Netherlands, November 27–31, Amsterdam.

Life-Cycle Analysis and Assessment in Civil Engineering: Towards an
Integrated Vision – Caspeele, Taerwe & Frangopol (Eds)
© 2019 Taylor & Francis Group, London, ISBN 978-1-138-62633-1

Risk analysis of chemical industrial complex using parallel CUDA algorithms

M. Jeremiah & M. Torbol
Ulsan National Institute of Science and Technology, Republic of Korea

ABSTRACT

Systems under disaster conditions pose stringent operational conditions to components and sub-systems. This study uses the classic relationship between demand curves and capacity curves: both are derived from the failure analysis of mechanical components and structures. The testbed that is used to present the model is a chemical industrial complex in Ulsan, Republic of Korea. While the model can accept any natural disaster as input, earthquake was the chosen one within the framework of maximum likelihood estimation and probability of occurrence of earthquakes. The annual probability of occurrence of earthquake is used to determine hazards curve which serves as input to the model the seismic risk curve that: provides the analyst with decision indicators (DI), and leads to an improved benefit/cost analysis for any change and improvement to the system, such as: retrofit of the components and base isolation. The vulnerability of the subsystems due to damage is obtained with a quantitative and indicative failure criterion. The gain, when there is a preventive retrofit, is also obtained according to the state of the system. Furthermore, the gain, when it is a mitigating procedure, can also be obtained accordingly. The output of the study is an indicator for measuring the level of economic implication which will result from damage to components under consideration. The further away the curve is from the abscissa and ordinate, the higher the risk; thus, preventive or mitigating retrofit is employed. Eventually, the result helps analyst and government in planning, maintenance and design of reliability assessment program for major infrastructure like chemical plants, nuclear power plants, thermal plants and other complex systems. The novelty of the study is the deployment of parallel algorithm to expediting the analysis.

Acceptance criteria for structural design or assessment in accidental situations due to gas explosions

R. Hingorani & P. Tanner
Instituto de Ciencias de la Construcción E. Torroja (IETcc-CSIC), Madrid, Spain

C. Zanuy
Universidad Politécnica de Madrid (UPM), Madrid, Spain

ABSTRACT

In addition to the persistent situations, structures might be exposed to accidental events, which are among the most common causes of structural failure (Vrouwenvelder 2000). Given their specific characteristics, such as low occurrence probabilities and high potential failure consequences, the general structural design philosophy for such events differs from the usual approach adopted for persistent situations (Ellingwood 2006). Design for accidental situations is in particular implemented to avoid catastrophes, wherefore local failure, such as a member or component failure, might be acceptable if neither the whole structure nor an important part of it will collapse. However, the practical implementation of this principle is not straightforward, inter alia because objective acceptance criteria for the associated life safety risks are lacking, being the consequence that such risks are often ignored.

The present study, based on (Hingorani 2017), is a contribution to close this gap. Following the general approach developed in prior studies (Tanner & Hingorani 2015), explicit risk analysis is conducted to quantify implicitly acceptable life safety risks associated with the effects of gas explosions on reinforced concrete structures, based on the probability of structural collapse and its consequences. Acceptance criteria for such risks are deduced from the findings, which facilitate the adoption of rational decisions on the need and the choice of appropiate strategies to counteract the effects of gas explosions on building structures, and in particular on their robustness. The provision of key elements, upon which depends the stability of the structure, or a large part of it, may be one of these strategies. For this purpose, acceptable risks are translated into target values for the conditional failure probability $p_{ft|EX}$ of a structural member given the occurrence of a gas explosion. The $p_{ft|EX}$ are defined as a function of the area affected by the collapse of a particular structural member A_{col} (Fig. 1). Minimum

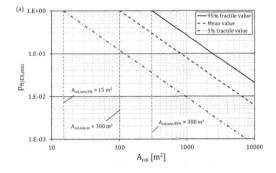

Figure 1. Minimum target value for conditional failure probabilities, $p_{ft|EX}$ ($T_{ref} = 50$ years for accompanying variable actions).

areas $A_{col,min}$ are obtained below which explosion-related safety measures would not be needed, i.e. where member design could be based on the persistent situation, without any further provisions for this accidental situation (Fig. 1). Beyond these threshold areas, diminishing $p_{ft|EX}$ with increasing A_{col} account for aversion to collapse events with larger consequences to persons.

REFERENCES

Ellingwood, B.R. (2006) Mitigating risk from abnormal loads and progressive collapse. *Journal of Performance of Constructed Facilities*, 20 (4), 315–323.

Hingorani, R. (2017) *Acceptable Life Safety Risks Associated with the Effects of Gas Explosions on Reinforced Concrete Structures*. Universidad Politécnica de Madrid (UPM). https://doi.org/10.20868/UPM.thesis.47772.

Tanner, P. & Hingorani, R. (2015) Acceptable risks to persons associated with building structures. *Structural Concrete*, 16 (3), 314–322.

Vrouwenvelder, T. (2000) Stochastic modelling of extreme action events in structural engineering. *Probabilistic Engineering Mechanics*, 15, 109–117.

Safety requirements for the design of ancillary construction equipment

P. Tanner
Instituto de Ciencias de la Construcción E. Torroja (IETcc-CSIC), Madrid, Spain
CESMA Ingenieros, S.L., Madrid, Spain

R. Hingorani
Instituto de Ciencias de la Construcción E. Torroja (IETcc-CSIC), Madrid, Spain

J. Soriano
CESMA Ingenieros, S.L., Madrid, Spain

ABSTRACT

Current structural standards like the Eurocodes do not provide a coherent framework for design or assessment of structures under temporary use such as ancillary construction systems. In the light of comparatively high failure rates on record (Ratay 2009), this situation requires improvement, especially with a view to avoid gross human errors. In addition, appropriate target reliability levels for temporary structures are needed (Caspeele et al. 2013).

Regarding this challenge, some basic principles were recently formulated (JCSS 2015). It was suggested that the fundamental basis for choosing the levels of safety for temporary structures or structures under temporary use shall not be different from those applied to permanent structures and should be fixed taking account of both, possible failure consequences and relative costs for risk-reduction measures. Moreover, in view of the important consequences the failure of structures under temporary use might entail, it is felt that there is no meaningful reason to choose a priori lower safety levels for such structures just because of their temporary use conditions (JCSS 2015).

Taking into account these considerations, target reliability levels for structural members under temporary use conditions were recently suggested (Tanner et al. 2018). The reliability requirements seek to ensure the same acceptable risk levels per unit of time as for permanent structures in the current best practice, determined in prior studies (Tanner & Hingorani 2015). The results obtained show that the target reliability index for structural members rises significantly with declining risk exposure times (Figure 1). Conversely, the design values for variable actions may be lowered in keeping with the duration of construction, as illustrated in a case study, the analysis of a movable scaffolding system for the erection of a continuous prestressed concrete bridge girder.

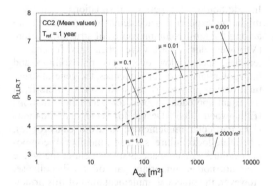

Figure 1. Target reliability index ($\beta_{t,LR,T}$) for members in temporary structures versus the area affected by member collapse (A_{col}) based on the mean value of implicitly acceptable risks.

REFERENCES

Caspeele, R., Steenbergen, R. & Taerwe, L. (2013) An adjusted partial factor method for temporary structures compatible with the Eurocode framework. *Civil Engineering and Environmental Systems*, 30 (2), 97–114.

JCSS (2015) Document on design of temporary structures. In: *Draft presented at the 59th JCSS Meeting by Vrouwenvelder, T., with input from Faber, M., Høj, N. P., Sørensen, J. and Tanner, P., ETH Zürich*. Joint Committee on Structural Safety.

Ratay, R.T. (2009) *Forensic Structural Engineering Handbook*. New York, McGraw-Hill.

Tanner, P. & Hingorani, R. (2015) Acceptable risks to persons associated with building structures. *Structural Concrete*, 16 (3), 314–322.

Tanner, P., Hingorani, R., Bellod, J.L. & Sanz, D. (2018) Thoughts on construction risk mitigation and acceptance. *Structural Engineering International*, 28 (1), 60–70.

Life-Cycle Analysis and Assessment in Civil Engineering: Towards an Integrated Vision – Caspeele, Taerwe & Frangopol (Eds)
© 2019 Taylor & Francis Group, London, ISBN 978-1-138-62633-1

Target reliability indices for existing quay walls derived on the basis of the LQI criterion

A.A. Roubos
Port of Rotterdam, Rotterdam, The Netherlands
TU-Delft, Delft, The Netherlands

D.L. Allaix & R.D.J.M. Steenbergen
TNO, Delft, The Netherlands
Gent University, Gent, Belgium

K. Fischer
Matrisk GmbH, Affoltern am Albis, Switzerland

S.N. Jonkman
TU-Delft, Delft, The Netherlands

ABSTRACT

General frameworks for reliability differentiation have evolved over time and are mainly developed for new buildings. However, recommendations for existing quay walls are lacking. In this study target reliability indices for assessing existing quay walls were derived by economic optimisation and by evaluating the Life Quality Index criterion (LQI). In quay wall design, some dominant stochastic design variables are largely time-independent, such as soil and material properties. The influence of time-independent variables on the development of the probability of failure was taken into consideration in this study, because this affects the present value of future failure costs and the associated target reliability indices. The reliability indices obtained in accordance with the LQI acceptance criterion were a little lower than the target reliability indices derived by economic optimization. The target reliability indices obtained for existing quay walls depend on the consequences of failure and the remaining service life. If failure modes of a quay wall are largely time-invariant and already survived the first period of the service life, the residual probability of failure is lower for an existing quay wall compared to a new quay wall. Hence, this should be considered in the determination of target reliability indices. The method of approach to assess the development of reliability over time can also be used for evaluating target reliability indices of other civil and geotechnical structures.

REFERENCES

Allaix, D.L., Steenbergen, R.D.J.M. & Wessels, J.F.M. (2017) Target Reliability Levels for the Design of Quay Walls. Delft, The Netherlands, TNO.
Diamantidis, D. (2017) A critical view on environmental and human risk acceptance criteria. *International Journal of Environmental Science and Development*, 8 (1), 62–68.
Fischer, K. & Faber, M.H. (2012) The LQI acceptance criterion and human compensation costs for monetary optimization – A discussion note. In: *LQI Symposium in Kgs*, Lyngby, Denmark.

Fischer, K., Virguez, E., Sánchez-Silva, M. & Faber, M.H. (2013) On the assessment of marginal life saving costs for risk acceptance criteria. *Structural Safety*, 44, 37–46.
Holický, M. (2011) The target reliability and design working life. *Safety and Security Engineering*, IV, 161–169.
ISO 2394 (2015) General Principles on Reliability for Structures. Geneva, Switzerland, International organization for standardization.
ISO 13822 (2010) *Bases of Design of Structures – Assessment of Existing Structures*. Geneva, Switzerland, International Organization for Standardization.
JCSS (2001) Probabilistic Model Code. Part 1. Joint Committee on Structural Safety. www.jcss.byg.dtu.dk.
Jonkman, S.N., Gelder van P.H.A.J.M. & Vrijling J.K. (2003) An overview of quantitative risk measures for loss of life and economic damage, Elsevier. *Journal of Hazardous Materials*, A99, 1–30.
Rackwitz, R. (2000) Optimization – the basis of code making and reliability verification. *Structural Safety*, 22, 27–60.
Rackwitz, R. (2006) The effect of discounting, different mortality reduction schemes and predictive cohort life tables on risk acceptability criteria. *Reliability Engineering & System Safety*, 91, 469–484.
Rackwitz, R. (2008) *The Philosophy Behind the Life Quality Index and Empirical Verification*. Joint Committee of Structural Safety.
Roubos, A.A. & Grotegoed, D. (2014) Urban Quay Walls. SBRCURnet. Delft, the Netherlands.
Roubos, A.A., Steenbergen, R.D.J.M., Schweckendiek, T. & Jonkman, S.N. (2018) Risk-based target reliability indices of quay walls. *Structural Safety*.
Steenbergen, R.D.J.M. & Vrouwenvelder, A.C.W.M. (2010) Safety philosophy for existing structures and partial factors for traffic loads on bridges. *Heron*, 55, 123–139.68.
Steenbergen, R.D.J.M., Sýkora, M., Diamantidis, D., Holický, M. & Vrouwenvelder, A.C.W.M. (2015) Economic and human safety reliability levels for existing structures. *Structural Concrete*, 16, 323–332.
Sykora, M. & Holický, M. (2011) Target reliability levels for the assessment of existing structures. In: *Proceedings of the ICASPII*, pp. 1048–1056.
Sýkora, M., Diamantidis, D., Holicky, M. & Jung, K. (2017) Target reliability for existing structures considering economic and societal aspects. *Structure and Infrastructure Engineering*, 13–1, 181–194.

MS5: Early BIM for life-cycle performance
Organizers: P. Schneider & P. Geyer

Design-integrated environmental performance feedback based on early-BIM

A. Hollberg & I. Agustí-Juan
Chair of Sustainable Construction, ETH Zurich, Switzerland

T. Lichtenheld
Chair of Building physics, Bauhaus University Weimar, Germany

N. Klüber
Fraunhofer Institute for Microstructure of Materials and Systems, Germany

ABSTRACT

Architects largely define the environmental impact a building will cause throughout its lifetime in early design stages. Therefore, Life Cycle Assessment (LCA) would be ideally used to optimize the environmental performance in these stages. Conventional LCA tools based on BIM require detailed information only available in detailed design stages when it is too late for major changes. Therefore, a novel tool based on simple 3D models and a parametric LCA method is developed. The cloud-based tool called CAALA is designed as a plug-in for 3D CAD software and provides the results in real-time (see Figure 1). The applicability of this early-BIM approach is validated using three case studies, each highlighting different aspects. The results show that CAALA provides meaningful information to architects and clients, is easy to learn, and applicable in early design with a reasonable effort. Finally, it can be concluded that holistic environmental performance optimization in the most important early design stages is now practically feasible.

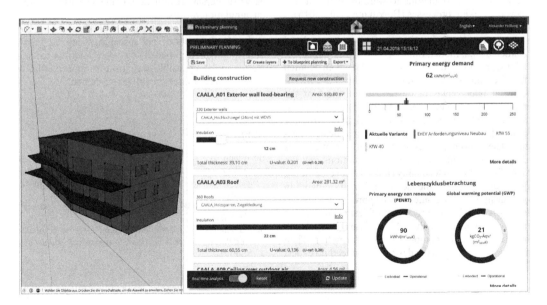

Figure 1. Graphical user interface of CAALA (left: geometry in Sketchup, middle: input of boundary conditions, materials and HVAC systems, right: output of results).

Life-Cycle Analysis and Assessment in Civil Engineering: Towards an Integrated Vision – Caspeele, Taerwe & Frangopol (Eds)
© 2019 Taylor & Francis Group, London, ISBN 978-1-138-62633-1

Seamless integration of simulation and analysis in early design phases

A. Zahedi & F. Petzold
Chair of Architectural Informatics, Technical University of Munich, Munich, Germany

ABSTRACT

Decisions taken at early phases of architectural design have a significant effect on the planning steps for the entire lifetime of the project as well as the performance of the building throughout its lifecycle (MacLeamy, 2004; Steinmann, 1997). Especially in the early stages of planning, fundamental decisions are already made that decisively determine the eco-logical properties of a building.

It has been shown that there is a lack of holistic and flexible models and novel methods for the use of digital building models to control complexity and make decisions transparent (Borrmann et al., 2015).

One major advantage of Building Information Modelling (BIM) lies in the direct reuse of data hold by the model for diverse analysis and simulation tasks, as e.g. the Life Cycle Assessment (LCA). This means that BIM could bring forward the planning and decision-making processes (Borrmann et al., 2015).

However nowadays, the use of analytical and simulation tools typically requires a mostly completed and detailed design model. Since such analysis and calculations, are mostly dependent on the data from detailed model design, these analytical procedures occur typically at later stages of design, where most of the important design-decisions have already come to pass.

The architect nowadays almost exclusively uses his know-how for evaluating and comparing design variants. The reasons for this are the insufficient process integration of supporting software solutions and the lack of required model quality (levels of development), due to uncertainty or the lack of information depth in early design phases. Some overall approximate simulation tools do exist for early stages of design but they are mostly uncoupled with the CAD tools and incapable to handle vague and incomplete input information. Then again, the architect as the designer is indeed no expert in various fields of analysis and simulations needed for evaluating a design alternative. Therefore, understandable evaluation of performance, costs, economics and constructability of different design variants are mainly lacking in these early stages.

To address this insufficiency, the research unit (Removed for the purpose of Blind Review) is dedicated to development of methods that allow the evaluation and assessment of alternative building design variants, which may also be partially incomplete and vague.

The above mentioned research group is fairly in the beginning of its work and this paper publishes the first results concerning one of its part projects. The focus of this part project in the above mentioned research group is twofold, one to establish a generic dialog-based interface to various simulation tools, and the other to develop methods for visual exploration and easy to understand evaluations of design variants based on simulation results.

As the contents of this paper will be describing the preliminary results of our research, using a generic description of the aimed dialog-based interface to simulation tools, a possible scenario for requesting LCA in early stages of design will be explained.

REFERENCES

Borrmann, A., König, M., Koch, C. & Beetz, J. (2015) *Building Information Modeling: Technologische Grundlagen und industrielle Praxis*. Herausgeber, Springer-Verlag.

MacLeamy, P. (2004) *The Future of the Building Industry: The Effort Curve*.

Steinmann, F. (1997) *Modellbildung und computergestütztes Modellieren in frühen Phasen des architektonischen Entwurfs*.

Early-design integration of environmental criteria for digital fabrication

I. Agustí-Juan, A. Hollberg & G. Habert
Chair of Sustainable Construction, ETH Zurich, Zürich, Switzerland

ABSTRACT

The building industry is a traditional sector, with high environmental impacts and low productivity compared to other industries. Research in digital fabrication is beginning to reveal its potential to improve the sustainability of the building sector (Agustí-Juan & Habert, 2017). However, new evaluation methods are needed to quantify the actual reduction of environmental emissions compared to conventional construction. The Life Cycle Assessment (LCA) method (ISO, 2006) is commonly employed to evaluate the environmental performance of buildings. Recent developments in LCA integration in CAD and BIM software have improved the evaluation of environmental impacts during design. However, these tools have still limitations, such as the visualization of results and the evaluation of early design stages. Moreover, they are only partially applicable to digital fabrication, because of differences in the design process. In contrast to the conventional design process, digital fabrication begins with the definition of material, functional, structural, etc. parameters and fabrication constraints, which optimization defines the final geometry. Therefore, this paper presents a design-integrated method for simplified LCA of digitally fabricated architecture. The evaluation method is divided into four design stages according to the degree of information available and geometry definition of the building element. The LCA is integrated in Grasshopper, a visual scripting interface that allows the manipulation of parametrized geometry and the extraction of data from the 3D model in Rhinoceros. In each design stage, by defining materiality and functionality, the Global Warming Potential (GWP) of the digitally fabricated building element can be assessed and compared to conventional construction. Specifically, in design stages 1 and 2, the tool provides a target value for the designer, while in design stages 3 and 4 a direct quantitative comparison is provided. Furthermore, the tool displays a

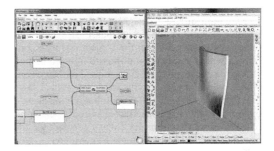

Figure 1. Design-integrated LCA of digital fabrication.

real-time visualization of the environmental comparison using a color scale (Fig. 1). The method grants continuous feedback for the designer and provides a basis for decision-making and project optimization based on environmental criteria. Finally, the paper presents the evaluation of the case study "Mesh Mould", a digitally fabricated complex concrete wall that does not require formworks (Hack et al., 2017). The results provide a comparison of embodied impacts between digital fabrication and conventional construction (kg CO_2 eq./m^2). The results prove the applicability of the method and highlight the environmental benefits that digital fabrication can provide.

REFERENCES

Agustí-Juan, I. & Habert, G. (2017) Environmental design guidelines for digital fabrication. *Journal of Cleaner Production*, 142, 2780–2791.

Hack, N., Wangler, T., Mata-Falcón, J., Dörfler, K., Kumar, N., Walzer, A.N., Graser, K., Reiter, L., Richner, H., Buchli, J., Kaufmann, W., Flatt, R.J., Gramazio, F. & Kohler, M. (2017) Mesh mould: An on site, robotically fabricated, functional formwork. In: *High Performance Concrete and Concrete Innovation Conference*, Tromsø, Norway.

ISO (2006) *14040: Environmental Management-Life Cycle Assessment-Principles and Framework.*

Efficient management of design options for early BIM

H. Mattern & M. König
Chair of Computing in Engineering, Ruhr-University Bochum, Germany

ABSTRACT

Investigating different design options represents an iterative process in the design phase of a building. Especially at early stages, the project client is confronted with a vast amount of possibilities combined with complex restrictions and unique boundary conditions which might heavily influence a building's Life Cycle Assessment (LCA). While studying diverse options is highly appreciated from an ecological point of view as well as from the client's perspective, the resulting model structures might be complex.

The gradually increasing implementation of Building Information Modeling (BIM) replaces conventional methods dominating the design, construction and use phase of a building. In the context of early design stages, the use of BIM offers promising features due to easily configurable models and reusable information. To guarantee information exchange processes between the high number of involved parties, the open data format "Industry Foundation Classes" (IFC) have been established. The scope of IFC4, however, does not cover the investigation and management of design options. Instead, explicit models need to be exchanged.

In this paper, an IFC-based management of design options is presented. The considered options evolve at early design phases and may influence a high number of building objects. Different option classes are defined in accordance with the appearance during the planning process as well as the effect on other building elements. Subsequently, the developed categories are compared to the existing structure and content of the IFC model. The management of models covering multiple options is implemented based on the concept of objectified relationships. By providing different options within one consistent model, a repeated creation of objects is avoided. Furthermore, the use

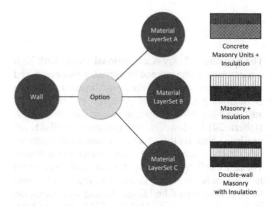

Figure 1. Example of considering different layers and materials for a single wall.

of consistent identifiers is supported as a repeated creation of building elements may lead to inconsistent data. To provide a transparent overview on the structure of the resulting model and the effect of single options, a graph-based representation is proposed.

REFERENCES

Ismail, A., Nahar, A. & Scherer, R. (2017) Application of graph databases and graph theory concepts for advanced analyzing of bim models based on ifc standard. In: *Proceedings of the EG ICE 2017*.

Jaly-Zada, A., Koch, C. & Tizani, W. (2015) Ifc extensions for design change management. In: *Proceedings of the 32nd CIB W78 Conference 2015*.

Koch, C. & Firmenich, B. (2011) An approach to distributed building modeling on the basis of versions and changes. *Advanced Engineering Informatics*, 25, 297–310.

The energy grey zone—uncertainty in embedded energy and greenhouse gas emissions assessment of buildings in early design phases

H. Harter, P. Schneider-Marin & W. Lang
Institute of Energy Efficient and Sustainable Planning and Building, Technical University of Munich, Munich, Bavaria, Germany

ABSTRACT

With buildings in Europe scheduled to become nearly zero-energy by 2020 during their use phase, an increasing input of building materials, especially insulation materials, is needed. Therefore, the energy and greenhouse gases spent on construction, refurbishment and demolition, also known as "grey" or embedded energy (EE) and greenhouse gases (EGHG), play an increasingly important role in the overall consumption of buildings. At the same time, the use of Building Information Modelling (BIM) is gaining momentum opening up opportunities to provide information about EE and EGHG at early project stages, when the possible influence on the life cycle of the built environment is greatest.

Currently there are no possibilities to estimate EE and EGHG early, since calculations require detailed knowledge about the building, which is usually only available later in the process. Therefore, this study develops a methodology to accurately predict EE and EGHG to be included in BIM processes. Additionally, the uncertainty of the predictions is calculated to enable the best possible assessment. This information helps planners to decide at an early stage between different designs and construction materials to find the best ecological solution.

The methodology is presented on a specific case study. The realized project is known in detail, such that all information is provided beforehand, which simplifies the assessment and verification of the methodology. The results for EE and EGHG of different design and construction solutions are compared amongst each other and with the final results to derive values about the uncertainty of the estimated EE and gain knowledge about the influence of the choices made in early project stages.

The study shows the driving factors for the amount of EE and EGHG and uncertainties in calculations in the first three Levels of Development (LODs), e.g. see Figure 1 for LOD 120. BIM methods help in efficiently exploring project variations, which have to be linked to LCA methods for assessing the best possible option concerning the sustainability of buildings. These methods have to be automated in the next steps to enable a much more efficient assessment of the uncertainty. The study concludes that EE can be predicted at early project stages only with great uncertainty. It helps to assess these uncertainty to be included in recommendations to clients.

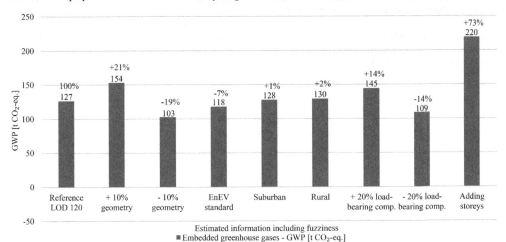

Figure 1. EGHG of LOD 120 – including uncertain information.

Life-Cycle Analysis and Assessment in Civil Engineering: Towards an Integrated Vision – Caspeele, Taerwe & Frangopol (Eds)
© 2019 Taylor & Francis Group, London, ISBN 978-1-138-62633-1

The early BIM adoption for a Contracting Authority: Standard and methods in the ANAS approach

A. Osello, N. Rapetti & F. Semeraro
Dipartimento di Ingegneria Strutturale, Edile e Geotecnica, Politecnico di Torino, Italy

ABSTRACT

In recent years, Building Information Modelling (BIM) has played a key role in order to improve the planning process and the entire Life-Cycle of buildings and infrastructures. Despite the high level of expertise reached for buildings applications, further research is still needed in the infrastructure field.

For this reason, the public company ANAS, which is the first Italian Contracting Authority for infrastructure projects, decided to arrange a partnership with Politecnico di Torino, in order to implement the BIM methodology into their actual practices. This is due to the BIM capability of forecasting construction criticalities and extremely high-cost operations and maintenance activities since the early stages of design.

The research activity was focused on the implementation of BIM methods and tools for an infrastructure preliminary project, providing standards and rules to obtain a suitable model for ANAS purposes.

The first step of the research was the identification of the main relevant datasets of the early stage of infrastructure design, which belong to several specific disciplines such as environment, geology, hydraulic, geotechnics, etc. Considering the large amount and the heterogeneity of data, it was necessary to define a structured database able to collect, share and exchange information, through the use of different parametric models.

The Contracting Authority has the necessity to introduce a clear set of information contents in their contracts, in order to obtain visible and positive results, in terms of model delivery, since the procurement stage. For this reason an Employers Information Requirements (EIR) has been developed.

This work revealed that main limitations to the spread of BIM into the early stage of the planning process are the immaturity of the Italian market, the software limitations and finally, the significant effort to evaluate real cost and time benefits of early BIM adoption.

REFERENCES

Bradley, A., Li, H., Lark, R. & Dunn, S. (2016) BIM for infrastructure: An overall review and construction perspective. *Automation in Construction*, 71, 139–152.

Eastman, C., Liston, K., Sacks, R. & Liston, K. (2008) BIM handbook: A guide to building Information Modelling. Hoboken, John Wiley & Sons, Inc.

EUBIM (2017) *Handbook for the Introduction of Building Information Modelling by the European Public Sector Strategic Action for Construction Sector Performance.* Technical Report, EU BIM Task Group.

McAuley, B., Hore, A.V. & West, R. (2012) Implementing Building Information Modeling in Public Works Projects in Ireland. In: Proceedings of the 9th European Conference Process Model, pp. 589–596.

McGrawHill (2014) *The Business Value of BIM for Construction in Major Global Market: How Contractors Around the World Are Driving Innovation with Building Information Modeling.* Technical Report, McGrawHill.

Succar, B. & Kassem, M. (2015) Macro BIM adoption: Conceptual structures. *Automation in Construction*, 57 (May), 64–79.

*Life-Cycle Analysis and Assessment in Civil Engineering: Towards an
Integrated Vision – Caspeele, Taerwe & Frangopol (Eds)
© 2019 Taylor & Francis Group, London, ISBN 978-1-138-62633-1*

How to make decision-makers aware of sustainable construction?

H. Kreiner, M. Scherz & A. Passer
*Working Group Sustainable Construction, Institute of Technology and Testing of Construction Materials,
University of Technology, Graz, Austria*

1 INTRODUCTION

Current environmental decisions and objectives
(COP21 Paris Agreement, EU Roadmap 2050) clearly
emphasize the need to counter current climate change
developments. The Intergovernmental Panel on Climate Change (IPCC) and related groups around the
world (such as the Austrian Panel on Climate Change
(APCC)) address future climate change, risks and
environmental impacts.

Due to the enormous masses of material and energy,
the implementation of sustainable development principles plays a key role in construction industry. While
buildings represent essential objects for humans, the
construction industry causes a variety of environmental impacts (e.g. the construction industry accounts for
up to 40% of global energy and for up to 50% of global
greenhouse gases). Reducing these impacts, including climate change, are important issues on the global
agenda.

2 OBJECTIVE

Due to multi-criteria requirements in sustainable construction complexity in decision-making process is
increasing. Limited awareness of systemic effects may
lead to imprecise and/or wrong decisions in early
design stages. At current there is a lack of appropriate
methods in order to manage a multi-criteria decision-making process based on a systemic approach. In this
article a methodological approach for the support of
decision-making in the early design stage is presented.

3 METHODOLOGY

Based on a grid of sustainability evaluation criteria a systemic approach is applied and implemented
in a three-level process model. Primarily, individual
decision-maker goals and quality levels for each sustainability criterion are defined. Secondly, by causal
loop analyses possible trade-offs and synergies can
be highlighted simultaneously. Impacts of these decisions can further be communicated transparently to
the decision-maker. Finally, to ensure the holistic
process implementation of decision-maker goals the
method of maturity assessment is applied in third
level of process model. The final step enables a holistic quality assurance over the whole decision-making
process.

4 CONCLUSION

The simplification of complexity in decision-making
by visualization the holistic impacts should lead
to more awareness towards implementing sustainable construction. The presented approach supports
decision-makers in promoting sustainable construction and steering the life-cycle performance of
buildings.

5 OUTLOOK

One important next step is the implementation of
the proposed process model in building information
modeling (BIM).

Therefore, it is necessary to identify the required
BIM-parameters according to individual sustainability
criteria. Within the comparison of the identified BIM-parameters and the parameters already implemented
in the ASI-property server it is possible to analyze the
gap of required BIM-parameters.

Life-Cycle Analysis and Assessment in Civil Engineering: Towards an Integrated Vision – Caspeele, Taerwe & Frangopol (Eds)
© 2019 Taylor & Francis Group, London, ISBN 978-1-138-62633-1

Information exchange scenarios between machine learning energy prediction model and BIM at early stage of design

M.M. Singh, S. Singaravel & P. Geyer
Department of Architecture, KU Leuven, Leuven, Belgium

ABSTRACT

The design process of a built facility is an evolutionary and iterative process incorporating various modelling and analysis activities and exploration of design space to select the best alternative. The expanding need of sustainable design and construction has made energy efficiency one of the most important factors during the building life-cycle.

Building Information Modelling (BIM) can facilitate energy performance analysis by reducing re-modelling efforts to develop Building Energy Model (BEM) in this iterative process. The design model doesn't contain all the information required to perform energy simulation at the beginning of the design process, making it difficult for the designers to follow a deterministic approach for performance assessment. It will require to generate samples of unknown information which can be used to assess the performance of proposed solution with probability (Cecconi, Manfren, Tagliabue, Ciribini, & De Angelis, 2017). Furthermore, the analysis of designed solutions will require modelling and simulation of energy model for each alternative which will require more computational resources (Geyer & Schlüter, 2014).The research work is focused at the issues of information exchange between BIM and energy simulation tool to analyze alternatives with unknown information at early stage of design.

The research work analyzes various scenarios of information exchange at the different level of details that link to an approach of Machine Learning (ML) energy prediction model with BIM data to facilitate energy-efficient design solutions. ML model are useful for design space exploration at the early stage of the design because of reduced time and modelling efforts (Van Gelder, Das, Janssen, & Roels, 2014). At any level of detail, information states are distinguished by the labels "available", "developing" and "unknown". This will be used to identify "fixed parameter", "constraint" and "unknown parameters" to develop energy model. The alternatives will be generated by defining constraints based on information which will be developed at that level of detail. At any level of detail, the generated alternative will be analyzed for energy performance using ML model. For unknown information, Monte Carlo method will be used to generate samples. This will cause uncertainty in the energy performance of design alternative. The uncertainty of energy performance of alternatives is depicted with mean, maximum and minimum value of heating load which is subject to range of values for unknown parameters. The generated alternatives will be stored using built-in feature of BIM tools to manage design solutions. The research will be useful in using the energy prediction model at the early stage of design and design space exploration.

REFERENCES

Cecconi, F.R., Manfren, M., Tagliabue, L.C., Ciribini, A.L.C. & De Angelis, E. (2017) Probabilistic behavioral modeling in building performance simulation: A Monte Carlo approach. *Energy and Buildings*, 148, 128–141. https://doi.org/10.1016/j.enbuild.2017.05.013

Geyer, P. & Schlüter, A. (2014) Automated metamodel generation for Design Space Exploration and decision-making – A novel method supporting performance-oriented building design and retrofitting. *Applied Energy*, 119, 537–556. https://doi.org/10.1016/j.apenergy.2013.12.064

Van Gelder, L., Das, P., Janssen, H. & Roels, S. (2014) Comparative study of metamodelling techniques in building energy simulation: Guidelines for practitioners. *Simulation Modelling Practice and Theory*, 49, 245–257. https://doi.org/10.1016/j.simpat.2014.10.004

Life-Cycle Analysis and Assessment in Civil Engineering: Towards an
Integrated Vision – Caspeele, Taerwe & Frangopol (Eds)
© 2019 Taylor & Francis Group, London, ISBN 978-1-138-62633-1

A multi-LOD model representing fuzziness and uncertainty of building information models in different design stages

J. Abualdenien & A. Borrmann
Chair of Computational Modeling and Simulation, Technische Universität München, Munich, Germany

ABSTRACT

The design of a building is a complex process in which a solution is developed iteratively in a way that fulfills the objectives and boundary conditions of the multiple design and engineering disciplines. The management of the early design phases is critical to circumventing a substantial amount of rework (Ballard and Koskela, 1998), increased costs and reduced productivity (Kolltveit and Grnhaug, 2004). Architects and engineers have a very high influence on the building design and consequently on its performance. However, during the early phases, information about the project activities and executions is insufficient to support the computational tools (Kolltveit and Grnhaug, 2004).

The planning process in the *Building Information Modeling* (BIM) approach starts from a coarse model and gradually refines it via multiple steps, described as *levels of development* (LOD). The principle of LOD describes the different stages of the project life-cycle by providing definitions and illustrations of BIM elements at the different stages of their development (BIMForum 2017). The LOD scale increases iteratively from a coarse level of development to a finer one. Consequently, the associated characteristics' quality of the exchanged model elements is increased. In this paper, the abbreviation *LOD* represents the composition of both the model's *Level of Geometry (a.k.a Level of Detail)* and *Level of Information (semantics)*.

To our knowledge, there is no formal approach for maintaining multiple levels of development throughout the design phases. Neither is there a formal definition of a building component's level of development nor is there an explicit description of the fuzziness of its geometric and semantic information. At the early design stages, where only rough information is available, a BIM model appears precise and certain which can lead to false assumptions and model evaluations, for example, in the case of energy efficiency calculations or structural analysis. This paper presents a multi-LOD meta-model to explicitly describe each individual LOD's requirements, which makes it possible to check the consistency of the geometry as well as the topologic and the semantic coherence across the different LODs.

The approach used to realize the multi-LOD meta-model is heavily based on the existing *Industry Foundation Classes* (IFC). IFC is an ISO standard, which is integrated into a variety of software products (Liebich et al., 2013). Thereby, the meta-model ensures a data format that is vendor-neutral, offering high flexibility and applicability.

The aim of this research is to provide a methodology for explicitly describing the building elements' multiple levels of development as geometric and semantic requirements, our approach takes into consideration the possible uncertainties. The multi-LOD meta-model can be depicted as a high-level interface, which simplifies the query and analysis of the elements and their relationships across the multiple LODs. The model introduces two layers, *data-model level* and *instance level*, which offers high flexibility in defining per-project LOD requirements and facilitates formal checking of their validity, such as requiring specific information to support the different *Embodied Energy* calculations.

REFERENCES

Ballard, G. & Koskela, L. (1998) August. On the agenda of design management research. *Proceedings IGLC*, 98, 52–69.

BIM Forum (2017) 2017 Level of Development Specification Guide.

Kolltveit, B.J. & Grnhaug, K. (2004) The importance of the early phase: the case of construction and building projects. *International Journal of Project Management*, 22 (7), 545–551.

Liebich, T., Adachi, Y., Forester, J., Hyvarinen, J., Richter, S., Chipman, T., Weise, M. & Wix, J. (2013) Industry Foundation Classes IFC4 Official Release. Online: http://www.buildingsmart-tech.org/ifc/IFC4/final/html.

Intelligent substitution models for structural design in early BIM stages

D. Steiner & M. Schnellenbach-Held
Institute for Structural Concrete, University of Duisburg-Essen, Essen, Germany

ABSTRACT

The design of buildings is significantly influenced by the concept of the bearing structure in addition to aesthetical and functional aspects. Because of this correlation, the factors time and cost are substantially affected by the configuration of the supporting framework and included structural elements. In early design stages a challenge arises from the small number of known design parameters. That leads to a limitation of applicable design principles to basic rough approaches like simplified formulae or vague engineering knowledge. At the same time, the building design process and further project processing require the relevant structural parameters like general realizability, structural element types, component geometries, rough material demands etc. Consequently, systems for a holistic support of the preliminary design are required for an earliest possible consideration of structural aspects in the design process. The involved highly complex decision-making process based on little information necessitates an adequate approach for engineering knowledge integration (Schnellenbach-Held & Albert 2003).

For this purpose, intelligent substitution models involving engineering expert knowledge for structural pre-design are developed as a decision-making support in early design stages. The fundamental concept of this new approach is based on detail-level dependent fuzzy knowledge bases and related inference systems. Therefore, adaptive levels of development (ALoD) and correlating information requirements are identified (see Tab. 1, Fig. 1). The developed models take effect as transfer functions, that determine necessary information for the next development level from limited and uncertain data. Generation of the expert knowledge and the inference rules is demonstrated by an example. Further assistance for the substitution models is given by complementary optimization tasks and modification management techniques. The support of preliminary design stages through the developed intelligent systems results in integration of structural engineering in early planning phases. This allows a harmonization and an efficiency increase of the process with avoidance of costly design cycles and providing realizable and economic structures.

Table 1. Developed basic detailing system for structural preliminary design.

ALoD	Designation	Content
0	Blackbox	Environmental conditions
1	Floor plan	Room and grid arrangements
2a	Position plan	Idealized structural elements
2b	Possibility plan	Suitability of construction types
3	Pre-design plan	Dimensioned structural solutions

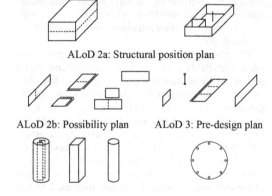

Figure 1. Adaptive Levels of Development for structural preliminary design.

REFERENCE

Schnellenbach-Held, M. & Albert, A. (2003) Integrating knowledge based systems with Fuzzy logic to support the early stages of structural design (in German). *Bauingenieur*, 11/2003, 517–524.

MS6: Performance of concrete during life-cycle
Organizers: J. Li, X. Gao & X. Ren

Remaining life prediction of aged bridges under salt and carbonation damages by concrete core tests

A. Miyamoto
Professor Emeritus, Yamaguchi University, Ube, Japan
(Visiting Professor, EPFL, Lausanne, Switzerland)

ABSTRACT

In recent years, the lifetime management of civil infrastructure has become one of the important issues in sustainable development in world societies. Civil infrastructure such as a bridge needs to be regularly assessed to make sure the continuity of the life of the bridge. Many efforts are now under way to ensure the longevity of the existing bridge through the improvement of the structural performance such as structural safety, remaining life, and so on, and management activities based on effective maintenance plans. An integrated lifetime management system for civil infrastructure in Japan, particularly bridges, becomes crucial based on the large number of aged bridges that need to gain attention. If the remaining life of an aged bridge is to be maximized, it is necessary to assess the structural performance of the bridge regularly. On the other hand, making a decision to remove an aged bridge is also an option. In order to make such a decision, it is important to make appropriate safety evaluation and remaining life prediction.

The bridge management system (J-BMS), which integrated with a Bridge Rating EXpert system (BREX), is one of the useful methods to assume the remaining life prediction of an existing concrete bridge. The BREX is a system that is designed for evaluating the present performance of the target bridge. The outputs are load-carrying capability and durability of each structure member. The input data for rating the concrete bridge are the technical specification of the target bridge, environmental conditions, traffic volume, and other subjective information that can be obtained through detailed visual inspection results. Evaluation results were thus obtained as the soundness level of the remaining life. This information processing approach makes it possible to deal with cases involving a large number of influencing factors. However, the BREX system result needs to be verified. To verify this system, concrete cores were extracted from some parts of the target bridge to conduct the carbonation and chloride ion tests. The result of carbonation and chloride ion tests will be used to assume the remaining life prediction.

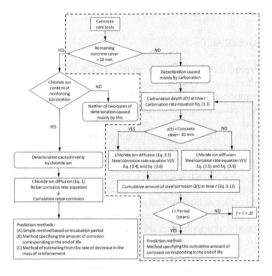

Figure 1. Flowchart of remaining life prediction.

This study aimed to evaluate the deterioration process of concrete cores extracted from two aged bridges (approximately 72 years old in service). Considering the most dominant factor affecting the deterioration of the bridge, between carbonation and chloride attack, based on the location of the bridge. In particular, propose a method of remaining life prediction in a case where the deterioration factor is caused mainly by chloride ion and the deterioration factor is caused mainly by carbonation (see Fig. 1).

REFERENCES

Miyamoto, A. & Motoshita, M. (2015) Development and practical application of a bridge management system (J-BMS) in Japan. *Civil Engineering Infrastructures Journal*, 48 (1), 189–216.

Widyawati, R., Takahashi, J., Emoto, H. & Miyamoto, A. (2014) Remaining life prediction of an aged bridge based on concrete core test. In: *2nd International Conference on Sustainable Civil Engineering Structures and Construction Materials*, 2.

Life-Cycle Analysis and Assessment in Civil Engineering: Towards an Integrated Vision – Caspeele, Taerwe & Frangopol (Eds)
© 2019 Taylor & Francis Group, London, ISBN 978-1-138-62633-1

Service life prediction of concrete under freeze-thaw deicing salt attack with intermittent dry periods

C. Thiel & C. Gehlen
Technical University of Munich, Munich, Germany

F. Foestl
Ed. Zueblin AG, Munich, Germany

ABSTRACT

Freeze-thaw-deicing salt attack (FTDSA) can lead to internal and external damage in concrete structural components. In order to plan, build and run buildings economically and ecologically, there is a need for research to develop a reliable probabilistic model that predicts the damage evolution of concrete structures under FTDSA. While current design methods distinguish only between constant and moderate moisture supply to concrete surfaces, the usual case of an alternating wet-dry exposure is not considered. Therefore, the effect of intermittent drying periods (IDPs) on the progress of freeze-thaw attack on concrete was investigated and possibilities for taking the results in service life prediction into account shown.

Concrete made with Portland cement and without entrained air voids and a *w/c-ratio* of 0.45 was exposed to continuous freeze-thaw cycles (FTC) at an age of 56 d. After 6 initial FTC, different cycle series were interrupted up to three times by a dry period which varied in duration (1, 2 and 7 days), temperature (5 and 20°C) and the duration of the subsequent capillary suction period (1 day in 3% sodium chloride solution, 2 days in water). The evaluation was based on the degree of external and internal damage, the water and chloride content as well as the changes in the pore system of the concrete. A series exposed to continuous freeze-thaw-cycles was run for comparison. The results show that the amount and rate of scaling are significantly reduced due IDPs and this is almost exclusively due to the lower amount of capillary pore water in the uppermost millimeters of the concrete near-surface zone. Furthermore, IDPs and following exposure to NaCl solution lead to an increase in chloride concentrations and penetration depths with a reduction in the freezing point of the pore water. As expected, chlorides penetrate into the interior of the concrete during drying and subsequent capillary suction. In addition, coarsening of the pore structure as observed for the continuous series was less pronounced when concrete was

subjected to IDPs. The role of IDPs on the evolution of scaling can be taken into account in:

- Modified performance tests to account for real exposure conditions
- Engineering models to improve the accuracy of service life prediction
- Numerical models to verify theoretical assumption

In current performance tests, FTDSA is accelerated by the use of a pessimal NaCl concentration (1–3% NaCl), high minimal temperature of up to −20°C and a constant moisture supply. The evolution damage is overestimated for concrete components used in the different climate conditions in Europe. FTDSA attack can be lessened by increasing the minimum temperature, reducing the number of freeze-thaw cycles, decreasing the thickness of the ice layer or introducing IDPs. However, real field behavior is only accurately simulated by introducing IDPs. Nevertheless, results from such modified performance tests could be used as compliance tests for new materials exposed to less severe conditions (e.g. vertical building components of bridges) rather than FTDSA with continuous moisture supply.

Reducing the severity of attack probably reduces the ability to distinguish between different concrete types and increases the scatter of the results. However, introducing a weather function in probability durability design can account for all kinds of realistic climatic conditions (i.e. leaching of salts due to long summers, very low or very high humidity conditions, different durations and temperatures of dry periods, etc.). Including the effect of IDPs in probabilistic service design concepts could optimize raw material use in an ecological and cost-efficient manner.

Furthermore, including the IDPs in numerical models could provide valuable proof of theoretical assumptions and therefore significantly contribute to a better understanding of damage evolution induced by FTDSA.

Life-Cycle Analysis and Assessment in Civil Engineering: Towards an Integrated Vision – Caspeele, Taerwe & Frangopol (Eds)
© 2019 Taylor & Francis Group, London, ISBN 978-1-138-62633-1

Effect of choice of functional units on comparative life cycle assessment of concrete mix designs

D. Kanraj
Department of Civil Engineering, University of Toronto, Canada

C.J. Churchill
Department of Civil Engineering, McMaster University, Canada

D.K. Panesar
Department of Civil Engineering, University of Toronto, Canada

ABSTRACT

The purpose of this study is to evaluate the effects that the choice of functional unit has on the results of comparative life cycle assessments (LCA) for concrete. When defining a LCA analysis, the functional unit provides the basis for quantification of all inputs and outputs and allows for comparison of LCA results based on equivalent performance of different processes or products (ISO 14040, 2006). Physical units such as mass or volume have been used as functional units for LCA analysis pertaining to concrete (Celik et al., 2015, Chen et al., 2010). Though physical units are convenient for measuring the overall environmental impact of any concrete construction, it does not satisfy the requirements for definition of a functional unit (Van den Heede and De Belie 2010).

This study uses properties of concrete such as volume, compressive strength (CS), and time to first repair (TFR) to define three functional units: volume, binder intensity with compressive strength as the performance indicator (BI-CS), and binder intensity with TFR as the performance indicator (BI-TFR). The effects of using the three functional units (volume, BI-CS, and BI-TFR) on the results of the comparative LCA were studied.

To consider the effect of cement replacement, four mix designs with varying levels of cement replacement by fly ash (0%, 25%, 35%, and 50%) were compared with each other based on a cradle to grave comparative LCA analysis.

When volume is used as the functional unit, the level of cement replacement and in turn the cement content influences the LCA results. In general, it was observed that the higher the level of cement replacement the lower the LCA results.

Human toxicity (non-cancer) and ecotoxicity impact categories were found to capture the effect of transportation of fly ash in the LCA analysis. As a result, an increase in percentage replacement of fly ash which results in an increase in transportation flows leads to a higher LCA result.

When BI-CS is used as the functional unit, it was again observed that the level of replacement influenced the LCA results. The trend of results for GWP, human toxicity non-cancer, and ecotoxicity are similar for when volume is used as the functional unit. However, the final trends of results which incorporate all impact categories are different from when volume is used as the functional unit. The reason for this result is that the binder intensity increases with increase in cement replacement due to: (i) decrease in 28 day compressive strength, (ii) cement and fly ash being treated as equivalent binders in terms of environmental performance and binder performance.

When BI-TFR was used as the functional unit, the LCA results were observed to reduce with increase in cement replacement. This is due to the increase in TFR obtained by increase in percentage replacement of cement by fly ash. This increase in TFR translated into reduction in binder intensity which is used as the functional unit, and hence, decreases the LCA results.

When the results of the comparative LCA performed using the three functional units were compared, it was observed that the analysis resulted in differing order of increasing LCA result depending on the functional units used in the study. Hence, the choice of functional unit determines the results of the comparative LCA to a significant extent.

REFERENCES

Celik, K., Meral, C., Gursel, A., Mehta, P., Horvath, A. & Monteiro, P. (2015) Mechanical properties, durability, and life-cycle assessment of self-consolidating concrete mixtures made with blended portland cements containing fly ash and limestone powder. *Cement and Concrete Composites*, 56, 59–72.

Chen, C., Habert, G., Bouzidi, Y. & Jullien, A. (2010) Environmental impact of cement production: detail of the different processes and cement plant variability evaluation. *Journal of Cleaner Production*, 18, 478–485.

International Standard Organization (2006) (R2016) *ISO 14040: Environmental Management-Life Cycle Assessment-Principles and Framework.*

Van den Heede, P. & De Belie, N. (2012) Environmental impact and life cycle assessment (LCA) of traditional and 'green' concretes: Literature review and theoretical calculations. *Cement and Concrete Composites*, 34, 431–442.

Evaluation of crack width in reinforced concrete beams subjected to variable load

T. Arangjelovski, G. Markovski & D. Nakov
Faculty of Civil Engineering, University "Ss. Cyril and Methodius", Skopje, R. Macedonia

ABSTRACT

Appearance of crack width in flexural reinforced concrete elements are expected under serviceability loads during service life of the structure, not only from permanent long-term loads but also from variable short-term loads.

In this paper, the influence of loading histories of variable (imposed) actions on the behavior of reinforced concrete beams especially the long-term crack width was analyzed. For the evaluation of long-term effects (effects due to creep and shrinkage), quasi-permanent combination of actions was used to verify the reversible limit state.

An experimental program and analytical research was performed to compare the experimentally obtained results of crack width and results of proposed calculation models given in EN 1992-1-1 Eurocode 2 and in the fib Model code 2010. For two specific loading histories, of series of beams D and E, the quasi-permanent coefficient ψ_2 was defined using the quasi-permanent combination of actions. These loading histories were consisting of long-term permanent action G and repeated variable action Q. The variable load was applied in cycles of loading/unloading for 24 and 48 hours in the period of 330 days. A total of eight reinforced concrete beams, dimensions 15/28/300 cm were tested. Four beams were made of concrete class C30/37 and four beams of concrete class C60/75.

At the start of the experiment at concrete age of $t = 40$ days, first the beams were loaded by the permanent load G which doesn't cause cracks in the section, then the variable load Q was applied and which causes cracks in the beams. First the crack width w_{G+Q} ($t = 40$) was measured approximately in the middle of the span, and then after unloading at the level of permanent load G crack width w_G ($t = 40$) was measured.

Because of the type of loading histories (repeated loading and unloading) the diagram of the measured crack w during time t has a form of an area defined by the limits of permanent load G and by G and variable load Q. One representative diagram was given for beam $D1$ on the Figure 1.

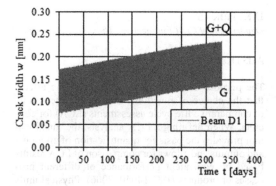

Figure 1. Diagram crack width w – time t for beam D1 concrete class C30/37.

Table 1. ψ_2 factors for series of beams D and E made of ordinary concrete C30/37 and high-strength concrete C60/75.

Series	Permanent action G kN	Variable action Q kN	ψ_2 factor	Crack width w mm
D (C30/37)	4	7.6	0.70	0.13
E (C30/37)	4	7.6	0.85	0.15
D (C60/75)	4	7.6	0.55	0.05
E (C60/75)	4	7.6	0.70	0.07

For the reversible serviceability limit state, quasi-permanent combination of actions was used to verify the experimental crack width at the level of permanent load G using Eurocode 2 and Model Code 2010 crack models. The results are given in the Table 1.

REFERENCE

Arangelovski, T. (2011) *Time-dependent Behavior of Reinforced High-strength Concrete Elements Under Action of Variable Loads*. Doctoral Dissertation, Skopje, University "St. Cyril and Methodius" and SEEFORM Doctoral Studies-DAAD Program, English Version.

Steel corrosion in ASR deteriorated concrete affected by de-icing salts

Y. Kubo, S.H. Ho & S. Kikuchi
Kanazawa University, Ishikawa, Japan

Y. Ishikawa
*Central Nippon Highway Engineering Nagoya Company Limited, Kanazawa Branch,
Road Inspection Section, Ishikawa, Japan*

ABSTRACT

In Japan, a large number of concrete structures have been constructed during the high economic growth period (1960 to 1975). A lot of early deterioration of concrete structure has been reported when concrete structures were affected by various severe conditions and it resulted in the occurrence of degradation of concrete structure due to early deterioration such as Alkali-silica reaction (ASR), chloride induced corrosion and other deterioration. In Japan, the standard countermeasure method is established for chloride induced corrosion which is known as one of the typical causes of concrete structure deterioration.

On the other hand, the countermeasure has not established for the concrete structures affected by alkali silica reaction since the deterioration mechanism of ASR is very complicated and it is not easily performed to remove or to reduce the deterioration factors such as reactive aggregates, alkali, and moisture in a concrete with the current techniques of repair. The proper repair method for ASR has not been established.

Furthermore, after the prohibition on the use of the studded tire, it has been reported that some concrete structures have been deteriorating significantly in a short term due to chloride induced corrosion affected by de-icing salts for the safety of traffic in cold region where a large amount of de-icing salts was used during the winter. It is known that the de-icing salts accelerate the chloride induced corrosion and promote Alkali silica reaction simultaneously by supplying the alkali (sodium) and moisture. Therefore, it is considered that the combined deterioration with Alkali silica reaction and chloride induced corrosion can occur easily where ASR deteriorated concrete is affected by de-icing salts. The real situation of steel corrosion in ASR deteriorated structure is not precisely clarified in existing concrete structure affected by de-icing salts.

Figure 1. Example of an abutment deteriorated by ASR.

In this study, the penetration of chloride ion and the steel corrosion in the concrete was investigated on the ASR deteriorated concrete abutment affected by de-icing salts. The influence of de-icing salts and alkali silica reaction on the steel corrosion was investigated by the in-situ survey and laboratory test. In the in-situ survey, the crack density and the surface water content of concrete were measured and the chipping investigation was conducted to take samples for chloride ion analysis and determine the corrosion grade of steel bars in order to clarify these influence of de-icing salts and Alkali silica reaction on the steel corrosion.

As the result, the influence of de-icing salts was not small since the chloride ion content was high and chloride ion permeation is large near concrete surface. It is considered that the increase of crack density as ASR progress may result in promoting water, oxygen and the chloride ion penetration supplied from de-icing salts and making steel bars easy to corrode in concrete structures. Therefore, the influence of de-icing salts was not small since the chloride ion content was high and chloride ion permeation is large near concrete surface. It was clear that cracks caused by Alkali silica reaction promote the penetration of chloride ions, which may promote chloride induced corrosion.

Life-Cycle Analysis and Assessment in Civil Engineering: Towards an Integrated Vision – Caspeele, Taerwe & Frangopol (Eds)
© 2019 Taylor & Francis Group, London, ISBN 978-1-138-62633-1

Analysis of the influencing factors for shear capacity of reinforced concrete beam-column joints

X. Gao, D. Xiang, Y. He & J. Li
Department of Structural Engineering, Tongji University, Shanghai, China

ABSTRACT

Reinforced concrete frame structure is a widely used structure form. The beam-column joints are the key and the vulnerable component in the seismic design of the RC framed structures. In this paper, the relevant test results of the three typical beam-column joints e.g. interior joint, exterior joint and joint with floor slab were collected and analyzed. The main influencing factors such as concrete compressive strength, stirrups of the joints, longitudinal reinforcement of the beam, and floor slabs and the reinforcement bars were investigated.

The shear bearing capacity provided by concrete to the beam column joints is investigated through the calculation of the shear capacity of the joints with different design standards such as Chinese standard (GB 50010-2010), European code CEB-FIP, European code2, American standard ACI352. Concrete is a significant factor that bears the shearing force of the joints. Increasing concrete strength or the area of the joint core can effectively improve the shear capacity of the joints.

There exist upper and lower limits of the volume stirrup ratio for the joint core and the threshold has been derived. Meanwhile, a relationship between the stirrups ratio and the reinforcement ratio of the inter-facing beam has been presented. An optimum design for RC framed structure should consider the relationship to achieve the most economical and reasonable design.

The ACI standard always does not consider the contribution of axial load to the shear capacity of the joints. With the increase of axial compression ratio, the shear capacity provided by axial load gradually increases according to the other three standards When the axial compression ratio changes from 0 to 0.3 the results calculated by the CEB-FIP standard show that the axial load provided 66% of the total shear capacity. The CEB-FIP code uses the shear-friction model to calculate the shear strength of the joints and the

friction force is mainly determined by the normal stress. Therefore, the greater the axial load, the larger the shear capacity is. Comparing the calculated results as per the different standards, it is easy to find that the shear capacity of joints calculated according to ACI is the most conservative among the four standards for the shear strength provided by axial load.

The relationship between the shear compression ratio of the spacing joint with the floor slab and the beam longitudinal reinforcement was diagramed. It is found that the floor slab can increase the shear capacity of the joint core. When longitudinal reinforcement ratio of the beam was in the regular range, the floor slab can improve the shear capacity of the joints significantly. A parameter J_s can be expressed by Equation (1). When J_s is equal to 0.125, the shear capacity of slab can provided 60% of the total shear capacity. When J_s is equal to 0.25, the shear capacity shared by the floor slab reaches up to 45% of the total shear capacity. For the joints with less longitudinal reinforcement of the beam, the effect of slab reinforcement is larger due to the relatively larger reinforcement of the slab.

$$J_s = \frac{f_y \left(A_s' + A_s \right)}{b_j h_j f_c} \left[1 - \frac{0.8 h_b}{H_c - h_b} \right] \quad (1)$$

where f_y is the yield strength of the beam longitudinal reinforcement, A_s' and A_s is the sectional area of the rebar in the compressive and tensile zone at the beam end, respectively, z is the length of the internal force arm, H_c is the distance between the bottom of the column and the loading points, h_b is the height of the section of the beam.

The effect of the floor slabs on the shear capacity of the joints cannot be ignored that can provide at least 35% of the total shear capacity of the joints.

The existence of the floor can affect the failure mode of joints. Therefore, proper contribution of the floor slab to the joint should be considered to reduce the stirrup ratio in the joint core.

Study on physical property and durability of concrete with high content of blast furnace slag

K. Eguchi & Y. Kato
Tokyo University of Science, Chiba, Japan

D.N. Katpady
Research and Development Division, Infratec Co. Ltd., Kagoshima, Japan

ABSTRACT

Currently, deterioration of environment on the global scale has become a problem. In the construction industry, by blending cement with industrial waste such as ground granulated blast furnace slag (hereinafter referred to as GGBS) and fly ash, from a view point of effective use of resources and reduction of CO_2 emissions, there are striving to reduce the environmental burden. GGBS is known to exhibit latent hydraulicity and when used in combination with cement, the water impermeability and the amount of binding chloride ions increases. GGBS concrete with a replacement ratio of about 50% is widely used in Japan. However, when the replacement ratio of GGBS in cement is 70% or more, since initial strength and neutralization resistance are lowered, high volume GGBS concrete is hardly used. The properties of concrete with such high replacements of GGBS are seldom studied and there are many unclear properties. Hence requires a thorough understanding of the behavior of such concrete for possible application. In this study, we performed tests for compressive strength, porosity, heat of hydration, pH of pore solution, surface air permeability and chloride ions ingress of concrete substituted with high percentage of GGBS. In this paper, those replaced with GGBS by 70% or more are regarded as high volume GGBS concrete. Ordinary cement only was used as binder to make OPC concrete and blended with GGBS with different replacement ratios (from 50 to 90%).

Figure 1 shows the compressive strength test results of concrete subjected to water curing. Specimens using GGBS had 28 day strength lower than OPC at any replacement ratio. B60 and B70 had similar strength to OPC at 56 days and 182 days curing, respectively. For the replacement ratio of 80% or more, the strength after 28 days was markedly lower than that of 70% or less, and thereafter the strength increase was very low.

The relationship between the compressive strength and the porosity is shown in Figure 2. Generally, it is believed that the compressive strength increases when voids are filled by hydration products. As can be seen from the results, the strength increase for

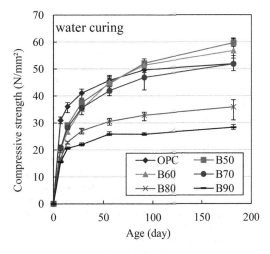

Figure 1. Compressive strength.

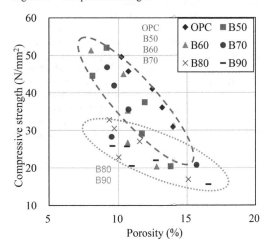

Figure 2. Relationship between strength and porosity.

any replacement ratio was observed for OPC and low volume GGBS concrete. But for high volume GGBS concrete, the increase in strength due to the decrease in porosity was small.

Life-Cycle Analysis and Assessment in Civil Engineering: Towards an Integrated Vision – Caspeele, Taerwe & Frangopol (Eds)
© 2019 Taylor & Francis Group, London, ISBN 978-1-138-62633-1

Estimation of compressive strength at early age using electric conductivity

L.L. Chacha Costa
Graduate School of Engineering and Science, Department of Architecture and Civil Engineering, Shibaura Institute of Technology, Tokyo, Japan

T. Shibuya & T. Iyoda
Department of Civil Engineering, Shibaura Institute of Technology, Tokyo, Japan

ABSTRACT

The techniques using electrical conductivity to estimate compressive Strength of concrete is effective in tunnel construction as concrete structures are greatly affected by demolding time. Results from this study present a methodology for estimating compressive strength at early age using electric conductivity by carrying out two main tests; compressive strength of different concrete mixes, using ordinary Portland cement (OPC), blast furnace slag type B (BB) and chemical admixture (A) as an accelerator, water cement ratio 50% (W/C) and monitoring the performance at temperature 5°C–35°C. Electric conductivity test to measure the phenomena of increase in conductivity ratio was conducted. The analysis revealed a model for predicting the compressive strength of concrete mixes at early age (24 hours). From the analysis and the results, there is a relationship between electrical conductivity and liquid water content in the pore in other words, the conductivity ratio and the compressive strength. When compressive strength was estimated from conductivity for OPC concrete at the atmospheric temperature of 28°C, it was confirmed that the estimated strength is slightly lower than the actual strength. So, it is possible to estimate the compressive strength with electrical conductivity. Electrical conductivity method proves to be an effective and reliable method for assessing various characteristics of concrete especially concrete strength development at early age. Moreover being a non-destructive test. Therefore using these techniques will provide indispensable techniques for efficient constructions of concrete structures.

REFERENCES

Japan Society of Civil Engineers: Formulation Specification for Concrete Established in 2012 [Construction Manual], p. 152.

Kumar Mehta, P. & Monteiro, P.J.M. (2006) *Concrete Microstructure, Properties, and Materials*. 3rd edition. New Delhi, McGraw Hill Education.

Madhavi, T.C., et al. (2016) Electrical conductivity of concrete. *ARPN Journal of Engineering and Applied Sciences*, 11 (9), 1–9.

Soroka, I. (2004) *Concrete in Hot Environment, 2004 Modern Concrete Technology Series*.

Takafumi, I. & Takeshi, I. (2016) Study of mechanism of compressive strength estimation using electrical conductivity, repair of concrete structures. *Upgraded Upgrade Papers Report*, 16, 227–232.

Takehiro, M. et al. (2017) Influence of hydration reaction of concrete not yet solidified on electrical resistance measured by DC four electrode methods. *Annual Concrete Engineering Proceedings*, 39 (1).

Takeshi, I. & Ota, M. (ACF 2016) Study of monitoring technique of hardening process concrete using electrical conductivity. In: *The 7th International Conference of Asian Concrete Federation*.

Life-Cycle Analysis and Assessment in Civil Engineering: Towards an Integrated Vision – Caspeele, Taerwe & Frangopol (Eds)
© 2019 Taylor & Francis Group, London, ISBN 978-1-138-62633-1

Investigating the effect of improving the low grade recycled aggregate by carbonation

A.A. Abdulkadeer
School of Engineering, Shibaura Institute of Technology, Japan

N. Matsuda
Tokyo Techno Company, Japan

T. Iyoda
Shibaura Institute of Technology, Japan

ABSTRACT

A potential solution for both sustainability in demolition and increased service life is repurposing concrete that is taken out of service as recycled concrete aggregate (RCA). This material allows for a more economical aggregate source for towns that are located far from any virgin aggregate sources. Recycling existing concrete for use as aggregate in new construction has gained consideration in recent years because of high costs of disposal of waste concrete and shortage of virgin aggregate in Japan, insufficient of natural aggregate has become more of great concerned. According to an investigation carried out in 2012 by the Ministry of land, Infrastructure and transportation of Japan (MLIT), the amount of construction waste generated in Japan is approximately 31 million tons annually. Also in Japan, Japanese Industrial Standard (JIS) categorized RCA into three classifications which are High-quality (H) RCA (JIS A 5021:2011), Medium-quality (M) RCA (JIS A 5022:2012), and Low-quality (L) RCA (JIS A 5023:2012), the classification is based on absolute dry density and water absorption. The H class of RCA has density above 2.5 g/cm^3 and water absorption of 3% and below. The M class of RCA has density of over 2.3 g/cm^3 and water absorption 5% and below. The L class of RCA has density bellow 2.3 g/cm^3 and water absorption below 7%. The High-quality (H) RCA could be used as natural aggregate in concrete, yet the high cost and high energy required in manufacturing of H has limited its utilization in concrete. On the contrary, the M and L type of RCA have relatively significant volume of adhered mortar and they could be manufactured with low cost and also required less energy. The problem of cost and quality has challenged many researchers to look in-depth for more sustainable solution for the problem. In this research experimental studies were carried out by using the carbonation technology to improve the quality of low recycles aggregate. In RCA, the adhered mortar contains calcium hydroxide which is a product of hydration, when carbon dioxide is injected into the adhered mortar through accelerated carbonation, calcium carbonated is produced as a result of reaction between the injected carbon dioxide and calcium hydroxide in the adhered mortar. The produced calcium carbonate that is large in volume would filled the pore in the adhered mortar which would lead to densification of RCA, and also improved the quality of RCA. This would increase the strength and the permeability resistance of concrete with RCA and other mechanical properties of RAC.

This paper presents the result of three different condition of experiment to evaluate the effect of RCA with carbonation and without carbonation at the different water cement ratio. Both the properties of RCA and RAC were examined. firstly, the concrete was cast at different water cement ratio (30%, 50% and 70%), the concrete was recycled in R30, R50 and R70. The result shows the high strength was at the lowest water cement ratio. R30 and R70 were partially mixed (3:70, 5:5 and 7:3) it was found out that at 5:5 the compressive strength was the highest. The second condition was the carbonation of RCA produced from concrete of different water cement ratio of 35%, 45%, 55% and 65%. the properties of RCA (relative density and water absorption) was significantly improved but the water cement ratio of the original concrete has no significant effect on the carbonation. The last was carbonation of the low quality (M and L) RCA which shows a tremendous improvement on both properties of RCA and RAC. the properties of low recycled aggregate can be enhanced by carbonation and also The compressive strength and the splinting strength of recycled aggregate concrete is the dependent and the strength of the parental concrete. carbonation of RCA can also reduce the drying shrinkage of RCA.

REFERENCE

Saeki, T. & Yoneyama, K. (1992) Change in strength of mortar due to carbonation. *Journal of the Japan Society of Civil Engineering*, 451, 68–78 (Japan).

Influence of ASR expansion on concrete deterioration observed with neutron imaging of water

Y. Yoshimura
Tokyo Institute of Technology, Tokyo, Japan
RIKEN, Saitama, Japan
Topcon Corporation, Japan

M. Mizuta, H. Sunaga & Y. Otake
RIKEN, Saitama, Japan

Y. Kubo
Kanazawa University, Ishikawa, Japan

N. Hayashizaki
Tokyo Institute of Technology, Tokyo, Japan

ABSTRACT

Deterioration of concrete such as salt attack and alkali silica reaction (hereafter ASR) relies on the water content. In case of ASR, for example, alkali silicate gel expands on absorption of water and ASR does not occur under non-water conditions. Since the water content or water distribution in concrete may affect the deterioration process level of the concrete, it is possible that the measurement of water in concrete will be useful for evaluating the durability of the concrete. Concerning the measurement of water content of concrete, embedded-type humidity sensors have been used. Visual inspection is also one of the conventional measurements. These methods had been easily chosen, whereas the limited information obtained was a demerit. The process of changing cannot be continuously followed and it is difficult to confirm the reproducibility of the experimental results. However, there had been few methods of measuring water in concrete conveniently.

For the reasons mentioned above, we adopted neutron beam which has high transmissivity for concrete and sensitivity for hydrogen in order to measure the water content of concrete. RIKEN accelerator-driven compact neutron source, RANS was used as a neutron generator and the water penetration of concrete was visualized by neutron imaging. In this research, the water absorption tests of concrete were conducted and the water contents were measured by neutron transmission imaging at fixed intervals. The concrete specimens were made with varied ASR expansion rates and mix proportions. First of all, the specimens were dehydrated in a desiccator at a temperature of 50 degrees until their weight was evened. And the bottom surface of the specimen was soaked in water over one week and the absorbed water was visualized by neutron transmission imaging. Water content of the concrete was derived as the difference of neutron transmissivity between dry and wet conditions by image analysis.

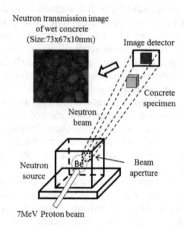

Figure 1. A schematic view of neutron imaging.

In this paper, we report the results obtained from evaluation for eight days by neutron imaging with a compact accelerator-driven neutron source about water penetration in a concrete specimen with ASR expansion.

(1) Water penetrating into concrete could be visualized by neutron imaging with a compact accelerator-driven neutron source.
(2) We performed neutron imaging on 5-cm-thick concrete, and the successful relationship between neutron transmissivity and water content for quantification was confirmed.
(3) The progress of water content and water penetration varied by the degree of ASR expansion. Also, the properties of water penetration were different between specimens cut from the same block. These results suggest that water penetrability is influenced by the degree of ASR deterioration.
(4) In water absorption tests performed twice, a specimen with ASR expansion strain 900 μ couldn't obtain repeatability regarding water penetration.

Life-Cycle Analysis and Assessment in Civil Engineering: Towards an Integrated Vision – Caspeele, Taerwe & Frangopol (Eds)
© 2019 Taylor & Francis Group, London, ISBN 978-1-138-62633-1

Cement-based geotechnical elements exposed to chemical attack—provisions for durability in standards

F. Wagemann & F. Schmidt-Döhl
Institute of Materials, Physics and Chemistry of Buildings, Hamburg University of Technology, Hamburg, Germany

A. Rahimi
Federal Waterways Engineering and Research Institute, Karlsruhe, Germany

ABSTRACT

Geotechnical elements made of concrete, mortar and cement slurry are frequently used for deep foundations or anchorages of hydraulic or other civil engineering structures. Depending on their application and the ground conditions, such foundation elements may be subject to chemical attack by soil or water. The type of attack varies, depending on the type of aggressive substance, its concentration and the flow velocity of the surrounding water, so that the construction material may corrode by expansion or decalcification. Such scenarios may result in considerable problems during the construction process if tried-and-tested geotechnical elements such as grouted anchors, for example, cannot be used to anchor construction pit walls. This may increase the cost of site preparation and impede construction.

The codes for the design and construction of common types of special geotechnical works in Germany contain only a limited amount of specific information on how to ensure the durability of geotechnical elements exposed to chemical attack. Designers are instead bound by the prescriptive provisions given in the standards for concrete (DIN EN 206-1:2001-07; DIN 1045-2:2008-08). However, when applied to geotechnical elements, these provisions are associated with uncertainties and must be treated with caution (Rigo & Unterderweide 2011).

The used elements in foundations and tiebacks for civil engineering works are grouted anchors, piles and jet-grouted elements. These elements differ with regard to their method of production and the cement-based materials used. Grouted anchors and grouted piles are executed with cement grout while concrete is used for piles. For elements produced by the jet-grouting method, a material consisting of the injected binder and the in-situ soil is produced.

The consequences of chemical attack to those elements depend on their production and structural performance. For example, attack with decalcification is very unfavourable for ground anchors and grouted micropiles as the loads are transferred via the surface of the grout body. It must be assumed that a reduction in the loadbearing capacity will occur.

The provisions given in standards are only partially able to ensure the durability of cement-based geotechnical elements. The technical rules for bored piles and displacement piles permit durable use in all XA exposure classes. However, the durability of micropiles and ground anchors is associated with uncertainties regarding different scenarios of chemical attack. Those elements are unfeasible under an intense chemical attack (XA3). Also elements built with the jet grouting method are associated with uncertainties under chemical attack. Depending on the design working life, the normative provisions are inadequate owing to the lack of available experience. In particular, the method cannot be used in soils containing sulphates.

REFERENCES

DIN 1045-2:2008-08. *Tragwerke aus Beton, Stahlbeton und Spannbeton – Teil 2: Beton – Festlegung, Eigenschaften, Herstellung und Konformität – Anwendungsregeln zu DIN EN 206-1.* Berlin, Beuth.
DIN EN 206-1:2001-07. *Beton Teil 1: Festlegung, Eigenschaften, Herstellung und Konformität.* Berlin, Beuth.
Rigo, E. & Unterderweide, K. (2011) Chemischer Angriff auf verpresste Anker und Pfähle. *Beton- und Stahlbetonbau,* 106 (5), 308–313.

*Life-Cycle Analysis and Assessment in Civil Engineering: Towards an
Integrated Vision – Caspeele, Taerwe & Frangopol (Eds)
© 2019 Taylor & Francis Group, London, ISBN 978-1-138-62633-1*

Shrinkage characteristics of ground granulated blast furnace slag high content cement

H. Mizuno
School of Engineering, Shibaura Institute of Technology, Tokyo, Japan

T. Iyoda
Shibaura Institute of Technology, Tokyo, Japan

ABSTRACT

In recent years, global warming has become a world problem. So reduction of carbon dioxide is required in all industries as against global warming. Focusing on the construction, Portland cement which is the most widely used construction material discharges large amount of carbon dioxide during the manufacturing process. Therefore, the use of mixed cement replacing admixture materials such as ground granulated blast furnace slag and fly ash with Portland cement is receiving a lot of attention. Mixed cement has advantages such as effect of protected penetrated chloride ion, prevent of ASR, enhancement of long-term strength, etc. However, it has been reported that drying shrinkage of cement with high content of ground granulated blast furnace slag becomes large. Cracks of drying shrinkage cause decreasing resistance of the durability on concrete structures. Therefore, countermeasure to reduce drying shrinkage are essential. So we focused on fly ash in this research. In the past researches, it has been reported to reduce drying shrinkage using fly ash. So it is expected that reduction of drying shrinkage can be expected, even in cement with high replacement ground granulated blast furnace slag. Therefore, in this study, mortar with fly ash added to blast furnace slag cement was made for the purpose of reducing drying shrinkage in blast furnace slag cement, and drying shrinkage were measured. In addition, since the drying shrinkage is closely related to the pore diameter and the pore volume of the mortar, the measurement of the pores was carried out from a mercury intrusion porosimetry. And the drying shrinkage is reduced in the mortar having the high replacement rate of the ground granulated blast furnace slag. We also confirmed which influences drying shrinkage in three component mortar.

REFERENCES

Civil Engineering Society Construction Guidelines for Concrete Using Blast Furnace Slag Fine Powder. ISBN 4-8106-0190.
Fishman, K. (2007) Supervision material design of concrete structure. *Ohmsha*, 7, 104–157.
Kato, Y., Tsunoda, T., Watanabe, T. & Umemura, Y. *Materials and Construction of Kashima Publishing Association*.
Katsuro, K. (2003) Effects of water cement ratio and curing conditions on concrete drying shrinkage. *Annual Concrete Engineering*, 25 (1).
Kohei, E. (2011) Experimental study on shrinkage characteristics and durability of ternary concrete using blast furnace slag fine powder and fly ash 66th Annual Scientific Lecture of JSCE V-262.
Takahashi, Y. Study on the durability of three component cement adjusted for content of various admixtures. *39th Society of Civil Engineers Kanto Branch Technical Research Presentation*.

Life-Cycle Analysis and Assessment in Civil Engineering: Towards an Integrated Vision – Caspeele, Taerwe & Frangopol (Eds)
© 2019 Taylor & Francis Group, London, ISBN 978-1-138-62633-1

Performance evaluation on chlorine ion immobilizing ability of concrete using calcium aluminate aggregate and additive

Y. Nakanishi
School of Engineering, Shibaura Institute of Technology, Tokyo, Japan

S. Ito
Cement & Special Cement Additives Research Department, Omi Plant, Denka Co. Ltd., Niigata, Japan

T. Iyoda
Shibaura Institute of Technology, Tokyo, Japan

ABSTRACT

Japan is facing the sea on every side. It has been reported that there are many deteriorated concrete structures due to salt damage. In recent years, in order to improve the resistance to salt-damage, materials have been research and developed using calcium aluminate. In this research, we focus on the aggregate ($CaO \cdot Al_2O_3$ aggregate) and admixture ($CaO \cdot Al_2O_3$) for making the specimens for improving resistance of salt damages. Compressive strength test, permeability of chloride ion and observation of aggregate interface on SEM were carried out, using concrete on different replacement of calcium aluminate materials. In order to confirm the composition of the product, a cement paste specimen using the calcium aluminate material was prepared and the hydration product was clarified using XRD. As a result of XRD test, it was found that the calcium aluminate materials react with water and calcium hydrate ($Ca(OH)_2$) which is a cement hydration product. And the calcium aluminate materials (especially CA aggregate) added with salt, it was found that chloride ion was immobilized as Friedel's salt and hydration product was generated on the aggregate interface. We can see many hexagonal plate at the aggregate interface on observation of the aggregate interface by SEM. It was confirmed that the products are Friedel's salt and hydration products form calcium aluminate materials from previous studies. Although according to water permeability test, the interface of aggregate may be densified by these hydration products. It is suggested that the densification of the aggregate interface may decrease the permeability inside the concrete. Furthermore, the effect of shielding salt is expected under the severe salt environment on long term.

REFERENCES

Ito, T. & Iyoda, T. (2014) Proposal for Improvement of Salt Penetration Performance of Low Heat Portland Cement. Shibaura Institute of Technology Faculty of Engineering Department of Civil Engineering, Japan.

Ito, S., Morioka, M., Iyoda, T. & Maruyama, I. (2016) Modification effect of transition zone by calcium aluminate type aggregate. *Japan Society of Materials*, 65 (11), 787–792 (Japan).

Kawano, S. & Ujike, I. (1999) Study on change of permeability coefficient of concrete by drying. *Report on Concrete Engineering Annual Papers*, 21 (2) (Japan).

Masuda, T. & Iyoda, T. (2015) Grasp of Concrete Performance Using Aggregate and Admixture with Chlorine Immobilizing Ability. Shibaura Institute of Technology Faculty of Engineering Department of Civil Engineering, Japan.

Tahara, K., Miyaguti, K., Mori Oka, M. & Takewaka, K. (2011) Immobilization ability of hydration behavior and chloride ions of the cement hardened body in a variety of types of mixed the $CaO \cdot Al_2O_3$. *Cement Science and Concrete Technology*, 65. Japan.

Life-Cycle Analysis and Assessment in Civil Engineering: Towards an
Integrated Vision – Caspeele, Taerwe & Frangopol (Eds)
© 2019 Taylor & Francis Group, London, ISBN 978-1-138-62633-1

Effect of chloride and corrosion of reinforcing steel on heat transfer of reinforced concrete

P. Sancharoen, D. Im, P. Julnipitawong & S. Tangtermsirikul

Sirindhorn International Institute of Technology, Thammasat University, Pathumthani, Thailand

ABSTRACT

Chloride induced corrosion of reinforcing steel is one of the most significant deterioration problems of reinforced concrete structure. Inspection of chloride content of concrete or corrosion of reinforcing steel is resources consuming. Therefore, effect of chloride and corrosion of reinforcing steel on heat transfer of reinforced concrete was studied in order to determine possibility of using thermograph to detect chloride contaminated concrete and corrosion of reinforcing steel.

In this study, concrete mix proportions with different water to binder ratio and fly ash replacement were prepared. External infrared heater was applied to surface of reinforced concrete specimen. Specimen was submerged in sodium chloride solution. Also corrosion of reinforcing steel was accelerated by impressed current. Temperature profile inside concrete with and without reinforcing steel was monitored every 1 minute by embedded thermocouple and data logger. Temperature was recorded during heating and cooling cycle. One cycle was 4 or 8 hrs.

The results show that both of chloride and corrosion products significantly affected thermal properties of concrete. Results show that chloride in pore solution of concrete reduced specific heat of concrete. As a result, temperature of concrete increased faster. The maximum difference is around 2–3°C at same measurement time and depth.

Similarly, corrosion products cause faster heat transfer into concrete. This is contradict to literature reported that rust of reinforcing steel act as insulator. Because in this study, rust is formed inside concrete and confined in the interfacial zone between steel and concrete. So formed rust is dense and act as steel which as high heat conductivity and low specific heat. At high corrosion level and depth around reinforcing steel, maximum temperature difference due to corroded bar is almost 1.6°C at depth around reinforcing steel.

From the results, detection of chloride in concrete and corrosion of reinforcing steel by infrared camera can be possible based on difference of heat transfer of deteriorated concrete. Further study is ongoing.

REFERENCES

Akiyama, T., Ohta, H., Takahashi, R., Waseda, Y. & Yagi, J. (1992) Measurement and modeling of thermal conductivity for dense iron oxide and porous iron ore agglomerates in stepwise reduction. *ISIJ International*, 32, 829–837.

Bentz, D.P. (2007) Transient plane source measurements of the thermal properties of hydrating cement pastes. *Materials and Structures*, 40, 1073–1080.

Caré, S., Nguyen, Q.T., L'Hostis, V. & Berthaud, Y. (2008) Mechanical properties of the rust layer induced by impressed current method in reinforced mortar. *Cement and Concrete Research*, 38, 1079–1091.

Choktaweekarn, P., Saengsoy, W. & Tangtermsirikul, S. (2009) A model for predicting the specific heat capacity of fly-ash concrete. *Science Asia*, 35, 178–182.

Jamieson, D. & Tudhope, J. (1970) Physical properties of sea water solutions: thermal conductivity. *Desalination*, 8, 393–401.

Jamieson, D., Tudhope, J., Morris, R. & Cartwright, G. (1969) Physical properties of sea water solutions: heat capacity. *Desalination*, 7, 23–30.

Klieger, P. & Lamond, J.F. (1994) Significance of Tests and Properties of Concrete and Concrete-making Materials. ASTM.

Michel, A., Pease, B.J., Geiker, M.R., Stang, H. & Olesen, J.F. (2011) Monitoring reinforcement corrosion and corrosion induced cracking using non-destructive x-ray attenuation measurements. *Cement and Concrete Research*, 41, 1085–1094.

Nossoni, G. & Harichandran, R.S. (2011) Electrochemical mechanistic model for concrete cover cracking due to corrosion initiated by chloride diffusion. *Journal of Materials in Civil Engineering*, 26, 04014001.

Val, D.V., Chernin, L. & Stewart, M.G. (2009) Experimental and numerical investigation of corrosion-induced cover cracking in reinforced concrete structures. *Journal of Structural Engineering*, 135, 376–385.

Wong, H.S., Zhao, Y.X., Karimi, A.R., Buenfeld, N.R. & Jin, W.L. (2010) On the penetration of corrosion products from reinforcing steel into concrete due to chloride-induced corrosion. *Corrosion Science*, 52, 2469–2480.

Life-Cycle Analysis and Assessment in Civil Engineering: Towards an Integrated Vision – Caspeele, Taerwe & Frangopol (Eds)
© 2019 Taylor & Francis Group, London, ISBN 978-1-138-62633-1

Long-term performance of cementitious materials used in nuclear waste disposal—a case study at SCK•CEN

Q.T. Phung, S. Seetharam, J. Perko, N. Maes, D. Jacques, S. Liu & E. Valcke
Belgian Nuclear Research Centre (SCK•CEN), Belgium

T. Seeman
Belgian Nuclear Research Centre, Belgium
Ghent University, Belgium

G. Hoder
Belgian Nuclear Research Centre, Belgium
Technical University of Delft, The Netherlands

L. Lemmens & A. Varzina
Belgian Nuclear Research Centre, Belgium
Leuven University, Belgium

R.A. Patel
Paul Scherrer Institut (PSI), Switzerland

ABSTRACT

The possibility to assess the service life of structures made of cement-based materials for nuclear waste facilities is of great practical importance but raises new challenges. Unlike classical structures, the durability of waste disposal facilities needs to be predicted over hundreds up to thousands of years because of the long-lived activity of nuclear waste. Over such long-time period, the properties of cement-based materials evolve under repository conditions. Endogenous concrete processes and loading cause mechanical and physical degradation, whereas geochemical interactions with the host environment and with the waste matrices lead to chemical degradation (Jacques *et al.* 2013). All these degradation processes will undoubtedly change the design properties of cementitious materials and have consequences on the long-term performance including changes in mechanical performance, transport properties and microstructure. Therefore, a systematic approach is needed to establish the relationship between multiple processes at multiple scales, which could provide knowledge of long-term safety of nuclear waste disposal and suggestions on how safety functions of the repository components can be improved in terms of material and structural designs.

This paper presents an overview of a research program at the Belgian Nuclear Research Centre (SCK•CEN) concerning long-term durability of cementitious materials. The need for this research originates from the Belgian nuclear waste disposal concept, which proposes ordinary portland cement as the candidate binder for engineered barriers. Three main degradation mechanisms are considered: calcium leaching, carbonation and combined leaching/carbonation. Specifically, the effect of these mechanisms on microstructural changes and hence on physical and transport properties are discussed. The discussion concerns recent advances in experimental techniques and numerical tools in support of experimental programme. Novel experimental techniques cover accelerated degradation experiments for sound/fractured material, and transport experiments to determine water permeability and diffusion coefficient. Numerical tool covers recent implementation of lattice-Boltzmann based reactive transport code to handle micro-meso-continuum scale modelling.

The findings from this research program show how the degradation of cementitious materials might be affected by severe/coupled conditions. These might induce an evolution of the microstructure, cracking and changes of transport, chemical and mechanical properties of cementitious materials at different time and spatial scales. In conclusion, a significant amount of work is still needed to fully understand long-term degradation processes and to ensure ordinary Portland cement is a durable binder material for nuclear waste disposal structures. There is still a long way to reach the safest but economical solution for nuclear waste disposal. In SCK•CEN, other studies to better understand the long-term performance of concrete are still on-going including (i) approaches to evaluate coupled chemical (carbonation and leaching) degradation; (ii) to characterize the (free) gas transport in concrete; (iii) to obtain sufficiently good 3D microstructure of hardened cement paste; and (iv) to study the interaction between concrete – (radioactive) waste.

REFERENCE

Jacques, D., Maes, N., Perko, J. et al. (2013) Concrete in engineered barriers for radioactive waste disposal facilities – phenomenological study and assessment of long term performance. In: *15th International Conference on Environmental Remediation and Radioactive Waste Management – ICEM2013*. Brussels, Belgium.

Investigating the effect of surplus filler of asphalt plant on setting time and compressive strength of cement mortar

M. Kioumarsi
Associate Professor, Department of Civil Engineering and Energy Technology, OsloMet – Oslo Metropolitan University, Norway

H. Vafaeinejad
Graduated Master Student in Civil-Structural Engineering, Alaodoleh Semnani Education Institute of Semnan, Iran

ABSTRACT

One of the solutions taken to reduce the use of cement is the use of materials such as pozzolans and partial cement replacement in concrete (Hale et al., 2008, Kartini et al., 2012, Zareei et al., 2017). Kartini et al. (2012) in their study showed that that 10% replacement of cement by rice husk ash causes a loss in compressive strength and durability of conventional concrete and with a further increase of these materials, the loss is more tangible. According to Hale et al. (2008), replacement of 25% slag and 15% fly ash improves long-term properties of concrete such as compressive strength, permeability and chloride ion penetration.

There are still possibilities of using new materials and surplus fillers in concrete to reduce the cement consumption. One of these materials is surplus filler of asphalt plant. In this study, effect of using the surplus filler of asphalt plant on the cement paste setting time and compressive strength of cement mortar were studied with the aim of reducing the cement consumption. This filler is produced during the process of manufacturing asphalt as a secondary and surplus material. Since this material has no other application in the process of asphalt production, it is necessary to dispose of this product. The disposal of this material from the plant leads to environmental pollution. Effect of replacing of cement by surplus filler of asphalt plant with percentages of zero to 50% was investigated with the aim of reducing cement consumption. To this aim and in the first step, the optimum water content of the cement paste was obtained by using a 10 mm needle of Vicat apparatus. Then, the effect of adding filler in different amounts on the initial and final setting of the cement paste was evaluated. By constructing 110 cubic specimens with the size of 100 mm, with constant mix design of water and aggregate, the compressive strength of the cement mortar containing filler at 3, 7, 28, 56 and 90 days was obtained with specified replaced percentages of filler at 20°C. Moreover, the growth rate of compressive strength was examined in 0 and 20 percentages of replaced filler with cement at the ages of 1, 3, 7, 14, 28, 56 and 90 days with the constant temperature (35°C). According to the results, adding the filler is a factor of delaying the initial and final setting time. The filler increment in the cement mortar leads to a drop in the compressive strength. However, this reduction is negligible when the filler percentage is less than 15%. In accordance with ASTM C618 standard, the pozzolanic activity of the investigated filler is in good range.

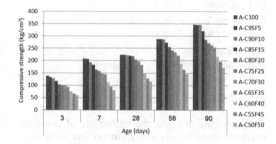

Figure 1. Comparison of compressive strength of specimens at different ages with different mixing at 20°C.

REFERENCES

Hale, W.M., Freyne, S.F., Bush, T.D. & Russell, B.W. (2008) Properties of concrete mixtures containing slag cement and fly ash for use in transportation structures. *Construction and Building Materials*, 22, 1990–2000.

Kartini, K., Nurul Nazierah, M., Zaidahtulakmal, M. & Siti Aisyah, G. (2012) Effects of Silica in Rice Husk Ash (RHA) in producing High Strength Concrete. *International Journal of Engineering and Technology*, 2.

Zareei, S.A., Ameri, F., Dorostkar, F. & Ahmadi, M. (2017) Rice husk ash as a partial replacement of cement in high strength concrete containing micro silica: Evaluating durability and mechanical properties. *Case Studies in Construction Materials*.

Effect of initial curing temperature on mechanical strength of concrete

M. Kioumarsi
Associate Professor, Department of Civil Engineering and Energy Technology, OsloMet – Oslo Metropolitan University, Norway

H. Vafaeinejad
Graduated Master Student in Civil-Structural Engineering, Alaodoleh Semnani Education Institute of Semnan, Iran

ABSTRACT

One of the importance factor, which can increase the concrete quality, is curing temperature. The curing temperature plays an important role in hydration and formation of the structure of a major part of the properties of Portland cement and cement products (Lothenbach et al., 2007, Lothenbach et al., 2008, Matschei and Glasser, 2010). The increment of the hydration temperature enhances the initial compressive strength; however, in the case of strength at higher ages, this temperature increment reduces the compressive strength compared to specimens kept at room temperature (Wang and Liu, 2011). To date, several studies have done to evaluate the effect of curing conditions on concrete behavior. According to author's knowledge, there is no work to find the optimum curing temperature in respect to different mechanical properties of concrete. This paper seeks to evaluate and compare different mechanical strengths of concrete with regard to curing temperature variations to obtain the optimum curing temperature for of each mechanical strength of concrete. To this end, the compressive, flexural and tensile strengths of the concrete specimens constructed at the ages of 3, 7, 28 and 90 days were investigated by changing the drying temperature of the concrete. In order to assess the effect of curing temperature on mechanical behaviors of concrete, concrete specimens using materials from concrete factory of Semnan municipality construction organization were tested.

According to the results, increasing in curing temperature enhances the compressive strength of 3 days specimens. Whereas, this increase in specimens at 7, 27, 56 days continues until the optimum curing temperature (15°C). The optimum curing temperature for the Brazilian tensile strength is obtained about 35°C. In flexural strength, alike compressive strength, the increase in curing temperature initially increased the flexural strength and then reduced the flexural strength at the age of 28 and 56 days. The optimum curing temperature for flexural strength was evaluated about 25°C.

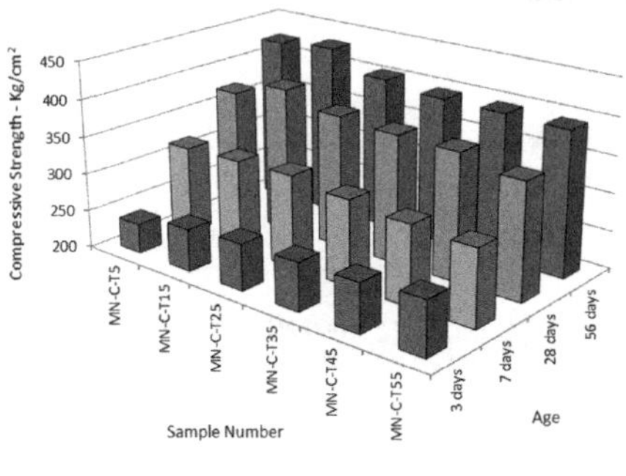

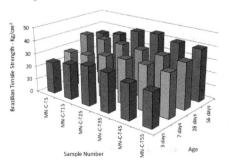

Figure 1. Comparison of compressive strength growth at different ages.

Figure 2. Comparison of tensile strength growth at different ages.

REFERENCES

Lothenbach, B., Matschei, T., Möschner, G. & Glasser, F.P. (2008) Thermodynamic modelling of the effect of temperature on the hydration and porosity of Portland cement. *Cement and Concrete Research*, 38, 1–18.

Lothenbach, B., Winnefeld, F., Alder, C., Wieland, E. & Lunk, P. (2007) Effect of temperature on the pore solution, microstructure and hydration products of Portland cement pastes. *Cement and Concrete Research*, 37, 483–491.

Matschei, T. & Glasser, F.P. (2010) Temperature dependence, 0 to 40°C, of the mineralogy of Portland cement paste in the presence of calcium carbonate. *Cement and Concrete Research*, 40, 763–777.

Wang, P.-M. & Liu, X.-P. (2011) Effect of temperature on the hydration process and strength development in blends of Portland cement and activated coal gangue or fly ash. *Journal of Zhejiang University-Science A*, 12, 162–170.

Life-Cycle Analysis and Assessment in Civil Engineering: Towards an Integrated Vision – Caspeele, Taerwe & Frangopol (Eds)
© 2019 Taylor & Francis Group, London, ISBN 978-1-138-62633-1

A sustainable approach to the design of reinforced concrete beams

A.P. Fantilli, F. Tondolo & B. Chiaia
Politecnico di Torino, Torino, Italy

G. Habert
ETH Zurich, Zurich, Switzerland

Reinforced concrete (RC) structures are generally designed to satisfy the ultimate and the serviceability limit states. In the specific case of RC beams, the current building codes only indicate how to comply with both the bearing capacity and the limitation of deflection, without considering the reduction of the environmental impact.

Conversely, in this paper, sustainability is taken into account in addition to the traditional limit states used to design beams in bending. As more eco-friendly cement-based mixtures can be attained by optimizing the performances and applying material substitution strategies, concretes made with fly ash are investigated herein. According to the Eurocode 2 requirements, a new procedure is proposed to design RC beams made with this kind of concrete.

As Fig. 1 shows, for beams attaining the same structural requirements, in absence of cement substitution (i.e., $S = 0$), the environmental impact BI of a beam increases with the initial strength f_{cA} of concrete. Nevertheless, the relative minimum of the curve BI-S moves towards higher S. Thus, the minim value of BI decreases when f_{cA} increases. It seems more convenient to use high strength concrete systems (i.e., with the highest f_{cA}), but with the maximum substitution rate of cement with fly ash.

In other words, when S is low, although the beam can be cast with a low amount of concrete (and steel, as well), the impact is high due to the high content of cement. Whereas, when S is high, the impact increases despite the low amount of cement (and low concrete strength), because large amounts of concrete and steel are needed.

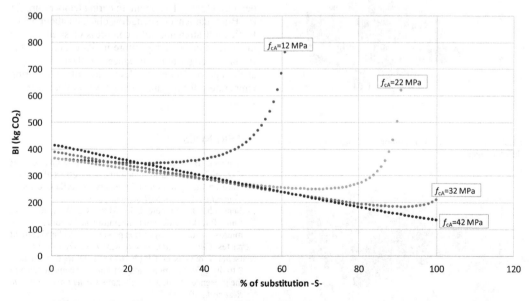

Figure 1. The impact of a reinforced concrete beam made with different concretes (having different f_{cA}).

MS7: Life-cycle engineering for hydraulic structures, levees, and other water related infrastructure
Organizers: F. den Heijer, J. Wessels, H. Yokota & M. Hoffmann

Life-Cycle Analysis and Assessment in Civil Engineering: Towards an Integrated Vision – Caspeele, Taerwe & Frangopol (Eds)
© 2019 Taylor & Francis Group, London, ISBN 978-1-138-62633-1

Asset management maturity for flood protection infrastructure: A baseline across the North Sea region

B. Gersonius & R. Ashley
IHE-Delft, Delft, The Netherlands

F. den Heijer & W.J. Klerk
Deltares, Delft, The Netherlands

P. Sayers
Sayers and Partners, Watlington, UK
Environmental Change Institute, University of Oxford, UK

J. Rijke
HAN University of Applied Sciences, Arnhem, The Netherlands

INTRODUCTION

North Sea Region (NSR) countries depend heavily on flood protection infrastructure, such as dykes, dams, sluices and flood gates. Many of these structures will reach their end-of-life between 2020 and 2050. Currently in all NSR countries, policies aim to maintain, renovate and adapt the existing infrastructure. This has lead to large-scale investment programmes, such as the UK flood and coastal erosion risk management investment programme and the Dutch flood protection programme.

Knowledge on where, when and how much to invest is of crucial importance for asset owners. This requires innovative approaches, such as using Life-Cycle-Costing to substantiate choices between maintenance, renovation and adaptation. It also requires looking for opportunities to connect with other complementary investments, such as for transportation, recreation and ecosystem restoration. The objective of the Interreg NSR FAIR project is to demonstrate improved approaches for investment planning. Initially, the project will analyse approaches in use across NSR countries, indicating the differences, advantages and disadvantages.

METHOD

From a desk study we have developed a comprehensive framework for asset management of flood protection infrastructure. The framework is comprehensive in that it incoporates and connects maintenance investment planning with strategic investment planning. We used the framework to analyse the maturity levels of asset management of various asset owners. The analysis was carried out in collaboration with the FAIR partners that own or operate assets, plan investments and set policies (e.g. RWS, LSBG).

RESULTS AND CONCLUSION

The result of the analysis is the baseline level of maturity for each NSR countries in terms of asset management approaches. These baseline levels will be used to identify the optimisation potential in the current approaches used by the asset owners, together with the preceived barriers for improved asset management. It is concluded from the analysis that there is ample potential in further mainstreaming asset management at the object-level and the network-level. Furthermore, we have observed that there is a clear difference in approach per country, which means that in the Interreg NSR FAIR project there is opportunity to share knowledge in a meaningful way, so as to enhance practice in the various NSR countries.

*Life-Cycle Analysis and Assessment in Civil Engineering: Towards an
Integrated Vision – Caspeele, Taerwe & Frangopol (Eds)
© 2019 Taylor & Francis Group, London, ISBN 978-1-138-62633-1*

Asset management of water and sewer networks, and levees: Recent approaches and current considerations

C. Curt & R. Tourment
Irstea, UR RECOVER, Aix en Provence Cedex, France

Y. Le Gat
Irstea, UR ETBX, Cestas, France

C. Werey
Irstea, UMR GESTE, Strasbourg, France

ABSTRACT

Infrastructure Asset Management (IAM) ensures that infrastructures deliver a certain level of service in a cost effective manner in both the short and long terms, with the service function depending on the infrastructure considered. This calls for integrated, sustainable, efficient rules and strategies to mitigate the consequences of ageing and possible obsolescence and which include life cycle approaches. IAM has advanced significantly since the beginning of the 1990s and is the subject of reference handbooks and European and national projects. Irstea has developed, or participated in the development, of various IAM methods and tools for different types of infrastructures (i.e. water and sewer networks, levees, etc.). Some of them are described in this paper:

– Analysis of flood protection systems based on functional and failure analyses; the scenarios identified can be analyzed in a probabilistic way;
– Approaches to assess performance: index based methods to assess levee structural performance (Tourment et al., 2012); failure prediction and impact valuation tools (Large et al., 2015), water network hydraulics and the analysis of water quality; long term analysis of deterioration;
– Multi-criteria methods: to help short-term prioritization at the pipe level (sewer); analysis of the positive and negative impacts of best practices (waste and rain water management);
– Approaches using both physical and monetary criteria: assessing the impacts of failures and those due to works (renewal or trenchless rehabilitation); cost benefit analysis for the different infrastructures (Werey et al., 2015);
– A Spatial Information-Reference System (SIRS Digues) that allows storing all levee systems data including their dating in GIS based software.

The article also aims at emphasizing the position of infrastructures with respect to LCM (Life Cycle Management):

– All the approaches developed place performance and risk at the heart of development to guarantee the performance of the service function in a cost effective manner;
– LCM of the infrastructures studied relies on cyclic approaches with account taken of short, medium and long-term temporalities;
– The IAM process involves three main dimensions whatever the infrastructure: operational, informational and governance issues;
– IAM performance depends on available, reliable, pertinent data and information throughout the lifecycle. The "informational capital" should be studied in close connection with the physical capital;
– IAM should explicitly take into account constraints and opportunities related to urban management and coordination with the renovation works of other infrastructures and urban planning decisions;
– A "single work/system" management rationale can be defined for levee, *vs.* a "pooled work" one in the case of meshed water networks. Sewer pipes to some extent have an intermediate status as inspection is possible but not performed on whole networks.

Finally, the desire to integrate sustainability (social, environmental and economic) and multi-functionalities (recreational activities on levees for instance) aspects in IAM is growing rapidly and globally. Encouraging results have been produced by recent studies on the topic.

REFERENCES

Large, A., Le Gat, Y., Renaud, E., Tomasian, M., Elachachi, S.M. & Breysse, D. (2015) Decision support tools: Review of risk models in drinking water network asset management. *Water Utility Journal*, 10, 45–53.

Tourment, R., Peyras, L., Vuillet, M., De Massiac, J. C., Allouche, A., Nicolas, L., Casteigts, C. & Delaunay, C. (2012) Digsure method: Decision support indicators and GIS tool for levees management. In: *2nd Eur. Conf. on Flood Risk Management FLOOD Risk 2012*, Rotterdam, NLD.

Werey, C., Rulleau, B. & Rozan, A. (2015) Valuation of social and environmental externalities from sewer networks: experiences and perspectives for asset management. *IWA LESAM Conference*, Yokohama, Japan.

Life-Cycle Analysis and Assessment in Civil Engineering: Towards an Integrated Vision – Caspeele, Taerwe & Frangopol (Eds)
© 2019 Taylor & Francis Group, London, ISBN 978-1-138-62633-1

Working towards one and the same objective: How serious gaming makes storm surge barriers more reliable

M. Schelland
TNO, Delft, The Netherlands

J.N. Huibregtse & M. Walraven
Rijkswaterstaat, Rotterdam, The Netherlands

E.C.J. Bouwman
Delta-Pi, Vught, The Netherlands

ABSTRACT

Rijkswaterstaat is the executive agency of the Ministry of Infrastructure and the Environment dedicated to promoting safety, mobility and the quality of life in the Netherlands.

Changes in policy, organisation and the relation to market, and the resignation of knowledge holders have put the retention of knowledge under pressure in the past years. For crucial objects such as the storm surge barriers, a loss of knowledge is not acceptable and is hence addressed in multiple ways.

This paper describes one instrument among others to help both employees and managers to structurally anchor the crucial knowledge in a sustainable manner: the workshop "Failure probability and PDCA". This workshop is developed for all teams who work on the management and maintenance of the Rijkswaterstaat storm surge barriers.

The Rijkswaterstaat storm surge barriers are managed and maintained in a risk-based way to achieve the joint objective, that is, reliable barriers. The principle here is that those actions which contribute the most to the barrier's performance (in terms of failure probability) and thus to the protection of the hinterland are carried out. The involved teams are organised according to a Plan-Do-Check-Act (PDCA) cycle. In practice the management and maintenance steps lie with various people and, in many cases, even with different parts of the organisation. This, in combination with an unruly reality, makes the work dynamic and complex. There is thus always the risk that the joint objective is lost out of sight. Besides all the necessary technically relevant expertise, soft-skills concerning team work and an eye for the role and position of the other are crucial elements in this cooperation.

Cooperating for the joint objective of reliable storm surge barriers is the point of focus of the workshop. All participants, who work on all levels from very detailed technical level to higher management level, are challenged to recognise their own contribution towards this joint objective.

In the workshop, we strengthen the cooperation between participants by:

- providing participants with insight into the importance and set-up of the failure probability performance and into the connected PDCA cycle;
- challenging them to recognise their own role in the PDCA cycle and those of their colleagues in this;
- offering a varied program in which technical content, cooperation, much interaction, creativity, fun and attention for each other are combined.

The workshop has had a positive effect on the participants' understanding of the processes in which they work and their joint objectives. The concept "failure probability" has become tangible and participants can more easily make the connection between their own work, the failure probability as an instrument and the reliability of the storm surge barrier. Participants know better from each other what is needed for a smooth cooperation and meet each other earlier in the process to coordinate. The enthusiasm for the workshop and in particular the game "Flood barrier in control?!" is, two years after the first time it was played, still very strong.

It has led to good results to pay attention to not only technical content but also explicitly to the processes shaping the way of working, to the mutual stresses and differences and to the joint objective. The playful forms which have been chosen have inspired the participants in new ways.

At the IALCCE 2018 conference, the game "Flood barrier in control?!" is available to be played by the conference visitors. Are you also coming to maintain your asset? We wish you much fun and inspiration!

Life-Cycle Analysis and Assessment in Civil Engineering: Towards an Integrated Vision – Caspeele, Taerwe & Frangopol (Eds)
© 2019 Taylor & Francis Group, London, ISBN 978-1-138-62633-1

A framework for assessing information quality in asset management of flood defences

W.J. Klerk
Deltares, Delft, The Netherlands
Department of Hydraulic Engineering, Delft University of Technology, Delft, The Netherlands

R. Pot
Fugro N.V., Leidschendam, The Netherlands

J.M. van der Hammen
Nelen & Schuurmans, Utrecht, The Netherlands

K. Wojciechowska
Deltares, Delft, The Netherlands

ABSTRACT

For asset management of flood defences a pivotal aspect is the quality and accessibility of available data and information. In the management of flood defences in The Netherlands, safety assessments are conducted every 12 years. To assure the quality and consistency of these safety assessments, flood defence engineers have to make use of a comprehensive toolbox and guidelines (WBI 2017). The safety assessment is one of the main starting points for the different asset management processes such as day-to-day maintenance and reinforcement-/reconstruction. However, there is no clear method available to assess whether the quality of underlying information is sufficient as a basis for decision making.

In this study a framework has been developed that consists of methods for an assessment of the quality of information, an assessment of the use and accessibility of information, and a rational framework for assessing the costs and benefits of obtaining additional information. Whereas many frameworks for information management start from the information itself, here the starting point is the extent to which the behavior of a flood defence is understood, in relation to the decision or management process considered. The Data-Information-Knowledge-Wisdom pyramid (Rowley, 2007) is used to derive different score categories for information quality as shown in Table 1. Here every flood defence can be rated on a 1-5 star scale indicating the level of information quality. A similar table was developed for assessing the use of information, mainly focusing on sharing and accessibility of information, as this is often a major problem in day-to-day practice. Additionally, to aid improvements in information quality a framework for Value of Information analysis is provided. This gives both simple and advanced guidelines for assessing the potential value of new information.

Table 1. Score performance card for information quality with descriptions for every score.

Score	Description
* *Insufficient data available*	The quality and usability of data is not assessed, insufficient and not validated.
** *Validated data available*	Quality of the available data is known, described and validated with different independent sources.
*** *Insight in flood defence behaviour*	Based on the information and knowledge of the flood defence a sensitivity analysis has been carried out that has revealed the major uncertainties.
**** *Capable of informed decision making*	The relevant uncertainties for decision making have been reduced and do not (significantly) influence the decisions to be taken.
***** *Optimal insight*	Based on a Cost-Benefit analysis of the different options for additional data collection the economically optimal insight into dike strength has been obtained.

The overall framework has been applied to a set of case studies where it was found applicable. Major finding was that the level of effort needed to carry out the analysis differs per case. For instance, in the case of dunes the adequateness of available information was quite clear, whereas for a regional flood defence a more thorough analysis was needed.

As the nature of the star rating framework is relatively qualitative, it is also of importance to assess whether it provides stable results. In order to verify this, the framework was tested in four workshops with in total nearly 100 participants, both individually and in groups. It was found that the framework provides stable results in general and its performance is considerably better when the analysis is carried out in groups rather than individually.

REFERENCE

Rowley, J. (2007) The wisdom hierarchy: Representations of the DIKW hierarchy. *Journal of Information Science*, 33, 163–180. doi:10.1177/0165551506070706

Fundamental study on effective utilization of caisson in breakwater

E. Kato, Y. Kawabata & K. Uno
Port and Airport Research Institute, MPAT, Yokosuka, Japan

H. Yokota
Hokkaido University, Sapporo, Japan

ABSTRACT

To control costs to maintain and update port and harbor structures under the condition of tight budget grounds, effective and appropriate utilization of existing stock is required in Japan in recent years. Following the performance-based design currently adopted in the technical standards for port and harbor facilities in Japan, when extend service period of an existing structure or convert it to another use, it must be appropriately confirmed that the performance of that structure and its individual structural members satisfy the required performances through its newly design service period MLIT et al. (2018).

This study investigated the results obtained by applying the current performance criteria to caissons in five breakwater facilities (A-E), designed based on previous technical standards. Caissons of course vary in terms of their shapes, dimensions, and expected wave conditions. However, the result of investigation indicates that when current design criteria are applied to caissons designed using previous methods, some or all of their members (i.e., bottom slabs, footings, outer walls, and partitions walls, see Figure 1) may fail to meet the current required performance. In addition, it was demonstrated that caisson members besides partition walls are especially likely to fail to meet serviceability requirements.

Based on the result of investigation and the present maintenance techniques, issues related to performance evaluation of caisson members, repair and reinforcement of a caisson, and ensuring performance throughout the design service period were discussed. Discussion follows the simple decision flow diagram shown in Figure 2 for making usability determinations for existing structural members Takano et al. (2017).

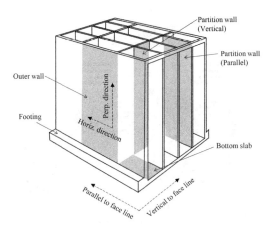

Figure 1. Caisson members and direction for design.

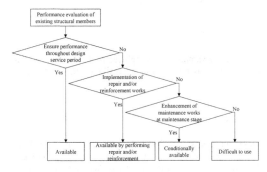

Figure 2. Decision flow diagram for utilization of existing structural members (Takano et al., 2017).

REFERENCES

MLIT, et al. (2018). *Technical Standards and commentaries for Port and Harbour Facilities in Japan.* (in Japanese).

Takano, et al. (2017) Study on considerations in design work for reformation of existing port facilities protective facilities and mooring facilities in port areas. *Technical notes of National Institute for Land and Infrastructure Management*, 944 (in Japanese).

Efficient investments in the waterway Danube with vessel traffic analysis and fairway optimization

M. Hoffmann & A. Haberl
Institute of Transportation, Vienna University of Technology, Austria

C. Konzel, T. Hartl, S. Simon & M. Simoner
viadonau – Austrian Waterway Company, Austria

ABSTRACT

Waterways are part of the trans-European transport network allowing a cost-efficient and environmentally friendly transport of passengers and goods. For managing the waterway Danube providing a sufficient level of reliability and availability a waterway asset management system (WAMS) has been established in Austria since 2015. Currently this software tool contains all necessary information on physical fairway availability providing viadonau with the means for an efficient planning and implementation of measures from dredging to fairway alignment. The current research project aims at implementing actual vessel traffic information and fairway utilization in the planning process. To achieve this goal anonymized transponder data with vessel trajectories, vessel type and draught loads are collected automatically in the WAMS – database. With a grid-based counting algorithm traffic flow and density are analyzed for one and two-way traffic during high and low water periods. The WAMS – tool is capable of generating heat maps and cross-sectional traffic distribution charts allowing for a comparison between provided fairway path and availability with actual utilization. Larger counting cells around ports and berths enable further analysis regarding site importance and length of stay. Further developments will also include a search option for optimal fairway alignment based on curve radii and fairway depth. This functionality will also be a good basis for a future route planner for an optimization of vessel loading and estimation of expected transport costs. The paper gives an insight into the WAMS system with focus on methodical aspects and actual findings of vessel traffic analysis as basis for a customer-oriented fairway optimization and an efficient use of funds. After implementation and testing of these new functionalities selected features and automatically updated results will be made available for the public in River Information Systems (RIS), navigational charts and notes to skippers.

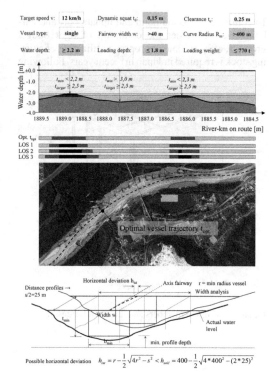

Figure 1. Example for a simplified route search algorithm on the waterway for maximizing available fairway depth.

REFERENCES

Haselbauer, K. (2016) *Transport Infrastructure Asset Management*. PhD-Thesis, Vienna University of Technology.

Hoffmann, M. & Hartl, T. (2016) Actual developments, treatment options and future waterway asset management implementation for the waterway Danube. In: *Peer reviewed proceedings IALCCE2016*, Delft.

Risk based inspection of flood defence dams: An application to grass revetments

W.J. Klerk
Deltares, Delft, The Netherlands
Department of Hydraulic Engineering, Delft University of Technology, Delft, The Netherlands

K.L. Roscoe & A. Tijssen
Deltares, Delft, The Netherlands

R.P. Nicolai
HKV Consultants, Lelystad, The Netherlands

J. Sap
Witteveen+Bos Consultants, Rotterdam, The Netherlands

F. Schins
Rijkswaterstaat Zee & Delta, Goes, The Netherlands

ABSTRACT

In The Netherlands, inspection and maintenance are essential for maintaining stringent flood protection standards. Flood defences are assessed every twelve years to ensure they meet their risk-based safety standards, which are given as legally-binding maximum failure probabilities. In between assessments, flood defence managers are subject to risk-maintenance requirements: they must ensure that the failure probability of the defence does not increase above its safety standard. However, there is no prescribed methodology to meet this requirement.

In this study we developed a method which enables flood defence managers to derive inspection strategies using visual inspection data that will allow them to meet their risk-maintenance requirements. Most of the inspections in the Netherlands are done based on visual inspection using the Digigids system, a database of reference pictures that can be used to determine a qualitative score (poor, fair, decent or good) for a flood defence. By connecting these scores to failure models it is possible to translate visual inspections to a failure probability and decide whether maintenance interventions are necessary.

We applied the method to the assessment of grass revetments on the outer slope of the Oesterdam, one of the dams of the Dutch Delta Works. We considered damage to the grass revetment due to the burrowing of small animals. We related each Digigids score to a category of the grass quality (closed, open, or fragmented sods), which served as input to a failure model. The failure probabilities for both closed and open sods were relatively small, whereas fragmented sod results in extremely large failure probabilities.

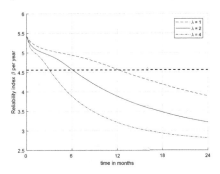

Figure 1. Failure probability in time for three degradation rates, related to the safety standard (horizontal line).

Therefore an inspection scheme should aim to prevent fragmentation of the grass.

For the deterioration of the grass quality (i.e. rate of holes over time) we assumed a homogeneous Poisson process. The rate of this process – the degradation rate – is of critical importance; however, there is no available degradation data with which to estimate it. We investigated the sensitivity of the inspection frequency by considering multiple degradation rates (see Figure 1), and found that the inspection frequency was highly sensitive to the choice of rate.

The application presented in this paper illustrates how inspection intervals can be derived to ensure the risk-maintenance requirements are met. However, a lack of data on degradation rates introduces substantial uncertainty in the required inspection frequency. Future efforts should focus on recording degradation observations. In the meantime, conservative values of the rate can be used, and be refined annually as degradation data is collected.

Comparative analysis of the reliability levels in hydraulic structures using partial safety factors and full probabilistic methods

A. Tahir & C. Kunz
Bundesanstalt für Wasserbau (BAW), Karlsruhe, Germany
(Federal Waterways Engineering and Research Institute)

ABSTRACT

Currently the design of new and the analysis of existing structures are conducted through partial safety factors derived from semi probabilistic methods (BAW 2016). To assure the safety throughout the life cycle of the structure these methods and safety factors are calibrated for a reliability index β of 3.8. Currently the calibration in Eurocode considers basic component (beams, columns etc.) and common structural arrangements such as buildings. Whereas hydraulic structures differ in several aspects including loads, their distribution, environmental conditions and geometrical requirements, therefore achieving the same safety levels need to be investigated. To assess the reliability level for hydraulic structure an exemplary reinforced concrete ship lock structure is selected along with its respective loading cases and limit state functions (bending moments with normal forces and shear) indicated in Figure 1.

A semi probabilistic design was conducted with partial safety factors from Eurocodes and German codes (DIN 19702 2013; DIN EN 1992-1 2010) which corresponds to a target reliability β of 3.8. Thereafter the calculated area of steel was used as an input in full probabilistic reliability analysis considering the same failure mechanism and loading cases. Full probabilistic reliability analysis was performed using First Order Reliability Method (FORM) (Rackwitz and Fiessler 1978) which provided reliability index β and sensitivity analysis for each load case and parameter. The results for each load case and limit state function are provided in the Table 1.

For the investigated ship lock comparative analysis of results indicated that the existing safety factors lead to a slight over design of the component with dependence on design conditions and load type. Coefficient of earth pressure, model uncertainties and water level are most sensitive parameter in case of the reinforced concrete ship lock structure. The presented methodology can be adapted for existing structures with inclusion of field data and verification of safety factors for hydraulic structures.

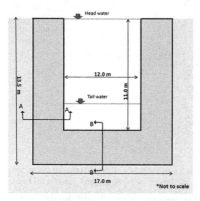

Figure 1. Generalized geometry of half frame ship.

Table 1. Reliability index (β) for load cases and limit states.

Section A-A Reliability Index β		Cases (figure 1)	
		water in chamber	No water in chamber
Moment	Waterside	5.00	–
	Earthside	4.47	2.98
Shear	–	4.56	3.07

Section B-B		Cases (figure 1)	
		water in chamber	No water in chamber
Moment	Waterside	3.97	–
	Earthside	4.07	5.26

REFERENCES

BAW (2016) *BAWMerkblatt: Bewertung der Tragfähigkeit bestehender, massiver Wasserbauwerke (TbW)*, Bundesanstalt für Wasserbau (BAW), Abteilung Bautechnik, Karlsruhe.

DIN 19702 (2013) Massivbauwerke im Wasserbau – Tragfähigkeit, Gebrauchstauglichkeit und Dauerhaftigkeit. *DIN 19702*, Beuth Verlag, Berlin.

DIN EN 1992-1 (2010) Eurocode 2: Bemessung und Konstruktion von Stahlbeton- und Spannbetontragwerken. *DIN EN 1992-1-1*, Beuth Verlag, Berlin.

Rackwitz, R., & Fiessler, B. (1978) Structural reliability under combined random load sequences. *Computers & Structures*, 489–494. doi:10.1016/0045-7949(78)90046-9.

A maintenance management plan for port mooring facilities based on cost-benefit analysis—a case study

Y. Kawabata, E. Kato & Y. Tanaka
Port and Airport Research Institute, Yokosuka, Japan

H. Yokota
Hokkaido University, Sapporo, Japan

ABSTRACT

Port mooring facilities play a key role in cargo handling and are thus effectively maintained over their service lives to avoid breakdowns that can negatively influence port activities. In Japan, the documentation of a global maintenance and management plan of a port as well as a single maintenance management plan for the facility is required. The superposition of the maintenance strategies of several facilities, on the contrary, is not necessarily the best solution for the global maintenance strategy of a port. Especially in the context of a limited budget for maintenance, having a single maintenance management plan for each facility does not necessarily produce a satisfactory result for global port maintenance management. When the deterioration is assessed to be severe, the operation of the facility may be terminated, resulting in loss of benefit. Berth dues as a benefit are crucial for the operation as well as the management of the mooring facilities of a port so that cost-benefit analysis with constrained conditions should be considered.

This paper presents a practical decision-making framework for the global maintenance management planning of port mooring facilities, using the net present value (NPV) as an indicator of a cost-benefit analysis (Yokota et al. 2017). A case study analysis is performed using a real port with five mooring facilities. The extended service life and the annual budget limit are set as constrained conditions. Combining the strategies of each facility, the best global maintenance scenario to minimize the LCC or maximize the NPV of the port is sought with the constrained conditions.

As a result, the NPV is found to be a promising indicator that leads the facilities toward preventive maintenance while the optimization of only the LCC tends to lead the maintenance strategy of the facility toward corrective maintenance to reduce cost which fails in NPV optimization. The NPV is also robust against the extended service life of the facility and the annual budget limit. Since these factors are difficult to be determined, the practical decision-making framework presented in this study is beneficial for documenting the global maintenance and management plan of ports.

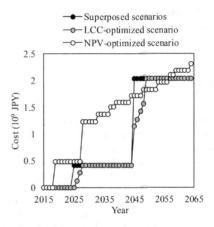

Figure 1. LCC of superposed maintenance scenarios & LCC/NPV-optimized global scenario (Annual budget limit = 0.9 billion JPY).

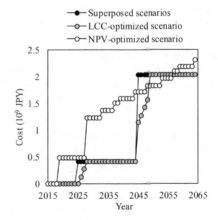

Figure 2. Effect of the annual budget limit on NPV.

REFERENCE

Yokota, H., Hashimoto, K. & Tani, T. (2017) Sensitivity analysis on repair prioritization for mooring facilities. In: *Life-Cycle of Engineering Systems: Emphasis on Sustainable Civil Infrastructures*, Bakker, Frangopol & van Breugel (Eds.), pp. 567–574.

Life-Cycle Analysis and Assessment in Civil Engineering: Towards an Integrated Vision – Caspeele, Taerwe & Frangopol (Eds)
© 2019 Taylor & Francis Group, London, ISBN 978-1-138-62633-1

Sustainable and future-proof port infrastructure

J.G. de Gijt
Delft University of Technology, Delft, The Netherlands

E.J. Broos & C. Bosschieter
Port of Rotterdam Authority, Rotterdam, The Netherlands

P. Taneja
Delft University of Technology, Delft, The Netherlands

H.E. Pacejka
Gemeente Rotterdam, Rotterdam, The Netherlands

ABSTRACT

Ports facilitate the domestic and international trade of goods and development of ports is vital for a nation's economy. However, ports today are facing a multitude of challenges related to ship sizes and economics of scale, competition with other ports and uncertainty in cargo volumes, and safety and environmental issues. Physical port infrastructure such as quay walls and jetties require large investments and therefore should be future-proof, i.e. they should be functional over a long-term future and they should be sustainable, integrating, ecological and economic principles in all aspects including development, operations and management.

This paper discusses the following key issues, citing numerous examples from the port of Rotterdam:

– current and future global developments that can have impacts on port development
– evolution of ship sizes in relation to dimensions of port infrastructure, with a focus on container ships
– developments in quay wall design and construction including new material applications and construction methods
– life cycle analysis of port infrastructure
– environmental issues related to ports

This paper concludes by presenting some ideas over what the port of the future will look like, and mentions some likely developments related to design and construction of future port infrastructure. Recommendation over further studies will also be given.

REFERENCES

de Gijt J.G. & Ham van N.A.V. (1996) Results of measurements of container crane loads versus design loads at the Amazone harbour quay wall of the port of Rotterdam. In: *International Harbour Congress*, Antwerp, 1996.

de Gijt J.G. (2004) Measurements and FEM-calculations for quay walls in Rotterdam. Kaimauer Sprechtag TUHH, Hamburg, 2004.

de Gijt, J.G. (2010) A History of Quay Walls, Past, Present and Future, PhD, Delft University of Technology.

de Gijt, J.G., Louwen, A.P. & Voogt, H. (2016) Implementation of life cycle costing in port infrastructure. In: *IILS Conference*, Delft.

Luijten, C.J.M., Andriessen, C.A., de Gijt J.G., Zwakhals, W., van Ewijk, F. & Broos, E.J. (2010) Preliminary results of research into carbon foot print of port infrastructure. In: *Port of Rotterdam, Port Infrastructure Conference*, Delft.

Luijten, C.J.L., de Gijt, J.G., Said, M. & Bouwheer, C.H.J. (2016). From LCA to LCC I infrastructure reducing CO_2 emissions in infrastructure. *LCC Conference*, Delft.

Maas, T., de Gijt, J.G. & van Heel, D.D. (2011) Comparison of quay wall design in concrete, wood, steel and composites with regard to CO_2-emission and the life cycle analysis. *MTEC*, Singapore.

McKinsey, How Container Ships Could Reinvent Itself for the Digital Age. https//www.mckinsey.com/industries/traveltransport and logistics.

van der Valk, R., de Gijt, J.G., Jonkers, H.M., Pacejka, H.E. & Vellinga, T. (2017) *A Fibre Reinforced Polymer Quay Wall Feasibility Study*, MSc Thesis, TU Delft.

van Heel, D.D., Maas, T., de Gijt, J.G. & Said, M. (2011) Comparison of infrastructure designs for quay wall and small bridges in concrete, steel, wood and composites with regard to the CO_2 emission and the life cycle analysis. *MTEC*, Singapore.

Winter, R., de Gijt, J.G., Jonkers, H.M., Pacejka, H.E., Kassapoglou, C. & Jonkman, S.N. (2017) *Feasibility Study: FRP Jetties: Investigating the Technical- and Economic Feasibility and Sustainability Aspects of Fiber-reinforced Plastic Jetties*, MSc, TU Delft.

Robustness as a decision criterion for construction and evaluation of ship lock chambers

C. Kunz
Bundesanstalt für Wasserbau, Karlsruhe (BAW), Germany
(Federal Waterways Engineering and Research Institute)

ABSTRACT

For the evaluation of new and existing hydraulic structures, i.e. ship locks, considering robustness became a meaningful tool for decisions. In present codes robustness is considered either by strategy recommendations or by construction rules and therefore only qualitative. Based on a reference with quantitative criteria worked out for bridges (Pötzl 1996) an approach for a robustness evaluation for ship lock chambers has been undertaken. In case of erection or replacement of ship locks two main types of construction for the lock chamber are suited: the reinforced concrete U-frame and the sheet pile wall structure with concrete base slab, Figures 1 and 2.

The heads are constructed in traditional way as a massive concrete structure. Both construction types have different shapes, cubing, safety fulfillments, construction costs and maintenance aspects. Eight Robustness criteria named by ROB_i ("ROBies") where the "i" numerates different criteria have been adapted respectively modified. They represent redundancy, ductility, monolithic construction, deformation capability, compactness, replaceability, adaptability and non-error-proneness and are determined with numerical values. The ROBies themselves have no physical properties but they are based on them. ROB is a fraction where a larger value means a better robustness.

ROBies are always relative and not absolute criteria. ROBies cover bearing capacity, serviceability, durability and range over lifecycle. The result shows clearly a better robustness for the RC U-frame type, Table 1.

Moreover, ROBies are suitable when judging existing hydraulic structures so as to support decision for the remaining lifetime and maintenance aspects.

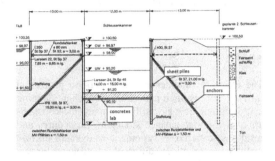

Figure 2. Cross section of a sheet-pile ship lock wall.

Table 1. ROBies for reinforced concrete U-frame and sheet pile lock chambers.

ROB_i	Description	RC U-frame lock chamber	Sheet pile wall lock chamber
ROB_1	Redundancy	5.94	3.00
ROB_2	Ductility	1.0	28.6
ROB_3	Monolithic	0.05	0.0005
ROB_4	Deformation cap.	2.1	1.4
ROB_5	Compactness	1.08	0.30
ROB_6	Replaceability	231.0	131.0
ROB_8	Non-error-pron.	43.1	23.5

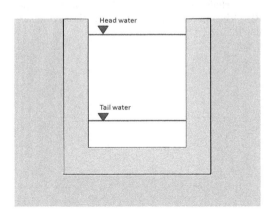

Figure 1. Cross section of a reinforced concrete U-frame ship lock chamber.

REFERENCE

Pötzl, M. (1996) *Robuste Brücken – Vorschläge zur Erhöhung der ganzheitlichen Qualität.* Vieweg Verlag, Braunschweig.

Life-Cycle Analysis and Assessment in Civil Engineering: Towards an Integrated Vision – Caspeele, Taerwe & Frangopol (Eds)
© 2019 Taylor & Francis Group, London, ISBN 978-1-138-62633-1

Improvement of the planning process of flood protection assets by using experiences with operation and maintenance—Hamburg case study

P. Fröhle, N. Manojlovic, S. Shaik, P. Jordan & L. Seumenicht
Institute of River and Coastal Engineering, Hamburg University of Technology, Hamburg, Germany

M. Schaper, J.-C. Schmidt & M. Roth
Hamburg State Agency of Streets, Bridges and Water (LSBG), Hamburg, Germany

ABSTRACT

The effort for maintenance of the flood protection infrastructure is often underestimated when planning and designing new assets. In Germany, the consideration of the maintenance strategies, procedures and results is scattered. A systematic analysis of current maintenance practice in different federal states for various types of structures is required to comprehensively understand the process and assess the possible benefits for their design and planning in the future. This paper introduces a methodology for a systematic analysis of maintenance strategies. A catalogue of criteria is developed that describes the maintenance process from the a strategic as well as the operational perspective, assessing the organisational aspects (such as the relevant legislative frameworks and regulations, responsibilities or ownership), the conflict potential (within the institutions or in general) or describing the assets and the corresponding maintenance procedure in details.

The results of the analysis are used to identify the room for improvement of the maintenance process but also to assess the potential transferability of the experiences from the maintenance to other aspects of the asset management, in particular for planning and (re)design of the new assets or upgrading the existing ones.

The methodology has been applied to analyse the flood protection assets in the Free and Hanseatic City of Hamburg in Germany. The focus has been put on the flood protection gates, as they are considered to be the most sensitive parts in the flood protection line. A representative gate "Große Elbstraße", designed as a multiple level safety system has been selected for a detailed analysis.

The preliminary analysis could already identify a number of possible improvements towards optimised adaptive planning, design and upgrade of the flood protection assets. The main need for improvement is seen in the following aspects:

– documentation of the maintenance procedures shifting it from the paper-based to a web based system,
– more intensive and purposeful interactions with the other departments and units (planning, operation, IT) or cross-institutionally,
– working more closely with the planning department especially when designing and upgrading the complex systems and assets such as the multiple safety gates.

The study has been conducted a part of the INTERREG Vb project FAIR (http://www.northsearegion.eu/fair/).

Life-Cycle Analysis and Assessment in Civil Engineering: Towards an Integrated Vision – Caspeele, Taerwe & Frangopol (Eds)
© 2019 Taylor & Francis Group, London, ISBN 978-1-138-62633-1

Peer reviews as an asset management tool

B. Lassing
TNO, Delft, The Netherlands

P. van Poorten
Paul van Poorten Consulting, Zwolle, The Netherlands

M. Walraven, S. van Herk & B. Vonk
Rijkswaterstaat, Utrecht, The Netherlands

R. Windsor
Environment Agency, London, UK

ABSTRACT

Hydraulic structures like storm surge barriers are essential for water safety/flood protection. Hence it is important to keep them in operation in the most efficient way during its Life Cycle. A peer review is a powerful asset management tool. It was developed within I-STORM (network of professionals on storm surge barriers), based on a tried and tested method of WANO (World Association of Nuclear Operators).

A peer review is a bespoke and subjective assessment of an asset's performance and its management organisation, based on objectively collected observed facts. It is delivered by independent peers from both inside and outside the receiving (host) organisation in cooperation with counterparts at the asset. Performance objectives and their descriptive criteria are used to analyse facts and find fundamental overall problems and gaps to excellence.

The tool has been applied to storm surge barriers:

- unique constructions which are idle most of the time, but must operate during a storm surge with an extreme high reliability;
- essential in providing water safety to people and property during their life time.

After four peer reviews on storm surge barriers the tool was evaluated using a systematic acquisition and assessment of information to provide useful feedback to the process. Three evaluation strategies:

1. Literature study/Cross-case analysis
Peer reviews can be considered as in-depth, objective reviews of the host barrier's operations by independent peers from outside the host organisations. A peer review typically resulted in:

- Around 300–400 facts and observations,
- Around three 'Fundamental Overall Problems',
- Multiple gaps to excellence, areas for improvement, strengths and best practices,
- Several recommendations and suggestions.

2. Participant-oriented evaluation method

Interviews indicated that peer reviews (can) have multiple, to some degree interrelated purposes:

- Quality check & control
- Training & education
- Networking & knowledge sharing
- Communication

3. Cost-benefit analysis
The costs of each topic executed by a peer team were compared to those of a hired consultant. In reviewing the value of the peer reviews, it is advisable to consider the following steps:

1. Determine which purposes are considered important or valuable;
2. Evaluate to what degree these purposes are currently met;
3. Assess them against the peer review.

Some building blocks are provided to do this.

Combining different purposes makes peer reviews a very effective tool for asset management and dissemination of knowledge. It can be used on all kinds of assets operated, maintained and managed by people. It is valuable for an (asset) manager who is committed to continuous improvement.

It was concluded that different types of peer reviews can be distinguished:

Type	Hydraulic Structure	Team	Duration
advanced	critical complex	international	10 days
detailed	large complex	national	4 days
simple*	small object	regional	2–5 objects/day

*quick scan

REFERENCES

Lassing, B. & Van Meerveld, H. (2015) *Peer Reviews I-STORM: Evaluation, Cost-benefit Analysis and Advice for Broader Implementation of the Peer Review Concept (Within Rijkswaterstaat)*. Delft, TNO.

Morris, E. & Van Poorten, P. (2012) *Peer Review Process Document: Collecting and Analysing Data.* London, International Network for Storm Surge Barriers (I-STORM).

Expert interviews in long-term damage analysis for bottom and bank revetments along German inland waterways

J. Sorgatz & J. Kayser
Federal Waterways Engineering and Research Institute, Karlsruhe, Germany

H. Schüttrumpf
Institute of Hydraulic Engineering and Water Resources Management, RWTH Aachen University, Aachen, Germany

ABSTRACT

In order to promote inland waterway transport in Germany, it is reasonable to provide and ensure broad navigability on the Federal waterways network even for possibly large vessels. Bank revetments at German inland waterways are mainly secured by loose or grouted armour stones on a geotextile or granular filter layer. Usually, increasing loads require a new design of the revetments. However, due to ecological and economic reasons, maintenance is favoured over new constructions.

A research project at the Federal Waterways Engineering and Research Institute (BAW) aims to establish substantial knowledge on damage processes and critical slope conditions to complement current German design standards (GBB 2010) allowing for a prediction of long-term stability and maintenance measures in regard to actual traffic and individual safety requirements.

An important aspect of the above mentioned research is damage/failure analysis which is explored by interviewing experienced engineers and technicians. Expert interviews are a scientific method aiming for qualitative data collection. In human sciences, they are a widely accepted method (Gläser and Laudel 2010; Meuser and Nagel 1991). In engineering practice nowadays, expert judgements become increasingly popular if systematic (long-term) observations are hard to obtain.

Potential life-cycle based maintenance guidelines require a damage classification system complying to observations in nature. Hence, a classification proposed by Kayser (2015) that distinguishes optically between four damage categories (see Figure 1) was discussed in the expert interviews in order to evaluate its applicability in practice.

While large-scale models provide reliable results, they imply simplifications, are expensive, time-consuming and hence can only be carried out selectively. The interviews revealed the most significant damage patterns occuring and loads bank revetments

Figure 1. Damage classification system by Kayser (2015).

are subjected to. Based on that profound knowledge on damage causes and damage patterns, a full-scale embankment model has been erected that specifically allows observing armour stone displacements.

This paper outlines a systematic procedure to conduct expert interviews as a useful tool in engineering. In the context of the presented paper, expert interviews are conducted to explore bank revetment deterioration along German inland waterways and thus to develop an integrated maintenance process. Design and realisation of expert interviews are presented using the example of loose riprap embankments. The results of the expert interviews will be evaluated and their contribution to the research project discussed. It will be explained how the knowledge gained by expert interviews was used to design the physical model.

REFERENCES

GBB (2010) *Principles for the Design of Bank and Bottom Protection for Inland Waterways.* Bundesanstalt für Wasserbau.

Gläser, J. & Laudel, G. (2010) *Experteninterviews und qualitative Inhaltsanalyse als Instrumente rekonstruierender Untersuchungen.* 4th edition. VS Verlag für Sozialwissenschaften, Wiesbaden.

Kayser, J. (2015) FuE-Abschlussbericht: Entwicklung des Zustands von Deckwerken bei Absenkung des technischen Standards. Bundesanstalt für Wasserbau.

Meuser, M. & Nagel, U. (1991) ExpertInneninterviews – vielfach erprobt, wenig bedacht: ein Beitrag zur qualitativen Methodendiskussion. In: Garz, D. and Kraimer, K. (ed.) *Qualitativ-empirische Sozialforschung: Konzepte, Methoden, Analysen, VS Verlag für Sozialwissenschaften,* Wiesbaden, pp. 441–471.

MS8: IEA EBC Annex 72: Assessing life-cycle related environmental impacts caused by buildings
Organizers: R. Frischknecht, H. Birgitsdottir, T. Lützkendorf & A. Passer

Life-Cycle Analysis and Assessment in Civil Engineering: Towards an Integrated Vision – Caspeele, Taerwe & Frangopol (Eds)
© 2019 Taylor & Francis Group, London, ISBN 978-1-138-62633-1

Potential for interconnection of tools for cost estimation and life cycle assessment of partial carbon footprint in the building sector in Czechia

A. Lupíšek & M. Nehasilová
University Centre for Energy Efficient Buildings, Czech Technical University in Prague, Czechia

J. Železná & P. Hájek
Faculty of Civil Engineering, Czech Technical University in Prague, Czechia

B. Pospíšilová & M. Hanák
ÚRS Praha, Czechia

ABSTRACT

With increasing strictness of EU regulations on operational energy efficiency in buildings the relative share of embodied environmental impacts rises. Life cycle approach becomes a necessity for building design optimization, but the existing LCA tools so far did not find way to offices of most architects and there is a need for easy to use software.

The idea examined in the paper is that the already widely adopted tools for construction budget estimation could be used also for life cycle analyses.

The objective was to evaluate the actual readiness of the construction cost estimation tools and LCA tools available in Czechia for mutual interconnection in order that construction cost estimation and cradle-to-gate environmental impact assessment can be made at once. We worked in the following steps:

– We took the major national construction cost estimation software KROS (ÚRS Praha, 2018) and analysed the structure and quality of data it provides for construction materials, products and construction works.
– We reviewed databases of environmental impacts available on the Czech market and selected the most favourable data source for further work.
– We compared the extent, structure and contents of the construction costing and environmental impact databases.
– We proposed a process for allocating the amounts of GHGs from the environmental database to the bill of quantities and cost items used in the costing software.
– In order to evaluate the practical applicability, we made a case study of a residential house, took its cost estimation made in KROS and tried to allocate to each of its items the amounts of GHG using a procedure we drew up.

The cost estimation in the case study consisted of 447 items. The number of the cost estimation items that could be directly interconnected with their counterparts in the LCIA database was 382 (85%). We were not able to easily interconnect 65 items (15%), but we found a development of a method to interconnect them not easy, but feasible, and we presumed there would be no items left as unsolvable.

Based on the case study of a typical current Czech residential building we were guessing that the interconnection of the databases for cost estimation and life cycle assessment of carbon footprint in the near future is technically possible. For the cost estimations created just from the pre-defined items (no custom items) shall be possible to automate the calculation of the embodied amount of GHGs for phases A1-A3 (cradle-to-gate). Inclusion of more construction elements and more life cycle phases would be complicated, because at the moment the budgeting nor environmental database provided satisfactory data in the time of writing this paper. Besides development of the cut-off rules, further work would be necessary to develop a detailed method that would exactly define the rules for allocation of the environmental data to the items of the pricing database in order to ensure the allocation to be made consistently, including updates. We estimated that it would take 2–5 years to create an easy to use tool available for common designers and architects.

ACKNOWLEDGEMENTS

This work has been supported by the Ministry of Education, Youth and Sports within National Sustainability Programme I, project No. LO1605.

Building life cycle assessment tools developed in France

B. Peuportier & P. Schalbart

MINES ParisTech, PSL Research University, CES – Centre d'efficacité Énergétique des Systèmes, Paris, France

ABSTRACT

In France, LCA began to be applied in the building sector in the 1990s [Polster, 1995 and 1996] and a common framework was sketched in the European project REGENER in collaboration with German and Dutch partners [Peuportier, Kohler and Boonstra, 1997]. Twenty years later, LCA is being integrated in the next regulation planned for 2018. This communication presents the modeling issues addressed, international exchange and comparison of tools, and proposes perspectives for further work.

From the beginning of its development, the life cycle simulation model EQUER has been associated with energy simulation. The first studies showed the importance of energy consumption in the overall environmental balance of a building. A design tool should therefore account for the influence of materials and components on energy performance. Dynamic simulation was later applied to the electricity system [Herfray, 2012], including production and grid, in order to account for temporal variation of the production (according to the season, the day of the week, the hour). A consequential approach was compared to attributional LCA, considering marginal instead of average production processes [Roux, 2016a]. Due to the long life span of buildings, prospective aspects were also studied [Roux, 2016b].

The French manufacturers association has developed a database using simplified life cycle inventories (e.g. dioxins fluxes are added to other substances in a global VOCs group). The consequence of such simplification was studied [Herfray, 2010]. Sensitivity and uncertainty analyses have been performed [Pannier, 2017]. Allocation methods have been compared regarding recycling processes and local renewable electricity production. Eight tools were compared in the frame of the European thematic network PRESCO [Peuportier, 2004], and collaboration continued in the European projects Eco-housing, ENSLIC Buildings and LORE-LCA.

The IEA annex 72 will provide further opportunities to study improvement of practices, harmonisation of methods and to promote LCA in the construction sector.

REFERENCES

Donald, A.P. & Gee, A.S. (1992) Acid waters in upland Wales: Causes, effects and remedies. *Environmental Pollution*, 78, 141–148.

Duff, P.M.D. & Smith, A.J. (1992) *Geology of England and Wales*. London, The Geological Society.

Haria, A.H. & Shand, P. (2004) Evidence for deep sub-surface flow routing in forested upland Wales: implications for contaminant transport and stream flow generation. *Hydrology and Earth System Sciences*, 8 (3), 334–344.

Herfray, G. & Peuportier, B. (2012). Evaluation of electricity related impacts using a dynamic LCA model. In: *International Symposium Life Cycle Assessment and Construction*, Nantes.

Peuportier, B., Kohler, N. & Boonstra, C. (1997) European project REGENER, life cycle analysis of buildings. In: *2nd International Conference "Buildings and the environment"*, Paris.

Polster, B., 1995. Contribution à l'étude de l'impact environnemental des bâtiments par analyse de cycle de vie, PhD Thesis, Ecole des Mines de Paris.

Polster, B., Peuportier, B., Blanc Sommereux, I., Diaz Pedregal, P., Gobin, C. & Durand, E. (1996) Evaluation of the environmental quality of buildings – A step towards a more environmentally conscious design. *Solar Energy*, 57 (3), 219–230.

Roux, C., Schalbart, P. & Peuportier, B. (2016a) Development of an electricity system model allowing dynamic and marginal approaches in LCA – Tested in the French context of space heating in buildings. *International Journal of Life Cycle Assessment*, 2016 (8).

Roux, C., Schalbart, P., Assoumou, E. & Peuportier, B. (2016b) Integrating climate change and energy mix scenarios in LCA of buildings and districts. *Applied Energy*, 184, 619–629.

LCA benchmarks for decision-makers adapted to the early design stages of new buildings

A. Hollberg, P. Vogel & G. Habert
Chair of Sustainable Construction, ETH Zurich, Switzerland

ABSTRACT

When employing Life Cycle Assessment (LCA) in the building design process a number of difficulties arise. In early design stages, not all material information are available yet and assumptions are needed to replace missing data. Therefore, assumptions on these materials are necessary to bridge data gaps. Furthermore, designers often find it difficult to interpret the LCA results (Meex et al. 2018) and use them to improve the building. Existing benchmarks are usually based on a number of reference buildings (cf. König & De Cristofaro (2012), Wyss et al. (2014)). These benchmark on the building level inform designers whether a certain threshold is met or not, but it does not indicate how the building's environmental performance could be improved. Therefore, benchmarks on element level are needed in addition to the target values for the entire building.

This paper introduces a method to derive benchmarks for individual architectural elements (walls, roof, etc.). The method consists of four steps (see Figure 1). First, a structured building component catalogue with LCA data is needed, for example the Swiss "Bauteilkatalog" (Holliger Consult 2002). Second, data on the market share of materials or types of construction is matched with the component catalogue. The percentages of market share serve as weighting factor of the components in the catalogue in step three. The weighted mean values are calculated as values to replace missing data. Furthermore, benchmarks as target values based on the 5th percentile are calculated using a lognormal distribution. This target is chosen to provide an ambitious yet feasible benchmark.

The benchmarks are used in a case study based on a real building with a published LCA report. The specific values of the materials and components chosen in the case study are slightly (about 1%) lower than the benchmarks for primary energy non-renewable (PEnr) and substantially lower (22% lower) for Global Warming Potential (GWP). However, the benchmarks and even the minimum values corresponding to the currently best available technology are far from the global goal of 1t CO_2 per person.

In the future, the benchmarks should be implemented into LCA tools applied during design to provide direct feedback on the optimization potential to decision makers.

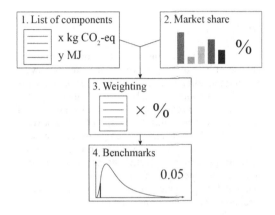

Figure 1. Methodological steps to calculate benchmarks.

REFERENCES

Holliger Consult (2002) "Bauteilkatalog." http://www.bauteilkatalog.ch/ch/de/Bauteilkatalog.asp.

König, H., & Lisa De Cristofaro, M. (2012) "Benchmarks for Life Cycle Costs and Life Cycle Assessment of Residential Buildings." *Building Research & Information* 40 (5): 558–80. doi:10.1080/09613218.2012.702017.

Meex, E., Hollberg, A., Knapen, E., Hildebrand, L. & Verbeeck, G. (2018) "Requirements for Applying LCA-Based Environmental Impact Assessment Tools in the Early Stages of Building Design." *Building and Environment* 133 (October 2017). Elsevier: 228–36. doi:10.1016/j.buildenv.2018.02.016.

Wyss, F., Frischknecht, R., Pfäffli, K. & John, V. (2014) "Zielwert Gesamtumweltbelastung Gebäude – Machbarkeitsstudie."

Principles for the development and use of benchmarks for life-cycle related environmental impacts of buildings

T. Lützkendorf & M. Balouktsi
Chair for Sustainable Management of Housing and Real Estate, Center for Real Estate, Karlsruhe Institute of Technology (KIT), Karlsruhe, Germany

ABSTRACT

Taking into account the use of resources and the adverse effects on the environment in the life cycle of buildings already from the early design stages is not only an important task, but also a challenging one. Benchmarks are required for both the formulation of design objectives within the scope of the task definition (client's brief) and the comparison and assessment of variants during the design process. Further fields of application are certification systems, funding programs and the signaling of an above-average environmental performance to third parties (i.e. tenants, purchasers, valuation professionals, banks). Demand for such benchmarks is expected to intensify even more in future because of the increasing use of life cycle assessment (LCA) methods and tools, supported by an increasing availability of related product data.

First of all, different types of benchmarks/performance levels are possible, naming – amongst other options – reference, limit or target values (Figure 1). To date, however, very few attempts have been made to define a systematic framework to set a multi-tier benchmarking/performance measurement system, representing different performance levels for different environmental indicators (indicator set). Furthermore, for deriving the benchmark/performance level values various possibilities exist. Besides the statistical evaluation of large data sets, there are also approaches involving technical feasibility or economic conditions as well as top-down approaches driven by political goals. The latter are based, in part, on scientific evidence on the capacity of the ecosystem.

Even when benchmarks are in place, for their proper application meeting certain comparability requirements is critical. Without a description of essential fundamentals and backgrounds, limit, reference and target values in combination with reference units can neither be adequately applied nor reliably interpreted. Benchmarks or benchmark systems of all kinds should

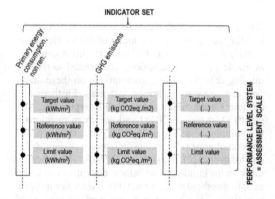

Figure 1. Illustration of a set of environmental indicators and a system of benchmarks/performance levels.

always be accompanied by a detailed description and explanation, including not only the description of the functional equivalent, but also the assumptions and system boundaries considered when developing the benchmark among other things.

In most cases, the use of a single assessment criterion and the focus on one reference or target value are no longer appropriate approaches considering the complexity of the assessment task. In future, all new buildings, and later the complete building stock, have to be "climate neutral" or have a net zero GHG emission balance regardless of their type and intensity of use. This uniform (political) goal now requires developing specific approaches for each building and use type.

The paper discusses all the above points with the end purpose of determining general rules for the development and application of a system of benchmarks. The results presented here represent a contribution to the new IEA EBC Annex 72 Subtask 1 and the work of WG 2 within the scope of ISO TC 59 SC17.

Life-Cycle Analysis and Assessment in Civil Engineering: Towards an Integrated Vision – Caspeele, Taerwe & Frangopol (Eds)
© 2019 Taylor & Francis Group, London, ISBN 978-1-138-62633-1

Assessing life cycle related environmental impacts caused by buildings (IEA EBC Annex 72)

R. Frischknecht
Treeze Ltd., Uster, Switzerland

ABSTRACT

Including environmental information in investment decisions on buildings is crucial given the consequential environmental impacts during a building's use phase of several decades and the important role of buildings with regard to greenhouse gas emissions.

The Annex 72 focuses on the assessment of different building use types, both new and retrofit during their entire life cycle. A building's embodied impacts from cradle to grave as well as operational impacts during the use stage are considered.

When including information on environmental impacts in investment decisions on buildings, efforts can be either focused on embodied or operational impacts. This gives rise to the question of the environmental optimum between gross zero operational energy buildings and minimal insulation buildings. Is it sensible to try to reduce energy consumption for heating and cooling to a level close to zero? Or do the environmental impacts of the additional material and equipment overcompensate the reduced environmental impacts during operation? In Annex 72 guidelines will be derived defining optimal ratios between investments in building materials and equipment and operational costs in the use stage.

Furthermore, methods will be established for the development of specific environmental benchmarks for different types of buildings to help designing buildings with a minimum life cycle based primary energy demand, greenhouse gas emissions and environmental impacts. Regional and national traditions will be captured to the extent feasible and necessary leading to regionally differentiated benchmarks. Case studies will be used to establish further empirical benchmarks and to validate the benchmarks defined based on the developed methods.

The environmental benchmarks will serve as important yardsticks and the guidelines for an environmental "optimisation" of buildings will help as an orientation for architects and planners, politicians, building owners, and investors in view of contributing to the fulfilment of the Paris agreement on the reduction of climate change.

Life-Cycle Analysis and Assessment in Civil Engineering: Towards an Integrated Vision – Caspeele, Taerwe & Frangopol (Eds)
© 2019 Taylor & Francis Group, London, ISBN 978-1-138-62633-1

Lessons learned from establishing environmental benchmarks for buildings in Switzerland

L. Tschümperlin & R. Frischknecht
Treeze Ltd., Uster, Switzerland

ABSTRACT

The Swiss technical bulletin SIA 2040 'SIA-energy efficiency path' defined target values for non-renewable primary energy demand and greenhouse gas emissions of buildings. They cover the impacts caused by the construction, operation and dismantling of the building as well as the mobility induced by the building. The target values are differentiated for several types of usage including residential buildings, office buildings and schools both for new and retrofit constructions.

Professional building owners asked to establish similar target values that cover the overall environmental impacts of buildings by using the ecological scarcity method. However hardly any experience regarding the environmental impacts of buildings was at hand. That is why over 30 buildings (residential buildings, school buildings, old people homes and office buildings, both retrofitted and new) were analysed with respect to the environmental impacts caused during construction, use and dismantling. They vary widely in terms of materialization and energy standards. The buildings were analysed regarding the overall environmental impacts and, to verify the results, regarding greenhouse gas emissions and the non-renewable primary energy demand.

A target value was derived from the level of environmental impacts of the current building stock and the need for reducing the actual environmental impacts in Switzerland to reach its policy-based target level. Complex electrical equipment and ventilation equipment are key components influencing the level of environmental impacts of a building. Similarly, buildings with a rather low energy efficiency using wood boilers for space heating may show a rather poor environmental performance. These aspects play a minor role when assessing primary energy consumption and climate change impacts. The presentation describes the approach chosen to quantify the environmental impacts and to define target values. Similarities and differences between the three indicators evaluated are highlighted.

A complementary case study further assessed the total environmental impacts of high tech and low tech buildings to reveal the reduction potential in environmental impacts due to improved building design. Three additional office buildings were analysed along with a high tech residential building and compared to the office and residential buildings assessed in the above mentioned study. The buildings vary widely in terms of materialization, technology level, energy standards and heating systems. From the comparison of the buildings insights were gained into the influence of a building's level of technology on its environmental impacts. It can be concluded that both buildings with more building technology as well as with a lean building technology can help in reducing the environmental footprint of buildings.

Life-Cycle Analysis and Assessment in Civil Engineering: Towards an Integrated Vision – Caspeele, Taerwe & Frangopol (Eds)
© 2019 Taylor & Francis Group, London, ISBN 978-1-138-62633-1

Belgian approach to mainstream LCA in the construction sector

D. Trigaux, K. Allacker & F. De Troyer
Department of Architecture, Faculty of Engineering Science, KU Leuven, Leuven, Belgium

W. Debacker & W.C. Lam
Unit of Smart Energy and Built Environment, VITO/EnergyVille, Mol/Genk, Belgium

L. Delem & L. Wastiels
Division of Sustainable Development and Renovation, Belgian Building Research Institute (BBRI), Brussels, Belgium

R. Servaes & E. Rossi
Public Waste Agency of Flanders (OVAM), Mechelen, Belgium

ABSTRACT

In recent years, various steps have been undertaken to stimulate the use of life cycle assessment (LCA) in the Belgian construction sector.

Firstly, an LCA method and expert model were developed to assess the environmental impact of building elements and buildings in a harmonized way (Allacker et al., 2018; De Nocker and Debacker, 2018). This method, called MMG ("Environmental profile of building elements"), is in line with current LCA standards and methods in Europe (CEN 2011; CEN 2013; EC-JRC 2011) and specifies the life cycle scenarios for the Belgian context. The MMG method allows an assessment both at the level of the individual impact indicators and as a single aggregated score, expressed in a monetary value and representing the external environmental cost. Concerning the environmental data, the MMG expert model uses generic data (Ecoinvent v3.3) which are adapted to the Belgian context.

Secondly, in the context of the Royal Decree laying down the requirements for placing environmental messages on construction products (Belgische Staatsblad, 2014), a national database with specific data for Belgian construction materials based on Environmental Product Declarations (EPDs) was established. In the near future these product specific data will serve for the assessment of building elements and buildings according to the MMG method.

Thirdly, the MMG expert model was translated to a publicly available web-based calculation tool, called TOTEM ("Tool to Optimise the Total Environmental impact of Materials") (OVAM et al., 2018). This tool is mainly oriented towards building designers, but can also be used by building clients, policy makers and other building professionals. The TOTEM tool should stimulate the use of LCA during the different stages of the design process and inform stakeholders on the environmental impact of design decisions.

In this paper, the current state of the Belgian approach to mainstream LCA in the construction sector is presented. An overview of these three initiatives is given and future developments are described.

REFERENCES

Allacker, K., Debacker, W., Delem, L., De Nocker, L., De Troyer, F., Janssen, A., Peeters, K., Van Dessel, J., Servaes, R., Rossi, E., Deproost, M. & Bronchart, S. (2018) *Environmental Profile of Building Elements [Update 2017]*. Mechelen, OVAM.

Belgische Staatsblad (2014) Koninklijk besluit tot vastelling van de minimumeisen voor het aanbrengen van milieuboodschappen op bouwproducten en voor het registreren van milieuproductverklaringen in de federale databank.

CEN (ed.) (2013) *EN 15804:2012+A1 Sustainability of Construction Works – Environmental Product Declaration – Core Rules for the Product Category of Construction Products*.

CEN (ed.) (2011) *EN 15978 Sustainability Assessment of Construction Works – Assessment of Environmental Performance of Buildings – Calculation Method*.

De Nocker, L. & Debacker, W. (2018) *Annex: Monetisation of the MMG Method [Update 2017]*. Mechelen, OVAM.

EC-JRC (2011) International Reference Life Cycle Data System (ILCD) Handbook – Recommendations based on existing environmental impact assessment models and factors for Life Cycle Assessment in a European context. Joint Research Centre (JRC) of European Commission – Institute for Environment and Sustainability (IES).

OVAM, SPW, Brussels Environment (2018) *TOTEM [WWW Document]*. Available from: www.totem-building.be [Accessed 23 February 2018].

Life cycle assessment benchmarks for Danish office buildings

F.N. Rasmussen & H. Birgisdóttir
Danish Building Research Institute, Aalborg University Copenhagen, Denmark

ABSTRACT

This study aims at developing and discussing a preliminary set of LCA reference values to serve as benchmarks for the Danish construction sector.

Benchmarks are derived from the LCAs of 16 office buildings, constructed between 2013 and 2017. The LCA tool used for the calculations (LCAbyg tool) operates with the Ökobau 2016 database.

Overall statistical benchmarks for a building's life cycle in a Danish context can be generated from the average results for electricity, heating and embodied impacts. The overall benchmark would thus be based on fixed benchmarks for each of the three constituting components (i.e. electricity, heating, embodied) illustrated as type 1 in Figure 1.

However, in the Danish energy regulation, electricity- and heating demands from a building are confined as one total energy demand. This allows for flexibility in the energy design of the building, because the sizes of the two demands (in $kWh/m^2_{heated\ floor\ area}/y$) are flexible as long as the total energy demand stays within a limit defined from regulation. This corresponds to type 2 in Figure 1. The type 2 benchmarks thus fall in line with the current regulatory focus and is considered most convenient for practitioners in the building sector. Table 1 shows the fixed embodied benchmarks.

If focus of regulation in the future is changed from energy design to life cycle design (i.e. taking the embodied impacts into account) a benchmark approach could be as outlined in type 3 of Figure 1. In this type, full flexibility of design is obtained by establishing one life cycle benchmark with flexible elements of both electricity-, heating- and embodied impacts.

The life cycle benchmarks established from current project are constituted by 2 components;

1) Statistically derived benchmarks of embodied impacts
2) Building specific benchmarks, based on regulation, for impacts from energy demand

Further studies are important to improve the validity of benchmarks from larger samples of case studies and

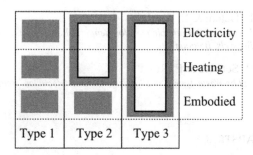

Figure 1. Schematic illustration of different types of life cycle benchmarks and the flexibility of the contributing components reflecting electricity-, heating- and embodied impacts.

Table 1. Embodied impacts for office building with service life 80 years, including life cycle modules A1-A3, B4, C3-4.

Benchmarks for embodied impacts		
GWP	in CO_2-eq/m²/y	5.3
ODP	in R11-eq/m²/y	$5.5*10^{-8}$
POCP	in C_2H_4-eq/m²/y	$2.6*10^{-3}$
AP	in SO_2-eq/m²/y	$1.3*10^{-2}$
EP	in PO_4-eq/m²/y	$1.7*10^{-3}$
PEnren	in kWh/m²/y	14.8
PEtot	in kWh/m²/y	19.2

to monitor the ODP and POCP impact categories in which notable variabilities were detected.

The life cycle benchmarks reflect contemporary, industrial, building practice. The foremost aim of the benchmarks is thus to increase knowledge and aspire front-runners in the sector to push in a more environmentally friendly direction. However, some political or value-based targets may become relevant when the construction sector has integrated the current benchmarks into practice.

Life-Cycle Analysis and Assessment in Civil Engineering: Towards an Integrated Vision – Caspeele, Taerwe & Frangopol (Eds)
© 2019 Taylor & Francis Group, London, ISBN 978-1-138-62633-1

Using BIM-based methods to obtain life cycle environmental benchmarks for buildings

A. García-Martínez, J.C. Gómez de Cózar & M. Ruiz Alfonsea
Instituto Universitario de Ciencias de la Arquitectura y la Construcción, Universidad de Sevilla, Seville, Spain

ABSTRACT

Life cycle assessment (LCA) methodology is an effective tool in determining the environmental impact of buildings. However, the LCA process requires extensive time and effort when applied to systems as complex as the life cycle of buildings. The development of Building Information Modeling (BIM) Platforms offers interesting opportunities to simplify the process for drawing up LCAs in the case of buildings. Recent research shows that the use of BIM-based methods leads to a significant reduction in time and effort which practitioners dedicate to applying the LCA methodology to entire buildings. LCA phases such as Life Cycle Inventory, traditionally extremely time-consuming, is almost self-operating when BIM resources are in use. The level of automation of LCA elaboration could be optimized by using a) the programming possibilities within the BIM platform framework; and b) the standardization of the models for analysis. In the last few years there has been significant progress in these two fields, based on a line of research proposed by the Master's in Innovation in Architecture: Technology and Design in the University of Seville. There has been a considerable development of programming software integrated in BIM. Many regions have encouraged countries to use BIM platforms when documenting the construction process.

This new conceptual framework provides the opportunity to assess the life cycle of numerous buildings in terms of the environment, obtaining specific environmental benchmarks for the different types. This communication describes a method proposed to obtain specific environmental benchmarks, – such as global warming potential and cumulative energy demand – for different types of dwellings using LCA methodology and BIM platforms. The application of this approach to different case studies is presented. The case studies selected for this research are those of housing located in different geographical and climate

contexts and using different constructive systems (e.g. masonry construction, wood construction, steel construction). Finally, this paper provides the benchmarks resulting from the application of the proposed method to the case studies.

The BIM models created, and represented in case studies, present major potential to carry out a complete inventory specifically detailing all the elements that intervene in the construction of a building. Depending on the precision and degree of definition of the model (LOD), the approximation to the real model can be very precise. The operation imposed by Revit, which classifies each element within a given constructive system, is fundamental when considering a BIM model as a potential case study for LCA. Assuming the use of unconventional construction systems not contemplated by Revit, the software has the potential to create specific families so that these elements can be modelled. The methodological proposal shown in section 2 allows the analysis to be linked directly to a BIM model.

Finally, the summary table below (Table 1) and those included in the Annex section, based on the ECB Annex 72 tables, show the benchmarks obtained to determine the environmental impact of the case studies relating to the constructive system and analysis.

Table 1. Benchmarks.

Indicator	Unit	P. Level 1 Reference Value: Worst-in-class	P. Level 2 Reference Value: Median	P. Level 3. Reference Value: Best-in-class
GWP	kg CO_2-eq./m^2	357.45	154.43	99.94
CED	MJ/m^2	3631.27	3122.44	4496.99

Life-Cycle Analysis and Assessment in Civil Engineering: Towards an Integrated Vision – Caspeele, Taerwe & Frangopol (Eds)
© 2019 Taylor & Francis Group, London, ISBN 978-1-138-62633-1

Development of a simplified methodology for creating embodied energy database of construction materials and processes in India

S. Palaniappan & V. Bindu Inti
Building Technology and Construction Management, Department of Civil Engineering,
Indian Institute of Technology Madras, Chennai, India

ABSTRACT

Evaluation of energy footprint and carbon footprint due to manufacturing of building materials, on-site construction and building operation is essential during the planning phase and the design phase of buildings to understand the sustainability aspects of built environment. This would help in choosing appropriate building plan, materials and construction methods and technologies to minimize the life cycle impacts of buildings. A comprehensive inventory database that is updated at regular intervals by the construction industry stakeholders is required to conduct such evaluation.

Several countries in Europe and North America have well established life cycle databases at the inventory and impact assessment levels. Examples are the German EPD system, Swiss KBOB list, Eco-invent database, GaBi database and the Inventory of Carbon and Energy (ICE) developed at the University of BATH, United Kingdom. Availability of similar database in India would be useful for the construction industry stakeholders to evaluate the sustainability performance of buildings during the pre-construction phase. India is one of the top five countries in the world in terms of energy consumption. The construction sector contributes to 8% of India's GDP. Significant growth is expected in the Indian construction industry in the next decade and there is a strong need for economic growth with optimal use of energy and materials. Methods used in the development of international life cycle inventory databases needs to be simplified before these methods are applied in India.

This study aims to develop a simplified methodology that would facilitate creating embodied energy database of construction materials and processes in India. The scope of this work is aligned with Activity 4.2 of IEA EBC Annex 72. The major phases of this study are: identification of materials and construction processes, review of methods used in LCA databases, documentation of user requirements from the perspective of material manufacturers and the development of a simplified method for establishing LCA database. The proposed method is expected to facilitate the creation of a transparent and reliable LCA database in India that is comprehensive and has the ability to evolve over a period of time in terms of structure and content.

Life-Cycle Analysis and Assessment in Civil Engineering: Towards an Integrated Vision – Caspeele, Taerwe & Frangopol (Eds)
© 2019 Taylor & Francis Group, London, ISBN 978-1-138-62633-1

The coupling of BIM and LCA—challenges identified through case study implementation

M. Röck & A. Passer
Working Group Sustainable Construction, TU Graz, Graz, Austria

D. Ramon & K. Allacker
Department of Architecture, KU Leuven, Leuven, Belgium

1 INTRODUCTION

To assess and improve the environmental performance of buildings over their entire life cycle the methodology of Life Cycle Assessment (LCA) is commonly accepted and increasingly used.

To conduct LCA on buildings a variety of information on the building, the components used for construction as well as data on its operation, is required. To provide and manage the variety of information required recent studies support the assessment using Building Information Modelling (BIM).

BIM is expected to enhance the applicability of LCA throughout the design process. The goal being to evolve the coupling of BIM and LCA and make its application feasible in everyday design practice.

2 OBJECTIVE

This study presents the findings of using BIM to conduct LCA acc. to EN 15978/15804 as well as in particular, experiences from testing the application of the Product Environmental Footprint (PEF) methodology using a BIM-LCA workflow. The identified potentials and challenges are discussed and conclusions drawn on requirements to improve the coupling of BIM and LCA and support integration of environmental assessment of buildings in the future.

3 METHODOLOGY

In order to identify generally applicable concepts of organizing building information, inventory and granularity of data used for assessments throughout the design process we reviewed literature, standards, and guidelines regarding conventions in LCA and BIM. Furthermore, the study discusses the potentials, challenges and learnings encountered during two BIM-LCA case studies from the PEF4buildings project and presents the workflow developed based on these case studies.

4 CONCLUSION

Multiple challenges for implementing a BIM-LCA workflow are present today, e.g.:

- Completeness and quality of BIM model regarding modelled/unmodelled elements and their Level of Development (LOD)
- Modelling and allocation of scenarios, which require different levels of granularity
- Differences in modeling structure in BIM and LCA making it difficult to automate direct data exchange between BIM and LCA software
- Harmonized aggregation and granularity of data and results, which are crucial for conclusive sensitivity and hotspot analysis as well as communication with designers and other stakeholders.

Potentials identified show that implementation of LCA and BIM could be supported by:

- Structuring building elements and sub-components hierarchically with specified levels of granularity which also supports results presentation on different levels of granularity
- Establishment of LCI as an intermediate step to support linking of detailed element and material quantities with respective LCA datasets and scenarios
- A parametric approach for establishing the LCI based on both BIM and LCA model to support easy changes –e.g. element's material composition and related scenarios.

5 OUTLOOK

To evolve the application of LCA to building as well as enhance the integration with BIM a **common definition** and **transparent documentation** are required regarding: a) **organization and structure** of LCI as well as; b) **granularity and scope** of both, BIM and LCA/PEF data. This includes consideration of the varying levels of development of BIM elements (e.g. during building design).

Life-Cycle Analysis and Assessment in Civil Engineering: Towards an Integrated Vision – Caspeele, Taerwe & Frangopol (Eds)
© 2019 Taylor & Francis Group, London, ISBN 978-1-138-62633-1

Environmental impacts of future electricity production in Hungary with reflect on building operational energy use

B. Kiss
Department of Mechanics Materials and Structures,
Budapest University of Technology and Economics, Budapest, Hungary

Zs. Szalay
Department of Construction Materials and Technologies,
Budapest University of Technology and Economics, Budapest, Hungary

E. Kácsor
Regional Centre for Energy Policy Research (REKK),
Corvinus University of Budapest, Budapest, Hungary

ABSTRACT

For environmental accounting of buildings Life Cycle Assessment (LCA) is a proper and more and more widespread method. The assessment of the embodied materials is well covered by constantly improving databases (Life Cycle Inventories). On the other hand, calculation methods for the operation phase of the buildings vary in the literature. Specifically the calculation of the environmental impact related to the use of the energy source raise important questions. Our results show that in many cases the energy source has the most significant influence on the environmental impact of a building. The consideration of future change in the electricity production mix is a key aspect in the modelling of the environmental impacts. We present a method for determining the future electricity production in Hungary and we investigate its influence on the specific environmental impacts. The method makes the dynamic modelling of the temporal variation of the electricity supply mix possible, that is later applicable to building LCA calculations. We also illustrate the results of this method on a specific future scenario and analyse the differences in the environmental impacts in compare to a static approach.

In this paper we investigated 132 different electricity mixes of different annual periods and future years. Although sophisticated methods are available for the calculation of both the energy use of the building and the electricity supply mix even on an hourly basis, in many cases only simplified calculations are made (early design stage, small buildings, etc.). Also, the use of hourly time-step raises some methodological issues (same meteorological data for energy and electricity market model should be used, shift in heat demand and electricity demand because of buffer tanks, etc.). In this paper we developed a simplified method to consider the most significant differences in the electricity supply mix (peak/off-peak and heating/non-heating periods) that is applicable with simple (seasonal, monthly steady state) energy calculations. In our model the difference in environmental impacts between heating and non-heating season turned out to be more relevant than the difference between peak and off-peak periods. We showed, that the improvement of the electricity supply mix strongly influences the environmental impacts in a specific future scenario. It is visible that the generally used static approach to calculate environmental impacts is inadequate in a long-term basis such as the operational phase of a building. However, with the improvement of the building energy performance and the shift to renewable energy sources in electric power generation, the difference may decrease, and more attention should be paid to embodied impacts.

Modular methodology for building life cycle assessment for a building stock model

Zs. Szalay
*Department of Construction Materials and Technologies,
Budapest University of Technology and Economics, Budapest, Hungary*

B. Kiss
*Department of Mechanics Materials and Structures,
Budapest University of Technology and Economics, Budapest, Hungary*

ABSTRACT

Life Cycle Assessment is a more and more widely used method for the environmental evaluation of buildings. However, there are ongoing discussions on how the particular calculations should be carried out, which method should be used for the environmental accounting and how the results should be interpreted. Since the main life cycle phases of a building are well distinguishable, a modular structure for an assessment is possible and appropriate to allow comparison between different methods in the above described elements of a building LCA. In this paper, we present a modular methodology for the environmental assessment and optimisation of buildings. The methodology is implemented in a parametric design tool where the building geometry can be modeled in 3D, then the energy performance is determined with the energy calculation module and the environmental impact is calculated with the LCA module. The life cycle impact minimization is possible through the optimization module. The implementation of this framework in a parametric environment is a great tool for optimizing in early design stage, in which case a much higher potential of improvement is possible because of fewer constraints. The real-time feedback of a parameter change helps the designer to find the most important variables and thus move to a better direction from the perspective of environmental impacts. The framework can be used for new buildings as well as for retrofit design. In the level of national energy regulations, there is a great potential to perform optimization for typical houses of the building stock, from which recommendations can be derived to support retrofitting subsidy programs. The framework is also capable for further extension with life cycle costing, that is also very important in the evaluation of new building design and retrofitting scenarios, however an optimization for life cycle cost would lead to

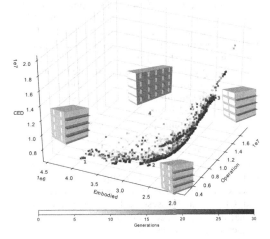

Figure 1. Visualization of the possible solutions and the specific developed geometries.

a different optimum than for environmental impacts. While it is not debatable that every building is unique, and the final choices of engineers are influenced by many factors beyond energy efficiency, such a tool can prove to be very useful in assisting the design process and may also contribute to significant savings in terms of cost. The method is demonstrated on a case study of a typical model of a house described by a typology of the Hungarian building stock. We present the application of the method for both new building design and energy retrofitting (Fig. 1). Such typological assessments facilitate the further modeling of a complete building stock (e.g. residential buildings), and with the help of optimization tools, the potential for environmental impact mitigation of the housing industry will be also possible.

Life-Cycle Analysis and Assessment in Civil Engineering: Towards an
Integrated Vision – Caspeele, Taerwe & Frangopol (Eds)
© 2019 Taylor & Francis Group, London, ISBN 978-1-138-62633-1

Effects of LCA impact categories and methodology on the interpretation of a building's environmental performance

L. Delem & L. Wastiels
Belgian Building Research Institute, Brussels, Belgium

K. Allacker
Department of Architecture, Faculty of Engineering Science, KU Leuven, Leuven, Belgium

ABSTRACT

Under impulse of the Product Environmental Footprint (PEF) initiative (European Commission 2013), CEN/TC350 wants to extend the current set of LCIA impact categories in the EN 15804 (CEN 2013) for the life cycle analysis (LCA) of construction products. The goal of the present study is to investigate the practical implementation and relevance of the additional impact categories in building life cycle assessment and to evaluate how their inclusion could change conclusions drawn from a (comparative) building LCA.

By means of a case study analysis, the life cycle environmental impact of six variants of an apartment building with the same lay-out and energy performance but different loadbearing structure is being assessed based on different impact categories (CEN and CEN+) and assessment methodologies (individual and aggregated). In comparison to the CEN impact categories (as required by EN 15804+A1 (CEN 2013)), additional CEN+ impact categories are considered (amended version of the EN15804 (foreseen 2018)). In addition, three different aggregation methods into a single score are considered for the evaluation: MMG (OVAM 2013), ILCD and ReCiPe.

Firstly, the results show that it is possible to use the additional indicators (*Ecotoxicity freshwater, Human toxicity cancer* and *non-cancer, Ionising radiation, Water resource depletion, Particulate matter, Land use Soil Organic Matter* and *Land use Biodiversity*) in building LCA as the corresponding methods and data are available. Concerning *Water resource depletion*, some incoherencies are detected in the regionalisation of input and output flows, leading to significant errors in the results. Concerning *Land use Biodiversity*, an inconsistency between LCI data and the characterisation factors is identified, leading to negative indicator results.

Secondly, the results indicate that the inclusion of the CEN+ impact categories does not drastically change the conclusions concerning most impacting life cycle phases of a building nor the ranking of building variants (in a comparative building LCA). However, their inclusion/omission could lead to different conclusions concerning the most impacting materials or processes.

Thirdly, the choice of the method to aggregate the CEN and CEN+ impact categories to one single score influences the identification of the building with the lowest or the highest total life cycle impact and of the most important impact categories. Also the relative importance of operational energy use versus materials related impact is affected by the aggregation method, which can be of significant importance for determining the optimal insulation level from an LCA perspective. Nevertheless, the aggregated results often indicate similar trends where the CEN+ impact categories *Human toxicity cancer, Human toxicity non cancer* and *Particulate matter* are being relatively important on building level.

Finally, the results show that the toxicity indicators are dominated by metals. Seen the shortcomings of current LCIA toxicity models for metals, and the high variation in results related to the inclusion or exclusion of characterisation factors for metals, the inclusion of those indicators in building evaluation tools (for designers) is questionable and should be considered with care.

REFERENCES

CEN (2013) *EN 15804:2012+A1:2013 – Sustainability of Construction Works – Environmental Product Declarations – Core Rules for the Product Category of Construction Products.*

European Commission (2013) *PEF – Product Environmental Footprint (EU) No179/2013 Commission Recommendation on the Use of Common Methods to Measure and Communicatie the Life Cycle Environmental Performance of Products and Organisations.*

OVAM (2013) *Environmental Profile of Building Elements (MMG Report). Towards an Integrated Environmental Assessment of the Use of Materials in Buildings.*

Life-Cycle Analysis and Assessment in Civil Engineering: Towards an Integrated Vision – Caspeele, Taerwe & Frangopol (Eds)
© 2019 Taylor & Francis Group, London, ISBN 978-1-138-62633-1

Analyzing the life cycle environmental impacts in the Chinese building design process

W. Yang & Q.Y. Li
School of Architecture, Tianjin University, China

L. Yang
China Institute of Building Standard Design & Research, China

J. Ren
Tianyou Architectural Design Company, China

X.Q. Yang
School of Architecture, Tianjin Chengjian University, China

ABSTRACT

Building LCA needs detailed data of operational energy consumption, material and construction. Therefore, most case studies were conducted when the design schemes were well defined and developed. International researchers had made efforts in developing the methods and tools to improve building life cycle performances in the early design stage. Building Information Model (BIM) integrates multiple properties with 3D model which enables it to support information extraction and feedback during the design process and later in the building life cycle stages. In this contribution, the potential and methods of integration of BIM, LCA and Chinese building design process is discussed. First, the status quo of BIM and LCA of Chinese buildings are reviewed. Then, the advantages and limits of existing tools and databases are analysed with a case study. Finally, the need for developing the BIM enabled LCA tool and databases for Chinese building design workflows is underlined.

The current sustainable building standards in China lack the LCA indicators and benchmarks. Life cycle performances are seldom analyzed in the early stages of the design process, which could lead to failure in selecting the optimized solution. To overcome the problem, a LCA tool that can cope with the progressive description of buildings in the design process is needed.

This contribution tries to apply the BIM enabled LCA approach to assist the design process of green building in the Chinese context. A primary school building in north China is taken as the case study project. As there is so far no well-developed BIM-LCA tool for the early design stage of Chinese building, several existing tools with connection to the commonly used BIM software are selected to support the design optimisation and to cope with the level of development of the information model at each design stage.

IES<VE> is applied as energy simulation tool. Its LCA model and IMPACT database is used for the LCA

in the conceptual stage, in which predefined elements can be assigned to each surface of the concept model. Yet the link between IES<VE> and BIM is limited to the recognition of surface area of the model. When more detailed construction information is described in the model, another tool with better interoperability is needed.

In the developed design stage, Tally is used as the LCA tool. Its dataset is based on building materials that could be allocated to each construction layers of the BIM elements. As a Revit plugin, it is well integrated in the BIM software.

At the construction design stage, the complete information and material quantity can be deduced from BIM. The LCA results of the whole building is calculated with Tally and with a simplified LCA tool based on generic LCA data for major Chinese building materials.

Results show that:

(1) It is necessity to analyze the potential life cycle performances of the design solutions in each stage of the design process.
(2) The BIM-LCA methods and tools should cope with the level of development of the model and the parameters to be considered in each design stage.
(3) There are limits of data inconsistency and modeling gaps in the existing tools and databases.
(4) There is a need to develop the BIM-LCA tool for Chinese building design and assessment.

REFERENCES

Gervásio H., Santos P., Martins R. & Simões da Silva L. (2014) A macro-component approach for the assessment of building sustainability in early stages of design. *Building and Environment*, 73, 256–270.
IES<VE> IMPACT. https://www.iesve.com/software/ve-for-engineers/module/IMPACT-Compliant-Suite/3273.
Tally. http://choosetally.com/.

Review of existing service lives' values for building elements and their sensitivity on building LCA and LCC results

S. Lasvaux, M. Giorgi & D. Favre
HEIG-VD, LESBAT, University of Applied Sciences of Western Switzerland (HES-SO), Yverdon-les-Bains, Switzerland

A. Hollberg, V. John & G. Habert
Chair of Sustainable Construction, ETH Zürich, Zürich, Switzerland

ABSTRACT

The service lives of building elements and building integrated technical systems (BITS) can vary within a large interval depending of the data sources used. It can lead to inconsistencies when performing building related assessment (such as environmental life cycle assessment (LCA) or life cycle costing (LCC)). To that purpose, this study presents a method for the estimation of uncertainties arising from the building element service life in the context of building Life Cycle Assessment (LCA) and Life Cycle Costing analysis (LCC).

First, different building elements' and materials' service lives sources are collected in the international literature. To date up to 60 sources were collected. Because of the search method and keywords used, 41% of the sources come from Switzerland, 35% are from Europe and 24% are non-European sources (mainly from Canada and the USA). Examples of sources include standards, professional owners document, SL guidelines, research reports, scientific papers etc. Each original source has lifetime values for different level of details and are not comparable. For example, the lifetime can be defined for a full element e.g., a compact façade, or defined at the layer/material scale. To be able to gather the lifetimes of all literature sources, a SL database has been set up according to the Swiss calculation cost structure with three main levels of classification as a starting point.

Second, the values are statistically sampled, and probability distributions of service lives are derived to serve as input data for the LCA and LCC of buildings' case studies.

To illustrate the approach, LCA and LCC were conducted on four newly built Swiss dwellings. The related uncertainties arising for the variability of the service lives was evaluated with a Monte Carlo simulation. In this application, the calculations used service

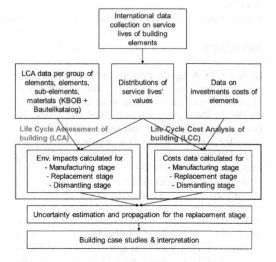

Figure 1. Methodology followed in this paper.

lives data taken at a lower level of details in the database so at the material scale. Other configurations could also be assessed e.g., using service lives data defined for each main group or group of elements. Figure 1 summarizes the methodology developed in this paper.

The outcome of the study showed that the greenhouse gas emissions or the costs uncertainties of the replacements represented between 6 and 16% of the total result of the buildings' LCA or LCC. Furthermore, the uncertainties estimation of the replacements allow to compare more accurately the buildings' environmental or economic performances. The study highlighted the interest of taking into account uncertainty estimation for the replacement phase, especially in the context of more energy efficient buildings.

Life-Cycle Analysis and Assessment in Civil Engineering: Towards an Integrated Vision – Caspeele, Taerwe & Frangopol (Eds)
© 2019 Taylor & Francis Group, London, ISBN 978-1-138-62633-1

Improving the reliability and specificity of an Input-Output-based Hybrid (IOH) method for computing embodied energy of a building

M.K. Dixit & V. Venkatraj
Texas A&M University, College Station, Texas, USA

ABSTRACT

Buildings consume approximately half of annual energy supply of globe in their construction and operation causing significant damage to the environment through resulting carbon emissions. A portion of this energy is used as life cycle embodied energy (LCEE) when buildings are initially constructed, maintained through their life cycle, and deconstructed. The LCEE includes the energy embodied in all products and processes used directly and indirectly over a building's life cycle. The operating energy (OE) is consumed in air-conditioning, heating, lighting, and operating building appliances. Both the embodied and operating energy use of buildings must be optimized to reduce their carbon footprint effectively. However, measuring and evaluating LCEE is more challenging and data-intensive than operating energy usage. Each of the existing embodied energy calculation methods suffers from the problem of either incompleteness, unreliability, or poor data quality. Although a consensus is lacking in literature over which method is apt, an input-output-based hybrid (IOH) method is considered relatively complete in terms of its system boundary coverage. However, an IOH method also requires improvements to offer reliable and product-specific results. Such improvements need extensive detailed data, which are not readily available.

In this paper, we propose a framework for improving the reliability and specificity of an IOH method using a tiered sensitivity analysis approach. We apply sensitivity analysis at various stages from an input-output (IO) model level to a product level to identify IO inputs that demonstrate significant impact over final embodied energy results. We then replace these IO inputs with reliable data in the IO model keeping other non-significant inputs unchanged. We use the same process to identify and disaggregate only the most influential industry sectors to improve the specificity of calculations. Because only the significant inputs are refined, the proposed IOH model does not required extensive amounts of data to improve reliability and specificity of embodied energy calculations.

Life cycle assessment of alternative masonry to concrete blocks

J. Dahmen & J. Kim
School of Architecture and Landscape Architecture, University of British Columbia, Vancouver, Canada

C.M. Ouellet-Plamondon
École de Technologie Supérieure, Université du Québec, Montréal, Canada

ABSTRACT

Concrete masonry is the second larger use of cement, after ready mix in North America. This life cycle assessment compares two emergent environmentally sustainable masonry blocks to conventional cement-based masonry blocks. The emergent stabilized soil block and alkali-activated blocks evaluated offer strength, durability and form is identical to that of conventional masonry blocks, enabling them to be used interchangeably with conventional masonry in developed and developing countries. The functional unit of this study is one hollow-celled masonry block. The emergent block materials, mix design and production process are from the emergent block are from actual operating masonry production. The system boundaries are from the raw materials extraction to storing at the manufacturing facility. The life cycle inventory analysis was performed with QuantisSuite, for its accessibility and transparency. Inventory data for the alkali activators sodium silicate and sodium hydroxide, along with other raw materials (aggregates, water, fuel and energy) were taken from Ecoinvent. Because cement production varies widely from country to country, cement data was also taken from U.S. Life Cycle Inventory (USLCI), which accurately reflects manufacturing practices in the U.S. where the blocks are produced. The environmental impact of each type of masonry block is compared to a base case consisting of conventional concrete block and architectural concrete block, according to the midpoint and endpoint indicators. The results indicate significantly improved performance of the emergent masonry blocks in the areas of embodied carbon. The stabilized soil block shows improvement of health and ecological indicators. The convention concrete block remained the most water efficient. The results are projected at the building, and neighborhood scale (Figure 1). The service life of the emergent block must be verified, as

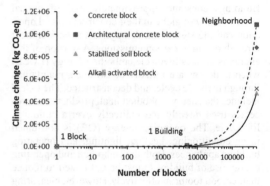

Figure 1. Climate change impact at the block, building and neighborhood level.

changing the reference study period as the most impact of masonry and concrete product. Both embodied energy and embodied carbon can be reduced further as efficiency is realized during the manufacturing process, which is still in an early stage of development, and energy source of the cement plant. Sourcing alkali activators as byproducts would also drive down water consumption required to produce alkali-activated block. These findings offer possible solutions for mitigating environmental impacts and achieving sustainable low carbon built environment.

REFERENCES

Dahmen, J., Kim, J. & Ouellet-Plamondon, C.M. (2018) Life cycle assessment of emergent masonry blocks. *Journal of Cleaner Production,* 171, 1622–1637.

Häfliger, I.-F., John, V., Passer, A., Lasvaux, S., Hoxha, E., Saade, M.R.M. & Habert, G. (2017) Buildings environmental impacts' sensitivity related to LCA modelling choices of construction materials. *Journal of Cleaner Production,* 156, 805–816.

MS9: Multi-hazard resilience assessment in a life-cycle context
Organizers: J. Ghosh & J.E. Padgett

Life-Cycle Analysis and Assessment in Civil Engineering: Towards an Integrated Vision – Caspeele, Taerwe & Frangopol (Eds)
© 2019 Taylor & Francis Group, London, ISBN 978-1-138-62633-1

Resilience assessment of highway bridges under multiple natural hazards

Y. Dong
Department of Civil and Environmental Engineering, The Hong Kong Polytechnic University, Kowloon, Hong Kong, China

D.M. Frangopol
Department of Civil and Environmental Engineering, ATLSS Engineering Research Center, Lehigh University, Bethlehem, PA, USA

ABSTRACT

Highway bridges are important infrastructure components for the safety and functionality of any society. The immediate damage of highway bridges under hazard effects can disrupt transportation systems. It is of vital importance to assess the performance of bridges under natural hazards in order to aid the emergency response and recovery decision. During their service life, bridges are exposed to multiple natural hazards. Earthquake and hurricanes can cause severe damage to structures and produce large economic losses to the society. All these facts have received an increasing attention from researchers, insurers and policy-makers. Therefore, a comprehensive resilience and probabilistic loss assessment of highway bridges under multiple hazards is necessary.

The capability of civil infrastructure systems to maintain prescribed safety, security and flexibility levels, and to recover from extreme events can be assessed using resilience. Resilience, related to the functionality of structural systems under extreme events and the recovery patterns, is becoming a paramount performance indicator within the hazard management process (Frangopol *et al.* 2017). The previous studies on resilience were mainly focused on the assessment of bridges under single hazard. Decò *et al.* (2013) and Dong and Frangopol (2015) computed the resilience of highway bridges under seismic hazard. Dong and Frangopol (2016) investigated the resilience of highway bridges under flood effects with different return periods. Relevant studies with respect to the resilience assessment of bridges under multiple hazards are scarce. Multiple hazards could bring much more disastrous consequences to the society than a single hazard (Padgett *et al.* 2009; Jalayer *et al.* 2011; Dong and Frangopol 2016).

In this paper, the resilience and long-term loss of highway bridges under multiple natural hazards are computed. The residual functionality and recovery functionality under specific recovery actions are identified. Vulnerability analysis is then used to predict the structural residual functionality under a prescribed hazard. Then, given the recovery actions, the time-dependent functionality of the bridges can be quantified. Herein, the uncertainties associated with natural hazard scenarios (e.g., earthquakes and hurricanes) and structural performance are considered within the evaluation process. Overall, in this paper, the resilience of highway bridges under multiple hazards (i.e., earthquakes and hurricanes) is investigated. The approach is illustrated on an example highway bridge.

REFERENCES

Decò, A., Frangopol, D.M. & Bocchini, P. (2013) Probabilistic seismic resilience of bridge networks. In: Deodatis, G., Ellingwood, B.R. & Frangopol, D.M. (eds.) *Safety, Reliability, Risk and Life-Cycle Performance of Structures and Infrastructures*. Boca Raton, FL, CRC Press. pp. 621–628.

Dong, Y. & Frangopol, D.M. (2015) Risk and resilience assessment of bridges under mainshock and aftershocks incorporating uncertainties. *Engineering Structures*, 83, 198–208.

Dong, Y. & Frangopol, D.M. (2016) Time-dependent multi-hazard life-cycle assessment of bridges considering climate change. *Journal of Performance of Constructed Facilities*, 30 (5), 04016034, 1–12.

Frangopol, D.M., Dong, Y. & Sabatino, S. (2017) Bridge life-cycle performance and cost: Analysis, prediction, optimization and decision making. *Structure and Infrastructure Engineering*, 13 (10), 1239–1257.

Jalayer, F., Asprone, D., Prota, A. & Manfredi, G. (2011) Multi-hazard upgrade decision making for critical infrastructure based on life-cycle cost criteria. *Earthquake Engineering & Structural Dynamic*. 40, 1163–1179.

Padgett, J.E., Ghosh, J. & Dennemann, K. (2009) Sustainable infrastructure subjected to multiple threats. In: *TCLEE 2009: Lifeline Earthquake Engineering in a Multi-Hazard Environment*. pp. 703–713.

*Life-Cycle Analysis and Assessment in Civil Engineering: Towards an
Integrated Vision – Caspeele, Taerwe & Frangopol (Eds)
© 2019 Taylor & Francis Group, London, ISBN 978-1-138-62633-1*

A discussion on the need for flexibility in infrastructure development

S. Torres & M. Sánchez-Silva
Universidad de los Andes, Bogotá, Colombia

ABSTRACT

Planning, design and operation of high-value infrastructure (i.e., infrastructure that cannot be replaced easily or at an acceptable cost) is central to society. Throughout the infrastructure's lifetime, managers and stakeholders have to face challenges that are difficult to define at the outset. Traditional over-design strategies are expensive, have a significant impact on the environment and are not sustainable. Thus, under uncertain exogenous conditions, success will depend on the ability to continuously adapt to requirements as the future unfolds. In other words, designs ought to account for provisions that can handle uncertain demands and cope with unplanned events (Cardin et al. 2015).

In response to the need of accommodating and managing future uncertain demands, concepts such as resilience, robustness, adaptability, flexibility, etc. within the engineering context have attracted attention recently. However, an exhaustive literature review shows that there is still a lack of rigorousness in their definition (Ross et al. 2008), (Fitzgerald 2012). This lack of consensus in the terminology has been, expectedly, transferred to the mathematical representation; quantifying the amount of flexibility in a system and assessing its value in practice is still an open question.

Making flexibility a central element in engineering design and operation is key to reduce risk and increase life-cycle value (Fitzgerald 2012). This claim is based on the fact that it is expected from a flexible system to have fast and cheap adaptation processes that increase its range of operating conditions and reduce its vulnerability.

This paper introduces a novel flexibility quantification index that attempts to capture the key behaviors of a flexible system. A system regarded as flexible must have the ability to be changed; the adaptation process has to be cheap and fast, and it must increase the perceived utility supplied by the system, i.e., the changes serve the purpose of incrementing the system performance under the new external conditions.

To exemplify the use of the flexibility index, a simple system that has to accommodate a demand that evolves in time is used to compare management strategies and system properties. The system modeled is a structure that has an initial capacity C_0 which costs K_0 to build. The capacity of the structure can be expanded by adding modules (modularity and scalability are typical examples of flexible design strategies), which share the same individual capacity C_i and construction cost K_i.

Two management strategies are compared: one where the systems is built with a large initial capacity that remains constant along the analysis period (20 years), and one where the capacity is expanded every time the ratio demand/capacity reaches certain threshold. The effects of the two strategies are analyzed in three scenarios where the initial construction cost can be higher for the more robust design, the cost can be equal, or where the cost for the flexible design are higher. The example demonstrated that the flexibility index captures adequately the balance between the benefits obtained and the incurred costs. It also showed that an adaptable system can generate value either by reducing costs when no extra capacity is needed or by keeping the capacity as close as possible to the demand.

REFERENCES

Cardin, M.-A., Ranjbar-Bourani, M. & De Neufville, R. (2015) Improving the lifecycle performance of engineering projects with flexible strategies: Example of on-shore LNG production design. *Systems Engineering*, 18 (3), 253–268.

Fitzgerald, M. (2012) *Managing Uncertainty in Systems with a Valuation Approach for Strategic Changeability*. MSc Thesis. Massachusetts Institute of Technology.

Ross, A.M., Rhodes, D.H. & Hastings, D.E. (2008) Defining changeability: Reconciling flexibility, adaptability, scalability, modifiability, and robustness for maintaining system lifecycle value. *Systems Engineering*, 11 (3), 246–262.

Life-Cycle Analysis and Assessment in Civil Engineering: Towards an Integrated Vision – Caspeele, Taerwe & Frangopol (Eds)
© 2019 Taylor & Francis Group, London, ISBN 978-1-138-62633-1

Modeling the resilience of aging concrete bridge columns subjected to corrosion and extreme climate events

H. Almansour, A. Mohammed & Z. Lounis
National Research Council Canada, Ottawa, ON, Canada

ABSTRACT

Global changes in temperature, precipitation and wind patterns threaten the integrity and functionality of reinforced concrete highway bridges. The expected continuation or acceleration of climate change can induce additional stresses that increase the risk of failure of critical components of aging bridges, such as piers, girders and deck-slabs. As a result of climate change increasing rates of chloride ingress into concrete and increasing rates of reinforcing steel corrosion are expected. This, in turn, will lead to reduced strength, stiffness and shortened service life of critical bridge components. The development of reliable structural evaluation tools that can provide quantitative or semi-quantitative estimates of the residual capacity of aging bridge structures under the effects of various damage mechanisms and climate change is of a major importance to bridge owners, engineers, and researchers. The estimation of the residual capacity of damaged bridge structures is a challenging task and requires a reliable simulation of the governing deterioration mechanisms, identifying the critical damage stages and their effects on flexural and shear stiffness and resistances, as well as ductility of materials and structural elements. The progressive changes of internal forces and stress distributions due to traffic and climatic loads combined with corrosion-induced deterioration raise many challenges to the evaluation of the residual capacity of reinforced concrete bridge elements. Experimental investigations show that reinforced concrete columns affected by reinforcement corrosion and loaded to failure by concentric or eccentric loads lose their strengths in successive stages. Each major sign of distress is related to major changes in the beam-column cross section, cracking and spalling, reduction of reinforcement cross-sectional area and ductility, reinforcement fracture or rebar bucking, and complete loss of concrete confinement. On the other hand, if the reinforcement corrosion is initiated and propagated, continuous local damage of the concrete cover, steel reinforcement and their bond could proceed in different stages. The effects of high levels of damage on the stiffness, resistance and ductility of the beam-columns could reach a critical threshold of their structural capacities. It is expected that climate change will lead to increased temperatures, which in turn will lead to increased rates of corrosion. The limited available knowledge on the time-dependent corrosion propagation in the context of climate change and its impacts on the properties of the concrete and steel reinforcement result in a very wide variability of the simulation results.

This paper presents an analytical model to evaluate the structural capacity of critical reinforced concrete beam-columns of highway bridges for different stages of deterioration induced by reinforcement corrosion and other hazards, including climatic loads and earthquake. In addition, the rapid resilience enhancement strategy using fiber reinforced polymers for their rehabilitation and strengthening. The proposed approach is based on a simplified nonlinear finite element model that is conceptually simple yet numerically efficient. The model was proven to be capable of capturing all stages of damage and the related residual capacities with acceptable accuracy. It is shown that the implementation of rapid strengthening methods using fiber reinforced polymers can enhance the resilience of critical bridge structures that ensures improved capacity and ductility and shorter traffic disruption time.

Uncertainty analysis and impact on seismic life-cycle cost assessment of highway bridge structures

S. Shekhar & J. Ghosh
Indian Institute of Technology Bombay, Mumbai, India

ABSTRACT

Uncertainty is ubiquitous in all sorts of modeling and experimental processes in science and engineering. Current research in seismic loss estimation considers the ground motion, material and structural modeling parameter, deterioration parameters, component repair strategies and its associated cost as individual random variables and propagates their uncertainty to estimate expected seismic life-cycle cost and associated confidence bounds. However, the quantification of each source of uncertainty towards total uncertainty is unknown. To take decisions related to field instrumentation of critical bridge components to aid life-cycle cost estimation, it is necessary to propagate and quantify the randomness associated with different input parameters.

Addressing these gap, this study first identifies the different sources of uncertainty that potentially affects the seismic vulnerability and loss estimation of highway bridge structures. Surrogate models are developed for each bridge component to estimate seismic demand which is used to estimate parameterized seismic fragility models conditioned on the ground motion intensity, bridge structural and modeling parameters and deterioration parameters. The contribution of each uncertainty source to the aggregate uncertainty around the mean estimated fragility is computed using a first order analysis method utilizing Taylor series expansion (Lei & Schilling 1994). The uncertainty bounds around the fragility estimates are used in conjunction with regional hazard information and repair strategies to evaluate the seismic life-cycle cost (Sebastiani 2016) and associated confidence bounds. The contribution of each parameters on seismic life-cycle estimates are also obtained. Figure 1 shows the 95% confidence bounds (represented in terms percentage change) around mean seismic life-cycle cost (*SLCC*) due to each of the

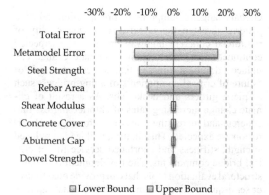

Figure 1. Partitioning of total confidence bounds around *SLCC* for bridge column in terms of various input sources of uncertainty (in percentage).

probabilistic input parameters for the bridge column in the form of a Tornado diagram. The metamodel uncertainty contributes the most followed by column rebar area and steel strength. For bridge system the upper and lower confidence bounds around the mean loss estimate is 22% lower and 30% higher with metamodel error, steel strength, column rebar area and dowel strength contributing most to the output uncertainty.

REFERENCES

Lei, J. & Schilling, W. (1994) Parameter uncertainty propagation analysis for urban rainfall runoff modelling. *Water Science and Technology*, 29 (1–2), 145–154.

Sebastiani, P.E. (2016) *Performance-Based Seismic Assessment for Life-Cycle Cost Analysis of Existing Bridges Retrofitted With Seismic Isolation*. PhD Dissertation. Sapienza, Italy, University of Rome.

Life-cycle based resilience assessment of bridges under seismic hazard

B. Sharanbaswa & S. Banerjee
Department of Civil Engineering, Indian Institute of Technology Bombay, Powai, Mumbai, India

ABSTRACT

There are growing worldwide recognition and interest among the engineering professionals and decision makers to build resilient and sustainable civil infrastructure systems. One approach towards meeting this goal is to analyze and design structures for possible extreme hazard scenarios considering structural deterioration (or, aging) during their service lives. Aging causes gradual degradation of bridge performance with time, and consequently, bridges exhaust their capacity much prior to their design lives. Therefore, assessment of bridge performance at various stages of its lifespan under possible extreme hazard scenarios demands life-cycle based resilience assessment.

The current research provides a broad assessment of seismic resilience of degrading bridges over their lifespans. A three span representative bridge designed according to the American Association of State Highway and Transportation Officials (AASHTO) bridge design specifications (AASHTO 2014) is analyzed for this purpose. Three dimensional (3D) finite element (FE) model of the bridge is developed to simulate its seismic behavior at pristine condition. Similarly, 3D FE model of the same bridge after 50 years of its lifespan is generated. Nonlinear time history analyses with 50 ground motions are performed to express seismic vulnerability of the bridge, at pristine and degraded (at 50-year) conditions, in the form of fragility curves. For evaluating seismic resilience of the bridge, losses due to bridge damage under earthquakes are estimated considering probable post-event recovery of the bridge. The loss model incorporated both direct (due to physical bridge damage) and indirect (due to bridge inaccessibility) losses. Realistic recovery functions for different bridge damage states are considered to account for post-event recovery of the damaged bridge. Finally, resilience evaluation integrated bridge performance (or, vulnerability), losses and recovery under seismic events (Cimellaro et al. 2010). Outcome of the study reveals the time-dependent variation of seismic resilience of the bridge under consideration. Figure 1 shows the variation of bridge seismic resilience, at pristine and degraded conditions, with ground motion intensity parameter PGA. As expected,

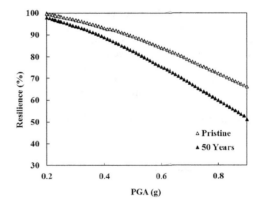

Figure 1. Lifetime seismic resilience of the bridge.

resilience values decrease with increasing PGA and structural deterioration. It is observed that 50 years resilience reduces more rapidly than that at pristine condition for higher ranges of PGA. A maximum reduction of 22% is estimated for 50-year deterioration at 0.9 g PGA. This reduction in resilience can be higher if structural conditions at 75 and 100 years after construction are considered. Uncertainties related to various analysis parameters (e.g. corrosion parameters, structural parameters, parameters related to loss and recovery models) may introduce some variation in the estimated resilience values. Such variations can be quantified through an extensive probabilistic study by conducting (i) sensitivity study to evaluate sensitivity of selected uncertain parameters on bridge seismic resilience and (ii) uncertainty analysis with sensitive uncertain parameters.

REFERENCES

AASHTO (2014) *LRFD Bridge Design Specifications*. Washington, DC, American Association of State Highway and Transportation Officials.

Cimellaro, G.P., Reinhorn, A.M. & Bruneau, M. (2010) Framework for analytical quantification of disaster resilience. *Engineering Structures*, 32 (11), 3639–3649.

Stochastic modeling of post-repair performance and integration into bridge life-cycle assessment

N. Vishnu & J.E. Padgett
Department of Civil & Environmental Engineering, Rice University, Houston, Texas, USA

ABSTRACT

A holistic bridge life-cycle assessment framework includes the impact of bridge failures due to truck loads as well as hazards, along with other life-cycle phases traditionally considered such as construction, maintenance, and end of life. This step is integrated in life-cycle analyses as costs incurred due to failure, such as repair and restoration of the bridge. Stochastic models such as renewal processes and specifically Poisson processes are used to effectively model this restoration during the bridge life-cycle. As shown in Figure 1(a), homogenous Poisson and other renewal models commonly assume that the bridge is completely restored to its as-built or pre-hazard event condition after repair (Nuti and Vanzi, 2003, Kumar and Gardoni, 2014). For deteriorating bridges, a non-homogenous Poisson process, as shown in Figure 1(b) is adopted with the capacity restoration assumed to be "pre-earthquake level" (Ghosh and Padgett, 2011, Padgett and Tapia, 2013). However, it is likely that some repair actions such as concrete patching and grouting for concrete bridge columns only restore partial capacity and functionality to the bridge. This paper formulates a stochastic methodology to model the uncertainty in the choice of the repair action as well as the uncertainty in post-repair bridge performance within a life-cycle formulation. The general formulation relieves restrictions of Poisson assumptions for bridge capacity and subsequent damage models while still assuming the hazard occurrence to be a Poisson process. A comparison of the life-cycle metrics in terms of embodied energy for a case study bridge column shows that the homogeneous Poisson model often underestimates the lifetime earthquake repair costs and the non-homogeneous model overestimates lifetime earthquake repair costs. This highlights the need to capture the uncertainty in the post-repair modeling of a bridge and opportunities to explore alternative stochastic processes to model hazard cost.

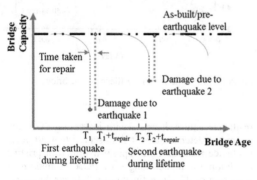

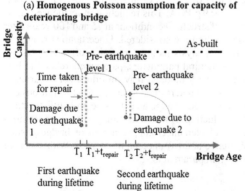

Figure 1. Poisson assumptions for bridge restoration after hazard.

REFERENCES

Ghosh, J. & Padgett, J. E. (2011) Probabilistic seismic loss assessment of aging bridges using a component-level cost estimation approach. *Earthquake Engineering & Structural Dynamics*, 40, 1743–1761.

Kumar, R. & Gardoni, P. (2014) Renewal theory-based life-cycle analysis of deteriorating engineering systems. *Structural Safety*, 50, 94–102.

Nuti, C. & Vanzi, I. (2003) To retrofit or not to retrofit? *Engineering Structures*, 25, 701–711.

Padgett, J.E. & Tapia, C. (2013) Sustainability of natural hazard risk mitigation: Life cycle analysis of environmental indicators for bridge infrastructure. *Journal of Infrastructure Systems*, 19, 395–408.

Life-Cycle Analysis and Assessment in Civil Engineering: Towards an Integrated Vision – Caspeele, Taerwe & Frangopol (Eds)
© 2019 Taylor & Francis Group, London, ISBN 978-1-138-62633-1

Probabilistic resilience assessment of infrastructure—a review

C. de Paor, L. Connolly & A. O'Connor
Roughan & O'Donovan Innovative Solutions, Dublin, Ireland

ABSTRACT

Resilience is typically defined as the capacity of a system, community or society potentially exposed to hazards to adapt, by resisting or changing in order to reach and maintain an acceptable level of function and structure (Bruneau *et al.*, 2003; Croope & McNeil, 2011; RESILENS consortium, 2016). This could be in response to a climate hazard (e.g. flooding, hurricane), natural hazard (e.g. earthquake), or a man-made hazard such as a cyber-attack or terrorism. Once 'rare' extreme weather events are now becoming increasingly frequent due to climate change, bringing with them potential to cause significant disruption to infrastructure networks. This has resulted in a need for our infrastructure to become more resilient (Hughes & Healy, 2014; Carey *et al.*, 2017). That is, a requirement for physical structures to withstand higher loading more frequently, for our emergency response organisations to react quickly, and for systems to resume normal operations quickly following a hazard or extreme event.

In this study, a quantitative methodology to determine resilience of infrastructure components, elements and structures to climate change and extreme weather change events is developed.

Traditional methods of quantitative resilience measurement consider system functionality, losses and recovery following an event deterministically. In this study, a relationship between loss and fragility is presented allowing resilience to be determined probabilistically through the use of fragility functions in cases where detailed structural data on system elements or components is available.

Fragility functions allow the user determine the probability that system response will exceed the performance limit state. They are essentially cumulative distribution functions indicating probability of limit state exceedance of a structure for a given hazard intensity measure (Guidotti *et al.*, 2016).

This methodology may be further expanded to a wider system or network.

REFERENCES

Bruneau, M., Chang, S.E., Eguchi, R.T., Lee, G.C., O'Rourke, T.D., Reinhorn, A.M., Shinozuka, M., Tierney, K., Wallace, W.A. & von Winterfeldt D. (2003) A framework to quantitatively assess and enhance the seismic resilience of communities. *Earthquake Spectra*. [Online] 19 (4), 733–752. Available from: doi:10.1193/1.1623497.

Carey, C., Clarke, J., Corbally, R., Connolly, L., et al. (2017) *Deliverable 6.3 – Report on benefits of critical infrastructure protection.* [Online]. (6.3). Available from: http://rain-project.eu/wp-content/uploads/2017/08/6.3_-FInal.pdf.

Croope, S. & McNeil, S. (2011) Improving resilience of critical infrastructure systems postdisaster. *Transportation Research Record: Journal of the Transportation Research Board*. [Online] 2234, 3–13. Available from: doi:10.3141/2234-01.

Guidotti, R., Chmielewski, H., Unnikrishnan, V., Gardoni, P., et al. (2016) Modeling the resilience of critical infrastructure: The role of network dependencies. *Sustainable and Resilient Infrastructure*. [Online] 1 (3–4), 153–168. Available from: doi:10.1080/23789689.2016.1254999.

Hughes, J.F. & Healy, K. (2014) *Measuring the Resilience of Transport Infrastructure*. [Online]. (February). Available from: http://www.nzta.govt.nz/assets/resources/research/reports/546/docs/546.pdf.

RESILENS Consortium (2016) *Methods for Resilience Assessment*. [Online] 138. Available from: resilens.eu/wp-content/uploads/2016/06/D2.2-Methods-for-Resilience-Assessment-Final.pdf.

MS10: Next generation asset management of civil infrastructure systems
Organizers: A. Michel, H. Stang, M.R. Geiker & M.D. Lepech

Life-Cycle Analysis and Assessment in Civil Engineering: Towards an Integrated Vision – Caspeele, Taerwe & Frangopol (Eds)
© 2019 Taylor & Francis Group, London, ISBN 978-1-138-62633-1

Comparison of optimization approaches for pavement maintenance and rehabilitation policies on road section and network level

V. Donev & M. Hoffmann
Institute of Transportation, Vienna University of Technology, Austria

ABSTRACT

The main goal in pavement management is to efficiently address various distress types with adequate maintenance and rehabilitation treatments (M&R). Common pavement management systems (PMS) aggregate short survey sections into long homogeneous sections aiming to reduce the amount of data and computational effort. However, this leads to loss of information, imprecise predictions and violations of condition thresholds contradicting ever more effective and accurate means of condition survey. The solution of this problem based on a novel life cycle approach allowing for an optimization of treatment type, timing and work zones is addressed elsewhere (Donev & Hoffmann 2017, 2018).

Furthermore, this paper shows that using optimization criterion based on the area between the do-nothing and the post-treatment performance curves of a weighted overall condition index (effectiveness area) leads to substantially more expensive M&R treatment strategies. Figure 1 shows that the strategy based on maximum effectiveness results in 62% higher average annual budget in comparison to the strategy based on minimum life cycle costs. In general, the strategy based on maximum effectiveness gives preference to heavy rehabilitation (larger area between the curves) and neglects maintenance treatments. Furthermore, the results of the optimization are different for different weights of the overall condition index. Prioritization under (tight) budget restrictions using the marginal cost-effectiveness approach will lead to even worse results, because fewer sections will receive more costly treatments until the budget is exhausted. This may result in an M&R backlog on roads with lower traffic volumes (if the effectiveness is multiplied by the traffic). Based on a simulation study the paper provides a short overview and comparison of both common and novel life cycle approach providing proof that this new approach yields far more practical and economic results under the same conditions (ceteris paribus).

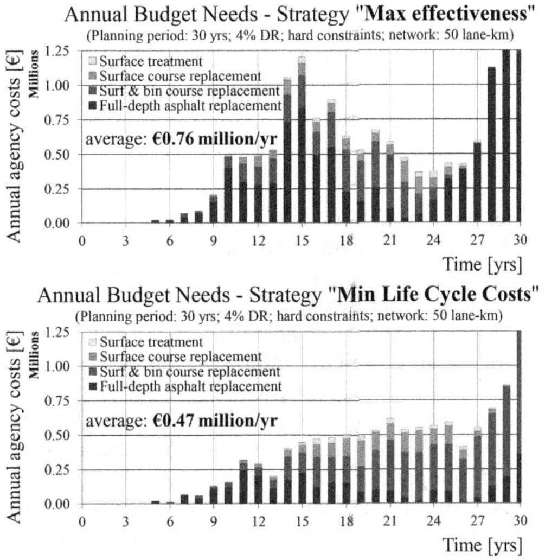

Figure 1. Comparison of annual budget needs without budget restrictions based on approaches using maximum effectiveness vs. minimum life cycle costs as optimization criteria.

REFERENCES

AASHTO (2012) *Pavement Management Guide*. 2nd edition. Washington, DC, American Association of State Highway and Transportation Officials.

Donev, V. & Hoffmann, M. (2017) Network-level optimization of pavement M&R activities, timing and work-zone length for multiple distress types. In: *World Conference on Pavement and Asset Management*, 12–16 June, Milan, Italy.

Donev, V. & Hoffmann, M. (2018) From cause-specific treatment selection on single road sections to work-zone optimization with user and external costs. In: *7th Transport Research Arena*, 16–19 April, Vienna, Austria.

Haas, R., Hudson, W.R. & Zaniewski, J. (1994) *Modern Pavement Management*. Malabar, FL, Krieger.

Weninger-Vycudil, A., Simanek, P., Haberl, J. & Rohringer, T. (2009) *Handbuch Pavement Management in Österreich* 2009 [Manual pavement management in Austria 2009]. Wien: BMVIT, Straßenforschung Heft 584 (in German).

Coupled mass transport, chemical, and mechanical modelling in cementitious materials: A dual-lattice approach

A. Michel, V.M. Meson & H. Stang
Technical University of Denmark, Kgs. Lyngby, Denmark

M.R. Geiker
Norwegian University of Science and Technology, Trondheim, Norway

M. Lepech
Stanford University, Stanford, California, USA

ABSTRACT

Deterioration of the civil infrastructure, such as bridges, tunnels, roads, and buildings, together with increasing functional requirements (*e.g.* traffic load and intensity) presents major challenges to society in most developed countries. A major part of the infrastructure is built from concrete and costs for maintenance, renovation, and renewing are growing and by now taking up a major part of concrete structure investments. While engineering tools and methods are well developed for the structural design of new structures, tools for assessing current and predicting the future condition of reinforced concrete structures are less advanced. Existing prediction tools are largely empirical, and thus limited in their ability to predict the performance of new material, structural, or maintenance solutions. As such, the inability to reliably assess the long-term future ramifications of to-day's design decisions poses a major obstacle for the design of reinforced concrete structures. A primary reason for the lack of reliable modelling tools is that deterioration mechanisms are highly complex, involve numerous coupled physical phenomena that must be evaluated across a range of scales, and often cut across several academic disciplines and faculties.

In this paper, a dual-lattice approach for coupled mass transport, chemical, and mechanical modelling for the deterioration prediction of cementitious materials is outlined (see Figure 1). Deterioration prediction is thereby based on coupled modelling of *i*) mass transport, *i.e.* moisture and ionic transport, in porous media, *ii*) thermodynamic modelling of phase equilibria in cementitious materials, and *iii*) mechanical performance including corrosion- and load-induced damages. The presented lattice approach is fully coupled, *i.e.* information, such as moisture content, phase assemblage, damage state, transport properties, *etc.*, are constantly exchanged within the model. While the general concept of the modelling framework and finite element method formulation for the fully coupled problem are outlined in this paper, a brief numerical example is provided to demonstrate the applicability of some aspects of the modelling framework. Selected results comprise ingress of chloride ions and moisture as well as charge balance in the center region of the cementitious beam domain taking into account the effect of cracks on the mass transport, see e.g. Figure 2.

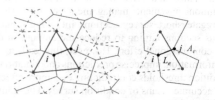

Figure 1. Structural (broken lines) and mass transport (full lines) lattice elements in dual-lattice mesh along with elemental dimensions for the mass transport lattice (right figure), after (Saka, 2013).

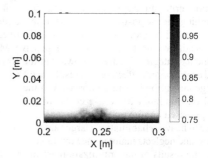

Figure 2. Distribution of relative humidity (RH) in cracked beam after 28 days of exposure.

REFERENCE

Saka, T. (2013) Dual-lattice-based simulations of coupled fracture-flow in reinforced concrete. In: *Proceedings of the 8th International Conference on Fracture Mechanics of Concrete and Concrete Structures, Framcos 2013*. Available at: http://findit.dtu.dk/en/catalog/249939251 [Accessed: 9 April 2018].

Measuring critical chloride contents in structures and the influence on service life modeling

C. Boschmann Käthler & U.M. Angst
Institute for Building Materials, ETH Zurich, Zurich, Switzerland

B. Elsener
Institute for Building Materials, ETH Zurich, Zurich, Switzerland
Department of Chemical and Geological Science, University of Cagliari, Cagliari, Italy

ABSTRACT

Given the major challenges arising from the continuous aging of infrastructures in all industrialized countries, there is an ever-increasing need for methods to reliably assess the condition of potentially corroding reinforced concrete structures. There exists currently no generally accepted method to assess the ability of a structure to withstand corrosion.

An important parameter in this regard is the critical chloride content (C_{crit}). Due to the lack of a reliable method to determine C_{crit}, this value is typically looked up in standards and text books (fib, 2006, SIA, 2011), based on empirical experience. In other words, the same values are used in the condition assessment of all kind of structures, irrespective of the materials used, site concrete properties, exposure history, age, etc.

This paper presents results obtained with a novel method that allows the determination of C_{crit} for individual engineering structures and reduces therefore current uncertainties in condition assessments (Angst et al., 2017). The presented case is a reinforced concrete tunnel in the Swiss alps, heavily exposed to road salts (XD3 exposure) during the winter half year. The age of the structure is 35 years. The tunnel consists of 16 elements, each of a length of 24 m. From the wall of two adjacent elements of the structure (labelled A & B), drilling cores ø 150 mm containing a steel bar were taken and tested in the laboratory. The results, depicted in Figure 1, indicate, that average values and distribution of C_{crit} may differ significantly from those given in the literature and in codes. Additionally, C_{crit} may differ significantly between individual structural elements. These results greatly influence the predicted time to corrosion initiation necessary to estimate the residual service life, as illustrated in Figure 2.

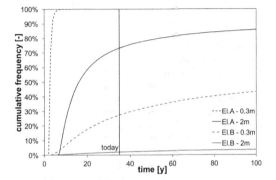

Figure 2. Probability of corrosion initiation for both elements (Black: Element A, Gray: Element B) at two different heights (Dashed: 0.3 m, solid: 2 m).

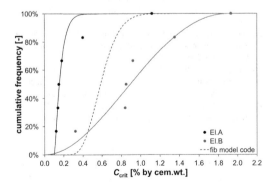

Figure 1. C_{crit} for the investigated structure, compared to C_{crit}-distribution of (fib, 2006).

REFERENCES

Angst, U.M., Boschmann, C., Wagner, M. & Elsener, B. (2017) Experimental protocol to determine the chloride threshold value for corrosion in samples taken from reinforced concrete structures. *Journal of Visualized Experiments*, 126.

fib (2006) *Model Code for Service Life Design*. Lausanne, International Federation for Structural Concrete (fib).

SIA (2011) SIA 269/2: Erhaltung von Tragwerken – Betonbau. Zürich, Schweizerischer Ingenieur- und Architektenverein.

A framework for modeling corrosion-related degradation in reinforced concrete

Z. Zhang & U.M. Angst
Institute for Building Materials, ETH Zurich, Zurich, Switzerland

A. Michel
Department of Civil Engineering, Technical University of Denmark, Kgs. Lyngby, Denmark

ABSTRACT

The corrosion of the reinforcement steel is the most common cause of degradation of reinforced concrete. However, most degradation models are empirical and based on the assumptions, such as homogeneous concrete structure and uniform corrosion products around rebar. They also intentionally ignore the penetration of corrosion products into the concrete matrix (Jamali et al. 2013). According to the degradation process, this paper proposes a framework for modelling corrosion-related degradation (see Figure 1).

This framework starts with establishing a microstructure for the steel-concrete composition material that must represent the real structure. Rather considering the composition material as a homogenous medium, we take into account the heterogeneity of concrete by using an "image-based local homogenization method" which has been employed to model mass transport (Zhang et al. 2018). Then, mass transport, including moisture, gaseous phases (dry air and O_2), and ions (especially OH^- and Fe^{2+}), must be well described. For the unsaturated condition, the transport separation of liquid and vapor should be considered (Zhang et al. 2015 & 2016). The multispecies transport can be adequately modelled by the Poisson-Nernst-Plank (PNP) equations which includes the electrical coupling between the different ions

Table 1. Relevant models for the proposed framework.

Stage	Option 1	Option 2
Structure	Homogeneous	Composite material
Moisture transport	Diffusion	Multiphase
Ion transport	Diffusion	PNP
Thermodynamic	GEMS	PHREEQC
Mechanics	Expansion	Poromechanics

(Samson & Marchand 1999). A simple way to consider the reactive transport is to use the reaction kinetic which can provide reasonable results (Stefanoni et al. 2018). For much complex chemical interactions and phase changes, a geochemical solver is suggested to perform chemical equilibrium calculations, such as GEM and PHREEQC. When solid phases precipitate in concrete, stresses are gradually generated which can be simulated by poromechanical considerations (Flatt & Scherer 2008). The relevant models in this paper are summarized in Table 1.

REFERENCES

Flatt, R.J. & Scherer, G.W. (2008) Thermodynamics of crystallization stresses in DEF. *Cement and Concrete Research*, 38, 325–336.

Jamali, A., Angst, U., Adey, B. & Elsener, B. (2013) Modeling of corrosion-induced concrete cover cracking: A critical analysis. *Construction and Building Materials*, 42, 225–237.

Samson, E. & Marchand, J. (1999) Numerical solution of the extended Nernst-Planck model. *Journal of Colloid and Interface Science*, 215 (1), 1–8.

Stefanoni, M., Zhang, Z., Angst, U. & Elsener, B. (2018) Corrosion-induced concrete cracking: bridging electrochemistry and engineering. *RILEM Technical Letters*.

Zhang, Z., Angst, U., Michel, A. & Jensen, M.A. (2018) An image-based local homogenization method to model mass transport at the steel-concrete interface. In: *6th ICDCS*. Leeds, UK.

Zhang, Z., Thiery, M. & Baroghel-Bouny, V. (2016) Investigation of moisture transport properties of cementitious materials. *Cement and Concrete Research*, 89, 257–268.

Zhang, Z., Thiery, M. & Baroghel-Bouny, V. (2015) Numerical modelling of moisture transfers with hysteresis within cementitious materials: Verification and investigation of the effects of repeated wetting-drying boundary conditions. *Cement and Concrete Research*, 68, 10–23.

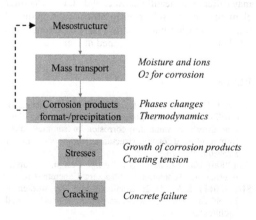

Figure 1. The proposed framework for modelling.

Life-Cycle Analysis and Assessment in Civil Engineering: Towards an Integrated Vision – Caspeele, Taerwe & Frangopol (Eds)
© 2019 Taylor & Francis Group, London, ISBN 978-1-138-62633-1

Design and maintenance of concrete structures requires both engineering and sustainability limit states

M.R. Geiker
Department of Structural Engineering, Norwegian University of Technology and Science, Norway

A. Michel & H. Stang
Department of Civil Engineering, Technical University of Denmark (DTU), Denmark

H. Vikan
The Norwegian Public Roads Administration, Norway

M.D. Lepech
Department of Civil and Environmental Engineering, Stanford University, California, USA

ABSTRACT

Reinforced concrete structures are – and will continue to be – a central part of the transportation infrastructure. Taking Norway as an example, the national and county transportation networks count about 17500 bridges and ferry quays, and about 2000 other load-carrying structures. Of these, 72% are concrete structures.

Huge sums are invested in the transportation infrastructure and the aging of the infrastructure is a recognized problem for the society. In the United States, 9.1% of bridges were recently reported as structurally deficient (ASCE, 2017), and a maintenance backlog of for Norwegian bridges and harbor structures has been estimated to 11–18 billion NOK (Sund, 2013, Sund, 2012). In 2012 Polder *et al.* (Polder *et al.*, 2012) published an overview maintenance expectations for Dutch reinforced concrete motorway bridges. They foresee that the number of bridges needing repair will increase by a factor of 2–4 within 20 years. It is their expectation that similar or larger repair needs will be faced in other European countries.

In addition to an aging infrastructure network and constrained budgets, additional challenges have arisen in recent years, most notably the contribution of infrastructure to global climate change. Thus, in response to the Paris Agreement, the Norwegian Transportation Plan 2018–2029 sets a reduction goal of 40% (from 1999 to 2030) for the emission of greenhouse gasses from the transport infrastructure (Norwegian Ministry of Transport and Communications, 2017), and by 2050, Norway is to be a low carbon emission society. In order to become a low emission society, measures must be taken to reduce emission of greenhouse gasses

during construction, operation and maintenance of the infrastructure.

Society depends on a safe and well-functioning transportation infrastructure. Environmental and economic demands stresses the need for innovation in materials, design and processes used for establishing and maintaining a sustainable transportation infrastructure. To support this we need digital frameworks for parametric and computational design.

Sustainable design and maintenance of reinforced concrete structures requires both engineering and sustainability limit states be considered. Reliable and valid multi-scale and multi-physics prediction models are needed for quantification of the environmental impact of the structures through their entire life cycle as well as design optimization and maintenance planning.

REFERENCES

ASCE (2017) Infrastructure Report Card.
Norwegian Ministry of Transport and Communications (2017) The Norwegian Transportation Plan 2018–2029, Meld. St. 33 (2016–2017) Report to the Storting (white paper).
Polder, R.B., Peelen, W.H.A. & Courage, W.M.G. (2012) Non-traditional assessment and maintenance methods for aging concrete structures – technical and non-technical issues. *Materials and Corrosion,* 63, 1147–1153.
Sund, E. (2012) Hva vil det koste å fjerne forfallet på riksvegnettet? Resultat av kartlegging. *Statens vegvesens rapporter No. 75.* Oslo, Norway, Statens vegvensen.
Sund, E. (2013) Hva vil det koste å fjerne forfallet på fylkesvegnettet? Resultat av kartlegging. *Statens vegvesens rapporter No. 183.* Oslo, Norway, Statens vegvensen.

Life-Cycle Analysis and Assessment in Civil Engineering: Towards an Integrated Vision – Caspeele, Taerwe & Frangopol (Eds)
© 2019 Taylor & Francis Group, London, ISBN 978-1-138-62633-1

Long-term planning within complex and dynamic infrastructure systems

M. Havelaar
Iv-Infra, Sliedrecht, The Netherlands

W. Jaspers
Bam Infra, Gouda, The Netherlands

A.R.M. Wolfert, G.A. van Nederveen & W.L. Auping
Delft University of Technology, Delft, The Netherlands

ABSTRACT

Cities are rapidly transforming, and infrastructures are not always resistant to these future developments. On a global scale cities are growing into mega cities, whilst counter urbanization causes significant population declines. To adapt infrastructures against these future developments, infrastructure asset management can be applied. Infrastructure asset managers require reliable insight in future developments, to make their intervention decisions strategically. This involves assessing multiple variables, and their underlying relations. Infrastructure systems can perform multiple functions, have various connecting interfaces, and are mostly situated in a dynamically changing environment. Therefore, intervention decisions are subjected to dynamic complexity, and uncertainty. However, asset managers predominantly make large intervention decisions based on decision support methods of a static nature, which provide insufficient insight in the dynamic complexity, and uncertainty, of future developments. Since asset managers perceive large challenges with decision-making, a decision support method able to incorporate all relevant complexity, and uncertainties, is inadmissible for infrastructure asset management.

This study aimed at providing substance to the challenges associated with large intervention decisions, by assessing dynamic complexity, and uncertainty, with multivariate simulation approaches as the Exploratory System Dynamics Modelling and Analysis approach

(ESDMA), and adapting strategies for large intervention decisions with the adaptation pathways approach. The approaches were applied to a case study, which is a highly schematized representation of the city of Amsterdam, with its interconnected infrastructure network.

The study showed that the proposed approaches have the potential to provide substance to the challenges associated with large interventions decisions on an infrastructure network level, by adapting them dynamically over time to uncertainty, and complexity. A framework could be developed including an adaptive set of future interventions, balanced on the basis of costs, target effects and side effects.

REFERENCES

Bhamidipati, S., Lei, T.V.D. & Herder, P. (2016) A layered approach to model interconnected infrastructure and its significance for asset management. *European Journal of Transport and Infrastructure Research (EJTIR)*, 16 (1), 254–272.

Haasnoot, M., Kwakkel, J.H., Walker, W.E. & Maat, J.T. (2013) Dynamic adaptive policy pathways: A method for crafting robust decisions for a deeply uncertain world. *Global Environmental Change*, 23, 485–598.

Pruyt, E. & Kwakkel, J. (2013) *Using System Dynamics for Grand Challenges: The ESDMA Approach*. Retrieved from Delft University of Technology.

UNDP (2017) *Probabilistic Population Projections based on the World Population Prospects: The 2017 Revision*.

Life-Cycle Analysis and Assessment in Civil Engineering: Towards an Integrated Vision – Caspeele, Taerwe & Frangopol (Eds)
© 2019 Taylor & Francis Group, London, ISBN 978-1-138-62633-1

Scheduling of waterways maintenance interventions applying queueing theory

F. Marsili & J. Bödefeld
Federal Waterways Engineering and Research Institute, Karlsruhe, Germany

H. Daduna
Department of Mathematics, University of Hamburg, Germany

P. Croce
Department of Civil and Industrial Engineering, University of Pisa, Italy

ABSTRACT

The geographical configuration of the German landscape presents an extended network of navigable rivers that has been exploited since the Roman age for the transportation of freights and it represents together with roads, rails and pipelines part of the ground-based traffic route network of the country.

Although the economies of using preexisting waterways are significant, efforts were required in order to improve their navigability. Thus in the 19th and 20th century, considerable expenditures have been made on the construction of locks, culverts, ports and weirs. Nowadays the waterways system has about 310 locks, 320 weirs, 450 culverts, 45 canal bridges, 2 ship lifts, 2 dams and about 1600 bridges. The fixed assets amount to approximately €50 billion. Several of these infrastructures have exceeded their design working life and are affected by deterioration phenomena: locks especially raise concerns because most of them have only one chamber and they are around 80–100 years old, showing evident signs of advanced degradation.

Resulting in a series system of infrastructures, the stall of only one lock may threaten the navigability of the entire waterway. For this reason inspections and maintenance – and in general the management of the asset – are of primary importance for the proper functioning of the entire network, also in view of the continuing transport growth foreseen in the upcoming years. However also during inspection and maintenance, the infrastructure should be put out of service; being these unavoidable actions, inoperative periods should be planned in such a way that their impact on the transportation of goods – the operability, or availability, of the network – is minimized.

The problem of scheduling maintenance interventions on waterways network can be approached with several methods; one way is to resort to queueing theory, a discipline originally born to manage telephone calls (Erlang 1909), but later largely applied whenever the study of waiting lines matters, like in computing and traffic engineering.

This paper especially proposes to model the waterways network as a system of queues with unreliable servers subjected to breakdowns and repairs, and to use simple explicit formulae in order to assess the performance of the system. An approximated equation has been used in order to compute the average queue size of one single lock undergoing inoperative periods. This approach allows the comparison of different maintenance policies, and eventually the choice of the optimal one. Results confirm that a maintenance strategy based on short and frequent interruptions has less impact on the operability of the network compared to those based on longer and sporadic intervention. Therefore, whenever the inoperative time devoted to the preparation of the intervention is short, and the maintenance intervention can be fractioned into smaller works, maintenance policies based on short but recurrent interruptions are preferable.

REFERENCES

Erlang, A.K. (1909) The theory of probabilities and telephone conversation, *Nyt Tidsskrift for Matematik* B, 20, 33–39. Rotterdam: Balkema.

Kleinrock, L. (1975) *Queueing Systems, Volume I: Theory*, John Wiley & Sons.

Sommer, J., Berkhout, J., Daduna, H. & Heidergott, B. (2016) *Analysis of Jackson Network With Infinite Supply and Unreliable Nodes*. Queueing Syst, Springer. pp. 181–207.

MS11: Life-cycle performance of structure and infrastructure under uncertainty
Organizers: M. Akiyama & D.M. Frangopol

Hazard analysis for bridge scour evaluation at watershed level considering climate change impact

D.Y. Yang & D.M. Frangopol
Department of Civil and Environmental Engineering, ATLSS Engineering Research Center, Lehigh University, Bethlehem, PA, USA

ABSTRACT

Scour is one of the major causes of bridge failures in the United States. Due to climate change and its associated extreme hydrological events, bridge scour becomes more severe and frequent. Therefore, hazard analysis for bridge scour evaluation must consider the detrimental effects of climate change. In this paper, a systematic approach is proposed to analyze the hazards causing bridge scour, in particular, flooding with extreme flow discharge. The future and/or ungauged discharge data are obtained by using hydrologic modeling at watershed level. The climate change impact is considered using projected climate data based on the global climate model (GCM) and different climate change scenarios. The proposed approach is applied to the Charles River watershed, Massachusetts, United States (see Figure 1).

As stream gauge data are no longer necessary, the proposed method is especially useful for scour evaluation of bridges without stream gauge data. More importantly, using hydrologic modeling, projected climate data can be incorporated into the analysis. Based on the proposed approach, this paper studies the changes in flow discharge of streams in the Charles River watershed in the 21st century. The GCM developed by Max Planck Institute for Meteorology and the climate change scenario of RCP 8.5 are used in the analysis (Taylor et al. 2012; IPCC 2014).

For the Charles River watershed under consideration, it is found from the analysis that the downscaled climate data from GCM and the selected climate change scenario indicate significant increase in temperature and a relatively mild change in precipitation pattern. Despite the less significant change in precipitation pattern, climate change is likely to affect the flow discharge within the Charles River watershed. The analysis shows that, as a result of climate change, the region is likely to suffer from more severe droughts and floods. The latter can dramatically increase the risk of scour-induced bridge failures. Therefore, bridge

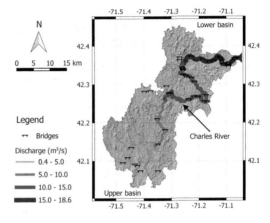

Figure 1. Flow discharge distribution in Charles River watershed in year 2050 under climate change scenario RCP 8.5.

management in this region must consider the detrimental effects caused by climate change.

The results in this paper provide useful input for climate change-informed risk assessment and mitigation with respect to scour- and flood-induced bridge failures (Mondoro & Frangopol 2018). The approach shown herein can also be applied to other regions.

REFERENCES

IPCC (2014) *Climate Change 2014: Synthesis Report.* Geneva, Switzerland, Intergovernmental Panel on Climate Change (IPCC).

Mondoro, A. & Frangopol, D.M. (2018) Risk-based cost-benefit analysis for the retrofit of bridges exposed to extreme hydrologic events considering multiple failure modes. *Engineering Structures*, in press.

Taylor, K.E., Stouffer, R.J. & Meehl, G.A. (2012) An overview of CMIP5 and the experiment design. *Bulletin of the American Meteorological Society*, 93 (4), 485–498.

Sensitivity analysis of time dependent reliability of RC members in general climate environment

D.-G. Lu, W.-H. Zhang & W. Wang
School of Civil Engineering, Harbin Institute of Technology, Harbin, Heilongjiang, China

ABSTRACT

Aging reinforced concrete (RC) members in buildings and bridges are subjected to the service loadings and aggressive environment, which may cause the degradation of structural resistance and affect the safety and serviceability of structures. Many existing RC structures in China have been in ill condition due to concrete carbonation and reinforcement corrosion. It is urgent to assess the durability, predict the remaining service life and monitor the health of these existing structures. The time-dependent reliability of deteriorated RC beams and columns in general climate environment of China is analyzed by four different approaches (Lu, Fan & Jiang 2012) in this paper.

In structural reliability analysis, there are many random variables and complex correlations among the random variables need to be considered. The influence of each random variable on the reliability is different. The key technique of investigation of the importance of each random variable to the structural reliability is the sensitivity analysis (Hohenbichler & Rackwitz 2007). To identify the most influencing parameters on the time dependent reliability indices of RC beams and columns in different actions, the time-variant

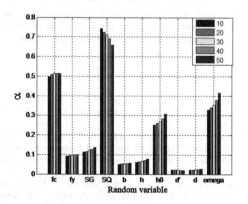

Figure 1. The change with time of importance vector α.

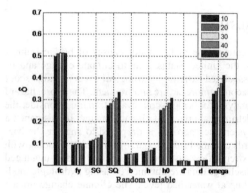

Figure 2. The change with time of importance vector δ.

sensitivity analyses for the time dependent reliabilities of RC members during their life cycles are performed using four reliability sensitivity measures (Lu & Li 2007). To investigate the changing rules of the time dependent reliability indices with the influencing parameters, the parametric analyses are further conducted for the time-variant reliability sensitivities of RC members. The analysis results can provide deep insights into the time-variation laws of dominating random variables for durability assessment and design for concrete structures.

REFERENCES

Hohenbichler, M. & Rackwitz, R. (2007) Sensitivity and importance measures in structural reliability. *Civil Engineering Systems*, 3, 203–209.

Lu, D.G. & Li, X.P. (2007) Probability importance analysis of seismic reliability for steel frame structures. *Journal of Building Structures*, 28, 156–164+178. (in Chinese)

Lu, D.G., Fan, X.P. & Jiang, W. (2012) Comparisons of four time-dependent reliability approaches for safety assessment of deteriorated concrete bridges. In: *6th International Conference on Bridge Maintenance, Safety and Management. Stresa, Lake Maggiore, Italy: IABMAS.* pp. 3442–3448.

Life-Cycle Analysis and Assessment in Civil Engineering: Towards an Integrated Vision – Caspeele, Taerwe & Frangopol (Eds)
© 2019 Taylor & Francis Group, London, ISBN 978-1-138-62633-1

Seismic fragility of corroding RC bridge substructures in marine environment

F. Cui
College of Civil Engineering, Shandong Jiaotong College, China
Department of Civil Engineering, The City College of the City University of New York, USA
School of Highways, Chang'an University, China

M. Ghosn
Department of Civil Engineering, The City College of the City University of New York, USA

ABSTRACT

This paper presents an improved reinforced concrete steel bar deterioration model that incorporates pitting corrosion and considers the change in after-cracking corrosion rate to assess the time-dependent seismic fragility of RC bridge substructures in marine environments.

As an application, the proposed model is implemented to conduct a probabilistic seismic fragility analysis of a three-span continuous box girder bridge accounting for uncertainties in bridge geometry, material properties, ground motion and corrosion parameters. The results show that the effect of chloride-induced corrosion cannot be neglected when performing the seismic fragility analysis of RC bridge substructures in marine environments. Additionally, the calculated time-dependent fragility curves indicate that there is a nonlinear accelerated growth of RC column vulnerability along the service life of highway bridges, especially after twenty-five years of exposure to chlorides. This highlights the importance of considering the joint effects of earthquakes and corrosion on bridge safety.

REFERENCES

Alipour, A., Shafei, B. & Shinozuka, M. (2011) Performance evaluation of deteriorating highway bridges located in high seismic areas. *Journal of Bridge Engineering*, 16 (5), 597–611.

Bastidas-Arteaga, E., Bressolette, P., Chateauneuf, A. & Sánchez-Silva, M. (2009) Probabilistic lifetime assessment of RC structures under coupled corrosion-fatigue deterioration processes. *Structural Safety*, 31 (1), 84–96.

Cao, C., Cheung, M.M.S. & Chan, B.Y.B. (2013) Modelling of interaction between corrosion-induced concrete cover crack and steel corrosion rate. *Corrosion Science*, 69, 97–109.

Choe, D.E., Gardoni, P., Rosowsky, D. & Haukaas, T. (2008) Probabilistic capacity models and seismic fragility estimates for RC columns subject to corrosion. *Reliability Engineering and System Safety*, 93 (3), 383–393.

Code for durability assessment of concrete structures (CECS) (2007) *Chinese Institute of Construction Standardization*. Beijing (China).

Cornell, C.A. (2002) Probabilistic basis for 2000 SAC federal emergency management agency steel moment frame guidelines. *Journal of Structural Engineering*, (4), 526–533.

Cui, F.K., Zhang, H., Ghosn, M. & Xu, Y. (2018) Seismic fragility analysis of deteriorating RC bridge substructures subject to marine chloride-induced corrosion. *Engineering Structures*, 155, 61–72.

Darmawan, M.S. & Stewart, M.G. (2007) Spatial time-dependent reliability analysis of corroding pretensioned prestressed concrete bridge girders. *Structural Safety*, 29 (1), 16–31.

Dong, Y., Frangopol, D.M. & Saydam, D. (2013) Time-variant sustainability assessment of seismically vulnerable bridges subjected to multiple hazards. *Earthquake Engineering Structural Dynamics*, 42 (10), 1451–1467.

Du, Y.G., Clark, L.A. & Chan, A.H.C. (2005) Residual capacity of corroded reinforcing bars. *Magazine of Concrete Research*, 57 (3), 135–147.

DuraCrete (2000) Statistical quantification of the variables in the limit state functions. The European Union – Brite EuRam III.

Ghosh, J. & Padgett, J.E. (2010) Aging considerations in the development of time-dependent seismic fragility curves. *Journal of Structural Engineering*, 136 (12), 1497–1511.

Guo, Y., David, T. & Yim, S. (2014) New model for estimating the time-variant seismic performance of corroding RC bridge columns. *Journal of Structural Engineering ASCE*, 141 (6), 1–12.

Liu, T. & Weyers, R. (1998) Modeling the dynamic corrosion process in chloride contaminated concrete structures. *Cement and Concrete Research*, 28 (3), 365–379.

Melchers, R. (2004) Pitting corrosion of mild steel in marine immersion environment-part 1: maximum pit depth. *Corrosion*, 60 (9), 824–836.

Picandet, V., Khelidj, A. & Bellegou, H. (2009) Crack effects on gas and water permeability of concretes. *Cement and Concrete Research*, 39 (6), 537–547.

Thoft-Christensen, P. (2000) Stochastic modeling of the crack initiation time for reinforced concrete structures. In: *Proceedings of the 2000 Structures Congress and Exposition ASCE. Pennsylvania (USA).*

Val, D.V. & Melchers, R.E. (1997) Reliability of deteriorating RC slab bridges. *Journal of Structural Engineering*, 123 (12), 1638–1644.

Vidal, T., Castel, A. & François, R. (2004) Analyzing crack width to predict corrosion in reinforced concrete. *Cement and Concrete Research*, 34 (1), 165–174.

Vidal, T., Castel, A. & François, R. (2007) Corrosion process and structural performance of a 17 year old reinforced concrete beam stored in chloride environment. *Cement and Concrete Research*, 37 (11), 1551–1561.

Vu, K.A.T. & Stewart, M.G. (2000) Structural reliability of concrete bridges including improved chloride-induced corrosion models. *Structural Safety*, 22 (4), 313–333.

Yuan, Y., Jiang, J. & Peng, T. (2010) Corrosion process of steel bar in concrete in full lifetime. *ACI Materials Journal*, 107 (6), 562–567.

Life-Cycle Analysis and Assessment in Civil Engineering: Towards an Integrated Vision – Caspeele, Taerwe & Frangopol (Eds)
© 2019 Taylor & Francis Group, London, ISBN 978-1-138-62633-1

Time-dependent structural reliability analysis of shield tunnels in coastal regions

Z. He & M. Akiyama
Waseda University, Tokyo, Japan

C. He
Southwest Jiaotong University, Chengdu, China

D.M. Frangopol
Lehigh University, Bethlehem, PA, USA

ABSTRACT

Deterioration of reinforced concrete (RC) structures due to chloride-induced reinforcement corrosion is a common problem in an aggressive environment. Especially for shield tunnels in coastal regions, RC segmental linings are subjected to the coupling effects of highly concentrated aggressive agents and high hydrostatic pressure. Aggressive agents diffuse under concentration gradients and/or permeate with liquids due to the high hydrostatic pressure in the linings (Jin et al. 2013; Zhang et al. 2016). This process often leads to premature steel corrosion and concrete cracking.

Since shield tunnels in coastal regions undergo more complex and rapid deterioration processes during their life-cycle, tunnel structures require higher structural performance and durability in aggressive environments; in order to accurately evaluate the structural performance of a shield tunnel during its life-cycle, the coupling effects of aggressive agents and high hydrostatic pressure on the deterioration of the shield tunnels need to be taken into consideration.

Over the past few decades, several attempts (Fagerlund 1995; Funahashi 2013 & Lei et al. 2015) using corrosion testing of the RC components and on-site monitoring have been focused on the qualitative durability assessment and lifetime prediction of underwater tunnels. The very few existing studies on marine environmental hazard assessments and deterioration processes of shield tunnels, especially considering the coupling effects, were not generally oriented to accurately assess the time-dependent structural performance of these tunnels.

In this paper, the chloride transportation in the segmental linings is examined with an emphasis on the impact of hydrostatic pressure. Two transport approaches of chloride ions including the diffusion process and the advection process are suggested, and thus the time to corrosion initiation is predicted. Next, considering that the deterioration of shield tunnels depends on structural locations and their surroundings, a Monte Carlo Simulation is used in conjunction with a time-dependent failure probability estimation associated with the deteriorated structural performance due to the steel corrosion.

Based on the results reported, the following conclusions can be drawn. Under the higher hydrostatic pressure and/or higher chloride hazard, chloride ions can permeate more quickly in the segmental linings; this generally results in a higher failure probability of undersea shield tunnels during their life-cycle. Meanwhile, the ratio of water to cement also has a significant effect on the structural failure probability. To improve the structural durability of undersea shield tunnels, a lower ratio of water to cement is suggested.

REFERENCES

Fagerlund, G. (1995) Penetration of chloride through a submerged concrete tunnel. *Report TVBM-7077*, 1–17.

Funahashi, M. (2013) Corrosion of underwater reinforced concrete tunnel structures. *Corrosion*, 1–15.

Jin, Z., Zhao, T. Gao, S. & Hou, B. (2013) Chloride ion penetration into concrete under hydraulic pressure. *Journal of Central South University*, 20 (12), 3723–3728.

Lei, M., Peng, L. & Shi, C. (2015) Durability evaluation and life prediction of shield segment under coupling effect of chloride salt environment and load. *Journal of Central South University (Science and Technology)*, 46 (8), 3092–3099.

Zhang, Y., Li, X. & Yu, G. (2016) Chloride Transport in Undersea Concrete Tunnel. *Advances in Materials Science and Engineering*, 1–10.

Non linear structural analyses of prestressed concrete girders: Tools and safety formats

B. Belletti, F. Vecchi & M.P. Cosma
Department of Engineering and Architecture, University of Parma, Parma, Italy

A. Strauss
Department of Civil Engineering and Natural Hazards, Institute of Structural Engineering, University of Natural Resources and Life Sciences, Vienna, Austria

ABSTRACT

The reliability analysis of structures is focused on the evaluation and on the prediction of the probability of reaching certain investigated limit states (i.e. the failure probability). The evaluation of the safety margins is a problem related to the degree of knowledge of the variability of factors that regulate the structural behaviour. For this reason, it is important to use a method to take into account these uncertainties rationally.

Traditional approaches simplified the problem by considering the uncertain parameters to be deterministic and accounted for them through the use of partial safety factors in the limit states. Such approaches not guarantee the required reliability and do not provide information on the influence of each parameters on reliability. So the main goal of this paper is the practical application of safety format methods to evaluate the structural response of prestressed concrete girders. Particular attention has been given to the fully probabilistic method that belongs to the highest approach in this context and in which the design resistance is evaluated by probabilistic analyses.

The Abaqus software has been used to develop deterministic and probabilistic nonlinear finite element analyses with the use of the PARC_CL 2.0 crack model implemented at the University of Parma.

Once the mechanical parameters influencing the structural behaviour of the beams are identified, they have been inserted in FReET, a software for statistical, sensitivity and reliability analyses. FReET generates random vectors used as input for probabilistic probabilistic analyses. The outputs of the probabilistic analyses (like peak loads) have been treated as random quantities and represented through a lognormal distribution type to estimate the design resistance R_d.

In particular the effects of the choice of the adopted failure criterion on the standard deviation of NLFEA results is treated in this paper. Moreover, a new method to reduce the number of non-linear analyses is proposed (Fractiles Based Sampling Procedure, FBSP).

With the aim to evaluate the use of mechanical parameters correlations reported in the Model Code 2010 (fib, 2013) in a probabilistic reliability analyses, a modified application of FBSP is also presented.

Finally, the design resistance obtained by FBSP has been compared with the results of fully probabilistic method (FP), the classical partial safety factor (PSF) method and the alternative Estimation of Coefficient of Variation method (ECOV).

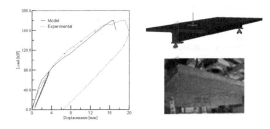

REFERENCES

Belletti, B., Scolari, M. & Vecchi, F. (2017) PARC_CL 2.0 crack model for NLFEA of reinforced concrete structures under cyclic loadings. *Computers and Structures*, 191, 165–179.

Cosma, M.P. (2018) *Fractiles Based Sampling Procedure: A New Probabilistic Method to Evaluate the Design Resistance of a Structural Element.* Master Thesis. Vienna, DIA University of Parma, IKI Boku University.

fib – International Federation for Structural Concrete (2013) *fib Model Code for Concrete Structures 2010.* Berlin, Verlag Ernst & Sohn.

Strauss, A., Krug, B., Slowik, O. & Novak, D. (2017) Combined shear and flexure performance of prestressing concrete T-shaped beams: Experiment and deterministic modeling. *Structural Concrete*, 1–20.

Novak, D., Vorechovsky, M. & Teply, B. (2007) Software for the statistical and reliability analysis of engineering problems and FReET-D: Degradation module. *Advances in Engineering Software (Elsevier)*, 72, 179–192.

Life-Cycle Analysis and Assessment in Civil Engineering: Towards an Integrated Vision – Caspeele, Taerwe & Frangopol (Eds)
© 2019 Taylor & Francis Group, London, ISBN 978-1-138-62633-1

Digitalization is the road to collaboration: A case study regarding the use of digital support tools on sewage and water pumping systems at Schiphol and Haarlem

W. Jaspers, L.S. van Duffelen & J.-J. de Jong
BAM Infra Asset Management, Gouda, The Netherlands

ABSTRACT

Requirements for infrastructure systems are constantly changing. This is caused by developments in mobility, climate, population, urbanization and incremental innovations. To ensure the performance of the infrastructure systems, contractors are hired as infrastructure asset managers. The contractor is responsible for applying technical and financial judgement to decide what interventions are needed to meet performance aims over the whole life cycle of the infrastructure systems. To improve their judgement, contractors are investing in digital support tools. The data obtained by these support tools provides valuable insight in all levels of decision-making (strategic, tactical, operational and executive). These insights can be used for day-to-day maintenance and management activities.

Furthermore, it can be used for strategic decision-making, which considers the technical and functional performance of the infrastructure network. By investing in digital support tools, the role of the contracting party is shifting from an executor to an advising partner. This new paradigm requires a restructuring of key-roles within currently used integrated contract types. This study focusses on two case studies in which the use of digital support tools led to added value on all levels of decision-making. The contractor acted as advising partner for the asset owner (the client). A transparent form of collaboration was initiated through which the client had access to all data at all times. Overall, the use of digital support tools resulted in a 20% decrease of total cost of ownership for both case studies.

Sensitivity analysis of fatigue lifetime predictions of pre-stressed concrete bridges using Sobol'-indices

D. Sanio
Ingenieurbüro Grassl, Düsseldorf, Germany

M.A. Ahrens & P. Mark
Institute of Concrete Structures, Faculty of Civil and Environmental Engineering, Ruhr-Universität Bochum, Germany

ABSTRACT

The fatigue of the pre-stressing steel in concrete bridges becomes relevant especially for existing structures. Here, the residual structural lifetime needs to be considered and can be predicted by Miner's rule based on load frequencies and stress ranges. Thereby, many different parameters need to be considered for the prediction. But in general, only a few of them are factual relevant. E.g. unsteady loads from traffic and the increasing density in time, the structural behavior – ruled by the structural dimensions, materials' stiffness, degradation (creep and shrinkage) and different others – and the fatigue behavior (shape of the S-N-curve) can control the structural lifetime significantly.

In cases of degraded structures measurement data from a monitoring can be helpful to increase the accuracy of a lifetime prediction. In general, experts chose the right parameters to be measured; this selection usually bases on experience. An objective assessment can be reached by sensitivity analyses. Here, variance-based sensitivity indices by Sobol' are a powerful tool to assess the impact of every parameter in the model. They are a measure of the influence of a single parameter's variance to the variance of the model output. Here, the first order sensitivity index S_i gives the direct impact of a parameter to the model. The total sensitivity index S_{Ti} is a measure for interactions, as well. This method is powerful to evaluate even non-linear models with strong parameter interactions, like the non-linear bearing behavior after cracking of the concrete.

This paper presents the methodology of the variance-based sensitivity indices – Sobol' indices. The results are evaluated for a reference structure – a 50 years old, pre-stressed concrete bridge in Düsseldorf, Germany. Specific characteristics like the non-linear bearing behavior of concrete after cracking are considered. Finally, the most relevant elements of a lifetime prediction can be identified and are shown in Figure 1.

The numerical evaluation reveals that especially time-dependent losses of the pre-stress due to creep and shrinkage and some specific types of traffic loads are relevant for the prediction. They are in strong

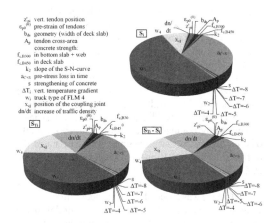

Figure 1. First-order S_i (top) and total sensitivity indices S_{Ti} (bottom left) along with quantified interactions (bottom right).

interactions with other parameters as well, what can be seen by the difference to total sensitivity index and the first order index ($S_{Ti} - S_i$). Additionally, the increase of the total traffic volume and the detailed position of the coupling joint were relevant as well. Consequently, a monitoring of these elements can be planned purposeful.

REFERENCES

Ahrens, M.A. (2012) Precision-assessment of lifetime prognoses based on SN-approaches of RC-structures exposed to fatigue loads. In: Strauss, A. et al. (eds.) *Proc. 3rd Int. Symp. on Life-Cycle Civil Engineering (IALCCE)*.

Sanio, D. & Ahrens, M.A. (2015) Identification of relevant but stochastic input parameters for fatigue assessment of pre-stressed concrete bridges by monitoring. In: M. Papadrakakis et al. (eds.), *Proc. 1st Int. Conf. on Uncertainty Quantification in Comp. Sci. and Eng. (UNCECOMP)*.

Sanio, D. (2016) *Accuracy of Monitoring-Based lifetime Predictions for Pre-Stressed Concrete Bridges Prone to Fatigue.* PhD Thesis. Germany. Ruhr-Universität Bochum.

A comparative study on fatigue assessment of orthotropic steel decks based on long-term WIM data

B. Wang & A. Chen
Department of Bridge Engineering, Tongji University, Shanghai, China

H. De Backer
Department of Civil Engineering, Ghent University, Zwijnaarde, Belgium

ABSTRACT

Fatigue is the problem that constantly bothers the durability and serviceability of orthotropic steel decks. The regular way based on S-N curves should be specifically discussed, since it is proven to be inconsistent with experiments and sometimes results in premature fatigue damage on actual bridges. On the other hand, a fatigue assessment method based on the fracture mechanics provides an alternative option, according to reports concerning welded joints in other industries.

This study compares the results from both methods. Firstly, a long-term weigh-in-motion data of the traffic flow on the Runyang Yangtze River Bridge in China was obtained. The analysis on traffic flow showed that the slow lane is more fatigue-prone as more damage was induced by larger accumulative total weights. Afterwards, the finite element model of the orthotropic steel deck is built, providing the influence surface of stresses at key positions. Consequently, the stress range histories induced by the traffic flow can be obtained, and different fatigue assessment methods were applied and compared.

The hot-spot stress method was firstly conducted in a deterministic way, and the results were given in Figure 1. It is, however, corresponding to the fatigue life when the test specimen fails, which is different from the real case that the fatigue problem severely reduces the serviceability of OSDs. As a contrast, the fracture mechanics method was applied by assuming an initial crack depth of 0.3 mm (see Figure 2), showing that the hot-spot stress method may not lead to conservative assessment. Furthermore, the results can be given in a probabilistic way with respect to a random initial crack depth. It will be helpful to further define the fatigue life as it tracked the entire fatigue cracking process. The results implied that there is a large chance to visually identify the fatigue cracks on the rib-to-deck welded joints on RYRB after a service time vary from 8.05 to 11.90 years, as the paint film damage will provide a clear sign at that time. It can be concluded that the fracture mechanics method was more applicable when combining with the inspection and maintenance works with respect to the life-cycle cost and performance. Future researches should focus on the detailed definition of the fatigue life of OSDs

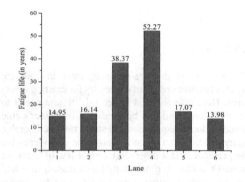

Figure 1. Fatigue life estimation using hot-spot stress method for each lane.

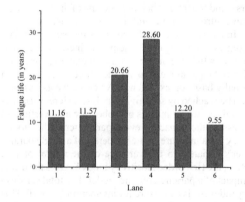

Figure 2. Fatigue life estimation with $a_0 = 0.3$ mm for each lane

with respect to its serviceability, together with the corresponding performance degradation model and assessment method.

REFERENCES

Kolstein, M.H. (2007) *Fatigue Classification of Welded Joints in Orthotropic Steel Bridge Decks*. TU Delft, Delft University of Technology.

Wang, B. (2017) *A Multiscale Study on Fatigue Mechanism and Life Estimation on Welded Joints of Orthotropic Steel Decks*. Ghent University.

Life-Cycle Analysis and Assessment in Civil Engineering: Towards an Integrated Vision – Caspeele, Taerwe & Frangopol (Eds)
© 2019 Taylor & Francis Group, London, ISBN 978-1-138-62633-1

Reliability analysis for geotechnical structures using iterative particle filter

T. Shuku
Okayama University, Okayama, Japan

I. Yoshida
Tokyo City University, Tokyo, Japan

ABSTRACT

With developments of sensor and computer technologies, large amount of data has been collected, stored and analyzed recently. This large amount of data can be used to model updates and parameter identifications in civil engineering practice.

The Bayesian updating provides coherent framework for assimilating data into models, in which prior distributions are updated with data. In Bayesian updating, however, posterior probabilistic model and the likelihood function need to be numerically evaluated. A numerical method for Bayesian updating includes the particle filter (PF, Gordon et al. 1993; Kitagawa 1996). The biggest advantage of the PF is its simple algorithm, and well-designed PF is faster than Markov Chain Monte Carlo (MCMC) method (Yoon et al., 2008). There have been some researches on the application of the PF to parameter identification (Shuku et al., 2012), model update (Yoshida and Sato 2002), and risk evaluation (Yoshida et al., 2009) in civil engineering.

The serious problem in the PF is "filter degeneracy" where all but one of the normalized weights are very close to zero. That means the statistical information in the particles becomes too low to be meaningful. The algorithm of the PF for accurate Bayesian updating to avoid filter degeneracy is necessary in reliability analysis in civil engineering.

This paper presents a new algorithm for Bayesian updating based on the PF called iterative particle filter with Gaussian mixture models (IPFGMM). The idea behind IPFGMM is to apply the Gaussian mixture model (GMM) as the proposal density and to introduce iterative to avoid filter degeneracy. The proposed method is demonstrated by application to parameter identification in two-degree-of-freedom (2DoF) building model and reliability analysis of a geotechnical structure using elasto-plastic finite element analysis.

REFERENCES

Bishop, C.M. (2006) *Pattern Recognition and Machine Learning*. Springer, 738 p.

Beck, J.L. & Au, S.-K. (2002) Bayesian updating of structural models and reliability using Markov chain Monte Carlo simulation. *Journal of Engineering Mechanics*, 128 (4), 380–391.

Ching, J. & Chen, Y.-C. (2007) Transitional Markov chain Monte Carlo method for Bayesian model updating, model class selection, and model averaging. *Journal of Engineering Mechanics*, 133 (7), 816–832.

Fan, Z., Ji, H. & Zhan, Y. (2015) Iterative particle filter for visual tracking. *Signal Processing: Image Communication*, 36, 140–153.

Gilks, W.R. & Berzuini, C. (2001) Following a moving target-Monte Carlo inference for dynamic Bayesian model. *Journal of Royal Statistical Society B*, 63 (1), 127–146.

Manoli, G., Rossi, M., Pasetto, D., Deiana, R., Ferraris, S., Cassiani, G. & Putti, M. (2015) An iterative particle filter approach for coupled hydro-geophysical inversion of a controlled infiltration experiment. *Journal of Computational Physics*, 283, 35–51.

Shuku, T., Murakami, A., Nishimura, S., Fujisawa, K. & Nakamura, K. (2012) Parameter identification for Cam-clay model in partial loading model tests using the particle filter. *Soils and Foundations*, 52 (2), 279–298.

Straub, D. & Papaioannou, I. (2015) Bayesian updating with structural reliability method. *Journal of Engineering Mechanics*, 141 (3), 040141134_1-040141134_13.

Schwarz, G. (1978) Estimating the dimension of a model. *Annals of Statistics* 6 (2), 461–464.

Yoon, C., Cheon, M., Kim, E. & Park, M. (2008) Road sign tracking using particle filter and Parzen window, SCIS & ISIS 2008: 2014–2017.

Yoshida, I. & Sato, T. (2002) Health monitoring algorithm by the Monte Carlo filter based on non-Gaussian noise. *Journal of Natural Disaster Science*, 24 (2), 101–107.

Yoshida, I., Akiyama, M. & Suzuki, S. (2009) Reliability analysis of an existing RC structure updated by inspection data. In: *Proceedings of 10th International Conference on Structural safety and Reliability*. pp. 2482–2489.

Yoshida, I. & Shuku, T. (2016) Particle filter with Gaussian mixture model for inverse problem, In: *Proc. 6th Asian-Pacific Symposium on Structural Reliability and its application (APSSRA2016)*. pp. 642–647.

Optimal inspection planning based on Value of Information for airport runway

Y. Tasaki & I. Yoshida
Tokyo City University, Setagaya, Tokyo, Japan

ABSTRACT

Optimal inspection planning of the airport pavement is studied based on Value of Information (VoI). The authors propose the methodology for optimal inspection planning of civil engineering structures based on VoI, which is one of decision making theory under uncertainty, and has been used in many fields. It can be interpreted to be expectancy of risk reduction or benefit by inspection information. The airport runway for the case study is composed of 100 units. The size of each unit is 21 by 30 meter. After constructing deterioration lines, namely their slope and intercept, for each unit separately considering spatial smoothness of the both parameters with Hierarchical Bayesian, we discuss optimal inspection timing of the airport runway based on the deterioration curves with VoI, and the applicability of the proposed method.

In order to evaluate life cycle performance of infrastructure, proper deterioration prediction is required. The prediction, however, involves a lot of uncertainty, so that efficient inspection to the structures is essential to the efficient maintenance.

The authors propose the methodology for optimal inspection planning of civil engineering structures based on Value of Information (VoI), which is one of decision making theory under uncertainty, and has been used in many fields. It can be interpreted to be expectancy of risk reduction or benefit by inspection information. An existing airport runway is studied as an application example, and the optimal inspection planning of the airport pavement is discussed based on VoI. Our target site is Naha Airport in southernmost main island in Japan, and it is shown in Figure 1. It is composed of 100 units. The size of each unit is 21 by 30 meter. The data about deterioration which are cracking, rutting, and roughness is obtained at each units by 3 to 8 times of inspection during around 28 years. Many researchers use Pavement Rehabilitation Index (PRI) for airport pavement maintenance in Japan. PRI is calculated based on the cracking, the rutting, and the roughness. The value of PRI is the higher, the soundness degree of pavement is the higher. Deterioration curves are obtained by the relation between PRI and year after repair at each units. After constructing deterioration curves for each units separately considering spatial smoothness of deterioration characteristics with Hierarchical Bayesian, we discuss optimal inspection timing of the airport runway based on the deterioration curves with VoI, and the applicability of the proposed method. We hope our method provides useful information for Risk-Informed Decision Making for the inspection planning of civil engineering structures.

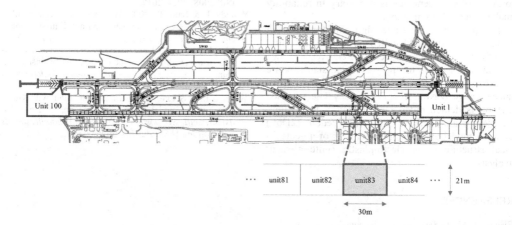

Figure 1. Naha airport.

A sampling-based approach to identifying optimal inspection and repair strategies for offshore jacket structures

R. Schneider & A. Rogge
Bundesanstalt für Materialforschung und -prüfung (BAM), Berlin, Germany

S. Thöns
Technical University of Denmark (DTU), Lyngby, Denmark

E. Bismut & D. Straub
Technische Universität München (TUM), Munich, Germany

ABSTRACT

Inspection and repair strategies for deteriorating structural systems can be optimized through risk-based inspection planning. Recently, Luque and Straub (2018) and Bismut et al. (2017) have developed an approach to perform such an optimization at the structural system level. The approach applies heuristics to define inspection and repair strategies at the structural system level and to reduce the number of possible strategies. For each defined strategy, the expected value of the service life cost of inspection, repair and failure is evaluated using a Monte Carlo approach based on simulated inspection and repair histories. A hierarchical dynamic Bayesian network (DBN) model of the deteriorating structural system forms the basis for computing the required repair and failure probabilities conditional on inspection and repair outcomes. The strategy that minimizes the expect value of the total service life cost is the optimal one in the subset of pre-selected strategies.

We adopt this methodology to optimize inspection and repair strategies for offshore jacket structures. As an alternative to the hierarchical DBN approach, the problem is here formulated such that the conditional repair and failure probabilities required for this analysis can be computed with structural reliability methods (SRM). The approach is implemented with subset simulation, which is a sampling-based SRM suitable for solving high-dimensional reliability problems. The application of subset simulation also enables the efficient generation of inspection and repair histories.

The approach is demonstrated in a numerical example considering a jacket-type frame subjected to high-cycle fatigue. The details of the underlying stochastic deterioration and structural model are documented in (Schneider et al. 2017). In the current case study, we essentially varied the inspection interval Δt_I, the threshold on the annual failure probability p_{th} and the

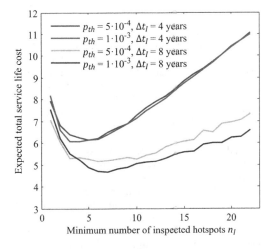

Figure 1. Expected value of the total service life cost as a function of the minimum number of inspected hotspots n_I, the inspection interval Δt_I, and the threshold on the annual failure probability p_{th}.

minimum number of inspected hotspots n_I. The results are shown in Figure 1.

REFERENCES

Bismut, E., Luque, J. & Straub, D. (2017) Optimal prioriization of inspections in structural systems considering component interactions and interdependence. In: *12th International Conference on Structural Safety & Reliability (ICOSSAR 2017)*, Vienna, Austria

Luque, J. & Straub, D. (2018) Risk-based optimization of inspection strategies in structural systems. *Structural Safety*, (under review).

Schneider, R., Thöns, S. & Straub, D. (2017) Reliability analysis and updating of deteriorating systems with subset simulation. *Structural Safety*, 64, 20–36.

Life-Cycle Analysis and Assessment in Civil Engineering: Towards an Integrated Vision – Caspeele, Taerwe & Frangopol (Eds)
© 2019 Taylor & Francis Group, London, ISBN 978-1-138-62633-1

Numerical durability simulation of reinforced concrete structures under consideration of polymorphic uncertain data

K. Kremer, P. Edler, S. Freitag, M. Hofmann & G. Meschke
Institute for Structural Mechanics, Ruhr University Bochum, Germany

ABSTRACT

Corrosion of the steel reinforcement and the (time variant) external loading are the dominant factors limiting the durability of reinforced concrete structures. Corrosive processes are influenced by time variant changes of the moisture and temperature field as well as by a coupling between evolving load induced cracks and the transport of corrosive substances into the structure.

The traditional method for estimating the lifetime of reinforced concrete structures is defining a concrete cover of the reinforcement based on exposure classes and using safety factors for loading and resistance. Numerical durability models, which consider the influence of uncertain design parameters and uncertain time variant structural actions, can be utilized to improve the prediction of the structural lifetime of reinforced concrete structures.

In this paper, the load bearing capacity and the cracking behavior of a reinforced concrete bridge structure is investigated by means of a finite element (FE) model. The used concrete material model is based on the model published in Meschke et al. (1998). For the steel reinforcement, an elasto-plastic model using the MISES criterion is applied. The reinforced concrete is modeled as a composite material using the MORI-TANAKA approach with an anisotropic formulation of the Eshelby tensor. Based on the internal damage variable, the crack widths at the reinforcement layers are calculated and by means of these crack widths the exposed lateral surface of reinforcement is computed as a durability measure of the structure.

In order to improve the durability of the initially designed structure, an optimization task is formulated to optimize the reinforcement layout. The durability objective function to be minimized is the mean value of the exposed lateral reinforcement surface taking uncertain material parameters and uncertain design parameters into account. The uncertainties of the design parameters (concrete cover of the reinforcement layers) are quantified by intervals considering imprecise reinforcement construction. The midpoints of the intervals are defined to be optimized, whereas the sensitivity of the imprecision is investigated by means of varying the radii of the interval design parameters. The Young's modulus is considered as a stochastic distributed uncertain a priori parameter, which also influences the tensile and compression strength of concrete. An accepted failure probability with respect to the load bearing capacity is considered as constraint of the optimization. The applied particle swarm optimization approach considering polymorphic uncertain data is based on the concept presented in Freitag et al. (2017).

For solving the optimization problem, the FE simulation model is replaced by two neural network based surrogate models, which are used to evaluate the objective function and the constraints of the optimization task, respectively. The obtained results show, that the consideration of imprecise design parameters is important for the durability assessment. The range of imprecision is very sensitive to the optimal concrete cover as well as to the values of the objective functions. Compared to a pure stochastic optimization, the optimal design differ up to 20%.

In further works, a meso scale model for the critical zones (hot spots) of reinforced concrete structures will be created to compute the interacting field variables (displacements, moisture and temperature) and the crack pattern at the critical structural sections and its influence to the corrosion process of the rebars.

REFERENCES

Freitag, S., Kremer, K., Hofmann, M. & Meschke, G. (2017) Numerical design of reinforced concrete structures under polymorphic uncertain conditions. In: Bucher, C., Ellingwood, B. & Frangopol, D., (eds.) *Safety, Reliability, Risk, Resilience and Sustainability of Structures and Infrastructure, Proceedings of the 12th International Conference on Structural Safety and Reliability (ICOSSAR 2017): 1535–1542, Vienna.*

Meschke, G., Lackner, R. & Mang, H. (1998) An anisotropic elastoplastic-damage model for plain concrete. *International Journal for Numerical Methods in Engineering*, 42, 703–727.

MS12: Life-cycle redundancy, robustness and resilience indicators for aging structural systems under multiple hazards
Organizers: F. Biondini & D.M. Frangopol

Life-Cycle Analysis and Assessment in Civil Engineering: Towards an Integrated Vision – Caspeele, Taerwe & Frangopol (Eds)
© 2019 Taylor & Francis Group, London, ISBN 978-1-138-62633-1

Influence of seismic hazard on RC buildings' resilience based on ANN

G. Bunea
Faculty of Civil Engineering and Building Services, Technical University "Gheorghe Asachi" of Iaşi, Iaşi, Romania

F. Leon
Faculty of Automatic Control and Department of Computer Engineering, Technical University "Gheorghe Asachi" of Iaşi, Iaşi, Romania

G.M. Atanasiu
Faculty of Civil Engineering and Building Services, Technical University "Gheorghe Asachi" of Iaşi, Iaşi, Romania

ABSTRACT

This paper analyzes the precision of the Artificial Neural Networks ANN prediction for a set of 3D building models subjected to various seismic actions. The ANN was trained based on the results of the Nonlinear Time-History Analysis performed with SAP2000 for 243 reinforced concrete frame structures. The finite element models were obtained by varying six structural parameters. From the total number of parameter combinations there were considered only the realistic ones. The database of seismic actions used in the ANN training includes 14 seismic recordings with different Peak Ground Accelerations PGA, each characterized by four parameters: PGA, Peak Ground Velocity PGV, Peak Ground Displacement PGD and Spectral Intensity SI.

The analysis presented in this paper implies modifying some input parameters by selecting values different than the ones used in the ANN training, in order to test the ANN accuracy. These values are changed both for the input structural parameters and the input seismic action parameters. The accuracy is tested by comparing the results of the Finite Element Analysis FEA, representing the desired results, with the output parameters given by the ANN. The output parameters were selected so that the global damage of the building after the seismic action could be assessed. Thus, five output parameters were considered, respectively the fundamental vibration periods of the undamaged and damaged structures, the Final Softening Index FS, the maximum displacement d_{max} and the Maximum Interstory Drift Ratio IDR. Among them there are two global damage indices, namely the Final Softening Index FS (Dipasquale and Cakmak, 1989) and the

Maximum Interstory Drift Ratio IDR (Rodriguez-Gomez and Cakmak, 1990), which can be used in assessing immediately the building damage level after an important seismic event.

The results confirm that the ANN thus created using the results of the Nonlinear Time-History Analyses can be used in assessing the damage level of reinforced concrete frame structures subjected to seismic actions. In this respect, the difference between the desired results given by the FEA and the approximated output parameters of the ANN is acceptable for four of the five output parameters considered in the analysis. The only output parameter which cannot be well approximated by the ANN, according to the results, is the Final Softening Index FS. In this case, the difference between the desired and the approximated output parameters could easily lead to a change of the damage level, taking into account the values provided in Ghobarah (2004).

REFERENCES

DiPasquale, E. & Cakmak, A.S. (1989) *On the Relation Between Local and Global Damage Indices. Technical Report NCEER-89-0034.* New York, National Center for Earthquake Engineering Research.

Ghobarah, A. (2004) On drift limits associated with different damage levels. In: *Proceedings of the International Workshop Performance-Based Seismic Design: Concepts and Implementation, Bled, Slovenia.* pp. 4321–4332.

Rodriguez-Gomez, S. & Cakmak, A.S. (1990) *Evaluation of Seismic Damage Indices for Reinforced Concrete Structures. Technical Report NCEER-90-0022.* Buffalo, New York, National Center for Earthquake Engineering Research NCEER.

Evaluation of road network performance considering capacity degradation on numerous links

H. Nakajima & R. Honda
The University of Tokyo, Tokyo, Japan

ABSTRACT

Consideration of multi-hazards is essential for the maintenance management of road network. Road network would be vulnerable to disasters when maintenance and management level is low. As prevention of an extreme loss in road network performance, the problem of removal of links is mainly examined in various research and practical works. In the past natural disasters, however, most of roads suffered capacity degradation caused by partial damage. We discuss the road network performance degradation due to the large number of links suffering capacity degradation.

Probability characteristics of network performance and risk quantification

We conducted simulation of grid networks, assuming user equilibrium state in the evaluation of network capacity. The expected value of performance of the network was lowered as the more links suffered capacity reduction. The simulation also shows that tail part of the probability distribution of the network performance did not decrease monotonically. The tail consists from several phases, which are dependent on the geometry of damaged links (Figure 1). The phases observed in the tail part of the probability distribution exhibited different characteristics when the capacity reduction level of damaged roads is different. It indicates that the maintenance level could affect the tail risk of the network performance. Such characteristics should be taken into consideration in the quantification of risk. We compared several risk quantification indices, to compare the risk of different maintenance level. It was found that Conditional Value at Risk (CVaR) exhibit consistent with the level of maintenance.

Identification of critical links in the road network

Applicability of the presented scheme was verified by applying to the road network in the Shikoku Area in Japan, and obtained the results consistent with the cases of grid networks. For practical purposes, we present a scheme to evaluate the critical links. Application to Shikoku Road Network reveals that the critical

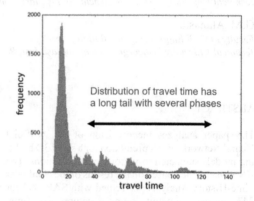

Figure 1. When links suffered severe degradation, probability distribution of network performance exhibited a long tail with several phases.

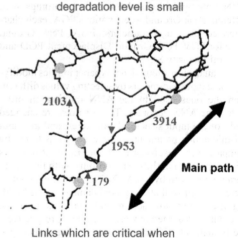

Figure 2. Critical links in the road network altered depending on the degradation level, thus, maintenance level, because characteristic of network performance also altered.

links and their priority alter according to the level of capacity reduction of network, thus, maintenance level (Figure 2).

Life-Cycle Analysis and Assessment in Civil Engineering: Towards an Integrated Vision – Caspeele, Taerwe & Frangopol (Eds)
© 2019 Taylor & Francis Group, London, ISBN 978-1-138-62633-1

Probabilistic life-cycle resilience assessment of aging bridges and road networks under seismic and environmental hazards

L. Capacci & F. Biondini
Department of Civil and Environmental Engineering, Politecnico di Milano, Milan, Italy

ABSTRACT

Structures and infrastructure systems are exposed to detrimental effects of natural and man-made disasters along with aging and deterioration processes over the system lifetime (Biondini & Frangopol 2016). Communities are hence called to cope with the impact of disruptive events, such as earthquakes, without suffering disproportionate sudden consequences and long-term effects with respect to the hazard exposure (Bruneau *et al.* 2003).

Road transportation systems are critical to ensure quick deployments of relief aids and resources to repair the surrounding facilities in the post-event emergency response. Indeed, widespread and severe damage of bridges and other vulnerable system components can cause direct monetary losses associated with maintenance and repair interventions, as well as indirect economic consequences induced by lack of network connectivity and downtime (Bocchini & Frangopol 2011, Capacci *et al.* 2016).

For aging bridges and transportation road networks exposed to seismic and environmental hazards, resilience depends on the time of occurrence of the seismic event (Biondini *et al.* 2015). In addition, the harmful effects of seismic damage and structural deterioration, as well as the beneficial effects of repair actions and post-earthquake recovery processes, are affected by significant uncertainties to be considered in the resilience assessment. Therefore, system functionality and seismic resilience should be formulated as time-variant probabilistic performance indicators under a life-cycle perspective (Capacci *et al.* 2018).

A probabilistic framework for life-cycle seismic resilience assessment of aging bridges and transportation road networks is presented. The time-variant fragilities and damage probabilities of the deteriorating bridges in the network are assessed through nonlinear incremental dynamic analysis and Monte Carlo simulation with reference to several limit states, from damage limitation up to collapse. The role of the seismic scenario is considered in terms of earthquake magnitude and epicentral distance within a seismic area source. The post-event seismic damage and the progressive restoration of bridge seismic capacity due to repair actions are related to traffic limitations and vehicle restrictions considering the uncertainty in the recovery process. Traffic analysis is hence carried out for each combination of bridge traffic restrictions in order to assess in probabilistic terms the network functionality profiles, the corresponding resilience levels, and a damage-based measure of life-cycle resilience. The proposed framework is applied to reinforced concrete bridges exposed to chloride-induced corrosion and simple road networks with a single bridge or two bridges in series under different earthquake scenarios. The applications show the importance of a life-cycle oriented approach to probabilistic assessment of seismic resilience of aging infrastructure systems.

REFERENCES

Biondini, F., Camnasio, E. & Titi, A. (2015) Seismic resilience of concrete structures under corrosion. *Earthquake Engineering & Structural Dynamics*, 44 (14), 2445–2466.

Biondini, F. & Frangopol, D.M. (2016) Life-cycle performance of deteriorating structural systems under uncertainty: Review, *Journal of Structural Engineering*, 142 (9), 1–17.

Bocchini, P. & Frangopol, D.M. (2011) A stochastic computational framework for the joint transportation network fragility analysis and traffic flow distribution under extreme events. *Probabilistic Engineering Mechanics*, 26 (2), 182–193.

Bruneau, M., Chang, S.E., Eguchi, R.T., Lee, G.C., O'Rourke, T.D., Reinhorn, A.M, Shinozuka, M., Tierney, K., Wallace, W.A. & Winterfeldt, D.V. (2003) A framework to quantitatively assess and enhance the seismic resilience of communities. *Earthquake Spectra*, 19 (4), 733–752.

Capacci, L., Biondini, F. & Titi, A. (2016) Seismic resilience of aging bridges and transportation networks. In: *8th Int. Conference on Bridge Maintenance, Safety and Management* (IABMAS2016), Foz do Iguaçu, Brazil, 26–30 June, 2016.

Capacci, L. & Biondini, F. (2018) Role of the earthquake scenario on life-cycle seismic resilience of aging bridge networks. In: *16th European Conference on Earthquake Engineering*, Thessaloniki, Greece, 18–21 June, 2018.

Life-Cycle Analysis and Assessment in Civil Engineering: Towards an Integrated Vision – Caspeele, Taerwe & Frangopol (Eds)
© 2019 Taylor & Francis Group, London, ISBN 978-1-138-62633-1

Innovative methodology of assessing the residual structural safety margin of reinforced concrete structures—application to cooling towers

N.C. Tran & C. Toulemonde
EDF R&D, Moret-Loing-et-Orvanne, France

F. Beaudouin & C. Mewisse
EDF R&D, Chatou, France

ABSTRACT

The management of the long term operation of critical infrastructures of civil engineering is always an important challenge for society. Many efforts and progresses have been done during the last three decades in both two areas: modeling as well as monitoring. However, we are still far from mastering the ageing of the structures and assessing precisely their residual safety margin. There are different difficulties for achieving it.

Firstly, even having progresses, numerical models are not be able to predict precisely the evolution of ageing pathologies whether it is creep, AAR or corrosion under real environmental conditions.

Secondly, the impact of some loads is not easy to evaluate (for example: settlement, wind. . .).

Thirdly, damage behavior of concrete could be simulated for small structural elements but is difficultly applied for big structures in reality (cooling tower, bridge, dam. . .). The main reasons are the difficulty of computation convergence and the difficulty in taking into account the damage accumulation due to real loading history which is not easy to assess.

Fourthly, there are more and more rich monitoring data for critical structures. A good computation model has to be able to be fit based on these data for a continuous improvement of modelling.

Fifthly, a modelling could give us images of the behavior of structure. But what decision maker needs is a quantitative index indicating the residual safety margin of the studied structure.

Finally, a civil engineering structure has a high level of uncertainty in terms of material properties, steel reinforcement position, environmental impacts. . . This has to be considered in evaluation of the safety margin.

The work presented in this paper aims to propose an integrated approach able to solve the five first difficulties mentioned above. The approach will be presented directly through an application: cooling towers of nuclear power plants and a fuel plant in France. These hyperbolic shaped towers have been built during the last decades of the twentieth century with an expected lifetime of forty years. The management of the long term operation and the optimization of the maintenance program primarily rely on an extensive set of periodic monitoring data: soil settlement, cracking patterns and geometry deformation. Destructive and non-destructive examinations have been implemented for characterizing ageing of reinforced concrete and the amplitude of the thermo-hydro mechanical loading. EDF approach being turned towards continuous improvement of safety and effective maintenance, the methodology is proposed to respond to the EDF engineering services' need: having an analysis tool assessing the safety margin of towers based on the above monitoring data and the knowledge acquired in construction phase. The methodology aims to calculate, for each tower, a risk index based on failure analysis of reinforced concrete.

The first version (corresponding to the mechanical module) and the second version (integrating the decision module) were presented in 2013 (Tran et al. 2013) and 2017 (Tran et al. 2017). This paper present the whole methodology including all three modules which corresponds to a new version currently under development.

The approach of this methodology could be extended for other critical civil engineering structures.

REFERENCES

Tran, N.C., Crolet, Y., Toulemonde, C., Schmitt, N., Courtois, A. & Genest, Y. (2013) Cooling tower fleet ranking strategy, In: *Symposium 2013 of IASS*, Poland.

Tran, N.C., Crolet, Y., Toulemonde, C., Schmitt, N., El-Yazidi, A., Courtois, A., Genest, Y. & Moriceau, S. (2017) Innovative methodology of ranking cooling towers based on structural safety margin. In: *Proceeding of 26th International Conference of Nuclear Engineering*, Shanghai, China.

*Life-Cycle Analysis and Assessment in Civil Engineering: Towards an
Integrated Vision – Caspeele, Taerwe & Frangopol (Eds)
© 2019 Taylor & Francis Group, London, ISBN 978-1-138-62633-1*

Life-cycle seismic performance prediction of deteriorating RC structures using artificial neural networks

S. Bianchi & F. Biondini

Department of Civil and Environmental Engineering, Politecnico di Milano, Milan, Italy

ABSTRACT

In recent years, many public authorities, private companies, and professional associations worldwide are dealing at local, national, and international level, with the condition rating of huge stocks of existing structures and infrastructure systems, including buildings, bridges, roads, railways, dams, ports, and other construction facilities (ASCE 2017). All such systems are at risk from aging, fatigue, and several deterioration processes due to chemical attacks and other physical damage mechanisms. The detrimental effects of these damaging phenomena can lead over time to unsatisfactory structural performance under service loadings or accidental actions and extreme events, such as earthquakes.

For Reinforced Concrete (RC) structures, damaging factors include the effects of diffusive attack from aggressive agents, such as chlorides, which may involve corrosion of steel reinforcement and deterioration of concrete. The direct and indirect costs associated with corrosion and related effects, particularly for RC buildings and bridges, are generally very high. It is therefore of major importance to promote a life-cycle approach to design, assessment, and management of RC structures under uncertainty (Frangopol 2011, Biondini & Frangopol 2016).

In this context, a proper planning of inspection and maintenance activities is crucial to prevent significant loss of structural capacity and to ensure a suitable level of system performance over the structure life-cycle. To inform the decision making process for the allocation of resources and assist designers and managers in the definition of operational policies and maintenance programs, automatic management systems, such as Bridge Management Systems (BMS), have been developed. These systems allows to elaborate monitoring and inspection data to assess the current state and predict the future condition of structures and infrastructures under uncertainty through probabilistic mathematical models, frequently based on the use of Markov chains.

Decision making processes for structure and infrastructure management are however based on data severely affected by aleatory and epistemic uncertainty and often need to incorporate engineering judgement and expert opinions. Soft computing techniques, including Artificial Neural Networks (ANNs), are particularly suitable to this goal since they can efficiently handle incomplete and subjective data (Flood 2008).

In this paper, a three-layer ANN is developed and trained to predict the life-cycle performance of deteriorating RC structures based on limited amount of information related to local damage of some components, typically obtained from the results of visual inspections. The proposed ANN is applied to the prediction of the life-cycle seismic capacity of a three-story RC frame under chloride-induced corrosion over a 50-year lifetime (Titi et al. 2018). The results of a probabilistic life-cycle analysis are used as training, validation, and test samples. The training datasets are formed to incorporate the results from several inspections carried out at different time instants over given observation time intervals, and to accommodate predictions over the remaining structural lifetime.

REFERENCES

ASCE (2017) *Report Card for America's Infrastructure.* Reston, VA, USA.

Biondini, F. & Frangopol, D.M. (2016) Life-cycle performance of deteriorating structural systems under uncertainty: Review. *Journal of Structural Engineering*, ASCE, 142 (9), F4016001, 1–17.

Flood, I. (2008) Towards the next generation of artificial neural networks for civil engineering. *Advanced Engineering Informatics*, 22 (1), 4–14.

Frangopol, D.M. (2011) Life-cycle performance, management, and optimization of structural systems under uncertainty: Accomplishments and challenges, *Structure and Infrastructure Engineering*, 7 (6), 389–413.

Titi, A., Bianchi, S., Biondini, F. & Frangopol, D.M. (2018) Influence of the exposure scenario and spatial correlation on the probabilistic life-cycle seismic performance of deteriorating RC frames. *Structure and Infrastructure Engineering*, 14 (7), 986–996.

Life-Cycle Analysis and Assessment in Civil Engineering: Towards an Integrated Vision – Caspeele, Taerwe & Frangopol (Eds)
© 2019 Taylor & Francis Group, London, ISBN 978-1-138-62633-1

Modelling of physical systems for resilience assessment

G. Tsionis, A. Caverzan, E. Krausmann, G. Giannopoulos, L. Galbusera & N. Kourti
Directorate for Space, Security and Migration, Joint Research Centre (JRC), European Commission, Ispra, Italy

ABSTRACT

Natural hazards, such as earthquakes, floods, or storms may have significant socio-economic impacts, e.g. injuries/casualties, disruption of services, reconstruction costs, etc. They can cause major accidents at hazardous industry (Natech accidents). Earthquakes in particular are a major threat to communities around the world and a number of events have shown that the consequences may be disproportionate to the damage of the built environment and that significant time and resources are needed to regain pre-event conditions. The paper focuses on the modelling of critical infrastructure and the built environment for assessing their resilience to natural and man-made hazards. It covers frameworks, tools and methods to design for resilience, inter-dependencies modelling, as well as the links among technological and societal systems.

The European Commission's Joint Research Centre (JRC) has developed a semi-quantitative methodology for Natech risk analysis and mapping which was implemented as a web-based software tool called RAPID-N (http://rapidn.jrc.ec.europa.eu) and includes all functionalities required for the assessment (natural-hazard assessment, industry damage severity and probability estimation, and risk assessment). The output is a risk summary report and an interactive risk map showing the scenario-specific impact areas. RAPID-N can determine the likelihood and severity of human impacts as well as of damage to neighbouring structures (e.g. power plants, ports, etc.). This helps to understand the risks of cascading effects that might hamper a speedy recovery.

Another useful tool for understanding resilience gaps in technological systems is the forensic analysis of past incidents. Using incident data available in the open literature, a study on the recovery of the power grid analysed how the characteristics of earthquakes, floods and space weather influence the grid's recovery time after a natural event. The analysis included other factors affecting power grid recovery time, e.g. the disruption of other critical infrastructures.

The JRC has embarked also on an effort to create practical toolkits in order to support the analysis of interdependencies at technological and economic level. To this end, it has developed GRRASP (Geo-spatial Risk and Resilience Assessment Platform) which is freely available and based on open source technologies. GRRASP follows a service-based approach to simulate interdependencies at technical level and an input-output based approach to provide an estimation of the economic impact of critical infrastructure disruption. The platform is continuously expanded with new functionalities in terms of models and data visualisation, leveraging the potential of GIS-based open-source technologies.

A project undertaken at the JRC addresses the seismic resilience of buildings across Europe. The first step was the review of fragility curves from literature and their qualitative evaluation in order to select the most appropriate ones for each building class and geographic region. A simplified quantitative risk analysis at country level was then performed to examine the effect of the detail of exposure data in large-scale seismic risk assessment. The case study showed that the readily available data from the Eurostat Census Hub can be used to assess with acceptable accuracy the seismic risk for all European countries. Lastly, an inventory of the housing stock in Europe and a classification of its seismic vulnerability at a regional level was performed. In the seismic-prone regions of Europe, the majority of buildings was designed without provisions for earthquake resistance or with moderate-level seismic codes.

A number of issues need to be further investigated and harmonised, including: boundaries in space and time for resilience assessment; all-hazards scenarios that consider cascading effects and interdependencies between systems; minimum data requirements in standardised format for model validation and assessment; calibration of recovery models and reduction of uncertainties through application to real-life systems; integration of engineering with organizational response and societal impacts.

Robustness analysis of 3D base-isolated systems

P. Castaldo, D. Gino & G. Mancini
Department of Structural, Geotechnical and Building Engineering (DISEG), Politecnico di Torino, Torino, Italy

ABSTRACT

The aim of the study consists of evaluating the seismic robustness of a 3D r.c. structure equipped with single-concave friction pendulum system (FPS) (Zayas et al. 1990) devices in reliability terms (Cornell 1968) considering different models related to different malfunctions of the seismic isolators.

The elastic response pseudo-acceleration corresponding to the isolated period is assumed as the relevant random variable and, by means of the Latin Hypercube Sampling technique, the input data have been defined in order to perform 3D inelastic time-history analyses.

In this way, bivariate structural performance curves (Castaldo et al. 2015) at each level of the r.c. structural system as well as seismic reliability-based design abacuses for the FP devices have been computed and compared in order to evaluate the robustness of the r.c. system considering different models related to the different failure cases analysed.

Contextually, the seismic robustness of the above-mentioned r.c. structural system has also been examined by considering both a configuration equipped with beams connecting the substructure columns and a configuration without these connecting beams (Figure 1) in order to demonstrate their effectiveness in improving the seismic robustness for a malfunction of a seismic device and provide very useful design recommendations for base-isolated structures equipped with FPS.

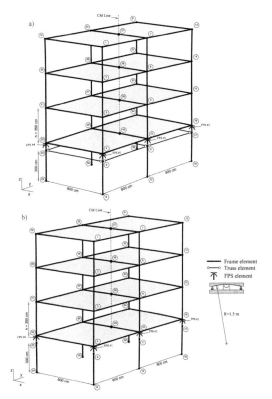

Figure 1. Base-isolated structure configurations with (a) and without (b) the beams connecting the substructure columns.

REFERENCES

Castaldo, P., Palazzo, B. & Della Vecchia, P. (2015) Seismic reliability of base-isolated structures with friction pendulum bearings. *Engineering Structures*, 95, 80–93.

Cornell, C.A. (1968) Engineering seismic risk analysis. *Bulletin of the Seismological Society of America*, 58 (5), 1583–1606.

Zayas, V.A., Low, S.S. & Mahin, S.A. (1990). A simple pendulum technique for achieving seismic isolation. *Earthquake Spectra*, 6, 317–333.

MS13: Advances in Structural Health Monitoring for real-world applications
Organizers: C.W. Kim, P.J. McGetrick, A. Cunha & T. Kitahara

Identification of damaged regions in dynamically loaded dams

M. Alalade & T. Lahmer
Institute of Structural Mechanics, Bauhaus-Universität Weimar, Weimar, Germany

F. Wuttke
Geomechanics and Geotechnics, Christian-Albrechts-University Kiel, Kiel, Germany

ABSTRACT

Dams are important structures employed in the generation of hydro-electricity, provision of water supply and for flood defense. However, a large percentage of dams in operation today are more than 50 years old and as such the true condition of its structural integrity is not fully known. The deterioration of dams may result both from short term effects (e.g earthquakes, impact loads, etc.) and from long term effects (e.g. aging, physical and chemical weathering, fatigue, etc). Though these effects may not lead to an immediate collapse/failure of the dam, it is necessary, in order to ensure optimum performance of these structures, to quickly identify regions which may jeopardize the safety of the dam. Unfortunately, the conventional methods are tedious, time consuming, expensive and inefficient. To tackle these challenges, various numerical methods and inverse analysis has been proposed, considering the coupling effects on various phenomena present in the operational life cycle of the dam.

Based on the successful application of Full Waveform Inversion (FWI) in geotechnical exploration and non-destructive testing (NDT), especially with subsurface structures our method aims to extends this application to the identification of damaged regions in dams. The material properties of the dam are deduced by reconstructing the wave propagation and its behavior as the wave field interacts with the dam. This is done by fitting the field response of the wave propagation recorded by sensors with a numerical model (forward model). The 2D elastic wave equation employed for the forward modeling is the stress-displacement formulation ($\tau - \mathbf{u}$). The reconstruction is an inverse problem which requires minimizing the data residuals δu or misfit between field data $\mathbf{u}^{exp}$ and the simulated data from our forward model $\mathbf{u}^{mod}$. The objective function $C_f(\mathbf{m})$ in Equation 1 which is to be minimized is the residual elastic energy in the residuals. That is the $L2$−norm of the error between the model response and the field or experimental data.

$$C_f(\mathbf{m}) = 0.5 \parallel \mathbf{u}^{mod}_{(m)} - \mathbf{u}^{exp} \parallel_{L2} \quad (1)$$

Using a starting model which corresponds with the material properties in the dam at its construction,

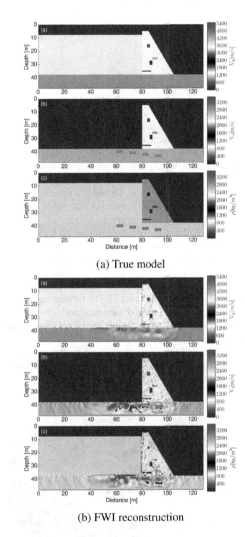

(a) True model

(b) FWI reconstruction

Figure 1. Distribution of seismic velocities and density in the air, water, foundation and dam structure (true and FWI model).

the anomalies/damages in the dam in the true model Figure 1(a) are deduded by FWI and obtained in (b) with the shear wave velocity, V_s model giving the best resolution of damages.

Installation and results from the first 18 months of operation of the dynamic monitoring system of Baixo Sabor arch dam

S. Pereira, F. Magalhães & A. Cunha
Construct – ViBest – Faculty of Engineering, University of Porto (FEUP), Porto, Portugal

J. Gomes & J.V. Lemos
National Laboratory of Civil Engineering (LNEC), Lisbon, Portugal

ABSTRACT

Hydroelectric power plants are responsible for nearly half the capacity installed in Portugal for energy production by renewable sources. This is the result of a continuous effort to decrease the country's dependence on fossil fuels, which in the hydroelectric case has been accomplished by both new infrastructure and an increase of installed capacity in existing infrastructure.

The Baixo Sabor hydroelectric power plant, located in the northeast of Portugal, includes a reservoir with a storage capacity of 630 million m^3 of water created by a concrete double-curvature arch dam, 123 m high (Figure 1). With the aim of studying Baixo Sabor arch dam dynamic properties and their evolution over time, in order to assess the dam's structural health and the effect of exceptional events in its behaviour, a continuous dynamic monitoring of the dam is being carried out by ViBest/FEUP and LNEC. The monitoring takes into account the variation of ambient and operational conditions, as well as the possible evolution of material mechanical properties.

The installed dynamic monitoring system involves 20 uniaxial accelerometers, 12 of them radially disposed in a gallery close to the dam crest, and the other 8 radially placed along two other galleries, being all of them connected to a set of digitizers distributed in the instrumented galleries and synchronized by GPS.

This paper introduces the dynamic monitoring system installed in the Baixo Sabor dam and the processing that the obtained data is subjected to. It presents as well the results obtained during the first eighteen months of operation of the referred monitoring system, between 01/12/2015 and 31/05/2017, including the reservoir-filling period.

The results include the characterization of accelerations amplitude and modal properties (natural frequencies, modal damping ratios and mode shapes), whose evolution over time is studied. The evolution of the 12-hour average for the first six modes natural frequencies is presented in Figure 2.

More over, the influence of reservoir water level and ambient temperature on natural frequencies is studied as well, and appropriate statistical models are applied in order to mitigate those effects and to obtain time series of natural frequencies suitable for detection of structural changes.

Figure 1. Baixo Sabor arch dam (EDP).

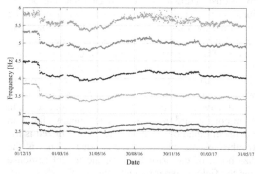

Figure 2. Time evolution of 12-hour average natural frequencies.

Life-Cycle Analysis and Assessment in Civil Engineering: Towards an Integrated Vision – Caspeele, Taerwe & Frangopol (Eds)
© 2019 Taylor & Francis Group, London, ISBN 978-1-138-62633-1

Monitoring of corrosive environment focusing on dew condensation in steel bridges

Z. Rasoli & K. Nagata
Nagoya Institute of Technology, Nagoya, Aichi, Japan

T. Kitahara
Kanto Gakuin University, Yokohama, Kanagawa, Japan

ABSTRACT

In Japan, the development of steel bridge infrastructure peaked during the period of high economic growth. In recent years, most of these bridges have reached a critical age. The number of bridges whose age will exceed 50 years will be over 43% of all bridges by 2023, and this number will increase significantly in the subsequent decade, as shown in Fig. 1. Additionally, major steel bridges are rather costly to construct, maintain, and rehabilitate. Preventive maintenance management has been introduced to handle this challenging issue. Corrosion is a major problem that causes the deterioration of steel infrastructure and affects the long-term mechanical performance and durability of existing aging steel structures. To carry out maintenance management as a countermeasure to environmental corrosion, it is important to initially inspect and survey these structures.

This paper proposes a methodology to accurately evaluate dew condensation, which is an important factor affecting the corrosion of bridges and deterioration of the protective coating. The evaluation of dew condensation caused by atmospheric temperature and relative humidity was carried out in Aichi prefecture using the weather research and forecasting (WRF) numerical weather prediction technique.

The simulation results revealed that dew condensation occurred in the entire prefecture and was significantly higher in the northeastern and northwestern regions, as shown in Fig. 2. Furthermore, the corrosion environment of a steel bridge was investigated on-site. The evaluation outcomes from the field measurements explicitly confirm the validity of the proposed method.

In this study, the girder temperature was assumed to be equal to the atmospheric temperature, although in reality, there were slight differences between them. Therefore, the steel girder temperature and atmospheric temperature differences were investigated via field measurements. Thus, it was found that the atmospheric temperature was lower than the girder temperature, and the differences varied by approximately 0.7°C–1°C.

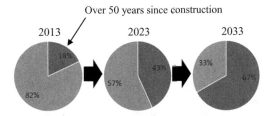

Figure 1. Change in the ratio of bridges over the age of 50 years.

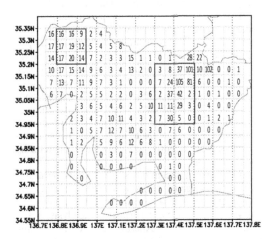

Figure 2. Evaluation results of dew condensation in Aichi Prefecture for January 2017.

REFERENCES

Ministry of Land Infrastructure Transport and Tourism (2014) *White Paper on Land Infrastructre and Transport in Japan.*
Nakashima, M. & Nagai, K. (2014) An investigation of road bridge maintenace system in japan in developed society. *Society for Social Management Systems Internet Journal,* 09.

Life-Cycle Analysis and Assessment in Civil Engineering: Towards an Integrated Vision – Caspeele, Taerwe & Frangopol (Eds)
© 2019 Taylor & Francis Group, London, ISBN 978-1-138-62633-1

Optimising circumferential piezoelectric transducer arrays of pipelines through linear superposition analysis

X. Niu
Department of Engineering Science, University of Greenwich, Chatham Maritime, Kent, UK
TWI Ltd., Cambridge, Cambridgeshire, UK

H.P. Chen
Department of Engineering Science, University of Greenwich, Chatham Maritime, Kent, UK

H.R. Marques
TWI Ltd., Cambridge, Cambridgeshire, UK

ABSTRACT

Ultrasonic guided wave inspection is one of the non-destructive testing (NDT) techniques available for the structural integrity evaluation of engineering structures. Compared with other NDT techniques, guided waves are able to propagate over tens of metres with a relatively high sensitivity to defects in the structure. For pipeline inspections, they enable identification of the location of defects on the pipe circumference using a circumferential array of piezoelectric transducers. In general, the circumferential array is clamped around the surface of the pipe. The general range for the sensitivity of the operation is 3%–9% reduction of the cross-sectional area (Mudge & Speck, 2004), depending on the signal-to-noise ratio. Therefore, optimisation of the transducer array design to enhance performance in pipeline inspections is still required. An optimised design aims to increase defect sensitivity without significantly increasing the manufacturing and testing costs.

In this paper, the initial steps towards optimisation of piezoelectric transducer arrays in pipes are investigated through a combination of finite element analysis (FEA), linear superposition analysis (LSA) (Marques, 2016) and experimental studies. A high mode purity for the $T(0,1)$ mode is a core objective of the optimisation. Here, an 8.18 mm thickness steel pipe of 4450 mm long and 219.1 mm in diameter is used for numerical simulations and the related empirical work. First of all, a transducer array of 24 transducers with an odd gap of 33 degrees generate torsional wave mode $T(0,1)$ with a group of flexural wave modes $F(n,2)$. The excitation signal is selected at the centre frequency of 35 kHz. The finite element models are simulated, and numerical simulation of dispersive propagation is evaluated through the related FE results. After that, to reduce transducer number

in each ring for signal transmission, a series of odd transducer spacings is investigated and the related prediction of guided wave propagation on the pipe is simulated effectively. Finally, the numerical FEA and LSA simulations are compared with the experimental results. Based on the numerical simulations and experimental studies, it can be found that the flexural wave modes $F(n,2)$ have a high contribution for the case of non-axisymmetric array with an odd spacing of 33 degrees using a centre frequency of 35 kHz excitation. With unregularly reducing the number of transducers, it is simulated that the energy is decreased at the non-transmission area. Also, the use of the theoretical modelling of LSA is an effective method to predict dispersive propagation of guided waves on a pipe. The LSA models show good agreement with the FEA simulations and later results in modelling, and also agree well the related experiment. Therefore, array design optimisation will lead to improvement in non-destructive evaluation of pipe defects using guided waves. It will also enhance effectively the health monitoring of structures and subsequently increase the safety of engineering infrastructure in operation.

REFERENCES

Lowe, M.J.S., Alleyne, D.N. & Cawley, P. (1998) The mode conversion of a guided wave by a part-circumferential notch in a pipe. *Journal of Applied Mechanics*, 65 (3), 649–656.

Marques, H.R. (2016) *Omnidirectional and Unidirectional Sh0 Mode Transducer Arrays for Guided Wave Evaluation of Plate-Like Structures*. PhD Thesis. Brunel University London.

Niu, X., Marques, H.R. & Chen, H.-P. Sensitivity analysis of circumferential transducer array with $T(0,1)$ mode of pipes. *Smart Structures and Systems*, accepted.

Life-Cycle Analysis and Assessment in Civil Engineering: Towards an Integrated Vision – Caspeele, Taerwe & Frangopol (Eds)
© 2019 Taylor & Francis Group, London, ISBN 978-1-138-62633-1

Particle Swarm Optimization for damage identification in beam-like structures

A. Barontini, M.-G. Masciotta, L.F. Ramos & P.B. Lourenço
ISISE, Department of Civil Engineering, University of Minho, Guimarães, Portugal

P. Amado-Mendes
ISISE, Department of Civil Engineering, University of Coimbra, Coimbra, Portugal

1 INTRODUCTION

Nowadays, developing cost-effective and automatic strategies for the maintenance of built environment is becoming essential, as many existing structures and infrastructures are close to the end of their service life (or over) and the new ones are growing in number, size and complexity. Structural Health Monitoring (SHM) is an ongoing field of research whose main aim is the implementation of strategies for the assessment of the health condition of a structural system and the prompt identification of damage.

To achieve a correct location and quantification of the damage extent, inverse model updating problem formulation is usually adopted, which consists in the minimization of an objective function defined in terms of discrepancies between the features extracted by operational modal analysis and those computed using a numerical or analytical model. Experience demonstrates that Particle Swarm Optimization (PSO) algorithms, as other metaheuristics, are suitable for the model updating despite a few well-known shortcomings. In the present paper, one of the most basic and well-known versions of the PSO algorithm by Shi and Eberart (Shi & Eberhart, 1998) is used to identify the location and extent of different damage scenarios in a clamped-clamped steel beam. The reference beam is numerically simulated in DIANA software to carry out the eigenvalue analysis, and the damage is introduced in the model through a reduction of the Young Modulus. The simulated scenarios are meant to reproduce the most expected damage conditions in the reference beam, namely damage close to the midpoint and damage at the beam clamps. The analysis allows to confirm

that even a basic version of the PSO is suitable for damage identification, although such a version is generally considered not efficient enough in the literature. The influence of parameter setting on the algorithm performance is also confirmed, especially in regard to the coefficients C_1 and C_2, whose values updated have been usually based on previous works performed on completely different classes of problems. Therefore, it is clear from the developed work how a proper parameter setting is pivotal to achieve an improvement in this field of research.

Besides the aforementioned aspects, the influence of the number of elements in the population on the algorithm performance is analyzed as well, demonstrating that increasing the population size does not imply an equivalent growth of the number of FEM analyses nor of the time required for the process.

Finally, the optimized algorithm instance resulting from all these analyses is tested over a set of more complex problems. The experiments carried out demonstrated that the PSO is a feasible way to face inverse problems for damage identification. A few questions are still open and worth of more research, namely the effect of noise polluted data and of problem dimension.

REFERENCE

Shi, Y. & Eberhart, R. (1998) A modified particle swarm optimizer. In: *1998 IEEE International Conference on Evolutionary Computation Proceedings. IEEE World Congress on Computational Intelligence (Cat. No. 98TH8360).* pp. 69–73. https://doi.org/10.1109/ICEC.1998.699146

Experimental investigation on crack detection using imbedded smart aggregate

C. Du
Delft University of Technology, Delft, The Netherlands
Shenzhen Graduate School, Harbin Institute of Technology, Shenzhen, China

Y. Yang & D.A. Hordijk
Delft University of Technology, Delft, The Netherlands

ABSTRACT

In the Netherlands, the maintenance of the existing reinforced concrete bridges is an important issue for the society. A large percentage of the bridges on the Dutch water and highway system are short span reinforced concrete bridges. To ensure the safety of these constructions, the health conditions of the bridges have to be evaluated. For reinforced concrete bridges, an important indication of the structure conditions is the widths and the distribution of cracks. This is usually assessed by visual inspection. Such inspection only works in presence of very large cracks, usually this means intervention is needed for the structure. Besides, such inspection is labour intensive. When the crack distributions can be obtained at an earlier stage with a more cost effective way, the bridge owners may have more economic options in terms of maintenance.

It has been recognized since long time that the wave transfer properties through solid reflect the change of properties inside the material. A common drawback for these methods are that the sensors have to be installed on the surface of the bridges, thus the accuracy of the measurement usually relies on the surface conditions of the structure, which often introduces more uncertainty when being applied on existing structures. Besides, such measurements usually demand considerable amount of preparation work and expertise from the operators, thus, always costlier for the bridge owners.

In this paper, an inspection strategy based on a new type of low cost sensor Smart Aggregate (SA) is introduced. For existing structures, they can be installed by casting them into the bored holes after obtaining concrete core samples. A scheme of such measurement is given in Figure 1. In such a monitoring scheme, the SA grid is arranged at the level of the neutral axis of the concrete bridge, for each measurement, one of the SA in the grid is used as a transducer marked as 'T' in Figure 1, another one or several SA's are used as receiver (marked as 'R'). With a mobile monitoring device, an electrical charge is given to the transducer, in the mean time, the receivers collect the responses of the

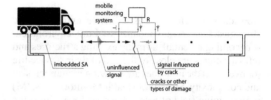

Figure 1. Scheme of structural assessment with SA arrays.

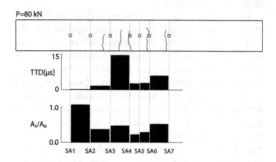

Figure 2. Travel time different and adjusted amplitude ratio between SA pairs when $P = 80\,kN$.

neighbouring SA's. The change of the wave properties is recognized when comparing the updated measurement with the reference one before damage occurs. The strategy enables a quick scan of the damage distribution of the target bridge without further disturbing its usage.

The inspection strategy is demonstrated by a scaled beam specimen casted in the lab. The damage of the beam was generated by applying a three-point bending test. SA's were imbedded in the specimen. Two parameter derived from the wave forms were used to indicate the cracking conditions, they are the travel time difference (TTD) and the amplitude ratio. Both uses the initial stage as reference. As shown in Figure 2, the comparison with the crack pattern and the two parameters showed good agreement.

*Life-Cycle Analysis and Assessment in Civil Engineering: Towards an
Integrated Vision – Caspeele, Taerwe & Frangopol (Eds)
© 2019 Taylor & Francis Group, London, ISBN 978-1-138-62633-1*

Development of a remote monitoring system with wireless power-saving sensors for analyzing bridge conditions

E. Sasaki, P. Tuttipongsawat & N. Sinsamutpadung
Department of civil Engineering, Tokyo Institute of Technology, Tokyo, Japan

H. Nishida
Development Center, OMRON Social Solutions Co, Ltd., Shiga, Japan

K. Takase
Social Infrastructure Monitoring Group, OMRON Social Solutions Co, Ltd., Tokyo, Japan

ABSTRACT

Recently, maintenance of bridges has been an important issue in Japan, and Ministry of Land, Infrastructure, Transport and Tourism (MLIT) enacted a ministerial ordinance in 2014 to have all the bridges inspected every five years. MLIT has also conducted the projects for the establishment of more efficient maintenance of infrastructures including bridges. Development of a bridge monitoring system is one of the on-going projects.

Our research group has been proposing a remote bridge monitoring system with wireless power-saving sensors. At present, the monitoring system is under operation for a typical type of a national highway steel bridge with RC deck under the project with MLIT. The proposed monitoring system consists of wireless sensor nodes and a server using 920MHz band, and can be installed to a bridge easily and quickly, which can reduce the application cost. Also, it is operated using our own high data synchronization technologies. The

sensor nodes can measure 3-axial acceleration, strain, temperature, humidity, and corrosive current of corrosion sensors. To measure strain, we developed a strain sensing technology using power-generating piezo-film sensors that can be applied for low frequency region.

The proposed monitoring system was developed based on the idea of "Characteristic Medical Chart" of the target bridge to enable road administrators easily understand conditions of the bridge. And the sensor arrangement in the monitoring system was customized considering possible deterioration scenarios of the target bridge, and the system covers damage evaluation using various indices including phase space property of measured data and the structural parameters identification. In actuality, the target bridge has cracks in RC deck and reduction of steel plate due to corrosion damage, and the indices have been analyzed to investigate the effects of the damage and temperature changes.

This paper illustrates the outline of the proposed monitoring system and shows the actual analyzed indices in the target bridge.

Investigation of Bayesian damage detection method for long-term bridge health monitoring

Y. Goi & C.W. Kim
Department of Civil and Earth Resource Engineering, Kyoto University, Kyoto, Japan

ABSTRACT

Aiming at efficient long-term health monitoring of bridges, this study proposes a novel damage detection method using vehicle-induced vibrations. Modal properties of bridges such as natural frequencies and mode shapes change as the structural integrity changes. In real bridges, however, modal identification requires subjective and troublesome operations for engineers to find the appropriate properties because of the uncertain external forces. Thus, the authors have investigated damage indicators directly derived from observed vibrations (Goi & Kim, 2017). Although the damage indicators enabled damage detection without identifying modal properties, they still require subjective decision on their model orders.

This study therefore investigates an efficient damage detection method adopting Bayesian regression (Bishop, 2006) to the regressive model representing bridge vibrations. The posterior distribution for the regressive coefficient derived from the vibration data. The posterior distribution provides damage-sensitive features based on its functional form of the probability distribution. Bayesian information criterion and stabilization diagram enable a straightforward choice of the model orders. The likelihood function is marginalized for the parameters. Bayesian hypothesis testing is formulated using a statistic called Bayes factor, which is defined as a ratio of marginalized likelihoods.

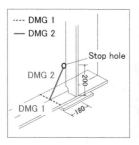

Figure 1. Artificial damage on the bridge deck.

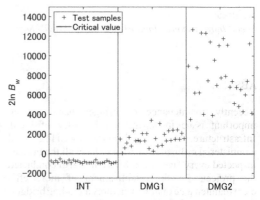

*INT: Intact bridge without damage.

Figure 2. Damage indicators monitored on the damage experiment.

The Bayesian hypothesis testing to detect changes in the damage-sensitive feature is proposed as the novel damage detection method.

Efficacy of the proposed method is assessed by a half-year monitoring experiment of a simply supported single span steel plate girder bridge. The proposed method efficiently detected the seasonal changes in the modal properties. Artificial damage imitating fatigue cracks are introduced to the monitored bridge as shown in Figure 1 after the half-year monitoring, and the feasibility of the proposed method is examined. Figure 2 shows an example of Bayes factors obtained from the monitoring on the damage experiment. The damage scenarios were clearly recognized by means of the proposed method.

REFERENCES

Bishop, C.M. (2006) *Pattern Recognition and Machine Learning*. New York, Springer.
Goi, Y. & Kim, C.W. (2017) Damage detection of a truss bridge utilizing a damage indicator from multivariate autoregressive model. *Journal of Civil Structural Health Monitoring*, 7 (2), 153–162.

A trial vibration measuring for evaluating performance of small bridge with small FWD system

H. Onishi
Faculty of Science and Engineering, Iwate University, Morioka, Japan

K. Ouchi
Graduate School of Engineering, Iwate University, Morioka, Japan

N. Kimura & D. Yaegashi
Graduate School of Integrated Arts and Sciences, Iwate University, Morioka, Japan

ABSTRACT

Recently, the aging and the deterioration of infrastructure is the one of the most important problems in Japan. The number of road bridges in Japan is almost 700,000. This number is too large to treat with existing techniques to investigate the performance or the conditions of each bridge.

There is a strong request to innovate the new method for efficient investigation to evaluate the performance of structures. In Japan, the most important member of road bridges is a deck slab because of the number of the case of the deteriorated bridge decks. Then we must spend the much more budget to repair the bridge decks. It is often said that the rate of spent budget for decks will be over 90% in Japanese highways. It is necessary to develop the new efficient investigation method for bridge decks.

Then we focused the small FWD system (Figure 2) and we achieved to develop the improved investigation

Figure 1. Measured bridge.

Figure 2. Small FWD.

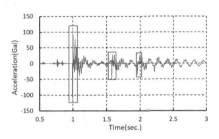

Figure 3. Measured acceleration.

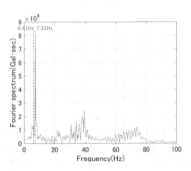

Figure 4. Fourier spectrum of acceleration.

method for bridge decks. And there is a necessary to improve or to extend this method to evaluate the performance of road bridges.

In this research, we tried to measure the vibration of bridge under impact load of small FWD system with extended new measuring method (Figure 3, Figure 4). And we also try to develop the new evaluation method for vibration data of bridges.

REFERENCE

Onishi, H., Santo, T., Chida, S. & Ouchi, K. (2016) A fundamental investigation to survey the state of RC bridge decks with vertical acceleration given by FWD test. *Life-Cycle of Engineering Systems: Emphasis on Sustainable Civil Infrastructure*, 697–704.

Structural health monitoring of a steel girder bridge utilizing reconstructed sparse-like system matrix

T. Mimasu, C.W. Kim & Y. Goi
Kyoto university, Kyoto, Japan

ABSTRACT

Structural health monitoring using the vibration data obtained from sensors installed on bridges has been developed for bridge condition assessments and maintenance. Wireless sensors are deemed as an indispensable sensing system in bridge health monitoring for their easy deployment and operation. However, saving power consumption of wireless sensor networks has been one of the technical issues to be solved especially for bridge health monitoring. Reduction of the radio communication between sensor nodes is necessary since major power consumption occurs during the radio communication between sensors.

This study aims to investigate a way to select sub-sensor groups so that the radio communication occurs inside the sensor group. This approach contributes to reducing the radio communication. Sparse-like system matrix is reconstructed utilizing the optimal sensor groups decided by correlations between sensors. The validity of the proposed approach is examined utilizing the data both from simulation and field experiment on a steel girder bridge. Multivariate auto regressive (MAR) model is used as a system identification method in time domain.

In the field experiment, artificial fatigue cracks at the girder end are considered by severing the lower flange and web plate. The cut on flange and cut on web are denoted as DMG1 and DMG2, respectively. Accelerations of the bridge are measured before and after introducing the damage, and the system identification utilizing obtained vibration responses are conducted. The sparse-like system matrix is built based on the optimal sensor group shown in Figure 1. The difference in frequencies identified from the original system matrix and reconstructed system matrix was

Figure 1. Example of the optimal sensor grouping decided by the correlations between sensors.

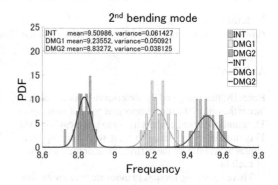

Figure 2. The histogram of frequencies identified utilizing reconstructed sparse-like system matrix.

small. As a result of system identification utilizing the sparse-like system matrix, probability distributions of frequencies identified from vibration data before and after the artificial damage changed as shown in Figure 2. The t-test which significance level is set to 5% was conducted utilizing the frequency distributions before and after the damage, and a significant difference was observed. Feasibility of damage detection utilizing sensor group was confirmed.

Future works will focus on system identification from wireless sensors utilizing sensor grouping method. It should be also examined if power consumption of wireless sensors reduces or not by utilizing sensor grouping method.

REFERENCES

Kim, C.W., Kawatani, M. & Hao, J. (2012) Modal parameter identification of short span bridges under a moving vehicle by means of multivariate AR model. *Structure and Infrastructure Engineering*, 8 (5), 459–472.

Nagayama, T., Moinzadeh, P., Mechitov, K., Ushita, M., Makihata, N., Leiri, M., Agha, G., Spencer, B.F., Fujino, Y. & Seo, J.W. (2010) Reliable multi-hop communication for structural health monitoring. *Smart Structures and Systems*, 6 (5–6), 481–550.

Damage identification of bridge structures using the Hilbert-Huang Transform

J.J. Moughty & J.R. Casas
Department of Civil and Environmental Engineering, Technical University of Catalonia (BarcelonaTech), Catalonia, Spain

ABSTRACT

Damage identification in bridges traditionally entails the assessment of modal parameters from accelerations; however, their application as damage features has found their performance to suffer under environmental and operational conditions (Moughty & Casas, 2017). Additionally, the non-stationarity of some vibration signals creates a problem when using standard modal techniques that employ Fourier-based transforms, as linear signal stationarity is assumed.

The Hilbert-Huang Transformation (HHT) (Huang et al., 1998) provides an alternative assessment method of signal frequency and energy that is suitable for non-stationary signals. Its instantaneous outputs of frequency and amplitude have been used for fault diagnosis of rotating machinery (Huang & Wu, 2008), but their application to bridge structures has been minimal due to the uniqueness of each structures and inadequate methods of Empirical Mode Decomposition (EMD). The present study aims to investigate the applicability of the HHT to non-stationary bridge vibrations from passing vehicles for the purpose of damage identification. To this end, an advanced method of EMD is employed and the resulting HHT outputs are quantified using a novel instantaneous damage parameter defined herein as *Instantaneous Vibration Intensity (IVI)*, presented in Equation 1, where $\alpha_i(t)$ and $\omega_i(t)$ are the HHT instantaneous amplitudes and frequencies, respectively.

$$IVI_i(t) = a_i^2(t) / \omega_i(t) \qquad (1)$$

The damage identification assessment of *IVI* as a damage sensitive feature is conducted on data

Table 1. Frequency percentage differences (Kim et al., 2014).

Mode	DMG2	RCV	DMG3
1st B. Mode	−2.67%	−0.13%	+0.31%
2nd B. Mode	+0.20%	−0.25%	−5.67%
3rd B. Mode	−0.21%	−0.87%	−9.05%
4th B. Mode	+0.22%	−0.72%	−6.87%
5th B. Mode	+0.58%	+0.19%	−0.16%

Table 2. *IVI* percentage differences at damage location sensor.

Damage Indicator	DMG2	RCV	DMG3
IVI	+10.31%	−25.56%	+36.40%

from a real bridge subjected to a progressive damage test under vehicle induced loading, which produces a highly non-stationary acceleration response. The results of the assessment found that IVI demonstrates damage detection and localization qualities, as portrayed in Figure 1, where the percentage difference between the cumulative *IVI* values obtained from baseline and damage states are presented at each sensor location. As a final comparison of *IVI*'s detection and localization capabilities, results are compared to modal frequencies obtained by Kim et al. (2014), as can be seen in Table 1 and Table 2.

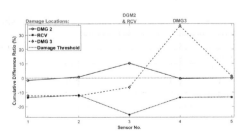

Figure 1. Evolution of distributed vibration intensity during progressive damage test under ambient conditions.

REFERENCES

Huang, N.E. & Wu, Z. (2008) A review on Hilbert–Huang transform: Method and its applications to geophysical studies. *Reviews of Geophysics*, 46, 1–23.

Huang, N.E., Shen, Z., Long, S.R., Wu, M.C., Shih, H.H., Zheng, Q., Yen, N.-C., Tung, C.C. & Liu, H.H. (1998) The empirical mode decomposition and the Hilbert spectrum for nonlinear and non-stationary time series analysis. *Proceedings of the Royal Society of London. Series A*, 454, 903–993.

Kim, C.W., Chang, K.C., Kitauchi, S., McGetrick, P.J., Hashimoto, K. & Sugiura K. (2014) Changes in modal parameters of a steel truss bridge due to artificial damage. In: *Proceedings of the 11th Inter. Con. on Structural Safety and Reliability, ICOSSAR, New York*. pp. 3725–3732.

Moughty, J.J. & Casas, J.R. (2017) A state of the art review of modal-based damage detection in bridges: Development, challenges, and solutions. *Applied Sciences*, 7 (5), 510.

Experimental studies on the feasibility of drive-by bridge inspection method using an appropriate vehicle model

S. Nakajima, C.W. Kim, K.C. Chang & S. Hasegawa
Department of Civil and Earth Resources Engineering, Kyoto University, Japan

ABSTRACT

Drive-by bridge inspection method is a method by which the bridg's dynamic characteristics or health status are identified from the dynamic responses of an inspection vehicle when passing over the target bridge. Compared with conventional direct measurement methods that require sensor deployment on the bridge, this method is thought to be more rapid, efficient and mobile, and especially to suit fast scanning needs. However, its practicability is verified in very limited numbers of field experiments; although several laboratory experiments were conducted, a difficulty was drawn in the identification of bridge frequencies among several frequencies induced by other factors such as vehicle vibrations and roadway surface roughness. Because of this, it needs an appropriate vehicle model that is sensitive to bridge vibrations but insensitive to other vibration sources.

In this study, a homemade trailer being tailored for capturing bridge vibrations more effectively was designed and assembled. This trailer's frequency could be tuned by changing the body mass and the suspension stiffness. Its frequency was tuned as high as 8 Hz, which was far away from most bridges' fundamental frequencies and therefore provided a wider sensible frequency band. After the assembly and tuning, the trailer was towed by a commercial car and tested on a 40-m long simply-supported bridge in field. Four artificial damage scenarios were introduced into the bridge, including an intact state, an artificial crack in the girder, recovery of the crack and an artificial freezing of a hinge support.

From the field moving vehicle tests, the bridge's fundamental frequency could be successfully identified (see Fig. 1) where INT denotes the intact condition or reference condition. Changes in the frequency caused by the fixing the support (DMG2) could be detected as shown in Figure 2, where DMG1 denotes the damage scenario applying artificial crack on lower flange and web plate of a bridge girder while RCV indicates recovering the artificial damage. This tendency is observed in the vibration response of the bridge. The feasibility and the reliability of the drive-by method using the homemade trailer was observed.

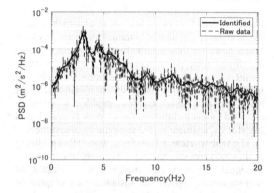

Figure 1. The PSD of the acceleration response in INT case.

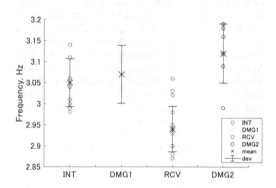

Figure 2. The identified bridge-related frequency.

REFERENCES

Kim, C.W. & Kawatani, M. (2009) Challenge for a drive-by bridge inspection. *Proceedings of the 10th International Conference on Structural Safety and Reliability ICOSSAR 2009*, 758–765.

Yang, Y.B., Lin, C.W. & Yau, J.D. (2004) Extracting bridge frequencies from the dynamic response of a passing vehicle. *Journal of Sound and Vibration*, 272, 471–493.

MS14: Monitoring of structures for informed-decision making
Organizers: A. Strauss & D.M. Frangopol

RAMS evaluation for the steel-truss arch high-speed railway bridge based on SHM system

Y.L. Ding & H.W. Zhao
Key Laboratory of Concrete and Prestressed Concrete Structures of Ministry of Education, Southeast University, Nanjing, China

A.Q. Li
Beijing Advanced Innovation Center for Future Urban Design, Beijing University of Civil Engineering and Architecture, Beijing, China

ABSTRACT

Long-span bridges are the vital projects on the high-speed railway lines. With the continuous construction of the high-speed railway network in China, the safe operation and routine maintenance of long-span high-speed railway bridges based on service performance become the challenges. Mostly long-span high-speed railway bridges in China are equipped with structural health monitoring (SHM) system to protect the daily operation of structure. The structural health monitoring system has assumed the important task of guiding the maintenance and management of the bridge structures since its birth. However, the existing evaluation methods of bridge structures are mainly based on the results of periodical inspection. The bridge health monitoring data has not been effectively applied to structural evaluation.

As a mature evaluation theory in railway engineering, Reliability, Availability, Maintainability and Safety (RAMS) theory can well integrate massive data of health monitoring system and be well applied to the evaluation, management and maintenance of long-span high-speed railway bridge structure. Using the Nanjing Dashengguan Yangtze River Bridge, a typical long-span steel-truss arch high-speed railway bridge on the Beijing-Shanghai high-speed railway,

Table 1. Grade for RAMS evaluation of bridge structure.

Grade	A	B	C	D	E
S	$1 \geq S \geq 0.9$	$0.9 \geq S \geq 0.8$	$0.8 \geq S \geq 0.65$	$0.65 \geq S \geq 0.4$	$0.4 \geq S \geq 0$

as the engineering background, this paper developed a RAMS evaluation method of serving state of the bridge structure based on the health monitoring system. According to the failure-risk, safety/availability, facility-maintenance of bridge members, the state evaluation method of each monitoring item is presented. The weights of each evaluation item in level 1 to 3 are determined, then service state of bridge structure can be grading evaluated. Finally, a new method of state evaluation for the existing steel-truss arch high-speed railway bridge has been developed.

REFERENCES

CENELEC (2007) *Railway Application-Specification of Railway Reliability Availability Maintainability and Safety (RAMS)*, EN 50126. Brussels.

Ding, Y.L., Zhao, H.W., Deng, L., Li, A.Q. & Wang, M.Y. (2017) Early warning of abnormal train-induced vibrations for a steel-truss arch railway bridge. *Journal of Bridge Engineering*, 22 (11), 05017011.

General Administration of Quality Supervision, Inspection and Quarantine of the People's Republic of China & Standardization Administration of the People's Republic of China (2008) Railway Application-Specification of Railway Re-liability Availability Maintainability and Safety (RAMS), GB/T 21562-2008. Beijing (in Chinese).

IEC (2007) *Railway Application-Specification of Railway Reliability Availability Maintainability and Safety (RAMS)*, IEC 62278-2002. Washington, DC.

Ministry of Railways of the People's Republic of China (2010) *Repair Rules for Railway Bridge and Tunnel, TG/GW103-2010*. Beijing (in Chinese).

Pratico, F.G. & Giunta, M. (2018) Proposal of a key performance indicator for railway track based on LCC and RAMS analyses. *Journal of Construction Engineering and Management*, 144 (2), 04017104.

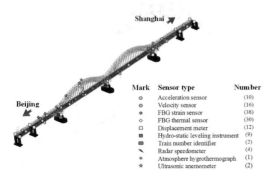

Figure 1. SHM system of the Nanjing Dashengguan Yangtze River Bridge.

Life-Cycle Analysis and Assessment in Civil Engineering: Towards an Integrated Vision – Caspeele, Taerwe & Frangopol (Eds) © 2019 Taylor & Francis Group, London, ISBN 978-1-138-62633-1

Crowd load prediction on pedestrian bridges using Fiber Bragg Grating sensors

K. Hassoun, J. Karaki, S. Mustapha, A. Kassir & Z. Dawy
Maroun Semaan Faculty of Engineering and Architecture, American University of Beirut, Beirut, Lebanon

H. Abi-Rached
Connected Experience Labs (CEL), General Electric Global Research (GRC), General Electric, USA

ABSTRACT

Unfortunately, high density human crowds or pedestrian flow has previously led to tragedies and or fatalities. In addition, crowd behavior as a reaction to an incident aggravates the complexity and disruption of human flow. Results of such situations include trampling and crushing. Therefore, it is important to monitor such crowd motion for danger warning and prevention. In this study, we are proposing a novel approach to provide a continuous crowd load prediction on pedestrian bridges, with particular focus on high density of crowds. The approach is based on continuous monitoring methods known as structural health monitoring (SHM). An important novelty in the approach is the choice of Fiber Bragg Gratings (FBG) Fiber Optic Sensors (FOS) over traditional monitoring instrumentations due to its numerous advantages. The approach also employs machine learning techniques to generate prediction models from training data gathered from FOS. The machine learning techniques applied were Support Vector Machines (SVMs) and exponential Gaussian Process (GP) regression applied on vibration data collected from FOS sensors mounted on a pedestrian bridge. The approach was validated using laboratory experiments on a real scaled bridge model. Different loading scenarios were investigated to predict the weight of individuals, groups as well as continuous flow of people activities. Current results demonstrated promising results capable of predicting the total load weight with an mean error of 10 kg and the activity type as fast or slow with an accuracy above 90%.

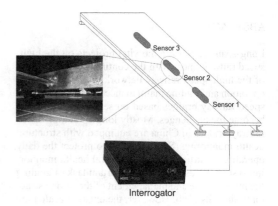

Figure 2. Experiment setup.

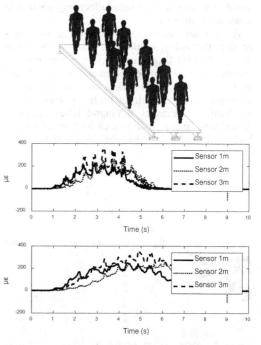

Figure 3. Ten persons – 726 kg (top) fast, (bottom) slow.

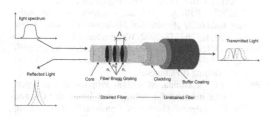

Figure 1. FBG structure and working principle.

A case study of structural monitoring as a control in the restoration process of heritage structures: The strengthening of the Vistabella Church's Tower

M. Llorens
Department of Mechanical Engineering and Industrial Constructions (EMCI), University of Girona, Girona, Spain

R. Señís & S. Pavón
Department of Architectural Technology (TA), Polytechnic University of Catalonia, Barcelona, Spain

B. Moreno
Smart Structural Sensing Technologies, S.L. (3STECH), Girona, Spain

The Sacred Heart Church of Vistabella (La Secuita, Spain) was designed by Josep Maria Jujol and built in the early twenties of the XX century. The slenderness of its steeple makes it extremely exposed to wind action, to the extent that its stability has been threatened since the construction started. At the end of 1934, the upper spire crumbled away, being rebuilt and reinforced with steel passive bars.

After some numerical approaches, it became clear that the lightness of such a masonry construction along with the feebleness of the connections between the structural members were at the basis of the problem. In order to provide additional vertical load without substantively modifying the structural scheme, a solution based on external prestressed bars, was implemented. The bars replace the existing passive bars, which have proven to be totally useless. The active system proposed allows to eliminate the masonry's tensile stresses under the action of the wind. The additional loading to be applied varies depending on the level (three levels in total), according to the gradual post-tensioning system proposed as a strengthening technique of the steeple. Therefore, it has been necessary to design a procedure able to introduce independent tensions at the different levels throughout the same tie and the corresponding monitoring system to control the tensioning process during the implementation (Fig. 1), based on the relevant references (Augenti et al. 2013, Leftheris et al. 2006, Masciotta et al. 2017, Mircam & Molins 2009, Van Balen & Verstrynge 2006). Monitoring the structure has allowed to know at all times the real situation, to quantify the applied efforts and, consequently, to make the most adequate decision in each moment.

The procedure, the theory that supports it, and the results of the follow-up are described in detail, these being the objective of the paper.

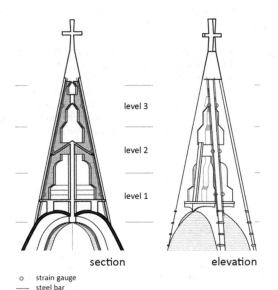
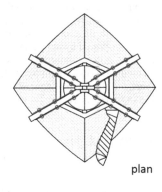

Figure 1. Position of the steel bars and strain gauges (monitoring system).

Life-Cycle Analysis and Assessment in Civil Engineering: Towards an Integrated Vision – Caspeele, Taerwe & Frangopol (Eds)
© 2019 Taylor & Francis Group, London, ISBN 978-1-138-62633-1

Levels of assessment for chloride model parameters of existing concrete structures

A. Strauss, C. Matzenberger & K. Bergmeister
University of Natural Resources and Life Sciences, Vienna, Austria

M. Somodikova
Brno University of Technology, Brno, Czech Republic

T. Zimmermann
City of Vienna, Building Inspection MA37, Vienna, Austria

ABSTRACT

The increasing age of important infrastructures, such as bridges, faces engineers in challenging tasks, since the safety index has to keep a certain level during the whole service live. Limiting factors are degradation processes, such as chloride ion migration, carbonation or freeze-thaw attack, but also load effects, a proper design, workmanship and maintenance. Today, the assessment of existing bridges mainly bases on visual inspections and hence, a subjective evaluation of the current performance. To deal with the increasing amount of aged reinforced concrete structures, more sophisticated assessment methods and degradation forecasting models has to be applied. Fib Model Code 2010 (fib 2013) distinguishes between different approaches for verification of limit states associated with durability: (i) probabilistic safety format, (ii) partial safety factor format, (iii) deemed-to-satisfy approach and (iv) avoidance of deterioration approach. In order to determine the safety level, only the fully probabilistic approach provides quantitative information and hence it is increasingly supported.

The microscopy thin oxidlayer on steel bars, which is built due to the alkalinity of the surrounding concrete, can be destroyed by the penetration of chloride ions dissolved in pore water and thus corrosion can occur. According to fib Bulletin No. 34 (2006), the Cl^- concentration at the depth of concrete cover c [mm] at time t [years], $C(c,t)$ [wt.-%/c] is calculated as:

$$C(c,t)=C_0+\left(C_{S,\Delta x} - C_0\right)\cdot\left[1 - erf\frac{c - \Delta x}{2\sqrt{D_{app}(t)\cdot t}}\right], \quad (1)$$

where C_0 [wt.-%/c] is the initial Cl^- content in concrete, $C_{S,0}/C_{S,\Delta x}$ [wt.-%/c] is the surface/substitute surface Cl^- content in depth of convection zone Δx [mm], and $D_{app}(t)$ [mm²/years] is the time dependent

migration coefficient and "erf" is the error function. According to fib Bulletin No.76 (2015), $D_{app}(t)$ can be determined as:

$$D_{app}(t)=k_e\cdot D_{app}(t_0)\cdot\left(\frac{t_0}{t}\right)^\alpha \text{ or}$$
$$D_{app}(t)=k_e\cdot D_{RCM}(t_0)\cdot\left(\frac{t_0}{t}\right)^\alpha, \quad (2)$$

where α [−] is the ageing exponent, t_0 [years] is the reference point of time, and D_{RCM} [mm²/years] is the diffusion coefficient based on the Rapid Chloride Migration (RCM) test method. The environmental variable k_e [−] that takes into consideration the effect of temperature on chloride ingress in concrete is defined as:

$$k_e = exp\left[b_e\left(\frac{1}{T_{ref}} - \frac{1}{T_{real}}\right)\right], \quad (3)$$

where b_e [K] is the temperature coefficient (proportional to activation energy), and T_{ref} [K] and T_{real} [K] are the reference temperature and the temperature of the structural element or the ambient air, respectively.

The objective of this article is to point out the methods for calibrating an applied model of chloride ion ingress, to increase the accuracy of predicting necessary interventions on concrete bridge structures.

REFERENCES

fib (2013) *fib Model Code for Concrete Structures 2010.* Berlin, International Federation for Structural Concrete.
fib Bulletin No. 34 (2006) *fib Bulletin No. 34: Model Code for Service Life Design.* Lausanne, Switzerland, Bulletin, Inter-National Federation for Structural Concrete (fib).
Somodíkova, M., Strauss, A., Ivan, Z. & Bretislav, T. (2018) PREPRINT: Quantification of pa-rameters for modelling of chloride ion ingress into concrete.

MS15: Probability-based service life design of reinforced concrete structures exposed to reinforcement corrosion
Organizers: G. De Schutter & S. Keßler

Life-Cycle Analysis and Assessment in Civil Engineering: Towards an Integrated Vision – Caspeele, Taerwe & Frangopol (Eds)
© 2019 Taylor & Francis Group, London, ISBN 978-1-138-62633-1

Evaluation of the chloride migration and carbonation coefficients of Belgian ready mixed and precast concrete for a performance-based design

P. Minne, E. Gruyaert & L. De Winter
Department of Civil Engineering, Technology Cluster Construction, KU Leuven, Ghent, Belgium

B. Craeye
EMIB Research Group, Faculty of Applied Engineering, University of Antwerp, Antwerp, Belgium
DuBiT Research Unit, Department of Industrial Sciences & Technology, Odisee University College, Belgium

R. Caspeele & G. De Schutter
Magnel Laboratory for Concrete Research, Ghent University, Ghent, Belgium

ABSTRACT

In the design of Belgian concrete structures, durability aspects are currently assessed based on a deemed-to-satisfy approach according to the standards NBN EN 206 (2014) and NBN B 15-001 (2012). As this design method hinders innovation, the trend worldwide is to move towards durability-based design of concrete structures. Within the Tetra project "DurO-Bet", a quantitative method has been developed to implement a performance-based design for durability in Belgium. For the damage mechanisms chloride ingress and carbonation, reference values have been set up in function of the exposure class, cement type, concrete cover, initial chloride content and curing time (NBN EN 206 2014, NBN B 15-001 2012, NBN EN 1992-1-1 2005). The reference values were deducted from available data and additional test results obtained within the "DurOBet" project, using a probabilistic approach.

In the study as presented here, twelve ready mixed and precast concrete mixtures were tested and evaluated with regard to their resistance to chloride and CO_2 ingress. The concrete mixtures were intended to be used in a specific environment EE2, EE3 or EE4 and for concrete strength classes C30/37 up to C50/60. For each of the concrete mixtures, the cement dosage, w/c-ratio, slump and 28-day mean compressive strength on cubes $150 \times 150 \times 150 \, \text{mm}^3$ were determined. In addition, for the damage mechanism chloride ingress the initial chloride content and migration coefficient of the concrete mixtures were determined according to NT BUILD 443 (1999), NT BUILD 208 (1996) and NT BUILD 492 (1999). For the damage mechanism carbonation, the carbonation coefficient was determined according to NBN EN 13295 (2004).

Concrete is considered to be durable if it can offer sufficient resistance to the environment to which it is exposed during its lifetime. In a deemed-to-satisfy- approach each concrete type has a specific combination of at least three durability requirements (maximum w/c-ratio, minimum cement content,

minimum concrete compressive strength class). The minimum air content can optionally be added.

For important projects, where a target service life and durability performances are required, a full-probabilistic or partial factor design is needed. The model Code for Service Life Design *fib* bulletin 34 (2006), the recent outcomes of *fib* bulletin 76 (2015) and the model of von Greve-Dierfeld & Gehlen (2016) discuss this full-probabilistic design approach for modelling chloride and CO_2 induced corrosion (depassivation) in uncracked concrete.

Within the Tetra project "DurOBet" tables, based on a full-probabilistic approach, were drawn up for the migration and carbonation coefficient.

Based on the deemed-to satisfy approach, it can be decided that some industrial concretes did not meet the durability requirements. This was mainly due to a slightly too large w/c-ratio. Based on the performance-based approach, all the concretes meet the durability requirements for carbonation, considering a design service life of 50 years. None of the concretes meet the durability requirements for chloride ingress (XD3 exposure class). From the tabulated values, it can however be concluded that these concretes can be used in a lot of other exposure classes where chlorides can induce reinforcement corrosion.

REFERENCES

fib bulletin 34 (2006) *Model Code for Service Life Design.* Stuttgart: Sprint-Digital-Druck. Fédération Internationale du Béton.

fib bulletin 76 (2015) *Benchmarking of Deemed-to-Satisfy Provisions in Standards: Durability of Reinforced Concrete Structures Exposed to Chlorides.* Germany, DCC Document Competence Center Siegmar Kästl. Fédération Internationale du Béton.

Von Greve-Dierfeld, S. & Gehlen, C. (2016) Performance-based durability design, carbonation Part 3: PSF approach and a proposal for the revision of deemed-to-satisfy rules. *Structural Concrete*, 17 (5).

*Life-Cycle Analysis and Assessment in Civil Engineering: Towards an
Integrated Vision – Caspeele, Taerwe & Frangopol (Eds)
© 2019 Taylor & Francis Group, London, ISBN 978-1-138-62633-1*

Probabilistic evaluation of service-life of RC structures subjected to carbonation

R.A. Couto & S.M.C. Diniz
Federal University of Minas Gerais, Belo Horizonte, Brazil

ABSTRACT

Current sustainability requirements for the built environment, have posed a great emphasis on the service-life prediction of reinforced concrete (RC) structures (Diniz *et al.*, 2016). Among the degradation mechanisms affecting RC structures, carbonation is one of the main causes of deterioration of the steel reinforcement. Considering that the concentration of CO_2 in the atmosphere in urban and industrial environments has increased over the last decades (Boden *et al.*, 2015), the deterioration of RC structures due to carbonation is a major issue. Additionally, due to the uncertainties in the variables related to the deterioration process, a probabilistic treatment of these uncertainties must be adopted in the service-life prediction of RC structures.

In spite of the relevance of this issue, durability of RC structures is treated via prescriptive procedures in most operational codes (e.g. ACI 318-14, NBR 6118:2014), i.e. specifying water/cementitious ratio (or concrete strength), cover thickness, as a function of the environmental exposure. On the other hand, other documents, e.g. fib Model Code (2010) and ACI 365 (2000) have paved the way for more advanced code formats which includes probabilistic assessment of the service-life of RC structures.

While a framework for such probabilistic assessment can be easily envisioned at present, the main difficulties in the implementation of such format arise from the selection of the models to be used in the calculation of the depth of carbonation and the statistical description of the attendant variables.

In this study, Monte Carlo simulation (MCS) is used for the probabilistic service-life prediction of RC structures, for the limit state of depassivation. Service-life predictions are obtained using three different models for the prediction of the carbonation depth. Different scenarios are assumed for the materials, construction practices and level of CO_2 concentration in the atmosphere.

Considering the high values of the predicted COV's of the service-life description according to the three considered models and estimated probabilities of failure, it is clear that a realistic treatment of the durability/service-life prediction of RC structures subjected to carbonation must be pursued within a probabilistic framework. To this end, it seems that the most important and challenging topic is the validation of carbonation models and the incorporation of the statistics of the corresponding model error into the probabilistic estimates.

ACKNOWLEDGMENTS

The authors would like to thank the Brazilian agencies CNPq ("Conselho Nacional de Desenvolvimento Científico e Tecnológico") and FAPEMIG ("Fundação de Amparo à Pesquisa de Minas Gerais") for the financial support provided.

REFERENCES

ACI Committee 365 (2000) *Service Life Prediction-State of the art report*, ACI 365.1R-00. Farmington Hills, Michigan, American Concrete Institute.

American Concrete Institute (2014) *ACI 318: Building Code Requirements for Structural Concrete*. Farmington Hills.

Associação Brasileira de Normas Técnicas (2014) *Projeto de Estruturas de Concreto – Procedimento* (NBR 6118:2014).

Boden, T.A., Marland, G. & Andres, R.J. (2015) *Global, Regional, and National Fossil-Fuel CO₂ Emissions*. Oak Ridge, TN, USA, Carbon Dioxide Information Analysis Center, Oak Ridge National Laboratory, U.S. Department of Energy.

Diniz, S.M.C., Padgett, J.E. & Biondini, F. (2016) Durability design criteria for concrete structures – An overview of existing codes, guidelines and specifications. In: *Life-Cycle of Engineering Systems: Emphasis on Sustainable Civil Infrastructure*. In: *Proceedings of the Fifth International Symposium on Life-Cycle Civil Engineering (IALCCE 2016)*.

Féderation Internationale du Béton – CEB-FIP (2011) *MC 2010: Model Code 2010*. Lausanne.

Life-Cycle Analysis and Assessment in Civil Engineering: Towards an Integrated Vision – Caspeele, Taerwe & Frangopol (Eds)
© 2019 Taylor & Francis Group, London, ISBN 978-1-138-62633-1

Durability design of concrete structures regarding chloride-induced corrosion by means of nomograms

A. Rahimi
Federal Waterways Engineering and Research Institute, Karlsruhe, Germany

ABSTRACT

The corrosion of reinforcing steel owing to the action of chlorides is an aspect of great importance as regards the durability of a large number of structures. Many types of civil-engineering structure are affected, including infrastructure such as bridges, tunnels and similar structures that are treated with deicing agents in winter for traffic-safety reasons and marine structures such as dams and locks. Chlorides penetrating concrete as far as the surface of the reinforcing steel and reaching a critical concentration at this point will damage the protective passive layer of concrete surrounding the rebars. This phase, known as the corrosion-initiation phase, does not itself cause any damage to the structure. It is only after the passive layer has been destroyed and when certain boundary conditions are present (moisture, oxygen ingress) that the reinforcing steel will begin to corrode (damage phase). Progression of this corrosion process can result in damage to the structural member, in some cases within a short period of time, with consequences for its serviceability and its load-bearing capacity.

The durability design of new structures has hitherto been performed in accordance with the prescriptive approach specified in standards by complying with certain minimum requirements for concrete composition and concrete cover to reinforcement based on experience but without considering the actual performance of the concrete and the structural members or the exposure.

For the purpose of assessing the residual service life of existing structures and planning repairs the prescriptive approaches are generally inadequate or cannot be applied owing to the lack of information on the composition of the materials. Performance-based design methods suitable for application in practice are not available.

A semi-probabilistic approach to the durability design of reinforced concrete members exposed to the action of chlorides was elaborated in *Rahimi (2017)* and simplified by developing design nomograms. The concept enables new structural members to be designed for durability and existing structural members to be restored by replacing damaged concrete; it can also be used to assess the residual service life of existing structural members. The concept is described briefly below. Details and explanations are given in *Rahimi (2017)*. The design concept was introduced in the Code of Practice *Dauerhaftigkeitsbemessung und -bewertung von Stahlbetonbauwerken bei Carbonatisierung und Chlorideinwirkung (BAW-MDCC)* ("Durability design and assessment of reinforced concrete structures exposed to carbonation and chlorides") published by the Federal Waterways Engineering and Research Institute, BAW, in November 2017.

REFERENCES

BAW-MDCC (2017) *Durability Design and Assessment of Reinforced Concrete Structures Exposed to Carbonation and Chlorides*. Published by the Federal Waterways Engineering and Research Institute, BAW, Karlsruhe, Germany (in German).

Rahimi, A. (2017) *Semi-Probabilistic Concept to the Service Life Design and Assessment of Concrete Structures Exposed to Chlorides*. Beuth, Berlin, DAfStb Heft 626 (in German).

Life-Cycle Analysis and Assessment in Civil Engineering: Towards an Integrated Vision – Caspeele, Taerwe & Frangopol (Eds)
© 2019 Taylor & Francis Group, London, ISBN 978-1-138-62633-1

Sensitivity analysis of a service life model linked to chloride induced corrosion focusing on the critical chloride content

G. Kapteina
HafenCity University Hamburg, Germany

ABSTRACT

To ensure the reliability of reinforced concrete structures linked to durability a full-probabilistic service life design can be applied. As the user of such models should have information of the model behavior a sensitivity analysis has been carried out to provide information to which extent the variables influence the calculation result. In the presented paper representative alpha values (α_R-value) have been evaluated to illustrate the influence of the variables on the calculation result (reliability index). The α_R-values have been evaluated for a concrete with a high and for a concrete with a low resistance against chloride ingress (initial calculations). Hereby the quantification has been chosen according to the model code. In both cases comparable high α_R-values have been calculated for the ageing exponent, indicating an important role of this variable.

As the influence of every single variable depends strongly on the quantification of all other variables, a further sensitivity analysis has been carried out multi-dimensionally. In this context special attention has been given to the critical chloride content C_{crit} as this variable is under discussion. According to the model code the current quantification is based on tests carried out on mortar electrodes in the laboratory. As this test setup leads to rather conservative values for C_{crit} the mean value has been increased by 25% to take in-situ conditions into account. Recent investigation

led to considerably higher values. The influence of higher critical chloride contents on the sensitivity of the variables has been examined by comparing the results of the initial calculations with the respective modified calculations. Hereby the modified calculation took an adapted quantification of C_{crit} into account, which is to be expected of higher relevance for in-situ conditions.

For concrete with a low resistance against chloride ingress the modification of C_{crit} leads to a very significant increase in the calculated reliability index and to significant changes of α_R-values. Due to the modification the influence of C_{crit} and $C_{S,\Delta x}$ becomes more important, whereat the variables linked to the chloride model itself have been significantly reduced. Due to the modification the importance of $C_{S,\Delta x}$ is even outrunning the importance of the ageing exponent. Basically, the same effects have been observed for a concrete with a high resistance, whereat the effects are less pronounced. The modification of C_{crit} for the high resistance concrete leads also to a reduction of the $\alpha_R(a)$, but the importance of the ageing exponent remains on a high level. The dominance of this variable within the modified calculation is plausible, as this variable is also related to the high material resistance. The sensitivity analysis demonstrates, that especially for concrete with a low resistance against chloride ingress the influence of the critical chloride content might become very important within service life design and should be considered more closely.

The importance of the size effect in corrosion of steel in concrete for probabilistic service life modeling

U.M. Angst
Institute for Building Materials, ETH Zurich, Zurich, Switzerland

ABSTRACT

Chloride-induced corrosion is a localized phenomenon, where corrosion initiates first at the weakest spot within an exposed steel surface area. This gives rise to a size scale effect, that is, the critical chloride content for corrosion, C_{crit}, is size dependent: the larger the size (e.g. rebar length), the lower becomes C_{crit}, and the lower becomes the scatter in C_{crit}.

This paper illustrates the impact of this size effect on probabilistic service life modeling. Examples are calculated using input data from laboratory studies and from engineering structures. The calculations highlight that the size effect greatly influences the outcome of probabilistic service life modeling. This is apparent from Fig. 1 that displays an example of a probabilistic corrosion prediction. Clearly, the chosen rebar length has an impact on the model outcome. The larger the considered size, the steeper decreases the reliability index and will thus earlier achieve the target reliability. Moreover, the figure shows that with C_{crit} determined on samples from a structure, a reasonable service life (t_{ini}) can only be achieved with relatively low target reliability indices (around $\beta = 0.5$) and for relatively short rebar lengths.

To consider the size effect in future model developments, materials science needs to be linked to structural engineering. Knowledge of the "relevant size" to be considered in corrosion predictions can only be obtained from structural considerations, taking into account the geometry of the structural member and the loading conditions.

Combining structural engineering with materials science also permits rationalizing the limit states. For *secondary reinforcement*, initiation of chloride-induced corrosion will typically lead to a serviceability limit state (SLS), in the form of rust stains and cracks becoming apparent at the concrete surface. For *primary reinforcement*, on the other hand, if corrosion initiates in a critical section of a structural member, an ultimate limit state (ULS) may be reached relatively fast. Suggestions for handling this are made and the need for further research is highlighted.

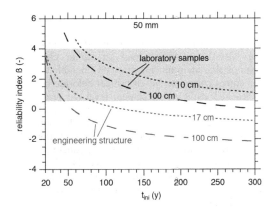

Figure 1. Reliability indices for full-probabilistic service life models considering C_{crit} for a cover depth of 50 mm. The calculations were run with different distributions of C_{crit} depending on selected rebar lengths (10 cm, 17 cm, 100 cm), based on data from laboratory samples (Angst and Elsener, 2017) and samples from a structure (Angst et al., 2017). The grey shaded area indicates common target reliability indices. t_{ini} = time to corrosion.

It is recommended that future developments of probabilistic service life modeling of chloride-induced corrosion take into account the size effect. This is considered important because clarity about limit states is crucial to justify the selection the target reliability index based on ISO standard 2394 (2015).

REFERENCES

(2015) ISO 2394:2015: General Principles on Reliability for Structures.

Angst, U.M., Boschmann, C., Wagner, M. & Elsener, B. (2017) Experimental protocol to determine the chloride threshold value for corrosion in samples taken from reinforced concrete structures. *Journal of Visualized Experiments*, e56229.

Angst, U.M. & Elsener, B. (2017) The size effect in corrosion greatly influences the predicted life span of concrete infrastructures. *Science Advances*, 3, e170075.

Life-Cycle Analysis and Assessment in Civil Engineering: Towards an Integrated Vision – Caspeele, Taerwe & Frangopol (Eds)
© 2019 Taylor & Francis Group, London, ISBN 978-1-138-62633-1

Evaluation of half-cell potential measurement and its impact on the condition assessment

S. Keßler
Centre for Building Materials, Technical University of Munich, Germany

ABSTRACT

Concrete structures exposed to chlorides due to deicing salts during winter road service or chlorides from marine environment are very vulnerable to reinforcement corrosion. When critical chloride content is reached at the reinforcement level the protective passive film on the reinforcement breaks down and reinforcement corrosion is initiated. The deterioration process continues with a local loss of cross section followed by cracking and spalling of the concrete cover when no repair measures are executed. The deterioration period to corrosion initiation and from corrosion initiation until concrete cracking covers a wide range depending on multiple factors.

Furthermore, no signs of active reinforcement corrosion are visible before cracking starts. Therefore, visual inspection does not lead to reliable information. Consequently, more elevated inspection tools such as half-cell potential measurement need to be applied supplemented by concrete cover measurement and chloride profiles.

This paper presents a case study of a real structure for updating the service life prediction with half-cell potential measurement in particular in view of the chosen threshold potential. For the sake of completeness the available concrete cover inspection data are considered as well.

Besides the probabilistic model the accuracy of the updated service life is dependent on the reliability of the used inspection method. The crucial point in the updating process with half-cell potential measurement is the definition of the so-called threshold potential, $E_{threshold}$ which divides the potential readings in active and passive probability density functions.

The reinforced concrete reference structure is an abutment wall located close to a highly frequented road in Munich built in 1961. Due to direct rain exposure and splash water the structures experience constantly drying and wetting. The splash water contains chlorides during winter time. Consequently, chloride-induced reinforcement corrosion is the decisive deterioration mechanisms. After 47 years in service the structure has a depassivation probability of 67%. The segment is subdivided into elements with the same size such as the grid size used. For each element the depassivation probability is recalculated with the assigned measured concrete cover. The update with half-cell potential measurement consists of the Bayes' approach since the half-cell potential measurement delivers qualitative information about the corrosion condition such as "corrosion indicated" or "no corrosion indicated". Furthermore, the a-priori depassivation probability as well as the reliability of the inspection result needs to be treated probabilistically.

Generally, the higher the potential threshold as higher is the corrosion detectability but as well the probability of false alarm. The probability of false alarms describes the probability to indicate a defect even there is no defect which leads to unnecessary repaired structures and needless expenses. A general recommendation of the threshold potential is not practical since the probability of detection and the probability of false alarm is more dependent on the active and passive probability density function than on a fixed threshold. The active and passive probability density functions are unique for each half-cell potential measurement. Therefore, the threshold potential has to be determined on the one hand under consideration of the probability density functions and on the other hand on the acceptable degree of probability of false alarm. The acceptable degree of probability of false alarm depends on the type of structure and shall be discussed with the structural engineer in charge.

Serviceability and residual working life assessment of existing bridges

J. Marková
Klokner Institute, Czech Technical University in Prague, Czech Republic

V. Navarova
College of Ceske Budejovice, Czech Republic

ABSTRACT

In the Czech Republic the existing bridges are categorized on the basis of visual inspections to seven categories. Several indicators of bridge state are defined in the standard CSN 73 6221. Category I is used for the best bridge condition while the load-bearing capacity of bridges classified to categories V to VII is considerably reduced from 60% up to 20%. However, such decision on actual load bearing bridge capacity is commonly accompanied by significant uncertainties.

Recently revised standard CSN 73 6222 provides supplementary guidance for determination of the load-bearing capacity and for estimation of the remaining working life of existing concrete bridges. Several bridge categories of prestressed and reinforced concrete bridges are distinguished and limiting values of crack width are recommended.

The probabilistic methods are applied for the verification of the reliability level of the bridge with respect to the serviceability limit states of crack width, Markova, Holicky (2010), Sykora et al. (2016).

The bridge may be considered as reliable if the condition $\beta \leq \beta_t$ is satisfied where the target reliability expressed in the reliability index β_t should not be exceeded during the bridge design working life.

The different crack width models recommended in the Eurocode EN 1992-1-1, draft prEN 1992-1-1 and Model Code are taken into account. The reliability analysis of a reinforced concrete bridge with respect to the serviceability limit states of crack width indicates that the uniform corrosion leads to a smaller reduction of the reinforcement area and higher reliability indices than the pitting corrosion.

The initial reliability of the bridge with respect to crack width is greater than the target value of the reliability index $\beta_t = 1,5$ for the verification of the limit states of crack width recommended in EN 1990. However, the diminishing area of reinforcement due to the reinforcement corrosion leads to decrease of the reliability index β with increasing working life of the bridge as shown in Figure 1.

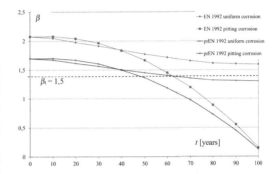

Figure 1. Variation of the reliability index β for uniform and pitting corrosion with time t for selected crack width limits w_{lim} considering crack width models given in EN 1992-1-1 and draft of revised prEN 1992-1-1.

The serviceability constrains recommended for the assessment of the residual working life of a bridge in current prescriptive documents should be further analysed and calibrated. Crack width models given in current codes should be harmonised.

It appears that the probabilistic assessment of existing bridges may facilitate the decision regarding their safety and serviceability, and indirectly contribute to a sustainable development.

REFERENCES

Marková, J. & Holický, M. (2010) Serviceability criteria in current codes. In: *Codes in Structural Engineering. Developments and Needs for International Practice*. Zagreb. pp. 887–894.

Sykora, M., Holicky, M., Markova, J. & Senberger, T. (2016) *Probabilistic Reliability Assessment of Existing Structures (Focused on Industrial Heritage Buildings)*. Prague, Czech Technical University in Prague, CTU Publishing House.

*MS16: Life-cycle maintenance and management for urban
infrastructures with big data*
Organizers: A. Chen, Y. Yuan & X. Ruan

Performance comparison for pipe failure prediction using artificial neural networks

S. Kerwin
Institute of Construction and Infrastructure Management, ETH Zurich, Zurich, Switzerland

B. Garcia de Soto
Institute of Construction and Infrastructure Management, ETH Zurich, Zurich, Switzerland
S.M.A.R.T. Construction Research Group, New York University Abu Dhabi, Abu Dhabi, UAE
Tandon School of Engineering, New York University (NYU), Brooklyn, NY, USA

B.T. Adey
Institute of Construction and Infrastructure Management, ETH Zurich, Zurich, Switzerland

ABSTRACT

Infrastructure managers must decide on the replacement timing of buried pipes in water distribution networks. These assets deteriorate resulting in failures and their associated consequences. In recent years, studies have investigated failure prediction models for pipes based on artificial neural networks (ANNs). These models are either generalized (i.e. trained with all pipe failures) or specialized (i.e. trained with certain pipe failures based on either pipe material or failure history) (Jafar et al., 2010, Harvey et al., 2014, Asnaashari et al., 2013). It is currently unclear whether infrastructure managers should develop several specialized ANNs or a generalized one.

In this study, the following steps were used to develop the ANN models for performance comparison:

1. Define level of specialization for ANN models
2. Choose output parameter
3. Choose input parameters
4. Determine ANN architecture
5. Vary ANN model parameters
6. Evaluate ANN performance

The following four models were developed: a generalized model for cast iron (CI) and ductile iron (DI) pipes; a specialized model for CI pipes, a specialized model for CI pipes with no previous failures and a specialized model for CI pipes, which had previously failed. CI failures were chosen for the comparison because of greater data availability.

The generalized model, with an architecture (i.e. number of neurons in layer) of 13-21-1, was trained with data from 5,824 failed pipes (2,468 CI failures, and 3,356 DI failures). The specialized model for CI pipes, with an architecture of 10-6-1, was trained using 2,468 CI failures.

Figure 1 and Figure 2 show the error histograms of one of the comparisons (generalized vs specialized for CI pipes). Overall, the study found minimal difference in performance between the generalized and specialized models.

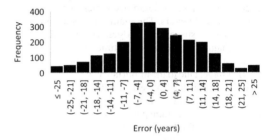

Figure 1. Error histogram of generalized model.

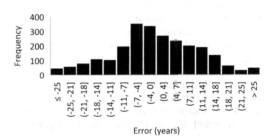

Figure 2. Error histogram of specialized model.

REFERENCES

Asnaashari, A., McBean, E.A., Gharabaghi, B. & Tutt, D. (2013) Forecasting watermain failure using artificial neural network modelling. *Canadian Water Resources Journal*, 38, 24–33.

Harvey, R., McBean, E.A. & Gharabaghi, B. (2014) Predicting the timing of water main failure using artificial neural networks. *Journal of Water Resources Planning and Management*, 140, 425–434.

Jafar, R., Shahrour, I. & Juran, I. (2010) Application of Artificial Neural Networks (ANN) to model the failure of urban water mains. *Mathematical and Computer Modelling*, 51, 1170–1180.

Life-Cycle Analysis and Assessment in Civil Engineering: Towards an
Integrated Vision – Caspeele, Taerwe & Frangopol (Eds)
© 2019 Taylor & Francis Group, London, ISBN 978-1-138-62633-1

Long-term mechanical reliability evaluation of the main cable of a suspension bridge

D. Wang, Y. Zhang, A. Chen & L. Li
Tongji University, Shanghai, China

H. Tian
Zhejiang Scientific Research Institute of Transport, Hangzhou, China

ABSTRACT

The carrying capacity of bridges inevitably deteriorate over time because of durability issues as well as the environmental factors. The main cable is an essential component of suspension bridges. The degradation of its mechanical properties with time is one of the key issues not only during the design phase but also in the service stage. Due to the importance of the main cable, a lager safety factor will be used at the beginning of the design to ensure the structural safety considering the degradation of structural performance. However, there is still a need for a method that can more accurately and truly evaluate the mechanical properties of a main cable in actual working environment.

The mechanical performance is evaluated by the multi-scale model in this paper which contains the steel wire scale, the strand scale, and the main cable scale. The negative influence of length effect and Daniels effect is considered in the wire scale and the strand scale. Based on the Monte Carlo Method, the statistical formulas of main cable's equivalent resistance frequency for different corrosion time are summarized into the time-dependent equivalent resistance probability model of the main cable as shown in Figure 1.

In order to study the stress distribution of main cable, the main cable units are divided into different modules according to their location and response (see figure 2). The stress of each main cable is taken as the analysis object to analyze the stress distribution rules of the main cable under temperature, wind and vehicle loads.

Figure 2. Module division of main cable.

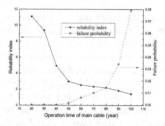

Figure 3. Reliability index and failure probability of main cable in different operation time.

Based on the reliability theory and Rice extrapolation theory, we can get the reliability index and the failure probability of the main cable under corrosion. By analyzing the results, we can come to following conclusion:

1) The mechanical properties of the steel wire or the main cable will be greatly attenuated after being corroded.
2) The current codes usually classify the corrosion grades of steel wires into four grades, which is not safe enough.
3) It is essential to install the dehumidification system on the main cable of suspension bridge.

REFERENCES

Council N. (2004) *Guidelines for Inspection and Strength Evaluation of Suspension Bridge Parallel Wire Cables*. Nchrp Report.
Cremona, C. (2001) Optimal extrapolation of traffic load effects. *Structural Safety*, 23 (1), 31–46.
Elachachi, S.M., Breysse, D., Yotte, S. & Cremona, C. (2006) A probabilistic multi-scale time dependent model for corroded structural suspension cables. *Probabilistic Engineering Mechanics*, 21 (3), 235–245.

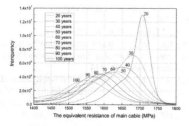

Figure 1. Resistance model of wires after considering Daniels effect.

Empirical Bayes-based Markov chain deterioration modelling for municipal sewer systems

P. Lin & X.X. Yuan
Department of Civil Engineering, Ryerson University, Toronto, ON, Canada
Ryerson Institute for Infrastructure Innovation, Ryerson University, Toronto, ON, Canada

R. Rashedi
Infrastructure Solutions Inc., Mississauga, ON, Canada

ABSTRACT

Municipal sewer systems are deteriorating in both structure integrity and performance with increasing service age. In this paper, the degradation of sewer structure condition is first discretized into five states based on Manual for Sewer Condition Classification (MSCC) developed by Water Research Center (WRc) in Great Britain; then a time-homogeneous Markov Chain model is developed to describe the transition between two successive condition states where the transition is represented by the transition intensity (also called transition rate), λ_i ($i = 1, 2, 3, 4$). The general idea is shown in Figure 1.

Based on the Markov Chain theory, the transition probability functions, defined as $p_{ij}(s,t) = Pr(X(t) = j | X(s) = i)$, where $0 \leq s \leq t$, are written as:

$$\mathbf{P}(s,t) = e^{(t-s)\mathbf{Q}} \quad (1)$$

where $\mathbf{P}(s,t)$ is the transition function in a matrix form with element $p_{ij}(s,t)$, and $\mathbf{Q}$ is the intensity matrix defined as:

$$\mathbf{Q} = \begin{bmatrix} -\lambda_1 & \lambda_1 & 0 & 0 & 0 \\ 0 & -\lambda_2 & \lambda_2 & 0 & 0 \\ 0 & 0 & -\lambda_3 & \lambda_3 & 0 \\ 0 & 0 & 0 & -\lambda_4 & \lambda_4 \\ 0 & 0 & 0 & 0 & 0 \end{bmatrix} \quad (2)$$

Small to medium municipalities usually do not have sufficient condition data to allow a confident estimation of λ_i. An empirical Bayesian approach is proposed to bypass this data shortage issue in estimation of the intensities, following two main steps: first, pool all available condition data to form a general database which is then used to determine the prior of λ_i using the maximum likelihood method; next, the Bayesian updating approach is utilized to find the posterior using municipality-specific condition data.

A case study is illustrated to elaborate the implementation of the Bayesian approach and the application of the developed Markov Chain model in prioritization and optimization of wastewater linear assets within the framework of risk-informed life-cycle cost management. A total of 3169 sewer condition inspections that are collected from numerous municipalities across Ontario, Canada is used as the general database; whereas the municipality-specific database contains a total of 228 sewer inspections from a small municipal S in Ontario. The λ_i values for municipality S are then determined and the developed model is used to predict the survival probability for a sewer asset with increasing service age, as shown in Figure 2.

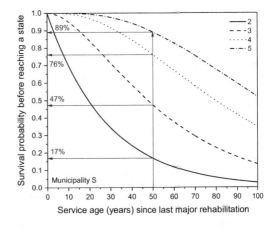

Figure 2. Survival probability before reaching a state against service age since last major rehabilitation.

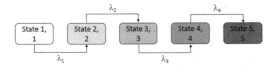

Figure 1. Deterioration of pipe structure conditions based WRc grading protocol.

Incremental launching construction of Chajiaxia Yellow River Bridge with data feedback

L. Zhou, J. Peng, Y. Wu & Z. Yin
Shanghai Urban Construction, Design and Research Institute, Shanghai, China

ABSTRACT

Incremental launching construction with the assistance of data feedback system is proven to be stable and accurate. For Chaijiaxia Yellow River Bridge, a curved cable-stayed bridge with two uneven pylons, this construction method is adopted. Elevation and layout of the bridge is illustrated in Figure 1.

For the construction process, 2 assembly platforms and 16 temporary piers are temporarily established, based on which the process are performed in 7 specific steps. Preliminary and refined analysis on different construction scenarios is implemented accordingly. Analysis results indicate that the stress state in the diaphragm near the support area is quite high. As a result, stiffening ribs in the lower chords and vertical supporting chords in the middle of the truss diaphragms are added to enhance the mechanical performance as shown in Figure 2.

The whole incremental launching system consists of hydraulic pump system, launching units and computer controlling system. Construction data feedback is integrated within this system. Four types of data are collected, the launching distance, the oil pressure, the girder displacement and environment temperature. The collected data is transmitted to the measurement

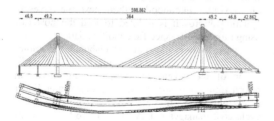

Figure 1. Elevation and layout of the bridge (unit: m).

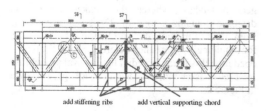

Figure 2. Component advancement based on construction scenario analysis.

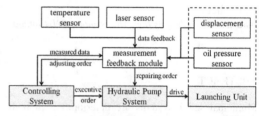

Figure 3. Incremental launching construction feedback system.

Table 1. Thresholds of collected data.

Collected data type		Threshold
longitudinal launching distance	different units	5 mm
	same unit	1 mm
vertical launching distance	same unit	4 mm
girder axis deviation		10 mm
longitudinal displacement of auxiliary piers		20 mm
altitude of the girder bottom		10 mm

feedback module and then to the power control module through the local area network. The whole process is illustrated in Figure 3.

During the incremental construction process, the synchronization of different units and pairing jacks in the same unit is ensured by the data feedback system. Once the measured data value exceeds its threshold, the computer controlling system will accordingly adjust the hydraulic pump system. The thresholds of the collected data are listed in Table 1.

Currently, Chaijiaxia Yellow River Bridge is under construction and is expected to open to public in the end of 2018.

REFERENCES

Arici, M. & Granata, M.F. (2007) Analysis of curved incrementally launched box concrete bridges using the transfer matrix method. *Bridge Structures*, 3 (3–4), 165–181.
Astaneh-Asl, A. & Black, R.G. (2001) Seismic and structural engineering of a curved cable-stayed bridge. *Journal of Bridge Engineering*, 6 (6), 439–450.
Hui, N., Wang, J.F., Zhang, Y.P., Zhang, Z.C. & Ya-Nan, Y.U. (2013) Study of incremental launching of space-curved butterfly-arch bridge. *Journal of Zhejiang University*, 47 (7), 1205–1212.

Bayesian formula based dynamic information updating of bridge traffic flow response probability model

X.J. Wang, X. Ruan & K.P. Zhou
Department of Bridge Engineering, Tongji University, Shanghai, China

ABSTRACT

Nowadays, with increasing amount of measuring instruments and data, continuous updating of information provides the basis for understanding structural characteristics and tracking the time-varying effects of the structure. Structural performance gradually deteriorates over time caused by the natural, man-made and other factors. And there is no appropriate code on the aging of the structure.

In this paper, the shortcomings of traditional probabilistic methods in data analysis of engineering structure from the perspective of information updating are discussed. Under the era of big data, only by constantly updating the understanding of the characteristics of the structure according to the newly added information can the value of the monitoring data be used to provide a better basis for structural assessment. On this basis, based on Bayesian theory and combined with the characteristics of engineering applications, a more practical bayesian formula for data updating and evaluation is deduced. The main conclusions are as follows.

By deducing the Bayesian formula, the posterior distribution formula expressed by the prior information and the sample information parameters is obtained, which greatly simplifies the complexity of the application of the Bayesian formula in the engineering field and provides a basis for the application and evaluation of the structural monitoring data. According to the measured WIM data, the data of backstay force are obtained. After fitting, it is found that its truncated distribution corresponds with the logarithm normal distribution. The traffic data of each group at different speeds and flows are respectively loaded, and the response data of each group is obtained for further update. From the updated results of each group, it can be seen that the updated posterior distribution probability density curve is located between the probability density curve of prior distribution and the test samples. In addition, the posterior distribution curve is closer to the prior distribution curve because the number of test specimens is less than that of the prior distribution.

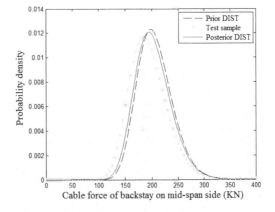

Figure 1. Flow–speed [200~300; 55~60].

After the update, the load level of vehicles responding posterior truncated distribution has been generally improved, which shows that the updated probability distribution is more in line with the actual situation of vehicle load development status today.

In summary, in this paper the necessity of information update is discussed, and a viable approach of information inheritance is proposed for massive monitoring data.

REFERENCES

Caspeele, R. & Taerwe, L. (2013) Numerical Bayesian updating of prior distributions for concrete strength properties considering conformity control. *J. Advances in Concrete Construction*, 1 (1), 85–102.

Law, S.S. & Li, J. (2010) Updating the reliability of a concrete bridge structure based on condition assessment with uncertainties. *Journal of Engineering Structures*, 32 (1), 286–296.

Wang, X.J. (2016) *Categorized Traffic Method Study Based on Loading and Traffic Characteristics of Traffic Flow.* D. Tongji University.

Zhou, K.P. (2015) *Information Updating Based System Reliability Assessment of String-Cable Arch Bridge in Operation Stage.* D. Tongji University.

Experimental study on the mechanical property evolution of main cable steel wires of suspension bridges under corrosion state

R. Ma, C. Cui, A. Chen & L. Li
Tongji University, Shanghai, China

H. Tian
Zhejiang Scientific Research Institute of Transport, Hangzhou, China

ABSTRACT

Corrosion of main cable wires of suspension bridges is a hot but difficult research topic worldwide. In this article, corrosion of main cable steel wires is studied by a current accelerated corrosion experiment. Corrosion morphologies of steel wires under different strain levels and corrosion time are observed and their mechanical properties are summarized. The experiment results could be used for structural design and long-term performance evaluation of suspension bridges. The results show that the modulus of elasticity does not change significantly with the increasing of corrosion time even under a higher strain level. It could be approximately assumed that both the strain levels and the corrosion time do not have much effect on the modulus of elasticity of the high-strength wires. Both the yield load and the ultimate load have an approximately linear relationship to the minimum diameter of cross section. Based on these relationships, experimental formulas are established to calculate yield load and the ultimate load according to corrosion time and strain level.

The relationship between yield load and minimum diameter of cross section can be seen in Figure 1.

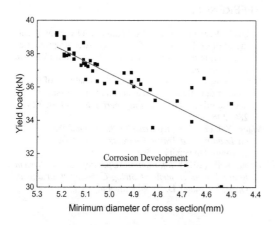

Figure 1. Relationship between yield load and minimum diameter of cross section.

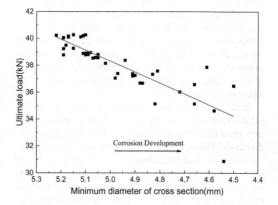

Figure 2. Relationship between ultimate load and minimum diameter of cross section.

The relationship between ultimate load and minimum diameter of cross section can be seen in Figure 2.

Linear relationship between minimum diameter of cross section and strain level is summarized as the following equation.

$$d_{min}(mm) = (5.5085 - 0.1645t) \\ + (3.0167 \times 10^{-5} \\ - 2.95e \times 10^{-5}t) \cdot \varepsilon \quad (1)$$

Thus, the final equations of yield load and ultimate load can be derived as follows.

$$p_y = (40.465 - 1.183t) + (21.693 \times 10^{-5} \\ - 21.21 \times 10^{-5}t) \cdot \varepsilon \quad (2)$$

$$p_u = (42.54 - 1.34t) + (24.58 \times 10^{-5} \\ - 24.04 \times 10^{-5}t) \cdot \varepsilon \quad (3)$$

ACKNOWLEDGMENTS

The authors would like to thank the National Natural Science Foundation of China (No. 51678437) and Major Science and Technology Special Program of Guizhou Province (No. [2016] 3013).

A risk management system of Hainan interchange merge area in maintenance zone: Risk analysis, identification, assessment and countermeasure

B. Liu, H. Yan & W. Zhao
Key Laboratory of Road Safety, College of Transportation Engineering, Tongji University, Shanghai, China

ABSTRACT

According to the present situation of frequent accidents in Hainan interchange merge area maintenance zone, this paper presents a risk management system. Accidents are investigated and analyzed, then a risk identification method based on the fault tree analysis (FTA) method is proposed to identify the risk factors with the combination of the work breakdown structure (WBS) and risk breakdown structure (RBS). A risk matrix based on the WBS-RBS is carried out and 16 sources of risk are analyzed and identified. Combining the method of FTA with the sources of risk, three main accidents of Hainan expressway maintenance are analyzed by the FreeFta Fault Tree analytical software. Further efforts, sources of risk are divided into three levels critical sources of risk, secondary sources of risk and non-critical sources of risk according to the sensitive coefficient values, and the rationality of the analytical method is elucidated. Finally, countermeasures of risk events are carried out and a framework of risk management is built. Initial result provides a reference for the risk management of expressway maintenance zone.

The highway administration, highway department and equipment maintenance department, from which accident data in expressway maintenance zone can be got, are investigated. There are three main accidents: accidents of maintenance work, accidents of vehicle collision and vehicles break into work area (Figure 1).

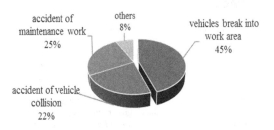

Figure 1. Statistics chart of accidents in expressway maintenance zone.

Table 1. The statistics of reasons for accidents.

The reason of accidents	Code	Vehicles break into working zone/%	Accidents of vehicle collision/%	Accident of maintenance work/%
Poor visibility of safety facilities	R1	12	0	0
Unreasonable layout of safety facilities	R2	32	0	0
Terrible road condition	R3	8	21	9
Motor vehicles and non-motor vehicle mixed together	R4	23	40	0
Driving infraction	R5	25	0	0
Un-closed expressway	R6	0	24	0
Bad visibility	R7	0	10	0
Unreasonable working zone layout	R8	0	15	0
Mechanical failure	R9	0	0	15
Improper operation	R10	0	0	23
Illegal operations	R11	0	0	18
Terrible management	R12	0	0	25
Unreasonable material placement	R13	0	0	10

REFERENCES

Arditi, D., Lee, D.-E. & Polat, G. (2007) Fatal accidents in night-time vs. day-time highway construction work zones. *Journal of Safety Research*, 38 (4), 399–405.

Ding, C., Huang, W. & Jiang, X. (2000) A method of fault isolation for spacecraft based on minimum cut sets rank. *Chinese Journal of Space Science*, 20 (1), 89–94.

Buckling analysis of a long span steel cable-stayed bridge

Y. Wei, Y.T. He & X. Ruan
Department of Bridge Engineering, Tongji University, Shanghai, China

L. Ding
China Design Group Co. Ltd, Nanjing, China

ABSTRACT

Buckling has always been one of the key problems in cable-stayed bridge design. For long span steel cable-stayed bridges, the main girder and the tower both have thin-wall sections, so the overall and local buckling stability of the structure will be more prominent.

It is not desirable for designers to see the instability damage of the structure before reaching its ultimate capacity, so the stability analysis must be carried out. Compared with concrete cable-stayed bridges, initial defects and residual stress of steel girder and steel tower will inevitably appear in processing and assembling process. Combined with the latter load action, the steel cable-stayed bridge is likely to have local buckling first, which forms weak area, and at last results in instability and failure of the whole structure.

Therefore, buckling analysis should be conducted. After the analysis, an eigenvalue, which is known as the buckling coefficient, is obtained and it can be used as the upper limit of structural ultimate capacity. Being a necessary process in the stability analysis of bridges, traditional buckling analysis generally adopts model based on beam element, which cannot consider local stability problem, and often get large buckling coefficient, whose referential significance to ultimate capacity is doubtful. Therefore, detailed study is conducted in this paper to solve these problems.

Based on a 500 m steel cable-stayed bridge, buckling analysis with beam-element model and hybrid model is conducted. The buckling coefficient and instability mode are obtained. The process of instability failure is studied through elastoplastic analysis.

From the analysis results, buckling coefficients of different load combinations are different. The coefficient for the same load combination but different

Table 1. Structural 1st order buckling coefficient under typical load combinations.

Load combination	Model	λ
Dead load	Beam-element model	4.74
	Hybrid model	5.05
Dead load + full vehicle load	Beam-element model	4.59
	Hybrid model	4.44

models is also different, which reflects the influence of different models on the global stiffness matrix.

Through the establishment of hybrid model, the local buckling of the main components can be considered, which helps to clear the cause of structural instability and makes it a good and comprehensive way to conduct structural analysis.

Although the analysis proves that the buckling coefficient can be used as upper limit to estimate the structural ultimate bearing capacity indeed, but the relationship between them still cannot be concluded very well. So, to get a more accurate and practical failure process and ultimate load, it is still necessary to consider the influence of nonlinearity, loading method, instability criterion and so on and do more in-depth research.

REFERENCES

Choi, D., Gwon, S. & Yoo, H. (2013) Estimation of inelastic buckling collapse loads for steel cable-stayed bridges. *IABSE Symposium Report*, 101 (13), 1–4.
Pedro, J.J.O. & Reis, A. (2016) Simplified assessment of cable-stayed bridges buckling stability. *Engineering Structures*, 114, 93–103.
Wu, B., Xiao, R. & Wei, L. (2009) Study on ultimate load capacity of steel pylon of cable-stayed bridge. *Journal of Shijiazhuang Railway Institute (natural science)*, 22 (2), 5–10.
Yoon, H. & Choi, D. (2009) Improved system buckling analysis of effective lengths of girder and tower members in steel cable-stayed bridges. *Computers and Structures*, 87, 847–860.

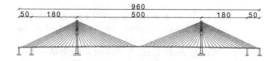

Figure 1. Overall arrangement of the steel cable-stayed bridge.

MS17: Life-cycle management as focus area within asset management
Organizers: J. Bakker, H. Roebers, M. Hertogh & J.F.M. Wessels

Life-Cycle Analysis and Assessment in Civil Engineering: Towards an Integrated Vision – Caspeele, Taerwe & Frangopol (Eds)
© 2019 Taylor & Francis Group, London, ISBN 978-1-138-62633-1

Quantifying the impact of variability in railway bridge asset management

P.C. Yianni
Asset Management Advisory, Jacobs, Wokingham, Berkshire, UK

L.C. Neves, D. Rama & J.D. Andrews
Resilience Engineering Research Group, Faculty of Engineering, University of Nottingham, University Park, Nottingham, UK

N. Tedstone & R. Dean
Network Rail, The Quadrant:MK, Elder Gate, Milton Keynes, Buckinghamshire, UK

ABSTRACT

Managing a portfolio of bridges can be a complex task considering the wide array of bridge configurations, material types and structural capability, as well as rising passenger numbers, increased freight usage and tightening budgets. To support bridge portfolio managers, decision support tools are becoming increasingly more important. There have been many studies to create bridge management models to support portfolio managers (Frangopol, Kallen, & van Noortwijk 2004, Miyamoto, Kawamura, & Nakamura 2000, Morcous, Rivard, & Hanna 2002). Although these models are very accurate and highly calibrated to the individual processes in the system, they can often overlook overarching complexities which limits their effectiveness as a tool for bridge portfolio managers.

This study investigates the effect of variability on railway bridge portfolio management, focusing on the aspects of variability which are human-induced, as they have most relevance for bridge portfolio managers. The factors have been identified through expert judgement and parametrised using historic records. The two factors which have been investigated in this study are 1) the variability introduced by using visual examinations, through to misdiagnosis of defects and the scheduling of incorrect maintenance actions and 2) the effect of imperfect repairs, with probabilities calculated from historic data. From there, the factors are incorporated into an existing railway bridge Whole Life-Cycle Cost (WLCC) model with simulations run under two scenarios: 1) the control scenario, which uses the industry standard policies and deterioration calibrated using historic data, and 2) the enhanced scenario, which uses the same base as the control scenario, but has been enhanced with probabilistic aspects to replicate the variability calculated from imperfect repairs and defect misdiagnoses.

The simulation results show that there is almost a two-fold increase in the cumulative WLCC of a typical bridge element over the simulation period of 100 years. This is a combination of: 1) dispensing with the assumption of perfect repairs, which inevitably means that the time between interventions is reduced as the condition uplift is no longer resulting in an "as new" condition every time and 2) misdiagnoses of defects, leading to incorrect maintenance actions being scheduled which means that the maintenance teams must return to carry out the appropriate maintenance action, increasing both the overall WLCC and operational burden.

Overall, it is not unexpected that the predicted financial and operational burdens increase when enhancing the WLCC model with human-induced variability. However, the quantification of this increase is significant as it shows how much of an effect that variability has on the WLCC of railway bridges. It is an important enhancement as it increases both the accuracy and usefulness of bridge management WLCC models.

REFERENCES

Frangopol, D. M., Kallen, M.-J. & van Noortwijk, J. M. (2004) Probabilistic models for life-cycle performance of deteriorating structures: review and future directions. *Progress in Structural Engineering and Materials, 6* (4), 197–212.

Miyamoto, A., Kawamura, K. & Nakamura, H. (2000) Bridge management system and maintenance optimization for existing bridges. *Computer-Aided Civil and Infrastructure Engineering, 15* (1), 45–55.

Morcous, G., Rivard, H. & Hanna, A. (2002) Modeling bridge deterioration using case-based reasoning. *Journal of Infrastructure Systems, 8* (3), 86–95.

Life-Cycle Analysis and Assessment in Civil Engineering: Towards an Integrated Vision – Caspeele, Taerwe & Frangopol (Eds)
© 2019 Taylor & Francis Group, London, ISBN 978-1-138-62633-1

Quantifying the Performance Age of highway bridges

Y. Xie & D. Schraven
Delft University of Technology, Delft, The Netherlands

J. Bakker
Rijkswaterstaat, Ministry of Infrastructure and Environment, Utrecht, The Netherlands

M. Hertogh
Delft University of Technology, Delft, The Netherlands

ABSTRACT

In this paper we propose a new parameter called Performance Age and the methodology to quantify it. This parameter provides a solid life cycle performance basis for bridge replacement. It is not only developed to give a new perspective to the replacement decision process, but also to support a strategic network management approach for civil infrastructures. The method draws on the principle that a bridge not necessarily has to be end of life at its estimated technical age, as long as it performs better than its expected use. For a bridge this idea could mean that it can reach a higher age than its assumed design life. We quantify Performance Age concept based on a hierarchical model, which consists of a translation of this expected performance with criteria and indicators. A prototype model was tested for three fixed concrete highway bridges managed by Rijkswaterstaat, an agency of the Dutch Ministry of Infrastructure and Environment. The results of this study show that Performance Age can be a useful parameter to predict the end of life of bridges.

Life-Cycle Analysis and Assessment in Civil Engineering: Towards an Integrated Vision – Caspeele, Taerwe & Frangopol (Eds)
© 2019 Taylor & Francis Group, London, ISBN 978-1-138-62633-1

A simplified approach to address uncertainty in Life Cycle Costing (LCC) analysis

Y. Sun & D.G. Carmichael
School of Civil and Environmental Engineering, The University of New South Wales, Sydney, Australia

ABSTRACT

Many probabilistic approaches have been proposed to deal with uncertainty in life cycle costing (LCC) analysis and include, for example, the use of probability density functions and fuzzy sets (Kishk 2004). However, probabilistic approaches are often criticized for their required knowledge of higher level mathematics and large data sets, while making the analysis complicated (Flanagan et al., 1987). The process of obtaining the required data can be costly and time consuming. Sun & Carmichael (2017) point out that three quarters of all probabilistic methods are either data-intensive or mathematically complicated. Such disadvantages deter people from including uncertainty in LCC calculations.

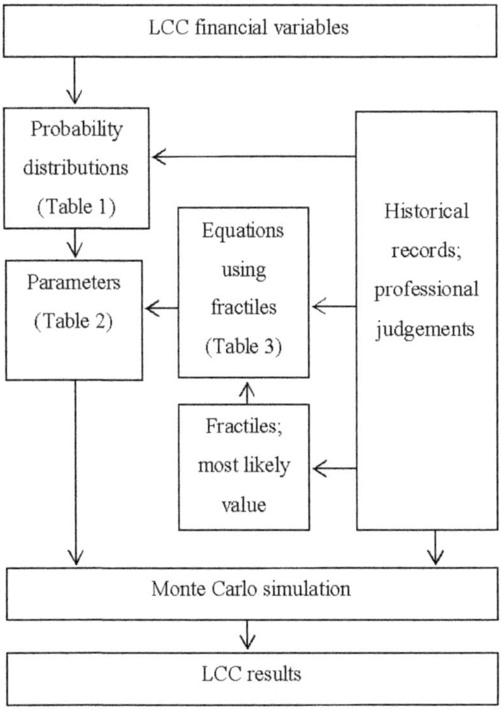

This paper proposes a simplified approach to address uncertainty (Figure 1). It summarizes five commonly used probability distributions describing the underlying LCC variables – cash flow, interest rate, timing of cash flows and asset lifetime – and reviews a range of methods, using fractiles, for estimating distribution key parameters, such as expected value and variance for a normal distribution, lowest and highest values for a uniform distribution, and shape and scale parameters for a Weibull distribution. With those estimated parameters, Monte Carlo simulation can be performed to obtain the distribution of present worth. A case study on a tourist centre building is worked in order to demonstrate the calculations and simulations involved. It is found that the proposed fractile approach is simple and flexible to implement.

The proposed approach will help to achieve a wider application of probabilistic LCC analysis to infrastructure. The paper will be of interest to anyone involved with life cycle costing analysis.

REFERENCES

Flanagan, R., Kendell, A., Norman, G. & Robinson, G.D. (1987). Life cycle costing and risk management. *Construction Management and Economics*, 5 (4), S53–S71.

Kishk, M. (2004). Combining various facets of uncertainty in whole-life cost modelling. *Construction Management and Economics*, 22 (4), 429–435.

Sun, Y. & Carmichael, D.G. (2017). Uncertainties related to financial variables within infrastructure life cycle costing: a literature review. *Structure and Infrastructure Engineering* published online 22 December 2017. https://doi.org/10.1080/15732479.2017.1418008

Figure 1. Incorporation of fractiles into Monte Carlo simulation for LCC.

Life-Cycle Analysis and Assessment in Civil Engineering: Towards an Integrated Vision – Caspeele, Taerwe & Frangopol (Eds)
© 2019 Taylor & Francis Group, London, ISBN 978-1-138-62633-1

Life cycle approach for sustainable pavement options for infrastructure projects

B. Czarnecki
Tetra Tech Canada Inc., Alberta, Canada

ABSTRACT

Majority of pavement design for new and reconstruction works in Alberta if for flexible structures with asphalt concrete surfacing. The design life of flexible pavements for municipal roadways is 20 years. The City of Red Deer requested alternative bid options for 2016 improvements and upgrades to a major transportation corridor servicing industrial and commercial developments and providing access to the city from a major highway. The project included the upgrade of two intersections to roundabouts including their approaches. The alternative options included asphalt concrete and Portland cement concrete. Life Cycle Cost Analysis was considered to allow for better understanding of the true costs of a roadway as opposed to considering only an initial cost of the pavement. The analysis included considerations for the differences in the design life of flexible and rigid pavements,

reduction in the volume of construction materials, durability and sustainability of the surfacing, maintenance needs over service life, and safety. Based on long term durability, extended service life to 40 years, life cycle costs, construction efficiency and less frequent disruptions to traffic due to future scheduled maintenance, the Portland cement rigid pavement option was selected for the upgrades. The initial construction costs were comparable for both options but the preservation costs over the life cycle were significantly lower for the Portland cement concrete option. Additional environmental benefits included lower total energy, reduced carbon dioxide emissions from vehicles operating on concrete pavements, reduction of urban heat island effect, and use of recyclable materials and industrial by-products. The paper discusses the structural performance of flexible and rigid pavements, durability and sustainability considerations and the rationale behind the selection process.

Comparison of truck fuel consumption measurements with results of existing models and implications for road pavement LCA

F. Perrotta, T. Parry & L.C. Neves
University of Nottingham, UK

T. Buckland, E. Benbow & H. Viner
TRL Ltd, UK

ABSTRACT

Life Cycle Assessment (LCA) is increasingly used to evaluate the impact of all lifecycle phases of road pavements on the environment. From the late '90s, this technique has continuously evolved and improved, however, there are still limitations and uncertainties in the framework. In this regard, Santero et al. (2011) showed that gaps still exist in the road pavement LCA methodology. More recently, Trupia et al. (2016) highlighted how existing models of the impact of the road pavement condition on vehicle rolling resistance and hence, fuel consumption (FC), can lead to very different results. This study presents a comparison between real measurements of truck fuel consumption from fleet manager's databases, and results of existing pavement models, MIRAVEC, a model recently developed within an ERA-NET ROAD action, funded by the 6th framework programme of the EU, and HDM-4, one of the most widely used models for estimating vehicle operating costs in road asset management.

The data are collected by standard sensors (SAE International 2016) installed on trucks and are provided anonymized by truck fleet managers that use these data for optimization of costs of their fleets.

In order to calibrate HDM-4, values of the parameters published in University of Birmingham (2011) and by Odoki et al. (2013) have been used.

No calibration has yet been performed on MIRAVEC. This is because MIRAVEC is more recent than HDM-4 and has been developed to fit the conditions of Europe (Benbow et al. 2013).

Results show that MIRAVEC performs slightly better than HDM-4 in predicting actual fuel consumption. However, both the models tend to underestimate the fuel consumption of light trucks while they tend to overestimate the fuel consumption of medium and heavy trucks.

In conclusion, the paper shows how far results of the considered models can be from reality (Figure 1) and opens a discussion of the implications of these differences on pavement LCA and strategic decisions of managers of the road infrastructure.

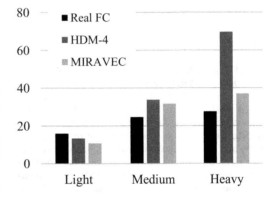

Figure 1. Comparison of fuel consumption estimates made by HDM-4 and MIRAVEC with real measurements for three types of trucks.

REFERENCES

Benbow, E., Brittain, S. & Viner, H. (2013) Potential for NRAs to provide energy reducing road infrastructure, Deliverable D3.1.

Odoki, J.B., Anyala, M. & Bunting, E. (2013) HDM-4 adaptation for strategic analysis of UK local roads. In: *Proceedings of the Institution of Civil Engineers – Transport*, 166 (2), 65–78. Available at: http://www.icevirtuallibrary.com/-doi/10.1680/tran.9.00026.

SAE International (2016) SAE J1939-71, Vehicle application layer – surface vehicle recommended practice. SAE International Standards. Available at: http://standards.sae.org/-j1939/71_201610/.

Santero, N.J., Masanet, E. & Horvath, A. (2011) Life-cycle assessment of pavements. Part I: Critical review. *Resources, Conservation and Recycling*, 55 (9–10), 801–809.

Trupia, Parry, T., Neves, L.C. & Presti, D.L. (2016) Rolling resistance contribution to a road pavement life cycle carbon footprint analysis. *International Journal of Life Cycle Assessment*, 972–985.

University of Birmingham (2011) Development of Socio-Economic Models for Highway Maintenance. Analysis of DfT Road Network Using HDM-4. WSP, (March).

Life-Cycle Analysis and Assessment in Civil Engineering: Towards an Integrated Vision – Caspeele, Taerwe & Frangopol (Eds)
© 2019 Taylor & Francis Group, London, ISBN 978-1-138-62633-1

Technical management risks for transport infrastructures along whole life cycle: Identification and analysis

D. García-Sánchez, J. Aurtenetxe & M. Zalbide
Fundación Tecnalia Research and Innovation, Bizkaia, Spain

R. Socorro
Acciona, Madrid, Spain

A. Pérez-Hernando
University of Cantabria, Cantabria, Spain

D. Inaudi
Smartec, Manno, Switzerland

ABSTRACT

RAGTIME is a Horizon 2020 project focused on development, demonstration and validation of an innovative infrastructures management approach. It lays out a whole system, based on standard multi-scale data models, able to facilitate a holistic management throughout the entire lifecycle of the infrastructure, providing an integrated view of risk based approach, implementing risk based models, resilient concepts and mitigation actions. It is structured in three modules: governance, finance and economic, and technical management.

The technical management module deals with technical risk assessment of the transport infrastructure approach based on an innovative 3D vector space, defined by following three variables: infrastructure component (bridge, tunnel, slope, open rail track and pavement), lifecycle stage (evaluation-planning, procurement-decision, design-project, construction, and operation and maintenance) and analysis parameter (green, resilient, social-inclusive, cost-efficient and safe-secure).

This paper will be focused on the very first step: the identification of threats, which impact on everyday technical management. The identification process of technical risks follows the inherent structure of a civil engineering project. Nevertheless, it is important to highlight that the present analysis defines the technical risks on infrastructures with a holistic and general perspective, determined based on the needs of the different agents involved and the technical framework in which the process is carried out.

During the identification stage, technical risks are identified and clustering as follow:

- Contractual: these risks are related to the conditions of the government capacity, public sector, private sector, markets, legal environment, public, welfare and/or security.
- Data: these cluster include the information needed for designing and calculating the adequate infrastructure dimensions.

- Design and calculations includes all risks to be account during the designing stage.
- Building and civil works: here are included those risks related to those originated because of building nature that can occur during construction.
- Unintentional hazards, natural disasters and intentional threats: all those exceptional risks that can occur with important consequences and high probability that have to be mitigated during exploitation stage.

After the identification of the technical risks, a final analysis has been made taking into account the transport mode, life cycle phases and stakeholders as criterion, setting two and leaving one as a variable. Thus, the objective of this analysis is to evaluate and preview the relative weight of each technical risk on each of the criteria. This first approach aims to identify the risks of higher index and to be able to guide in the management of these technical risks for mitigation actions design.

REFERENCES

CEDR (2013) *Conference of European Directors of Roads.*
Huddin, W., Hudson W. R. & Haas, R. (2012) *Public Infrastructure Asset Management*, Mc Graw-Hill.
Hughes, J.F. & Healy K. (2014) *Measuring the Resilience of Transport Infrastructure.* New Zealand: AECOM Ltd.
Jaafari, A. (1997) Concurrent construction and life cycle project management. *Journal of Construction Engineering and Management-Asce*, 123 (4), 427–436.
OECD (2015) *Towards a Framework for the Governance of Infrastructure.* Paris, OECD Publishing.
REFINET (2015) *Rethinking Future Infrastructure Network.* Coordination and Support Action – GA 653789.
UE (2014) *2014/24/UE Directiva del Parlamento Europeo y del Consejo de 26 de febrero de 2014 sobre contratación pública.* UE Publishing.

Life-Cycle Analysis and Assessment in Civil Engineering: Towards an Integrated Vision – Caspeele, Taerwe & Frangopol (Eds)
© 2019 Taylor & Francis Group, London, ISBN 978-1-138-62633-1

Computational framework for a railway bridge maintenance strategies affected by gradual deterioration

J. Fernandes, J.C. Matos & D.V. Oliveira
ISISE, School of Engineering, University of Minho, Guimarães, Portugal

A.A. Henriques
Faculty of Engineering, University of Porto, Porto, Portugal

ABSTRACT

Bridges are extremely important to social-economic development of a country once they are one of the most crucial components by providing crossings at critical locations that otherwise would add sig add significant travel time and cost (Markow and Hyman, 2009). However, bridges do not last forever. Whatever the materials that are used, the effects that cause deterioration to the bridge start to appear. Therefore, it comes up the necessity of applying several actions like designing, building, maintaining and even replacing the bridge in worst cases. These actions involve critical decisions due to the high cost of investments. In this way, the importance of minimizing life-cycle cost (LCC) and provide safe and efficient mobility transportation system users arises. In this way, several bridge management systems (BMSs) has been developed with the purpose of determining the optimal interventions to be executed over the bridge life-cycle (Mirzaei and Adey, 2014). A BMS is a decision-support tool that supplies analyses and summaries of the data, use models and algorithms to make predictions and recommendations (FHWA, 2010).

To keep the structures safe throughout their life, they require regular maintenance actions whose costs are generally supported by the operator. Accordingly, it becomes important to define strategies to maximize the societal benefits, derived from the investment made in these assets. This investment should be planned, effectively managed and technically supported by appropriate management systems. This process can be completed by improving the planning of maintenance strategies that will not only consist of defining the goals to be achieved, but that should also identify the investment needs and priorities based on an LCC criteria. The need to manage bridges in an efficient way have led to the development of bridge management systems (BMS) all over the world. The comparison between the BMSs can be found in a report presented by IABMAS – International Association for Bridge Maintenance and Safety (Adey et al., 2014).

In this paper, a deck of a railway bridge of Portugal will be analysed in terms of its reliability over the time being proposed a maintenance and rehabilitation optimal maintenance schedule.

REFERENCES

Adey, B., Klatter, L. & Kong, J. (2014) The IABMAS Bridge Management Committee overview of existing bridge management systems. In: *5th Int. Conf. on Bridge Maintenance, Safety and Management*, 2014.
FHWA (2010) Bridge Management Questionnaire Report, FHWA Office of Asset Management.
Markow, M. J. & Hyman, W. A. 2009. *Bridge Management Systems for Transportation Agency Decision Making – A Synthesis of Highway Practice. NCHRP Synthesis 397*, Transportation Research Board.
Mirzaei, Z. & Adey, B.T. (2014) Investigation of the use of three existing methodologies to determine optimal life-cycle activity profiles for bridges. *Structure and Infrastructure Engineering*, 1–26.

A many-objective optimization model for sustainable pavement management considering several sustainability metrics through a multi-dimensionality reduction approach

J. Santos
Department of Construction Management & Engineering, Faculty of Engineering Technology, University of Twente, Enschede, The Netherlands

V. Cerezo
IFSTTAR, AME-EASE, LUNAM Université, Bouguenais, France

G. Flintsch
Center for Sustainable Transportation Infrastructure, Virginia Tech Transportation Institute, Jr. Department of Civil and Environmental Engineering, Virginia Polytechnic Institute and State University, USA

A. Ferreira
Road Pavements Laboratory, Research Center for Territory, Transports and Environment, Department of Civil Engineering, University of Coimbra, Portugal

ABSTRACT

This study presents the development and application of a many-objective optimization (MaOO) framework for sustainable pavement management that combines a comprehensive and integrated pavement life cycle costs – life cycle assessment model that covers the whole pavement's life cycle, a multi-objective evolutionary algorithm and a multi-dimensionality reduction method. The development of this method is based on those proposed by Deb & Saxena (2006) and Brunet et al. (2012) and comprises three main steps. In the first step, the original many-objective optimization problem (MaOP) is decomposed into multiple bi-objective optimization problems (b-OOP), in which the life cycle highway agency costs (LCHAC) are optimized against each individual environmental metric separately. This procedure generates several b-OOP whose cardinality is determined by the number of environmental metrics considered. In the second step, a Principal Component Analysis (PCA)-based dimensionality reduction approach is applied. Next, a heuristic rule based on the analysis of the eigenvectors of the correlation matrix calculated from the assemble of Pareto-optimal solutions of the several b-OOP is adopted to identify the redundant environmental metrics (Deb & Saxena, 2006). It starts by retaining the first ordered principal components (PCs) such that their cumulative variance is greater than 95%. For the retained PCs, the environmental metrics with the most positive and negative components of the eigenvectors are kept in the analysis. The result of the reduction method is a MOO problem with a reduced set of objective functions (OFs) and a reduced solution space. Finally, in the third step the MOO module is applied in a reduced domain of OFs representing the retained environmental metrics and the economic OF.

The potentialities of the proposed framework are illustrated through a French case study consisting of identifying sustainable pavement maintenance and rehabilitation (M&R) strategies for a road pavement section that concurrently minimize the net present value of the life cycle costs incurred by the highway agency and nine life cycle environmental indicators. They are as follows: (1) climate change; (2) acidification; (3) eutrophication; (4) human toxicity; (5) abiotic resources depletion; (6) terrestrial ecotoxicity

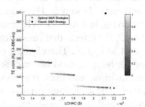

Figure 1. Pareto optimal sets of solutions in the objective space along with the M&R strategy corresponding to the French practice. Note: The fuzzy cardinal priority ranking of each non-dominated solution was normalized so that it falls into the range [0;1].

(TE); (7) particulate matter formation; (8) cumulative energy demand of renewable resources (CED-R); and (9) CED of non-renewable resources.

The results of the PCA determined that only two OFs (i.e., LCHAC and TE score) are necessary to be kept in the analysis, while the remaining ones can be discarded.

Figure 1 display the Pareto-optimal sets of solutions in the objective space, along with the M&R corresponding to the French practice. Each point in figure represents an Pareto-optimal nondominated pavement M&R strategy and its color the fuzzy cardinal priority ranking. According to this theory, the closer the ranking to 1, the better the solution is.

REFERENCES

Brunet, R. Guillén-Gosálbez, G. & Jiménez, L. (2012) Cleaner design of single-product biotechnological facilities through the integration of process simulation, multiobjective optimization, life cycle assessment, and principal component analysis. *Industrial & Engineering Chemistry Research*, 51 (1), 410–424.

Deb, K. & Saxena, D. (2005) On finding pareto-optimal solutions through dimensionality reduction for certain large-dimensional multi-objective optimization problems. In: *Proceedings of the IEEE congress on evolutionary computation (CEC2006), Vancouver*. pp. 3353–3360.

Expert-driven and data-driven risk-centred maintenance decision-making approaches for railway transport assets

F. Dinmohammadi
School of Engineering and Built Environment, Glasgow Caledonian University, Glasgow, UK
School of Engineering and Physical Sciences, Heriot-Watt University, Edinburgh, UK

ABSTRACT

The effective inspection and maintenance of railway transport assets has become increasingly important in recent years for almost all countries in the world. The consequence of a defect in rail sector can be catastrophic, and may lead to traffic delays and disruptions, passenger inconvenience, and risk to the health and safety of the crew. For this reason, preventive maintenance (PM) of railway assets plays an important role in network availability and reliability, passenger safety and comfort, as well as energy efficiency. Currently, the PM programs in rail sector are diverse from *periodic renewals* (carried out based on a standardised set of frequencies) to *prognostics and health management (PHM)* systems (that are based on monitoring of assets' conditions).

In recent years, several rail transport organisations have shifted towards using risk-centred approaches to optimise the maintenance activities of various rail transport systems on the network. Risk-centred maintenance (RCM) is an effective tool that links risk assessment to the maintenance decision-making process. This tool can help railway organisations to analyse, prioritise and optimize maintenance requirements based on the criticality of asset failures so as to focus on those areas having the greatest impact on the business. This enables asset managers to rank PM tasks in order of risk levels or by the size of opportunity for improvement in cost-effectiveness of existing maintenance practices.

This paper presents risk-centred approaches to optimize decision-making for the maintenance of railway transport assets, including rail infrastructure and rolling stock systems. In general, the RCM methodologies can be applied in two basic ways: qualitative (or expert-driven) and quantitative (or data-driven). As shown in Figure 1, for the identification and evaluation of systems failure modes and failure effects, some expert-driven tools such as safety integrity level (SIL) analysis, root cause analysis (RCA), fault tree analysis (FTA), reliability block diagram (RBD), and failure mode and effects analysis (FMEA) are used, whereas

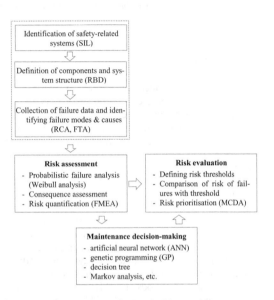

Figure 1. Architecture of RCM decision-making.

for the planning of maintenance tasks data-driven methods such as artificial neural network (ANN), genetic programming (GP), decision trees, etc. can be applied. For the purpose of clearly illustrating the proposed approaches, a number of case studies from rail operating companies are provided and the results will be discussed.

REFERENCES

Aven, T. (1992) *Reliability and Risk Analysis*. London, Elsevier Applied Science.

Dinmohammadi, F., Alkali, B., Shafiee, M., Berenguer, C. & Labib, A. (2016) Risk evaluation of railway rolling stock failures using FMECA technique: A case study of passenger door system. *Urban Rail Transit*, 2 (3), 128–145.

Dinmohammadi, F., Alkali, B. & Shafiee, M. (2017) A risk-based model for inspection and maintenance of railway rolling stock, In: *Risk, Reliability and Safety: Innovating Theory and Practice*, Taylor & Francis. pp. 1165–1172.

Life-Cycle Analysis and Assessment in Civil Engineering: Towards an
Integrated Vision – Caspeele, Taerwe & Frangopol (Eds)
© 2019 Taylor & Francis Group, London, ISBN 978-1-138-62633-1

Evaluation and application of AHP, MAUT and ELECTRE III for infrastructure management

Z. Allah Bukhsh, I. Stipanovic & A. Hartmann
Department of Construction Management and Engineering, Faculty of Engineering Technology,
University of Twente, Enschede, The Netherlands

G. Klanker
Rijkswaterstaat Ministry of Infrastructure and the Environment, Utrecht, The Netherlands

ABSTRACT

Infrastructure management renders a number of decision-making problems from asset's condition inspections to maintenance planning and resources optimization. Since management of infrastructure pertains to not only technical requirements but also to societal and economic developments, these decision problems have multiple and often conflicting objectives. Various methods of MCDA based on the decision theory and game theory are proposed to aid-in decision-making problems (de Almeida et al., 2015). Owing to the wide area of applications and extensive variation in MCDA methodology, the selection of appropriate MCDA method pertaining to the specific needs of infrastructure management and decision maker is a difficult task. In this paper, two synthesis-based methods (i.e. AHP and MAUT) and an outranking method (i.e. ELECTRE III) is applied on same maintenance decision making problem to evaluate them for their scalability, ease of use, risk consideration, and few other aspects.

To illustrate how the different methods of MCDA can be applied for the maintenance decision-making, we used data of twenty-two randomly chosen bridges from Netherlands road network. Using this raw provided data, we computed condition index on overall bridge-level, owner cost incurred due to maintenance activity, user delay cost and environmental cost for each of the bridge. Table 1 provides the final ranking of 22 bridges computed using AHP, MAUT and ELECTRE III. It is interesting to see that each of the MCDA method have ranked Bridge M highest, while this bridge does not have lowest condition index and lowest user delay cost. This is because MCDA methods systematically account for all the attributes involved in decision-making instead of ranking on the bases of single attribute only.

The final results of evaluation suggest that a) without a computerized tool the scalability of these methods is tedious task b) only MAUT considers the risk attitude of a decision maker c) AHP and MAUT both require the data to be converted to definite scale for

Table 1. Ranking of twenty-two bridges computed by AHP, MAUT and ELECTRE III.

Alternatives	AHP		MAUT		ELECTRE III
	Score	Rank	Score	Rank	Rank
Bridge A	0.061	18	0.972	22	17
Bridge B	0.048	14	0.748	13	13
Bridge C	0.048	13	0.826	15	14
Bridge D	0.026	3	0.350	3	2
Bridge E	0.032	5	0.588	7	7
Bridge F	0.060	16	0.854	16	18
Bridge G	0.068	22	0.865	18	21
Bridge H	0.043	12	0.793	14	11
Bridge I	0.036	9	0.492	5	5
Bridge J	0.050	15	0.873	19	16
Bridge K	0.026	4	0.333	2	4
Bridge L	0.033	8	0.671	9	8
Bridge M	0.022	1	0.141	1	1
Bridge N	0.040	11	0.738	12	10
Bridge O	0.026	2	0.383	4	3
Bridge P	0.061	19	0.694	10	15
Bridge Q	0.064	21	0.854	17	22
Bridge R	0.063	20	0.896	20	20
Bridge S	0.032	6	0.647	8	6
Bridge T	0.032	7	0.570	6	9
Bridge U	0.038	10	0.716	11	12
Bridge V	0.061	17	0.901	21	19

analysis, for instance, to Saaty scale of comparison and to utility functions respectively and d) unlike other two, ELECTRE works on preference structure and yields partial pre-orders.

REFERENCE

de Almeida, A.T., Ferreira, R.J.P. & Cavalcante, C.A.V. (2015) A review of the use of multicriteria and multi-objective models in maintenance and reliability. *IMA Journal of Management Mathematics*, 26 (3), 249–271.

Life-Cycle Analysis and Assessment in Civil Engineering: Towards an Integrated Vision – Caspeele, Taerwe & Frangopol (Eds)
© 2019 Taylor & Francis Group, London, ISBN 978-1-138-62633-1

Assessing approximation errors caused by truncation of cash flows in public infrastructure net present value calculations

R. Treiture, L. van der Meer & J. Bakker
Rijkswaterstaat, Ministry of Infrastructure and Water Management, The Netherlands

M. van den Boomen, R. Schoenmaker & R. Wolfert
Faculty of Civil Engineering and Geosciences, Delft University of Technology, The Netherlands

ABSTRACT   Life cycle costing analysis for public infrastructure assets is not a straight forward exercise and may be sensitive to input data and calculation approaches. In general alternatives are compared on net present value (NPV) of life cycle costs (LCC) over a bounded calculation horizon. However, alternatives may differ in life cycle characteristics. Issues to address include the chosen calculation horizons, the truncation method for the end of horizon cash flows, the discount rates and inflation rates. This study investigates the deviations of NPV calculations with horizons of 100, 200 and 300 years with an NPV calculation over an infinite time horizon. Three public infrastructure asset case studies with their specific LCC characteristics are considered: a bridge, a high way and dike revetment. A sensitivity analysis for the discount rate and differential inflation is performed. This study develops generic guidelines for approximation errors as a consequence of a premature truncation of cash flows in public infrastructure LCC calculations.

1 INTRODUCTION

Public sector NPV calculations for comparison of alternatives often use a standardised calculation horizon between 25 years and 100 years. Future cash flows are forecasted on a time scale and discounted to their present values. Discount rates used in Dutch public sector organisations are relatively low and range from 1% to 5%.

In the absence of a proper truncation method for infrastructure cash flows at the end of a calculation horizon, cash flows after the calculation horizon are neglected. This will lead to approximation errors in NPV calculations and possibly unfair comparison between different alternatives.

An alternative to a proper truncation method is no truncation of cash flows. Public infrastructure assets often have a function or service life that approximates infinity. This service life contains several future replacements.

The current study investigates the deviations of NPV's over truncated time horizons of 100, 200 and 300 years compared to the NPV over an infinite time horizon. For the NPV calculations over an infinite time horizon, less-known mathematical LCC-concepts have been used and adapted for inclusion of differential inflation.

Three case studies, a bridge, dike revetment and a highway, with different life cycle characteristics, underlie the derivation of a generic rule.

2 RESULTS

Upper and lower boundaries for deviations between bounded and infinite NPV calculations are given by common multiples of life cycles in chosen calculation horizons:

$$\% \ deviation = \left(\frac{1+f_d}{1+r}\right)^{aN} \cdot 100\% \qquad (1)$$

where f_d = differential inflation; r = real discount rate, aN = calculation horizon; N = life cycle of an asset and a = multiplication factor.

The execution of NPV calculations over an infinite calculation horizon is less intuitive for practitioners. Therefore, as a practical guideline, this study recommends for public infrastructure NPV calculations, to use calculation horizons of at least 100 years for discount rates of 3% to 5% as long a differential inflation is not involved. For other circumstances, upper and lower boundaries of deviations for different calculation horizons, discount rates and differential inflation rates can quickly be determined with Equation 1.

Life-Cycle Analysis and Assessment in Civil Engineering: Towards an Integrated Vision – Caspeele, Taerwe & Frangopol (Eds)
© 2019 Taylor & Francis Group, London, ISBN 978-1-138-62633-1

A survey of health monitoring techniques for the Dutch transportation infrastructure

J.F.M. Wessels, P.J. van der Mark & K.E. Bektas
TNO, Delft, The Netherlands

J. Bakker & M. van der Voort
RWS, The Netherlands

ABSTRACT

Transportation infrastructure assets are designed to meet pre-specified reliability, availability, maintainability and safety (RAMS) requirements. Electrical and mechanical components are critical elements with respect to the RAMS-criteria. An unplanned failure on the these components may have a direct impact on the availability of the primary function or safety provision of the asset. When it is not repaired on time, failures can escalate (as chain of failure), may result in a complete unavailability of the asset, unsafe situations, excessive maintenance costs, high societal impact with ending up high financial consequences to the public. A local event can develop into a global disturbance.

Maintenance of the assets to prevent local failures and avoid global disturbances are traditionally based on pre-existing knowledge of failure, combined with knowledge of current state of the asset. This practice has already moved to preventive maintenance, where the effectiveness is greatly influenced by the length of predetermined maintenance interval. The experiences showed that is not often the case. The maintenance schedule is often suboptimal meaning that they are done either too early or too late resulting in high costs, even societal impact.

The trend in asset management is to apply data-driven, predictive maintenance. Technological developments such as those using a myriad of low cost sensors and wireless communication techniques are used for monitoring the state of assets and in various fields, they have been in use to predict behaviour of assets and choose an optimal maintenance strategy. These technologies have proven already very positive results for the asset owners/managers (particularly in aviation, automation, production and process industries) especially when applied into (or onto) components which have measurable conditions.

The Dutch asset manager, Rijkswaterstaat, has decided to explore the opportunities these new technologies offer to develop new solutions for predictive maintenance, in order to reduce the probability of unplanned failures of ageing electrical and mechanical installations. The question to answer has become:

Which technologies fit to the Rijkswaterstaat assets and which steps need to be taken to include relevant technologies to the maintenance and organizational processes in order to move towards predictive maintenance practice?

To answer that is not only relevant for ageing assets but also understating the possibilities with built-in sensor technologies of new components (renovation) or in their contemporary assets (replacement).

This paper firstly analyse Rijkswaterstaat' systems (its assets, the characteristics and the state of the art of the maintenance strategy). It secondly scans other industries' predictive maintenance & monitoring techniques; their characteristics and similarities to the Rijkswaterstaat characteristics and (their possible applicability). The paper thirdly defines a roadmap in order to apply relevant techniques and provide a step-wise application and implementation strategy for Rijkswaterstaat and provides conclusion and follow-up.

In this paper, we have seen that the current risk-based maintenance approach of the Rijkswaterstaat and the data-driven approach are complementary. New opportunities of data driven predictive maintenance suggested a useful direction for Rijkswaterstaat as they can be added to a palette of techniques for performing maintenance.

This study is a part of the project "Vitale Assets. The paper here provides the results from several parallel activities. The activities are such as performing pilot projects, building the ICT infrastructure for predictive maintenance, an industry scan (based on literature of predictive maintenance practises in various industries (and countries), and defining a roadmap for implementation of these technologies in the Rijkswaterstaat maintenance and planning processes.

Considerations on the use of data to better predict long term replacement and renovation activities in transport infrastructure

J.F.M. Wessels & H. van Meerveld
The Netherlands Organisation for Applied Scientific Research TNO, Delft, The Netherlands

J. Bakker
Ministry of Infrastructure and Environment, Rijkswaterstaat, Utrecht, The Netherlands

K.E. Bektas
The Netherlands Organisation for Applied Scientific Research TNO, Delft, The Netherlands

ABSTRACT

A large number of assets is reaching the their 'end of life'. Large interventions like renovations and replacements open opportunities for improving the transport infrastructure network and land use. Such undertakings require a planned approach and early justification and securing of resources (Institute of Asset Management, 2015).

Rijkswaterstaat (RWS) has set up a program to address replacement and renovation needs. Part of this program is to forecast large interventions. The expected budgets are based on statistics, acceptable for the prediction of the total budget needed. The challenge is to provide accurate (long term) predictions for specific asset. An opportunity is recognized in a data-driven approach.

The goal of this research is to improve the long term prediction of 'end of life'. First insight in the complex mix of 'drivers' of 'end of life' decisions is gained, as end of life seems to be more like a decision than a status (Bakker et al, 2016). The findings are summarized in Table 1.

Table 1. Drivers for end of life of structures.

Driver: Functional performance of the asset

– The required functionality cannot be fulfilled by the assets and/or;
– The required functional performance cannot be delivered by the asset.

Driver: Physical performance and characteristics of the asset

– The asset cannot fulfil the technical (safety) requirements (which in practice also influences the functional performance)

Driver: Economical/Financial performance of the asset

– The maintenance costs are becoming too high (exceed the available budgets)
– Large investments (renovation/renewal) are more cost efficient than maintenance measures

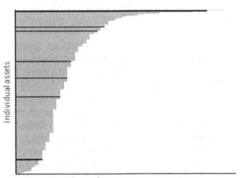

Figure 1. Number of measures per asset, differentiated to incidental (left) and periodic (right) measures.

With this knowledge data from 'RWS – DISK', a database containing the forecast of maintenance activities, is used for analysis. E.g. the number of measures per asset and the number of incidental and periodic measures are plotted. In the plots the asset being considered for replacement and/or renovation are highlighted (Figure 1).

The study indicated that data analysis already resulted in interesting insights but so far does not improve the prediction. The idea is that further data analysis from different available sources will bring more facts and figures to better understand the rational of end of life decision making. Hopefully this will lead to better informed decisions and opportunities to improve the land use and networks against lower costs.

REFERENCES

Bakker, J., Roebers, J.H. & Knoops, J. (2016) Economic End of Life Indicator (EELI). In: *Proceedings of 6th international conference IALCCE, Delft, Netherlands.*

Institute of Asset Management (2015) *An Anatomy of Asset Management*, Issue 3.

Life-Cycle Analysis and Assessment in Civil Engineering: Towards an Integrated Vision – Caspeele, Taerwe & Frangopol (Eds)
© 2019 Taylor & Francis Group, London, ISBN 978-1-138-62633-1

Life cycle cost analysis for short span bridges in Indiana

S.L. Leiva & M.D. Bowman
Purdue University, West Lafayette, Indiana, USA

ABSTRACT

Life cycle cost analysis (LCC) has been defined as a method to assess the total cost of a project. It is a simple tool to use when a single project has different alternatives that fulfill the original requirements. Different alternatives could differ in initial investment, operational and maintenance costs among other long-term costs. This research is focused on short span bridges (less than 30.00 m) which represents 90% of the NBI INDIANA bridge inventory.

Even when all the research efforts are focused on improving the life-cycle engineering, Frangopol and Soliman (Frangopol & Soliman, 2016) also underline the breach between these efforts and the real-world applications. It is important then to formulate the methods necessary to connect the decision-making process with all the life-cycle research efforts. Consequently, this study is concentrated on connecting de designers into the asset management process giving them the necessary tool to consider cost effective solutions based on LCCA since the conception of a new bridge alternative.

Two different span ranges are considered for the analysis along with different structural types and span configurations. Assumptions made are based on typical structure configurations and road conditions of the Indiana assets part of National Highway System (NHS) of the United States. Additionally, LCCA for each superstructure type is developed in order to categorize the most cost effective alternative. This research is only considering a deterministic analysis as a first approach to the problem.

It is assumed that in the case of highway assets of long service lives like bridges, it is likely to replace the structure in the same place over and over again rather than replace it in different locations each time. It can be assumed that each alternative will be indefinitely replaced, in other words in perpetuity. Equation 1 shows Ford's alternative (Ford *et al.*, 2012), where P_p is the present worth of LCCA in perpetuity, P is the life cycle cost of a single service life, i is the interest rate used and SL is the service life in years of each option. Using this equation, it is possible to compare different alternatives with different service lives in terms of life-cycle costs.

$$P_p = \frac{P}{(1+i)^{SL} - 1} \tag{1}$$

Recommendations on selecting different superstructure types, depending on the principal span length are made to designers based on cost effectiveness using BLCCA.

REFERENCES

Ford, K.M. *et al.* (2012) *Estimating Life Expectancies of Highway Assets. Volume 1: Guidebook.* National Academies Press. doi: 10.17226/22783.

Frangopol, D.M. & Soliman, M. (2016) Life-cycle of structural systems: recent achievements and future directions. *Structure and Infrastructure Engineering*, 12 (1), 1–20. doi: 10.1080/15732479.2014.999794.

Lifecycle management and replacement strategies: Two of a kind?

M. Zandvoort
Wageningen University, Wageningen, The Netherlands
Tauw BV, Deventer, The Netherlands

M.J. van der Vlist
Wageningen University, Wageningen, The Netherlands
Rijkswaterstaat, Utrecht, The Netherlands

R. Haitsma & E. Oosterveld
Tauw BV, Deventer, The Netherlands

ABSTRACT

Due to the end of lifetime of infrastructure assets, water managers need to develop strategies for their replacement or demolishment. We outline four distinct reasons in which replacements differ from new assets and analyze which particular considerations about replacement can complement lifecycle management (LCM) practices (Fig. 1). We conclude that these replacement considerations lead to additional technical design features for hydraulic structures and other assets. Complementing LCM with considerations about the new asset lifecycle after replacement informs planning approaches for developing adequate replacement strategies in the context of asset management.

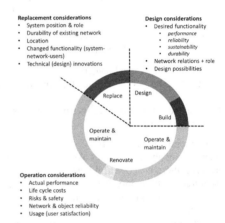

Figure 1. Considerations for three phases in the life cycle of a singe asset.

A holistic approach to Life Cycle Management at Rijkswaterstaat

J. Bakker & H. Roebers
Rijkswaterstaat, Utrecht, The Netherlands

ABSTRACT

Life cycle thinking takes an increasing role in both construction projects and in asset management. The term "Life Cycle Management" (LCM) is introduced as a principle for the management of Life Cycle Performance (LCP), Life Cycle Risks (LCR) and Life Cycle Cost (LCC) in a dynamic environment (figure 1, Fuchs et al. 2014).

In Asset Management processes typically 3 roles are differentiated (van der Velde et al. 2012):

1) Asset Owner: The legal owner
2) Asset Manager: Responsible for managing the assets
3) The Service Provider: Delivering services like maintaining the assets

Agreements between parties have a major impact on behavior of the parties at stake. If all targets are short-term, there is no fruitful basis for LCM. Traditional agreements are mostly based on time, prize and quality over a contract period. Unless this period is really long, these sort of agreements stimulate short term behavior. In figure 2 an "LCM-add-on" is proposed. Agreement on short term performance should

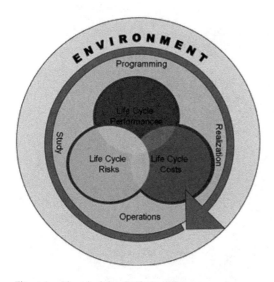

Figure 1. The principle of Life Cycle Management.

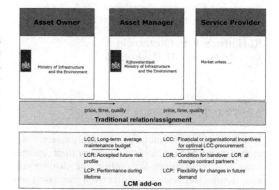

Figure 2. Roles in Asset Management and the LCM add-on (Fuchs et al. 2014).

be enriged by quantified, agreed and accepted effects on LCP, LCC and LCR.

Rijkswaterstaat is a Dutch government agency responsible for both new construction projects and the asset management of existing national infrastructure.

Many examples of "life cycle thinking" can be found in all processes at Rijkswaterstaat. However, there are many different perspectives on life cycle thinking. Life Cycle Management could be a binding approach to bring together all these different perspectives. This paper summarizes different life cycle perspectives at Rijkswaterstaat, and evaluates the possibilities for integral Life Cycle Management.

REFERENCES

Fuchs, G.H., Bakker, J.D. & Mante, B.R. (2014) A business case of the estimated profit of life cycle management Principles. In: *Proceedings of the 4th International Symposium on Life-Cycle Civil Engineering, IALCCE 2014.* Japan.

Van der Velde, J., Klatter, H.E. & Bakker, J.D. (2012) A holistic approach to asset management in the Netherlands. *Structure and Infrastructure Engineering*, Taylor & Francis.

Life-Cycle Analysis and Assessment in Civil Engineering: Towards an Integrated Vision – Caspeele, Taerwe & Frangopol (Eds)
© 2019 Taylor & Francis Group, London, ISBN 978-1-138-62633-1

Lessons learned from data analytics, applied to the track maintenance of the Dutch high speed line

R. Schalk
Mott MacDonald B.V., Arnhem, The Netherlands

A. Zoeteman
Section of Integral Design and Management, Delft University of Technology, Delft, The Netherlands

A. Núñez
Section of Railway Engineering, Delft University of Technology, Delft, The Netherlands

ABSTRACT

Life cycle performance and risk management are often mentioned as critical tasks for infrastructure managers. However, without proper data collection and analytics these tasks cannot be executed. This paper will demonstrate a case where a data analytics approach was deployed, when an unexpected phenomenon occurred on the Dutch High Speed Line (HSL-Zuid) and the paper will draw lessons learned from this case.

In November 2014, it was found that large sections of the HSL-Zuid were affected by a severe type of rolling contact fatigue (RCF). The RCF resulted in deep cracks on top of the rail. These damages were unexpected as the rails were only 5 years in operation and for these rails with proper maintenance it was expected to last about 20–25 years. In this case, resulting in about 20 km of rail replacements and additional grinding campaigns.

As the causes were unknown, the authors applied data analytics to evaluate the possible causes of the RCF. Several measurements of the infrastructure, maintenance and the rolling stock resulted in parameters which were combined in a model. This resulted

into two approaches; a bottom approach evaluating the affected sections to find similar parameter values among these and a top-down approach evaluating the whole track, looking for parameter values which could describe why certain sections were affected and others were not.

The outcomes of the analysis indicated that it is highly likely that one type of rolling stock was affecting the rails in the curves of the HSL-Zuid. As the track was designed at the highspeed sections for 220–300 km/h, this type of rolling stock was driving below design speed which results in different loading of the rails throughout the curves.

Lessons learned from this case do not only apply to the technical area of wheel/rail and vehicle/infrastructure interfacing, but also to the usage of data analytics itself and life cycle management. The paper will use this case study propose how data collection and analytics can be better embedded by (rail) infrastructure managers, in an early stage of development and use of infrastructure, and will suggest further scientific development for infrastructure data analytics.

Life-Cycle Analysis and Assessment in Civil Engineering: Towards an Integrated Vision – Caspeele, Taerwe & Frangopol (Eds)
© 2019 Taylor & Francis Group, London, ISBN 978-1-138-62633-1

Asset information management using linked data for the life-cycle of roads

B. Luiten & M. Böhms
TNO, Delft, The Netherlands

D. Alsem
Royal HaskoningDHV, Amersfoort, The Netherlands

A. O'Keeffe
Roughan & O'Donovan Consulting Engineers, Dublin, Ireland

ABSTRACT

From September 2016 to August 2018 a CEDR (Conference of European Directors of Roads) research has been conducted by the consortium INTERLINK into asset information management for European Roads life-cycle management using Linked Data. This research aims to improve interoperability between the European National Road Authorities (NRAs) and their stakeholders, mainly to explore the procurement of road asset information by NRAs for better operations, asset management and project management, to investigate suitable data structures for asset information, including BIM and GIS, to develop common principles for a European Road object type library (OTL) and to design and test a basic European Road OTL connected to open BIM/GIS standards. The CEDR INTERLINK project will design a managed European Road OTL as the basis of a common language to be used for developing high quality national Road OTLs for exchanging and sharing road asset data. This European Road OTL is supported by three pillars to optimize the success of subsequent implementation: 1. a quality open available standardized Technical Specification, 2. a suitable and sustainable Standardization Body, with a plan for development beyond this project and 3. acceptance in practice by the industry through engagement and dissemination. First the business and data needs have been investigated by INTERLINK through a literature review, interviews with stakeholders (NRAs, contractors, consultants, suppliers, academics, software developers and standardization initiatives in Europe) and an industry survey, which resulted in a approach for the European Road OTL. The common principles, based on the requirements, describe a hybrid solution, with semantically-rich data referencing more traditional document-based information, combining the strengths of currently applied BIM/GIS standards with Linked Data/Semantic Web technology. The proof-of-concept European Road OTL is tested via three trial cases, which demonstrated data transfers representing typical processes during assets' life-cycles in three European countries: Germany, Sweden and The Netherlands. A road map is delivered for development and implementation of the Road OTL from document based to object based road information management based on the European Road OTL open standard using Linked Data.

REFERENCES

BuildingSMART International (2016) Proof of the value of Linked Data for infrastructure, In: *leaflet presented at the building SMART Summit, South Korea, Sept, 2016.*

Böhms, M., Luiten, B., O'Keeffe, A., Stolk, S., Wikström, L., Weise, M. (as part of INTERLINK consortium) (2017) Principles for a European Road OTL. www.roadotl.eu/publications/ [Accessed on 13 October 2017].

INSPIRE Thematic Working Group Transport Networks (2014) *D2.8.1.7. Data Specification on Transport Networks-Technical Guidelines.* European Commission Joint Research Centre.

Institute of Asset Management (2016). *Asset Management – an Anatomy.* https://theiam.org/what-is-asset-manage ment/anatomy-asset-management [Accessed 27 September 2017].

Luiten, B, Böhms, M., Alsem, D., O'Keeffe, A. (as part of INTERLINK consortium) (2017) Asset Information management for European roads using Linked Data. In: *Proceedings of 7th Transport Research Arena TRA 2018, to be published at the TRA-2018, April 2018.*

O'Keeffe, A., Alsem, D., Corbally, R., van Lanen, R. (as part of INTERLINK consortium) (2017) Information management for European roads using Linked Data. In: *Investigating the Requirements.* https://roadotl.geosolutions.nl/ publications/ [Accessed 12 September 2017].

V-Con (2016) *International Ontologies for V-Con, 2016.* (Virtual Construction for Roads Project), Technical Specification Phase 3, April, 2016.

Life-Cycle Analysis and Assessment in Civil Engineering: Towards an
Integrated Vision – Caspeele, Taerwe & Frangopol (Eds)
© 2019 Taylor & Francis Group, London, ISBN 978-1-138-62633-1

Mechanisms for managing changes in construction projects

H.Ç. Demirel
Faculty of Civil Engineering and Geosciences, Delft University of Technology, Delft, The Netherlands

L. Volker
Architecture and Built Environment, Delft University of Technology, Delft, The Netherlands

W. Leendertse
Ministry of Infrastructure and Water Management, The Netherlands
Faculty of Spatial Sciences, Groningen University, The Netherlands

M. Hertogh
Faculty of Civil Engineering and Geosciences, Delft University of Technology, Delft, The Netherlands

ABSTRACT

Due to the extensive duration and the dynamic environment of construction projects, changes in contract conditions are inevitable during its implementation. Therefore, it is important that changes are actively managed during the project life cycle. However, contracts do not always offer effective ways within the standardised change procedures to cope with change events. We argue that changes do not have to be managed solely through the contract, but can also be managed by applying alternative kinds of dealing mechanisms.

The objective of the research presented in this article is to achieve a better understanding of the range of different contractual and non-contractual mechanisms for change management in public contracting of construction projects. To reach this objective, a literature review on change management in construction projects and especially on the dealing mechanisms used in this management has been conducted. Additionally, an in-depth case study through ethnographic

and action research into the modus operandi with regard to (unexpected) changes in the realization phase of a large scale PPP (Public Private Partnership) infrastructure project was conducted. This study reflects real life practice of change management and presents how various (contractual and non-contractual) dealing mechanisms are interactively employed in the actual management of changes. The comparison of literature and practice, subsequently leads to conclusions for researchers and practitioners about how to set up, manage and improve the practice of change management in public contracting.

The dealing mechanisms can be discussed as the following;

– Contract rules
– Relations
– Tools
– Organisational structure
– Knowledge
– Skills

MS18: Serviceability of underground structures
Organizers: Y. Yuan, E. Bilotta, H. Yu & Q. Ai

Life-Cycle Analysis and Assessment in Civil Engineering: Towards an Integrated Vision – Caspeele, Taerwe & Frangopol (Eds)
© 2019 Taylor & Francis Group, London, ISBN 978-1-138-62633-1

The effect of hydrogen embrittlement on durability of buried steel pipes

M. Wasim, M. Mahmoodian, D. Robert & C.-Q. Li
RMIT University, Melbourne Central, Australia

ABSTRACT

In civil engineering projects, mild steel because of its mechanical properties is commonly used in buildings, bridges, railways and underground structures and utilities like buried water, gas and oil pipelines, etc. The life cycle of the steel structures is seriously affected when subjected to various surrounding environments; because of their vulnerability to corrosion attack and hydrogen-induced embrittlement. Hydrogen embrittlement (HE) is a form of deterioration which can be associated to corrosion and corrosion-control processes. Several researchers have reported the damaging effect of hydrogen on the ductility of steel structures. However, there is a difference of opinion among researchers about its effect on the yield and failure strength which are the key parameters for the life cycle design of any steel structure. The current research is a contribution to previous studies on HE of low carbon steel and aims at quantifying the degradation of mechanical properties of mild steel due to HE in various simulated environments by laboratory-controlled experimentation. Mild steel specimens are immersed in various simulated corrosive environments of hydrochloric acid and soil solutions. For each corrosive environment, nine specimens for tensile and three specimens for hydrogen measurements are tested at three time periods (7, 14 and 28 days). From the experimental findings, loss of mechanical properties (yield and ultimate strength) and ductility are observed. Various characterization techniques such as SEM, optical microscope and XRF are used to provide the solid reasoning of HE effect on the durability of mild steel. This paper presents qualitative and quantitative explanations of hydrogen-induced damage on structures and materials made up of mild steel. Moreover, the time-dependent corrosion and hydrogen concentration relations can be modelled and can be used for more accurate prediction of the safe life of various steel structures.

Rapid detecting equipment for structural defects of metro tunnel

K. Wang & X. Yao
Department of Geotechnical Engineering, Tongji University, Shanghai, China

ABSTRACT

Structural defects are main factors affecting the safety and durability of metro tunnel. How to effectively obtain and correctly detect the tunnel defect has become a fundamental work of tunnel maintenance. It is also an essential precondition linking the evaluation of the service state. Surface defects and cross-section deformations are two kinds of pathogenic defects in the service life of metro tunnels. With the expansion of city metro, traditional manual inspection approaches cannot meet the increasing demand of inspection work on account of their low efficiency, long feedback cycle and human judgment error etc. Thus, it is urgent to develop rapid detection equipment to meet current needs.

This paper illustrates surface defect recognition technology and cross-section deformation measurement technology. For surface defects, the technology mainly uses the difference image method in image processing to identify and classify the defect (Table 1). For cross-section deformation, the paper set up a computational process based on the principles of transmission projection.

To realize automation and integrated detection of metro tunnel surface defects and cross-section deformation, a power drive cart was developed. The power drive cart is mainly loaded with acquisition system, power supply system and power control system (Figure 1). Using CCD camera as basic element for information acquisition, the cart carries out an interleaving integrated acquisition scheme. The equipment

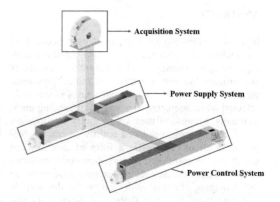

Figure 1. Schematic of power drive cart.

is able to acquire the service condition information of tunnel structure at a steady speed of 10 km/h. And the features of tunnel defect are extracted rapidly and intelligently by image processing software specialized developed for metro tunnel. The analysis result can be used for giving scientific judgment on the service state of metro tunnel structure through the setting criteria.

The proposed equipment has been applied in metro tunnels of Shanghai, Beijing, Guangzhou and Shenzhen, the testing results show that its technical performance fully meet the requirement of rapid detection.

Table 1. Features of different types of defect.

Type of defect	Water leakage	Segment damage	Cracks
Area size	Large	Medium	Small
Long axis length	Large	Medium	Large
Short axis length	Medium	Medium	Small
Long/Short	Large	Small	Large
Fill rate	Large	Large	Small

REFERENCES

Ai, Q. (2016) State-Oriented Maintenance Method of Tunnel Structure. PhD Thesis. Shanghai, Tongji University (in Chinese).

Ai, Q., Yuan, Y. & Bi, X. (2016) Acquiring sectional profile of metro tunnels using charge-coupled device cameras. *Structure & Infrastructure Engineering*, 12 (9), 1065–1075.

Huang, Y., Liu, X., Yuan, Y., Liu, C. & Wang, X. (2012) Auto inspection technology for detecting leakage in a shield tunnel. *Journal of Shanghai Jiaotong University*, 46 (1), 73–78 (in Chinese).

A state-oriented maintenance strategy of tunnel structure

Q. Ai
Department of Civil Engineering, Shanghai Jiao Tong University, Shanghai, China

Y. Yuan & X. Jiang
College of Civil Engineering, Tongji University, Shanghai, China

ABSTRACT

The surveys of tunnel structures show that many tunnels deteriorate in their early service lives. As important underground infrastructures, the inspections and maintenance costs of tunnel structures are huge, thus the research on scientific maintenance method of tunnel is becoming very necessary, urgent and forward-looking. Maintenance strategy focuses on the problem of how to allocate maintenance resource to improve maintenance effect in a better way, and it is of vital important in current situation.

A state-oriented maintenance strategy of tunnel structure is proposed in this paper. Comparing with traditional corrective maintenance and predictive maintenance, the major difference of proposed strategy is that it uses dynamic inspection schedule (as shown in Figure 1) and multi-state oriented maintenance rule, which means tunnel structure in different service states will be inspected by different frequencies and maintained by different levels of repair measure.

In order to investigate the proposed maintenance strategy with traditional ones, a mathematical framework of maintenance strategy optimization is formulated. Grid exhaustive optimization algorithm is applied to minimize the lifetime maintenance cost of tunnel structure under safety requirement. The optimal

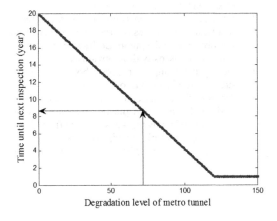

Figure 1. An aperiodic inspection schedule function.

Table 1. Optimal maintenance costs of different maintenance strategies.

Strategy No.	Characteristics of strategy	Maintenance cost
MS#1	Periodic inspection & control-limit maintenance rule	134
MS#2	Periodic inspection & multi-state oriented maintenance rule	38.5
MS#3	Aperiodic inspection & control-limit maintenance rule	124
MS#4	Aperiodic inspection & multi-state oriented maintenance rule	34

maintenance costs of different maintenance strategies are shown in Table 1.

The results of numerical investigation show that: 1) with the same maintenance rule, maintenance strategies using periodic inspection schedule and aperiodic inspection schedule get similar optimal maintenance cost; 2) when there are several repair measures available and minor or medium repair is more effective and cheap, the multi-state oriented maintenance rule is recommended in maintenance strategy, which will greatly reduce the maintenance cost. Thus, the state-oriented maintenance strategy is recommended, which will greatly reduce maintenance cost during the whole service life of tunnel structure.

REFERENCES

Do, P., Voisin, A., Levrat, E. & Iung, B. (2015) A proactive condition-based maintenance strategy with both perfect and imperfect maintenance actions. *Reliability Engineering & System Safety*, 133 (11), 22–32.

Grall, A., Bérenguer, C. & Dieulle, L. (2002) A condition-based maintenance policy for stochastically deteriorating systems. *Reliability Engineering & System Safety*, 76 (2), 167–180.

Liu, M. & Dan, M.F. (2004) Optimal bridge maintenance planning based on probabilistic performance prediction. *Engineering Structures*, 26 (7), 991–1002.

Changes in rheology of printable concrete during pumping process

Y. Yuan
State Key Laboratory of Disaster Reduction in Civil Engineering, Tongji University, Shanghai, China
Department of Geotechnical Engineering, Tongji University, Shanghai, China

Y. Tao & X. Wang
Department of Geotechnical Engineering, Tongji University, Shanghai, China

ABSTRACT

For printable concrete, the fluidity performance during pumping process directly affects the quality of printing. If the fluidity of the concrete becomes bad during transportation, it would be difficult to extrude. If the fluidity of the concrete becomes good, fresh concrete would not meet the requirements of printability. It is necessary to explore the changes of fluidity during pumping process.

Compared with slump test, Rheology can supply valuable and practical results regarding the properties of printable concrete.

IACR rheometer is used to measure the rheological parameters of printable concrete. The ICAR rheometer is based on the principle of the coaxial cylinders. The ICAR rheometer is shown in Figure 1.

The procedure employed is specifically designed to investigate the effect of pumping distance and shear time of rheometer on printable concrete.

The pump flow is fixed as 10 L/min. And the pipe diameter is fixed as 50 mm. The length of the pipeline varies from 0 to 8 meters. Six cases are adopted.

The results are shown below.

(1) The static yield stress of fresh concrete is 401.6 Pa after mixed. And the static yield stress of fresh concrete after pumping is 256.9 Pa. The static yield stress of fresh concrete drops obviously through the screw pump.

(2) After pumped for 2 meters, the static yield is 265.5 Pa. After pumped for 4 meters, the static yield is 246.1 Pa. After pumped for 6 meters, the static yield stress is 190 Pa. After pumped for 8 meters, the static yield stress is 177.6 Pa. Generally, the static yield stress of fresh concrete gradually decreases as the pumping distance increases.

(3) The plastic viscosity of fresh concrete is 38.7 Pa·s after mixed. And the plastic viscosity of fresh concrete after pumped is 22.3 Pa·s. The plastic viscosity of fresh concrete drops obviously through the screw pump.

(4) After pumped for 2 meters, the plastic viscosity is 26.3 Pa·s. After pumped for 4 meters, the plastic viscosity is 26.8 Pa·s. After pumped for 6 meters, the plastic viscosity is 26.1 Pa·s. After pumped for 8 meters, the static yield is 25.8 Pa·s. Generally, with the increase of the pumping distance, the plastic viscosity increases first and decreases then.

Following conclusions have been derived from the current research.

(1) The dynamic yield stress is basically zero. Tested fresh concrete is Newtonian flow.
(2) Because of the compression and shearing effects, rheological properties of fresh concrete changes sharply through the pump. Both static yield stress and plastic viscosity drop obviously.
(3) During the pumping process, both static yield stress and plastic viscosity increase first and decrease then. Based on the thixotropy theory, static yield stress and plastic viscosity increase due to significantly weaker shearing effect. Static yield stress and plastic viscosity both decrease due to the shear effect during transportation process.

Figure 1. ICAR Rheometer.

Settlement control, monitoring and analysis of utility tunnel on soft soil foundation

J. Huang, H. Wang & J. Wang
Shanghai Municipal Engineering Design Institute (Group) Co. Ltd., Shanghai, China

ABSTRACT

This paper takes Sangtiandao UTs in Suzhou Industrial Park as the background. Firstly, geological conditions of the construction site and the overall layout of UTs are introduced. Foundation pit scheme and settlement reduction measures are selected according to local conditions. Settlement monitoring points are then arranged on UTs and roads to observe the settlement after construction. Finally, the settlement data of UTs and roads are analyzed.

The post-construction settlement of roads and UTs increases with time, which can be classified as two patterns, that is the gradual change type and the two-stage type. During the first year after construction, the

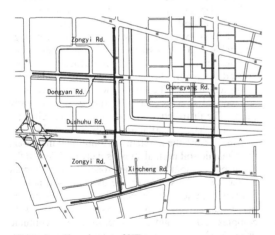

Figure 1. Plane layout of UTs.

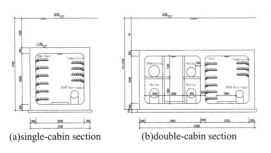

(a) single-cabin section (b) double-cabin section

Figure 2. Standard section of UT.

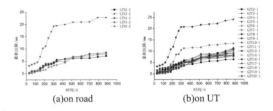

(a) on road (b) on UT

Figure 3. Settlement along Zongyi Rd.

settlement of the two-stage type almost reaches 80% of the total amount.

Settlement of UTs, the differential settlement between UTs and roads, the differential settlement of UTs, etc. are all within limits, which indicate that it is rational to use pipe pile foundations in the soft soil area for the settlement reduction. The construction method and technology will provide reference for similar projects.

REFERENCES

Hunt, D.V.L., Nash, D. & Rogers, C.D.F. (2014) Sustainable utility placement via multi-utility tunnels. *Tunnelling & Underground Space Technology Incorporating Trenchless Technology Research*, 39 (1), 15–26.

Jin, H. et al. (2013) *GB50911-2013 Code for Monitoring Measurement of Urban Rail Transit Engineering*. Beijing, China Architecture and Building Press. (in Chinese).

Qin, C. et al. (2008) *GB 50308-2008 Code for Urban Rail Transit Engineering Survey*. Beijing, China Architecture and Building Press (in Chinese).

Teng, Y. et al. (2011) *GB50007-2011 Code for Design of Building Foundation*. Beijing, China Architecture and Building Press (in Chinese).

Wang, D. et al. (2016) *JGJ8-2016 Code for Deformation Measurement of Building and Structure*. Beijing, China Architecture and Building Press (in Chinese).

Xiao, X. et al. (2006) *GB/T12897-2006 Specifications for the First and Second Order Leveling*. Beijing, China Surveying and Mapping Publishing House (in Chinese).

Xie, M., Huang, M., Lv, Y. et al. (2014) Experimental study on mechanical properties of marine soft soil in Fuzhou. *Chinese Journal of Underground Space and Engineering*, 10 (6), 1285–1292 (in Chinese).

Life-Cycle Analysis and Assessment in Civil Engineering: Towards an Integrated Vision – Caspeele, Taerwe & Frangopol (Eds)
© 2019 Taylor & Francis Group, London, ISBN 978-1-138-62633-1

Dynamic soil normal stresses on side wall of a subway station

Z.M. Zhang
Department of Geotechnical Engineering, Tongji University, Shanghai, China
Department of Civil, Architectural and Environmental Engineering, University of Napoli Federico II, Naples, Italy

Y. Yuan
State Key Laboratory of Disaster Reduction in Civil Engineering, Tongji University, Shanghai, China

E. Bilotta
Department of Civil, Architectural and Environmental Engineering, University of Napoli Federico II, Naples, Italy

H.T. Yu
Key Laboratory of Geotechnical and Underground Engineering of Ministry of Education, Tongji University, Shanghai, China

H.L. Zhao
Department of Civil Engineering, Shanghai University, Shanghai, China

ABSTRACT

Dynamic soil normal stress is still not clear or not certain for many engineering designers and science researchers. As one of the important loads, the accurate magnitude and distribution of dynamic soil normal stresses are not well known and remain a question worth discussing. Most research on this topic has focused on rigid or flexible retaining walls so far. Limiting-equilibrium Mononobe-Okabe type solutions is widely used in designing the above rigid or flexible retaining walls. Though Mononobe-Okabe method is widely applied around the globe, many researchers have presented doubts to it. Whether or not the Mononobe-Okabe method is really suitable for the evaluation of other structures may be debatable since the tests to define the method were made for cantilever walls retaining medium dense sand. For other structures and ground conditions, Mononobe-Okabe method has not been validated. Less research is conducted on other structures, such as underground subway stations. Obviously, these two kinds of structures have different mechanical characteristics, which will lead to their different dynamic soil normal stresses. To

extend knowledge about dynamic soil normal stresses, this study would investigate the characteristics of dynamic soil normal stresses for a subway structure experiencing sinusoidal ground motions. The numerical simulations were conducted on a scaled structure model. The results showed that the dynamic soil normal stresses experienced two stages: transient stage and stable stage. The dynamic soil normal stresses had obviously periodic features. The frequencies of the calculated stress time histories were same as the input sinusoidal waves. The distribution of peak dynamic soil normal stresses along the side wall was nearly linear. The distribution of real-time stresses changes with time. Their distribution differed from the peak distribution. As the acceleration amplitude increased, the dynamic soil normal stresses increased. When the frequency of input motions approached the dominant frequency of the soil layer, the dynamic soil normal stresses became much larger. These findings proved that dynamic soil normal stress distributions were complex and it might be not reasonable to continue using M-O method in some structures like subway stations.

284

Life-Cycle Analysis and Assessment in Civil Engineering: Towards an
Integrated Vision – Caspeele, Taerwe & Frangopol (Eds)
© 2019 Taylor & Francis Group, London, ISBN 978-1-138-62633-1

Shaking table model tests on tunnels at different depths

X. Zhao, R.H. Li, M. Zhao & L.J. Tao
Key Laboratory of Urban Security and Disaster Engineering of Ministry of Education,
Beijing University of Technology, Beijing, China

ABSTRACT

There are now increasing numbers of damage reports on mountain tunnels under strong earthquake including deep buried tunnels with high overburden. Failure mechanisms for deep buried tunnels and shallow buried tunnels are different. But the effects of high overburden during earthquake on tunnel have not been understood thoroughly. Few model tests have been applied on such conditions as the overburden is hard to be applied. A set of flexible rubber airbag loading equipment have been invented and used to apply overburden on top of surrounding rock in a rigid testing box to perform model tests on shaking table. Then, a series of model tests have been performed to study the dynamic response of tunnels with different overburden depth with the above test system. The rock condition around tunnel is designed as class V according to the China National Design Code for Mountainous Tunnels. A new similarity material with a main component of fly ash has been successfully created to simulate the surrounding rock to meet the requirement of the shaking table test. The results showed that

this testing system is feasible. Tests with prototypes of 25 m, 175 m and 275 m buried depths are performed by given different airbag pressure. The results show that, the acceleration amplifier factor increases from the bottom to top of the surrounding rock. This trend is amplified with the increasing buried depth. Maximum structure moments and displacements are found at the shoulder and foot of the tunnel which result in serious cracks in such areas. But failure modes of tunnels under different depth are similar. It is also found that, the additional moments from vibration decreases with the increase of buried depth. And more serious cracks developed are observed in rather shallow buried tunnel It could be concluded that, shallow buried tunnels are much more vulnerable than deep buried tunnels, while the vulnerability of deep tunnels are similar with different buried depth. The deep buried tunnels has a strong restraining effect on the tunnel structure and can restrain the de-formation of the structure under the earthquake, which is in favor of the overall stability of the tunnel. These results could be referred in designing tunnels considering seismic responses.

Life-Cycle Analysis and Assessment in Civil Engineering: Towards an Integrated Vision – Caspeele, Taerwe & Frangopol (Eds)
© 2019 Taylor & Francis Group, London, ISBN 978-1-138-62633-1

Two-dimensional finite element analysis of the seismic performance in complex underground structure

X.S. Cai, Z. Ye & Y. Yuan
Tongji University, Shanghai, China

ABSTRACT

This paper presents a study of dynamic responses of underground structures based on "World Expo" underground structure in Shanghai. Analysis is performed using the open source finite element framework OpenSees. A typical section of four-storey thirteen-span "World Expo" underground structure is adopted to analyze by using dynamic time-history analysis. 2D Opensees numerical analysis model is performed to compare displacement and acceleration of each soil layer. Shear force, moment of structure components, such as side walls, beams, columns and slabs, are analyzed to obtain the seismic response pattern of the structure. FEA (finite element analysis) model with super-structure and FEA model without super-structure are built to compare different responses during seismic response.

With the rapid development of cities, more and more large underground structures are under construction in seismically active areas, such as subway station, underground market and underground parking garage. In recent years, Underground structures seismic damage occur frequently, especially the Kobe earthquake, Chichi earthquake and Duzce earthquake, which has caused damage to underground engineering facilities. The seismic issues of underground structures have been highly valued by earthquake scholars all over the world.

Compared with traditional above-ground structure seismic issues, underground structures have different boundary conditions during earthquake. Underground structures are usually buried in soil or rock. In addition, length of tunnel in axial direction is far greater than the other two directions. Therefore, failure characteristics of underground structures during earthquake are as follows: (1) in the same region, seismic capability of the underground structure is superior to seismic capability of the super-structure. With the increase of buried depth, extent of structural damage decrease. With the increase of soil shear modulus, extent of structural damage decrease. (2) When high frequency waves occupy the main components in a seismic wave, the tunnels are prone to partial damage; (3) seismic duration is an important factor affecting the extent of damage of underground structures.

As shake91 is based on the frequency theory, the free field response seismic result of shake91 could be regarded as target solution and standard solution. It can be concluded that both Abaqus solution and OpenSees solution have a good match with the target solution.

Model with super-structure and model without super-structure were established to get soil response and structure response during seismic process.

From the calculated results, it can be concluded that it is conservative for underground structure design to neglect super-structure effects in dynamic cases. From shear force and moment results of side walls and columns, it can be concluded that columns section is critical section in complex underground structure during seismic process. Also more efforts are need in considering gravity in the FEA model, more cases should be analyzed to get the response patterns of different underground structures.

Shaking table test on shaft ingate of shield tunnel

J. Zhang
Department of Geotechnical Engineering, Tongji University, Shanghai, China

X. Tu, X. Zhang & F. Li
Department of AC Transmission Project, State Grid Corporation of China, Beijing, China

Y. Yuan
State Key Laboratory for Disaster Reduction in Civil Engineering, Tongji University, Shanghai, China

ABSTRACT

During the construction of typical shield tunnels, working shafts were always built at the two ends of the tunnels. Generally, shield tunnels have good structural performance against earthquakes, because the tunnel consists of segmental linings. However, the shaft ingates, where the tunnel initiates or ends, is prone to seismic hazard due to its sharp structural and geometric changes, and it could not by analyzed by any analytical methods yet. Most studies on this part of the shield tunnels were carried out by numerical methods (Zhao et al. 2012). This paper will present the work of a shaking table test, targeted on the seismic behavior of the shaft ingate.

A shear box is used as the model container. The operating space of the box is 4.5 m long, 3.5 m wide and 3.0 m high. The model soil is a combination of dry sand and saw dust. Two different kinds of model soils were applied to simulate the stratified ground. And dynamic triaxial tests were carried out on both the model and the original soils. The structure model consists of the shaft model and 20 tunnel ring models. The shaft model is designed according to the similitude ratio of mass. The inner and the outer diameters of the tunnel ring model are 58 cm and 53 cm, respectively. There are three steel rings embedded inside the model. And each ring model is divided into 8 segments by aluminum sheets. Figure 1 is the planar view of the model test.

The acceleration responses of the model soil are recorded in the free field model test. The acceleration responses of the tunnel and the shaft and the extensions of circumferential joints are monitored in the tunnel model test.

The boundary conditions of the test are validated by the acceleration response of the model soil in the free field model test.

According to the data acquired in the test, the working shaft and the shield tunnel have discrepant acceleration responses during earthquakes, especially in the frequency domain higher than 10 Hz.

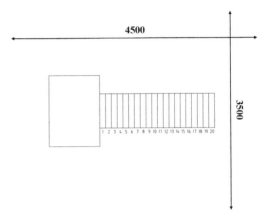

Figure 1. Planar view of the model test. (unit: mm).

The extensions of circumferential joints near the working shaft are much larger than that of the other joints. This could be explained by the non-uniform movements of the shaft and the tunnel during earthquakes. And in this very case, the working shaft could influence the adjacent 7 joints.

The results from the shaking table test are agree with the field observations by Koizumi (2009). The shaft ingate of shield tunnel is prone to seismic hazard. It is urgent to develop effective countermeasures.

REFERENCES

Koizumi, A. (2009). Seismic damages and case study for shield tunnel (Translated by W. Zhang, D. Yuan), 1st ed., China Architecture and Building Press, Beijing. (In Chinese)

Zhao W, He X, Chen W, et al. (2012). Analysis of seismic damage of segments and joints at the junction of shield tunnel and shaft. Chinese Journal of Rock Mechanics & Engineering, 31(2): 3847–3854. (In Chinese)

MS19: Circular economy to improve sustainability of infrastructure
Organizers: S. de Vos-Effting & R. Hofman

Are recycled and low temperature asphalt mixtures more sustainable?

M. Hauck, E. Keijzer, H. van Meerveld, B. Jansen & S. de Vos-Effting
TNO, The Hague, The Netherlands

R. Hofman
Rijkswaterstaat, The Hague, The Netherlands

ABSTRACT

Innovations are necessary to achieve reduction of both CO_2 emissions and use of fossil materials during construction and maintenance of roads. To enable selection of promising innovations and to facilitate material selection in green procurement, quantitative tools to assess both life cycle costs and environmental impacts are used.

In this research, life cycle costs and environmental product indicators of asphalt innovations were used to investigate the financial and environmental consequences of the strategic decision of the Dutch national road administration *Rijkswaterstaat* to use "green" porous asphalt (PA) top layers. Sustainable asphalt mixtures that were produced at low temperatures and/or with different degrees of reclaimed asphalt were compared to a reference hot mix PA.

Asphalt with recycled content turned out to have 4%–5% lower life cycle costs than the reference mixture and caused 12%–17% less CO_2 emissions. Low temperature mixes were estimated to cause 5%–10% less environmental impacts than the reference, but the costs increased with a few percent due to the addition of rejuvenators.

Life time expectancy is relevant for the outcomes of the comparison: if the sustainable asphalt mixtures would have a two year shorter life timer, the environmental and cost impacts would overshoot the reference impacts.

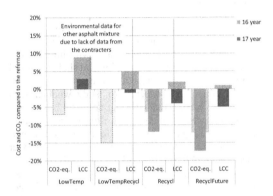

Figure 1. Comparison of the deltas of life cycle CO_2 equivalent (eq.) and costs (LCC) of four asphalt innovations relative to reference hot porous asphalt.

Suggestions for further research are

1) to monitor lifetime of sustainable pavements;
2) to decrease uncertainty in the environmental results by updating this study when more information on the asphalt environmental profiles is available; and
3) to investigate in cost and environmental assessment of other solutions for more sustainable asphalt, such as self-healing, rejuvenating, and reduced resistance and fuel consumption.

Life-Cycle Analysis and Assessment in Civil Engineering: Towards an Integrated Vision – Caspeele, Taerwe & Frangopol (Eds)
© 2019 Taylor & Francis Group, London, ISBN 978-1-138-62633-1

Relevance of the information content in module D on circular economy of building materials

K. Krause & A. Hafner
Resource Efficient Building, Ruhr University Bochum, Bochum, Germany

ABSTRACT

The waste legislation requires a high-quality utilization of materials in the construction industry in Germany. Thus construction waste with high-quality properties contributes to recycling and shall be reused extensively to replace fresh materials.

The aim of this paper is to study whether it is possible to present the recycling potential in the construction sector on building level or at product level by means of parameters of life cycle assessment. The existing system of the Environmental Product Declaration (EPD) is be used for this study.

According to DIN EN 15978 and DIN EN 15804, the LCA information is structured in the lifecycle sections of buildings and the materials used therein. This specifies the exact time in the lifecycle and the extent to which certain effects can be quantified from the use of buildings or the building products.

The LCA information is subdivided into the environmental impacts of the building up to the completion of a building (Module A), the environmental impacts (Module B) during its use and the expenses at the end of the building life cycle (Module C). An additional module D summarizes all benefits and loads outside the system boundaries. It declares the advantages or loads associated with raw material properties due to its reuse, recovery or recycling of a construction product.

For this paper a building have been analyzed by means of life cycle assessments and module D has nearly always shown negative values, which means benefits through recycling. Here more detailed investigations have been conducted to show from which material fraction and in which lifecycle stage the benefits occur. The paper will investigate the question whether this module offers the possibility to present recycling potentials at the end of a building life cycle.

Life-Cycle Analysis and Assessment in Civil Engineering: Towards an Integrated Vision – Caspeele, Taerwe & Frangopol (Eds)
© 2019 Taylor & Francis Group, London, ISBN 978-1-138-62633-1

PAPERCHAIN project: Establishment of new circular economy models between pulp and paper industry and construction industry to create sustainable infrastructures

M.S. Martín-Castellote, J.J. Ceprià-Pamplona, M.M. Pintor Escobar & E. Guedella-Bustamante
Technology & Innovation Division, ACCIONA Construction, Alcobendas, Madrid, Spain

ABSTRACT

The European Pulp and Paper Industry (PPI) generates 11 million tonnes of waste yearly. Most of them are currently burnt for energy recovery or used for landspreading, but around 1.5 million tonnes are still disposed. These wastes, mainly inorganic, are produced in the causticizing process or in waste-to-energy plants. If managed in a sustainable manner, they can become a valuable secondary raw material for other resource intensive industries such as construction. PAPERCHAIN, a research and innovation project funded by the European Commission under Horizon 2020 Framework Programme, addresses this potential resource to demonstrate from a technical, economic, social and environmental point of view the feasibility of applying Circular Economy Models between the PPI and the construction industry. In detail, the application of different waste streams for concrete filler, asphalt pavements, composite for slope stabilization and binder for soil stabilization, is presented as an example of cooperation between the PPI and the construction sector.

Life-Cycle Analysis and Assessment in Civil Engineering: Towards an Integrated Vision – Caspeele, Taerwe & Frangopol (Eds)
© 2019 Taylor & Francis Group, London, ISBN 978-1-138-62633-1

Introducing the circular economy in road construction

W.L. Leendertse
Faculty of Spatial Sciences, Infrastructure Planning, University of Groningen, The Netherlands
Ministry of Infrastructure and Water Management, Rijkswaterstaat, The Netherlands

M.E.M. Schäffner
Witteveen + Bos, Consulting Engineers, Deventer, The Netherlands

S. Kerkhofs
Ministry of Infrastructure and Water Management, Rijkswaterstaat, The Netherlands

ABSTRACT

In our current 'throwaway culture' we produce, use and throw away what is left if useless. A circular economy is aimed at reducing or even preventing waste. Instead of a single lifetime loop, the circular economy assumes that resources should be part of a closed system of subsequent loops, where the output of one loop is the input of the next loop. Currently, there is a growing political and societal pressure to reduce the use of basic materials and resources and to prevent creating waste. The road construction industry is infamous for its' major use of energy resources and materials, necessary for construction and maintenance. Under the above mentioned political and societal pressure also the road construction industry is strongly challenged to become circular.

But, how to design and realize a circular road for multiple material lifecycles? And what economical mechanisms can be used to stimulate circularity in road construction?

The paper discusses these questions by confronting theory about circular design and circular economy with a caseproject, InnovA58, a planned reconstruction of the A58 highway in the southern part of the Netherlands. This highway, including characteristic parts like pavement, earthwork, viaducts, bridges and fly-overs, will be designed circular to discover the accompanying challenges and dilemmas. The aim is to understand circular design in highway planning and to derive general design principles. The transition to a circular economy is a major social transition. Transitions do not happen automatically but can arise if various impulses coincide in a certain direction. Rijkswaterstaat has set up the InnovA58 project as an experimental environment in order to provide such an impulse for the construction industry in particular. In this paper the results (sofar) of InnovA58 are presented with the aim of raising a further debate about circular design and the transition to a circular economy in the construction industry and to challenge new mutually reinforcing impulses.

CEO & CAMO ontologies: A circulation medium for materials in the construction industry

E.M. Sauter
Geographical Information Management and Applications (GIMA) Programme, The Netherlands

R.L.G. Lemmens
Faculty of Geo-Information Science and Earth Observation (ITC), University of Twente, Enschede, The Netherlands

P. Pauwels
Department of Architecture and Urban Planning, Ghent University, Ghent, Belgium

ABSTRACT

The Circular Economy (CE) paradigm proposes a redesign of material flows by keeping materials in circulation, thereby avoiding waste and raw material extraction (Nasir et al., 2017). Pioneering CE initiatives suggest the creation of material passports to capture building component information. Spring & Araujo (2017) state product passports allow for (1) informed decisions about products' next reuse steps; (2) the creation of markets for reused or disassembled products; and (3) new service offerings around the product. Although Building Information Modelling (BIM) is currently suggested to store material passports, this technology does not address how to put material circulation in action: pose queries over information, handle large data amounts from disparate sources, make 'smart matches' between CE actors and their materials. This paper evaluates the potential of semantic technologies to facilitate such building material circulation in the construction industry within the CE context. It suggests material passports as a data source for the materials to be circulated and integrates them by means of Linked Data. Linked Data is selected for its ability to connect heterogeneous datasets, and its integrated query potential of building material profiles across the web.

Two ontologies are proposed, aiming to allow the combination and re-use of material data: the Circular Exchange Ontology (CEO) and the Circular Materials and Activities Ontology (CAMO). The ontologies allow to represent materials under a CE classification scheme and describe the elements required for material exchange to occur: actors, the activities they perform and the referents of these activities (resources, products, tools and waste). An evaluation of the ontologies is made by use case-driven SPARQL queries which follow specified triple patterns to retrieve data from semantic databases. Next, a decentralized data ecosystem is proposed to enable users to publish information and dynamically interact with repositories to exchange materials (Figure 1).

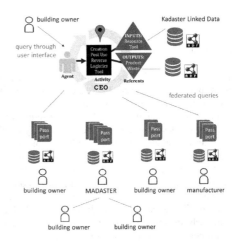

Figure 1. Architecture showing Linked Data repositories published by building owners and organizations following CEO/CAMO ontologies and querying by end user applications.

Use case #1 shows how Madaster material passport data is used to connect a ship builder needing waste beams (hardness >100 HB) with the building owner (http://ld-ce.com/query). Results indicate that these ontologies can foster a circulation medium between diverse construction industry resources. Future work considers validation of both ontologies in broader real-world settings with computer scientists and companies in the Benelux construction sector.

REFERENCES

Nasir, M. H. A., Genovese, A., Acquaye, A. A., Koh, S. C. L. & Yamoah, F. (2017) Comparing linear and circular supply chains: A case study from the construction industry. *International Journal of Production Economics*, 183, 443–457.

Spring, M. & Araujo, L. (2017) Product biographies in servitization and the circular economy. *Industrial Marketing Management*, 60, 126–137.

Life-Cycle Analysis and Assessment in Civil Engineering: Towards an Integrated Vision – Caspeele, Taerwe & Frangopol (Eds)
© 2019 Taylor & Francis Group, London, ISBN 978-1-138-62633-1

SUP&R DST: SUstainable Pavement & Railways Decision Support Tool

J. Santos
IFSTTAR, France

S. Bressi
DICAM, Universita' degli Studi di Palermo, Italia

V. Cerezo
IFSTTAR, France

D. Lo Presti
Nottingham Transportation Engineering Centre, University of Nottingham, UK

ABSTRACT

The importance of sustainability in transportation infrastructure has raised in response to the link between anthropogenic activity and global challenges, such as climate change, as well as in consequence of the ongoing development of models quantifying the social and economic impacts resulting from infrastructure development. Therefore, addressing the sustainability of transportation infrastructures requires exploring the environmental, social, and economic impacts of technological options while balancing the often conflicting priorities of different stakeholders, at an early design phase of the infrastructure delivery process. This is a typical multi-criteria decision-making (MCDM) problem, in which the decision-makers need to measure the sustainability through a set of meaningful, representative and quantifiable criteria, balance the relative importance of those criteria and determine the sustainability sequence of multiple alternative technologies for fostering transportation sustainability.

In order to help the decisions makers to efficiently address this challenging task, a decision support toll (DST) was developed in the scope of the training-through-research programme Sustainable Pavements & Railways Initial Training Network (www.superitn.eu). It consists of a computational platform that implements a conceptual framework developed to quantify sustainability. It comes with a set of sustainability indicators tailored to both road and railway systems as well as several objective and subjective weighting methods. Amongst those belonging to the last category, the DST includes a set of default weights derived from an Analytical Hierarchy Process (AHP)-based survey that engaged stakeholders from different sectors and from several European countries. At last, the Preference Ranking Organization Methodology of Enrichment Evaluation II (PROMETHEE-II) MCDM method is employed for prioritizing alternative road pavement and railway tracks solutions at the design stage.

The SUP&R DST is a freely available upon request (http://superitn.eu) and can be used at professional level, by professionals interested in advancing sustainability in transportation, as well as for educational purposes, to provide knowledge and educate on the use sustainability concepts and on what are the important issues to consider during the sustainable transportation decision-making process.

SPECIAL SESSIONS
SS1: Structural Health Monitoring and decision making for
infrastructures in multi-hazard environment
Organizers: M.P. Limongelli, J.R. Casas, M.G. Stewart & B. Imam

Life-Cycle Analysis and Assessment in Civil Engineering: Towards an
Integrated Vision – Caspeele, Taerwe & Frangopol (Eds)
© 2019 Taylor & Francis Group, London, ISBN 978-1-138-62633-1

Stochastic differential equations for modeling deterioration of engineering systems and calibration based on Structural Health Monitoring data

L. Iannacone & P. Gardoni

Department of Civil and Environmental Engineering, MAE Center,
University of Illinois at Urbana-Champaign (UIUC), Urbana, IL, USA
NIST-funded Center of Excellence for Risk-Based Community Resilience Planning, USA

ABSTRACT

The effects of aging and deterioration processes on the performance of engineering systems have become a growing concern over the past years. A generalized framework is required to model system deterioration in an accurate and efficient way, so that it is possible to estimate the current performance of the system and, at the same time, predict how it will respond to future demands that might be coming from regular service or shock occurrences. It is important to establish the models that describe the evolution of each of the state variables, i.e. the physical quantities that the system depends on. Data from Structural Health Monitoring (SHM) can be used for that purpose.

Jia & Gardoni (2018) developed a model that is able to account for the interaction between the different deterioration processes (both shock and gradual) by looking at the evolution of the state variables. The attractive feature of this work is that multiple deterioration processes can be included in the model for deterioration, the interaction between different state variables naturally comes into the framework by considering state-dependent processes, and obtaining the reliability of the system once the models for the state variables are available is a relatively straight-forward process (Gardoni 2017).

The models available in literature for the evolution of the state variables over time are characterized by high values of uncertainty concerning the form of the models and the values for the coefficients. Data from SHM can be used to (1) assess the current values of the state variables and then use such values to estimate the current reliability of the system, (2) gain insight into the underlying phenomena to develop better deterioration model forms, and (3) calibrate the deterioration models. Procedures available in literature rely on sim-

plifying assumptions such as the discretization of the time domain in order to operate on the single step increment of the state variables and, in addition, they generally are computationally expensive.

This paper proposes a stochastic model for the evolution of the state variables that does not define the stochastic process in terms of the change in the chosen unit of time (making the unknown variables independent from the chosen discretization) and is computationally more efficient. The proposed model is based on a system of Stochastic Differential Equations (SDE). The advantage of using this formulation is that it retains the qualities of the framework proposed by Jia & Gardoni (2018) (interaction between state variables, interaction between different deterioration processes, time-variant formulation for capacity and demand of the system), while at the same time, using a compact formulation for the evolution of the state variables, it allows to use the powerful tools available in the field of stochastic calculus (Grigoriu 2013) for both the simulation and the calibration of the processes. An example is provided to show the performance of the proposed model when combined with data from SHM.

REFERENCES

Gardoni, P. (2017) *Risk and Reliability Analysis: Theory and Applications: in Honor of Prof. Armen Der Kiureghian.* Springer.

Grigoriu, M. (2013) *Stochastic Calculus: Applications in Science and Engineering.* Springer Science & Business Media.

Jia, G. & Gardoni, P. (2018) State-dependent stochastic models: A general stochastic framework for modeling deteriorating engineering systems considering multiple deterioration processes and their interactions. *Structural Safety*, 72, 99–110.

Life-Cycle Analysis and Assessment in Civil Engineering: Towards an Integrated Vision – Caspeele, Taerwe & Frangopol (Eds)
© 2019 Taylor & Francis Group, London, ISBN 978-1-138-62633-1

Information requirements for effective management of an ageing transport network

J.H. Paulissen, S.H.J. van Es & W.H.A. Peelen
TNO, Delft, The Netherlands

H.E. Klatter
Ministry of Transport, Rijkswaterstaat, Utrecht, The Netherlands

On top of intensified use, strict regulations and public demand comes the effects of ageing and degradation of the structures (Klatter, 2016). It is expected that an increasing part of the ageing stock of structures will reach its end of service life due to mentioned influences. In the Netherlands Rijkswaterstaat (RWS) is the executive agency of the Ministry of transport responsible for three national networks; highways, waterways and water system. RWS established a program for replacement and renovation. This program, has amongst others the responsibilities to provide a forecast of the replacement need on an object level, to integrate replacement need in the investment programs and to coordinate execution programs and projects.

It has been recognized that these tasks require specific assessment procedures and reliable information on the assets which differs from those for regular maintenance. The technical indicators are not equal to those used in regular maintenance as described e.g. in (van Kanten, 2013). This paper gives a description of this information need and the assessment procedures in general and specifically for steel bridges.

The replacement program discriminates three time periods; short-term object oriented (10 years), mid-term issue oriented (10–15 years) and long-term (>15 years) statistically oriented. In first period in-depth technical information on an object level is needed, whereas in the second time period information on a set of structures, with the same issue is called for. The latter implies the need for models to correlate results obtain on individual structures over the set of structures in a network.

Traffic and structural safety is an important aspect for managing an ageing transport network. Also defining a cost-effective renovations with acceptable effects on the traffic, the surroundings and other stakeholders is challenging. The actual and future technical condition of the objects, as well as the loads they are exposed to are dominant parameters concerning replacement decisions. Establishing safe but not too frequent inspection intervals for (groups) of structures with a common issue is another important information need. These technical indicators cannot be seen

independent of other procedures and processes within RWS, therefore these are addressed in the paper as well.

For steel bridges RWS determines on an object and situation specific level which method of assessment is most suitable to address the specific concern. The assessment methods have an increasing degree of accuracy since they use an increasing amount of bridge specific data and models, e.g. from simple analytical models to advanced FE-models. The required information to perform such assessments as well as technical options to obtain the information are described in this paper.

Furthermore some improvements in the use of inspection and monitoring techniques to come to more realistic and/or cost-effective assessments are indicated. For instance it has been established that information on fatigue load (effects) could be used in combination with inspections with the goal to obtain a more optimal inspection interval, and reduce costs and traffic disruption. The same holds for implementation of permanent crack monitoring systems. Compared to the current situation it would be beneficial to consider measurements alongside with more detailed calculations at an earlier stage in processes with the aim to formulate optimal strategies in processes where the effort to justify the structural safety is step wise increased. Also continuous condition monitoring of critical elements can help during the time required to implement more permanent measures (e.g. strengthening).

REFERENCES

Klatter, H.E. & Roebers, H. (2016) Assessment of need for renewal on a multi-network level. In: *Proceedings of IALCCE 2016*.

H.F.B.F., 2015, *Technical and Economic Feasibility of Structural Integrity Monitoring for Fatigue Life Prediction of Orthotropic Bridge Decks, Proc. SHMII-7*, Torino, Italy 2015.

Van Kanten, W. List of condition indicators, 2013, Deliverable D2.1 of Collaborative project FP7-285119 http://trimm.fehrl.org/?m=3&id_directory=7539.

Life-Cycle Analysis and Assessment in Civil Engineering: Towards an
Integrated Vision – Caspeele, Taerwe & Frangopol (Eds)
© 2019 Taylor & Francis Group, London, ISBN 978-1-138-62633-1

ROC-based performance analysis and interpretation of image-based damage diagnostic tools for underwater inspections

M. O'Byrne & V. Pakrashi
*Dynamical Systems and Risk Laboratory, School of Mechanical and Materials Engineering,
University College Dublin, Ireland*
Marine Renewable Energy Ireland (MaREI), University College Dublin, Ireland

F. Schoefs
Research Institute of Civil Engineering and Mechanics (GeM)/Sea and Littoral Research Institute (IUML), CNRS UMR/FR, Université Bretagne-Loire, Université de Nantes, Nantes, France
IXEAD/CAPACITES Society, Nantes, France

B. Ghosh
Department of Civil, Structural and Environmental Engineering, Trinity College Dublin, Ireland

ABSTRACT

It is of practical importance for inspectors to have knowledge of the efficiency of Non-Destructive Testing (NDT) tools when applied commercially. It has become common practice to model the performance of NDT tools in a probabilistic manner in terms of Probability of Detection (PoD), Probability of False Alarm (PFA) and eventually by Receiver Operating Characteristic (ROC) Curves. Traditionally, these quantities are estimated from training data, however, there are often doubts about the validity of these estimates when the sample size is small. In the case of underwater inspections, the scarcity of good quality training data means that this scenario arises more often than not. Comprehensive studies around the on-site performance of image-based damage diagnostic tools have only recently been made possible through the availability of online resources such as the Underwater Lighting and Turbidity Image Repository (ULTIR), which contains photographs of various damages forms captured under controlled visibility conditions. This paper shows how meaningful information can be extracted from this repository and used to construct ROC curves that can be related to the on-site performance of image-based NDT methods for detecting various damage forms and under a range of environmental conditions. The ability to draw connections between image-based techniques applied in real underwater inspections with ROC curves that can be constructed on-demand provide the engineer/inspector with a clear and systematic route for assessing the reliability of data obtained from image-based methods. As a case study, the general approach has been applied to characterise the performance of image-based techniques for identifying instances of corrosion and cracks on marine structures. A discussion around how the results can be used for further analysis is provided. This includes looking at how the results can be fed into in the decision chain

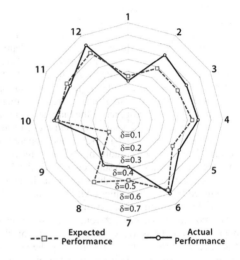

Figure 1. The expected and actual performances for each situation of detection.

and can be used for risk analysis, intervention and work scheduling, and eventually understanding the value of information.

REFERENCES

O'Byrne, M., Schoefs, F., Ghosh, B. & Pakrashi, V. (2013) Texture analysis based damage detection of ageing infrastructural elements. *Computer-Aided Civil and Infrastructure Engineering*, 28, 162–177.

O'Byrne, M., Schoefs, F., Pakrashi, V. & Ghosh, B. (2014b) Regionally enhanced multi-phase segmentation technique for damaged surfaces. *Computer-Aided Civil and Infrastructure Engineering*, 29, 644–658.

O'Byrne, M., Schoefs, F., Pakrashi, V. & Ghosh, B. (2018) An underwater lighting and turbidity image repository for analysing the performance of image based non-destructive techniques. *Structure and Infrastructure Engineering*, 14(1), 104–123.

Real-time monitoring as a non-structural risk mitigation strategy for river bridges

F. Ballio & G. Crotti
Department of Civil and Environmental Engineering, Politecnico di Milano, Milano, Italy

A. Cigada
Department of Mechanics, Politecnico di Milano, Milano, Italy

ABSTRACT

A monitoring system has been installed on the road bridge over the Po River at Borgoforte, Italy (Fig. 1) as a base for a non-structural risk mitigation tool. After a flood in year 2000 a bathymetric survey indicated the presence of a residual scour hole between the two central piers in the river, whose maximum depth was 15 meters; it is reasonable to assume that larger depths may have been reached during the flood event. Therefore, three piers were stabilized by adding four piles to each pier (piers on the left side of Fig. 1) and a rip-rap around the foundations. No structural reinforcement was applied to the fourth pier in the main stream (on the right side of Fig. 1), as it is situated close to the bank, where the bed level is higher and apparently more stable. However, hydraulic modelling of the river reach indicates that, for high stage conditions, the flow tends to concentrate on the right bank so that it cannot be excluded that also such pier could be significantly stressed during a future flood event.

The monitoring system is devoted to measure environmental conditions rather than the state of the structure; real-time variables recorded by the system are: water elevation, bed elevation at the pier, wind velocity, accumulation of floating debris. Measured values are taken as initial conditions for evolution scenarios of the actions by means of proper forecasting models and, through a structural model of the bridge, of the safety conditions of the bridge (Fig. 2), thus allowing to decide about the necessity of closing it to traffic in due time. Particular care is devoted to procedures for risk management, that is, a system of thresholds and consequent actions based on pre-determined (minimum) reaction time intervals along which evolution scenarios are forecasted. Fig. 2 shows the evolution of the actual state of the bridge during the year 2014 and the corresponding 14-hours worst case forecasts. Figure shows that, during a high water depth event in the winter, the bridge may have reached critical conditions, so that a procedure aimed at closing the bridge should have started.

The whole chain is able to prevent casualties and, by reducing loads connected with traffic, possibly also avoid collapse of the structure.

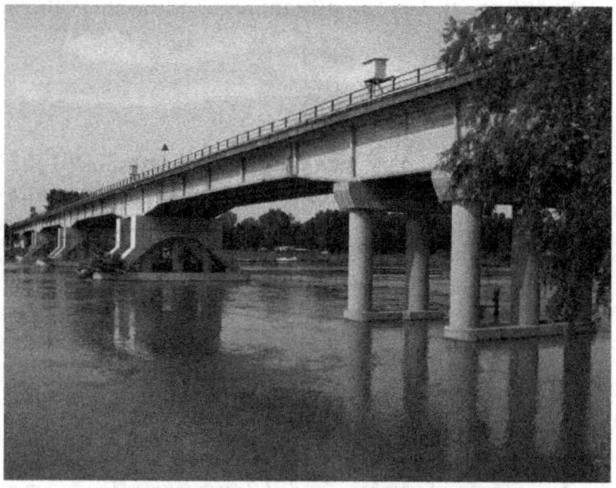

Figure 1. View of Borgoforte bridge from upstream.

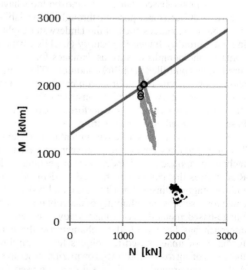

Figure 2. Structural state of the bridge along year 2014 in the N-M plane Black: real time values; grey: worst case 14-hours estimate.

Fiber optic sensing in an integrated Structural Health Monitoring system

R. Blin & D. Inaudi
Smartec SA, Manno, Switzerland

ABSTRACT

Designing a Structural Health Monitoring (SHM) system starts by identifying the risks associated with the specific structure and their probability. The risk analysis will lead to a list of possible events and degradations that can possibly affect the structure. Example of risks and uncertainties are corrosion, loss of pre-stressing, creep, subsidence of foundations, earthquake strike, unauthorized overloads, impact, inaccuracy of Finite Elements Models, poor building material quality and poor execution. The severity and probability of each risk will be classified using the usual risk analysis procedure to produce a ranking of risks. In this context, risks that are more likely to occur simultaneously or cascading will deserve special attention. Some risks will be retained, others will be addressed by regular inspection and the remaining will be dropped because of a low impact and/or probability. The result is a ranked list of risks that must be addressed by the Structural Health Monitoring system.

When selecting the best sensors for the specific risks associated with a given structure, it is often necessary to combine different measurement technologies. The Structural Health Monitoring system is addressing local properties, critical members, global performances and responses and network scales when several structures belonging to an infrastructure system are monitored. A few structures might present deficiencies which cannot be identified by visual inspection, modelling or standard survey. In these cases it is crucial to undertake appropriate remedial or preventive actions before it is too late. Having permanent and reliable long term monitoring data from a structure can guarantee the safety of the structure and its users.

It is of fundamental importance that a monitoring system is designed as an integrated system, with all data flowing to a single database and presented through a single user interface. The integration can be achieved

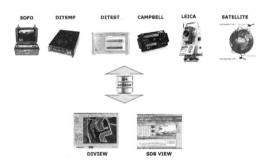

Figure 1. Integration of sensing technologies into a single database and user interface.

at several levels. The data management system must interface to all types of dataloggers, Figure 1.

To illustrate those concepts, we will provide examples where local and distributed optical fiber sensors are used to permanently monitor structures including bridges, buildings and pipelines.

REFERENCES

Del Grosso, A. & Inaudi, D. (2004) *European perspective on monitoring-based maintenance*, IABMAS International Association for Bridge Maintenance and Safety, Kyoto, Japan.

Glisic et al. (2003) *Health monitoring of a pipeline based on distributed strain and temperature measurements.*

Inaudi, D. (2009) *Integrated structural health monitoring systems for buildings*, Stanford.

Inaudi, D. & Church, J. (2011) *Paradigm shifts in monitoring levees and earthen dams: distributed fiber optic monitoring systems*. In: Proceedings, 31st USSD Annual Meeting & Conference, San Diego, California.

Jordan, A. & Papilloud E. (2015) *Penstock structural health monitoring*, Hydro 2015, Session 19 Gates and penstock, Bordeaux, France.

Structural and climate performance indicators in service life prediction of concrete bridges in multi-hazard environment

M. Kušter Marić & A. Mandić Ivanković
Faculty of Civil Engineering, University of Zagreb, Zagreb, Croatia

J. Ožbolt
Institute for Construction Materials, University of Stuttgart, Stuttgart, Germany

ABSTRACT

Adriatic coast is characterized by multi-hazard environment: high salinity of the Adriatic Sea, strong winds, high seismicity and extreme traffic load during summer tourist season. Case studies of Adriatic large span reinforced concrete arch bridges confirm chloride induced corrosion as the main degradation mechanism. In order to provide better bridge assessment and optimal maintenance planning, the coupled 3D chemo-hygro-thermo mechanical (CHTM) model was developed in order to enable a realistic simulation of reinforcement corrosion processes, taking into account influence and interaction of numerous material, structural, environmental and climate performance indicators (Fig. 1).

Climate performance indicators are of great importance to precisely determine environmental loads on structures (Table 1). Structural health monitoring, non-destructive testing and visual inspections are essential for model verification through comparison of numerical results and data obtained on existing bridges providing a base for more accurate prediction of the remaining service life of structures.

Table 1. Performance indicators considered in bridge service life prediction.

Material PIs	Source	Structural PIs	Source
Concrete		Structure geometry	●■
Tensile strength	●□	Long. reinforcement	●■
Compressive strength	●■□	Shear reinforcement	●■
Fracture energy	□	Confining reinforcement	●■
Young modulus	●■□	Concrete cover	●■
Poisson's ratio	●■□	Defects in concrete	■
Weight density	●	Crack width	■
Max aggregate size	●■□	Crack depth	■
Aggregate-Cement ratio	●	Deformation	■
Water-Cement ratio	●	Construction loads	●■◊
Water diffusivity	□◊	Wind load	●◊♦
Creep factor	□◊	Seismic load	●◊♦
Shrinkage factor	□◊	Traffic load	●■◊
Chloride diffusivity	□◊	Strains	●■◊
Coef. of chloride adsorption velocity	□◊	Environmental load	
Porosity	●□◊	Water in concrete	■□◊♦
Permeability	■□◊	Oxygen in concrete	◊♦
Electrical resistivity	■	Chloride in concrete	■□◊♦
Heat conductivity	□◊	Temp. in concrete	■◊♦
Heat capacity	□◊	Corrosion rate	■◊
Reinforcement		El. potential	■◊
Area	●■	Air temperature	♦
Young modulus	●□	Precipitation	♦
Poisson's ratio	●□	Wind rose	♦
Ultimate stress	●□	Wind speed	♦
Yield stress	●□	Relative humidity	♦
Hardening	●□	Tides	♦
		Freeze/thaw cycle	♦◊
		Wetting/drying cycle	♦◊

● Data obtained during design and construction
■ Data obtained from NDT and structure monitoring
□ Data obtained from experiments and destructive testing
◊ Data obtained from calculation or assumption
♦ Data obtained from environment and climate monitoring

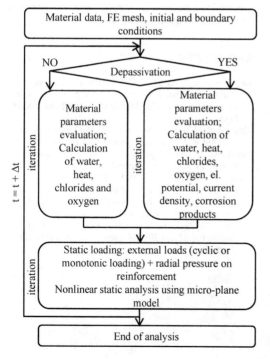

Figure 1. Alghoritam of the 3D CHTM model.

SS2: Climate adaptation engineering
Organizers: E. Bastidas-Arteaga, M.G. Stewart & Y. Li

Life-Cycle Analysis and Assessment in Civil Engineering: Towards an Integrated Vision – Caspeele, Taerwe & Frangopol (Eds)
© 2019 Taylor & Francis Group, London, ISBN 978-1-138-62633-1

Balancing payoff and regret: A bi-objective formulation for the optimal adaptation of riverine bridges under climate change

A. Mondoro & D.M. Frangopol
Department of Civil and Environmental Engineering, Engineering Research Center for Advanced Technology for Large Structural Systems (ATLSS Center), Lehigh University, Bethlehem, PA, USA

ABSTRACT

Adapting riverine bridges to a changing climate is essential for maintaining the functionality of civil infrastructure. Riverine bridges are vulnerable to damage due to flooding, including damage to decks, piers, and/or foundations. In order to manage vulnerable bridges, an accurate estimate of future flooding is required. However, the information regarding future events is not currently available; but rather, a set of flooding scenarios describes the potential climate changes.

Probabilities cannot be assigned to the scenarios in the set since the likelihood of climate change scenarios is unknown. This, along with the deep uncertainty in future economic scenario, complicated the decision making process. When decisions are made, the governing mentality is to maximize expected utility. However, in the face of deep uncertainty, there is also a desire to not choose a suboptimal solution; decision makers do not want to regret having made the wrong decision.

This paper discusses the application of Regret, i.e. a metric that corresponds to the feeling of loss and the opportunity lost by having made the wrong decision, in climate change adaptation. Regret is defined in terms of the economic regret associated with not choosing the best alternative, similar to the concept of opportunity loss. However, Regret may also include psychological regret, i.e. the desire for the decision makers not to perform poorly for their boss, their peers, and their community. In this paper, Regret is defined as the economic regret associated with a strategy.

Minimizing Regret is presented as an alternative to the traditional objective of maximizing utility (typically quantified through payoff). In order to account for the complexities of decision making under deep uncertainty, the adaptation optimization problem is posed in bi-objective form: maximize payoff and minimize Regret. The prosed bi-objective formulation results in a Pareto set of optimal solutions. The set represents solutions where one objective cannot be improved without negatively impacting the performance of the other. Once the Pareto set is determined, the decision maker, or group of decision makers, may then choose their preference after having preformed a diligent and unbiased search for information. The proposed framework is applied to an illustrative example for a typical riverine bridge located in the Northwest region of the United States.

ACKNOWLEDGMENTS

The support by grants from (a) the National Science Foundation (NSF) Award CMMI-1537926, (b) Commonwealth of Pennsylvania, Department of Community and Economic Development, through the Pennsylvania Infrastructure Technology Alliance (PITA), the U.S. Federal Highway Administration (FHWA) Cooperative Agreement Award DTFH61-07-H-00040, and (c) the U.S. Office of Naval Research (ONR) Awards N00014-08-1-0188, N00014-12-1-0023, and N00014-16-1-2299 is gratefully acknowledged. Opinions presented in this paper are those of the authors and do not necessarily reflect the views of the sponsoring organizations.

REFERENCES

Bell, D.E. (1982) Regret in decision making under uncertainty. *Operations Research*, 30(5), 961–981.
Espinet X., Schweikert A. & Chinowsky, P. (2017) Robust prioritization framework for transport infrastructure adaptation investments under uncertainty of climate change. *ASCE-ASME Journal of Risk and Uncertainty in Engineering Systems, Part A: Civil Engineering*, 3(1), E4015001.
Hallegate S. (2009) Strategies to adapt to an uncertain climate change. *Global Environmental Change*, 19(2), 240–247.
Hirabayashi Y., Mahendran R., Koirala S., Konoshima L., Yamazaki D., Watanabe S., Kim H. & Kanae S. (2013) Global flood risk under climate change. *Nature Climate Change*, 3(9), 816–821.
IPCC (2014) Climate change 2014: Synthesis report. *Contribution of Working Groups I, II and III to the Fifth Assessment Report of the Intergovernmental Panel on Climate Change* [Core Writing Team, R.K. Pachauri and L.A. Meyer (eds.)]; IPCC, Geneva, Switzerland.
Landman, J. (1993) *Regret: The persistence of the possible*. Oxford University Press.
Loomes, G. & Sugden, R. (1982) Regret theory: An alternative theory of rational choice under uncertainty. *The Economic Journal*, 92(368), 805–824.
Mondoro A., Frangopol D.M. & Liu L. (2017) Bridge adaptation and management under climate change uncertainties: A review. *Natural Hazards Review* (in press).

Modeling the climate change effects on storm surge with metamodels

A. Contento, H. Xu & P. Gardoni
University of Illinois at Urbana-Champaign, Champaign, USA

S. Guerrier
Pennsylvania State University, University Park, USA

ABSTRACT

Storm surge is an abnormal rise of the sea level above the astronomical tide consequent to the wind circulation around the eye of a hurricane. Several stretches of the U.S. East and Gulf coasts are vulnerable to this phenomenon that may cause a significant amount of damage. Moreover, the effects of climate change on hurricanes may lead to an increase in storm surge intensity and/or occurrences and consequently in the related damages. While hurricanes and the damage they directly generate have been the focus of many studies, storm surge that is responsible for a significant part of the total damage due to hurricanes has been far less studied (Lin et al. 2012). One of the reasons is that the accurate modeling of storm surge is computationally expensive and consequently it is difficult to get a high number of simulations to use for probabilistic studies. On the contrary, especially for analyses on the influence of climate change on the intensity of the storm surge and for the comparisons between possible climate change scenarios, multiple analyses are needed. Metamodels are a viable alternative to perform probabilistic analyses on storm surge (e.g., Jia & Taflanidis 2013). Usually, metamodels are trained using results coming from high-fidelity physics-based simulations. The accuracy of the predictions often depends on the number storm surge scenarios and locations used to train the models.

The metamodel for storm surge presented here is obtained combining a logistic regression model with a random field based on the Improved Latent Space Approach (ILSA) proposed in Xu & Gardoni (2018). In general, the latent space approach considers additional regressors as latent spaces of a random field and models the spatial correlation as a function of the differences between regressors at two locations. Differently from previous formulations (e.g., Risser & Calder 2015), Xu & Gardoni (2018) proposes models of variance and correlation scaling parameters that are functions of the regressors. The advantages of the presented model are that it can use data coming from historical records in addition to those obtained with high-fidelity simulations and it can give storm surge predictions in locations different from those of the data used for the training of the model. Moreover, the proposed model can be calibrated with a relatively small amount of data. In this way, having a limited number of simulations for each scenario of climate change would allow calibrating the model that could be used to perform comparisons between different climate change scenarios.

The model provides predictions of storm surge in a specific location in terms of the probability of the location of being wet (flooded) and an estimated value of water depth, as shown in Figure 1.

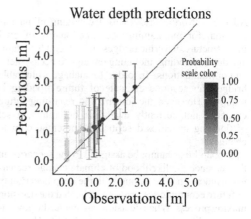

Figure 1. Prediction of water depth.

REFERENCES

Jia, G. & Taflanidis, A.A. (2013) Kriging metamodeling for approximation of high-dimensional wave and surge responses in real-time storm/hurricane risk assessment. *Computer Methods in Applied Mechanics and Engineering.* 261–262, 24–38.

Risser, M.D. & Calder, C.A. (2015) Regression-based covariance functions for nonstationary spatial modelling. *Environmetrics*, 26(4), 284–297.

Xu, H. & Gardoni, P. 2018. Improved latent space approach for modelling non-stationary spatial–temporal random fields. *Spatial Statistics*, 23, 160–181.

Impact of climate change on optimal wood pole asset management

A.M. Salman & Y. Li
Department of Civil Engineering, Case Western Reserve University, Cleveland, OH, USA

E. Bastidas-Arteaga
UBL, GeM, Institute for Research in Civil and Mechanical Engineering/
Sea and Littoral Research Institute, CNRS UMR/FR, Université de Nantes, Nantes, France

ABSTRACT

Overhead electric power distribution systems are supported by single-pole structures that are mostly wood poles. With over a hundred million wood poles in the US and millions more in other countries, their worth is in the billions. Wood poles are preferred over other materials because they are relatively cheaper to purchase, lighter and easier to transport, are easy to climb and are non-conductive, which makes them safer for utility workers. However, wood poles are susceptible to decay over time, which can significantly reduce their strength. Thousands of wood poles are condemned every year due to decay. The rate of decay depends on climatic conditions particularly relative humidity and temperature. As such, the future rate of wood pole decay is expected to change due to the impact of climate change. Considering the vast number of wood poles in use all over the world, it is imperative to study the potential impact of climate change on decay rate and come up with an optimum adaptation strategy. This paper presents a method for optimal wood pole asset management considering the impact of climate change on decay rate. Two IPCC emission scenarios are considered: RCP 4.5 and RCP 8.5.

Renewal theory is used to find the optimal pole replacement/reinforcement age in a network. The objective is to determine the optimum replacement (or reinforcement) age of the poles that will minimize the total maintenance cost (preventive and corrective) over a period of time. The optimization results for periodically chemically-treated poles are shown in Figure 1. The cost ratio on the vertical axis is the ratio of the cost of preventive maintenance to the additional cost incurred if failure were to occur. If the costs of replacement before and after failure are known (preventive and corrective maintenance costs), then the optimal replacement/reinforcement time can be read

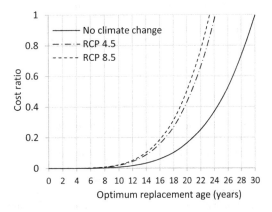

Figure 1. Optimization results for treated poles.

from Figure 1. For example, the cost of replacing a failed pole is estimated to be around $10,000 compared to the cost of reinforcement using fiber reinforced polymer (FRP) which is about $1,500. This means the cost ratio is about 0.18. Based on this ratio, the optimum reinforcement age without climate change is about 20 years which is much lower than the estimated mean service life of the pole of about 36 years. This implies that since the cost of reinforcement is much less than the additional cost if failure were to occur, the best strategy is to reinforce the pole much earlier in its life. For RCP 4.5 and RCP 8.5, the optimum reinforcement time for both scenarios is about 16 years.

Note that the results presented above are based on specific climate change scenarios and wood decay model which were selected to demonstrate the proposed method. Using different environmental and climate conditions could lead to different conclusions.

Life-Cycle Analysis and Assessment in Civil Engineering: Towards an Integrated Vision – Caspeele, Taerwe & Frangopol (Eds)
© 2019 Taylor & Francis Group, London, ISBN 978-1-138-62633-1

Climate change impact on safety and performance of existing and future bridges

A. Nasr, O. Larsson Ivanov & I. Björnsson
Division of Structural Engineering, Lund University, Sweden

J. Johansson
Division of Risk Management and Societal Safety, Lund University, Sweden

D. Honfi
Rise Research Institutes of Sweden, Sweden

E. Kjellström
Rossby Centre, Swedish Meteorological and Hydrological Institute, Sweden

ABSTRACT

Recent decades have seen an increased attention towards the threat of climate change to our built environment and not least our infrastructure. Accounting for the different ways in which potential climate change scenarios can affect our infrastructure is paramount in determining appropriate adaptation and risk management strategies. Accomplishing this aim will help safeguard the safety and performance of the existing as well as future structures while improving the resilience of the infrastructure these structures support. Overall, this will lead to a better preparedness and awareness, for all stakeholders involved, of some of the critical issues which require addressing for our society to be better equipped in taking on the challenges associated with climate change.

This paper presents the very initial findings of a research project that is concerned with establishing an improved management of the risks to our infrastructure, especially bridges, in light of a changing climate. The focus of this project is on existing bridge structures; however, valuable conclusions for the design of new bridges are drawn as well. This project aims to survey the potential climate change risks on bridges, prioritize them, assess and quantify the prioritized critical risks in a more elaborate manner, critically scrutinize the current design; inspection; and maintenance practices deriving insight from the attained results, and finally develop a framework for the

cost-effective adaptation of bridges and demonstrate its applicability to case studies.

Here, a preliminary survey of the climate change relevant risks on bridges is presented. Firstly, the relevant projected climate changes and the different emission scenarios are presented. Afterwards, an attempt to link these climate changes to some of their potential consequences on bridges is made. This is done by reviewing the published literature on the topic and looking at documented cases of bridge failure. In total, 19 different risks are highlighted in this paper, beginning with 4 risks that are presented in more detail; the accelerated degradation of structural and non-structural elements, higher flood levels and more frequent flooding, damage to pavements and railways, and the risk of higher scour rates. However, no inference about the criticality, i.e. prioritization, of these risks should be derived from the order and/or level of detail in which these risks are presented. The interplay between the different risks and how the occurrence of one risk may influence other risks is also briefly discussed. In addition, a short discussion illustrating some of the most important factors that should be considered for risk prioritization is given. It should be stressed that, it is not the intention of this paper to give an all-encompassing list of all the possible risks caused by climate change rather, the intention is to exhibit some of the most pertinent ones. More elaborate results are expected as the project progresses.

*Life-Cycle Analysis and Assessment in Civil Engineering: Towards an
Integrated Vision – Caspeele, Taerwe & Frangopol (Eds)
© 2019 Taylor & Francis Group, London, ISBN 978-1-138-62633-1*

A tool to evaluate effectiveness of climate change adaptation measures for houses subjected to coastal flood risks

A. Creach
Laboratoire ENeC, UMR, CNRS, Sorbonne University, France

M. Gonzva
Laboratoire Lab'Urba, Département génie urbain, SYSTRA & Université Paris-Est Marne-la-Vallée, France

E. Bastidas-Arteaga
Institut de Recherche en Génie Civil et Mécanique (GeM), UMR, CNRS, Université de Nantes, France

S. Pardo
Institut d'Economie et de Management de Nantes, Université de Nantes, France

D. Mercier
Laboratoire ENeC, UMR, CNRS, Sorbonne University, France

ABSTRACT

The V.I.E. index is a tool designed to assess the vulnerability of residential buildings for occupants facing flash floods. Its aims it to locate buildings that could be a trap for people in case of flood and where they can be drown.

In addition to the diagnosis, the index allows to compare the efficiency of four different adaptation strategies to reduce vulnerability and protect people: (i) protection, (ii) adaptation of houses, (iii) prevention and warning, and (iv) resettlement. This comparison is provided through a cost-efficiency analysis.

The whole method is designed to be a tool to help the decision process in vulnerability reduction strategies and human life protection against floods.

The goal of this talk is to discuss some points of the methodology used for the cost-efficiency analysis in order to enhance the method:

(i) The first point attempts to address different occurrences of flood, as the initial tool is only based on the most important event known. The goal is to compare the efficiency of the different strategies for several intensities of floods and to see at which intensity a particular strategy offers benefits to protect human life compared to the investment cost.

(ii) The second point focuses on the assessment of the mix of strategies that offers the best efficiency for a minimal investment, as the initial method only compare the interest of the strategy the one after the other. However, the best way to protect human life is a bit of all strategies. Which recipe appears to be the best?

(iii) The third point discusses the perspectives of a such vulnerability index to other infrastructures. The case of railway transport systems as infrastructures particularly vulnerable to floods is discussed to identify the opportunity of using an index approach similar to the V.I.E. index.

These results will be useful to propose a more robust tool that could provide recommendations that are more realistic for decisions makers.

Life-Cycle Analysis and Assessment in Civil Engineering: Towards an Integrated Vision – Caspeele, Taerwe & Frangopol (Eds)
© 2019 Taylor & Francis Group, London, ISBN 978-1-138-62633-1

Evaluating the effect of climate change on thermal actions on structures

P. Croce, P. Formichi & F. Landi
Department of Civil and Industrial Engineering, University of Pisa, Pisa, Italy

F. Marsili
Federal Waterways Engineering and Research Institute, Karlsruhe, Germany

1 INTRODUCTION

The evidence of climate change is widely accepted in the scientific community, and since the 1950s, many of the observed changes are unprecedented over decades to centuries (IPCC, 2013). Climate change potentially affects all regions of the world by alteration of natural processes, modification of precipitation patterns, melting of glaciers, rise of sea levels, etc. Whatever the warming scenarios and the level of success of mitigation policies, the impact of climate change increases because of the delayed impacts of past and current greenhouse gas emissions.

Current values of climatic loads given in the Eurocodes are commonly based on data series, usually spanning on time intervals of 40–50 years.

Although these data series are suitable enough for estimating characteristic loads (50 years return period), they are not adequately extended over the time to appreciate the potentially relevant effects of the climate change. For this reason, to assess possible changes in extremes it is necessary to rely on climate projections, resulting from appropriate Global or Regional Climate Models.

In the paper, focusing the attention on thermal actions on structures, an ensemble of 6 different Regional Climate Models (RCMs) outputs, derived in two different greenhouse gas emission scenarios, is analyzed to estimate future trends in extreme maximum and minimum temperatures. To take into account the uncertainties related to the internal climate variability, a new weather generator, proposed by (Croce et al., 2017), has been used.

2 CONCLUSIONS

In order to assess the impact of climate change on thermal actions on structures, a suitable procedure for the estimation of future trends in extreme temperatures starting from the analysis of climate model outputs has been presented.

The proposed method takes into account the three main sources of uncertainty affecting climate projections (emissions scenario, global climate model, internal variability). In particular, an ensemble of six different climate models run according different emission scenarios has been considered in the analysis and a new weather generator developed by the authors has been implemented to assess internal variability.

Each climate data series of the ensemble has been considered as an equally probable representation of future climate and then an extreme values analysis has been carried out for moving time windows forty years long to assess the trend in characteristic values of daily maximum and minimum temperature.

In this way, delta factor of changes for the characteristic values of maximum and minimum temperatures have been derived and their uncertainty range have been evaluated from the ensemble of factor of changes obtained according to a medium emission scenario (RCP4.5) and to most severe emission scenario (RCP8.5).

The results, presented in terms of confidence maps for the investigated region, confirm that the technique is very promising and can provide guidance for potential amendments of the current version of temperature maps present in technical standards.

In the future works, the proposed methodology will be extended also to different climate variables, obtaining information regarding other climatic actions potentially affected by the impact of climate change.

REFERENCES

Croce P., Landi F., Formichi P. & Castelluccio R. (2017) Use of weather generators to asses impact of climate change: thermal actions on structures. In: *Proceedings of the Fifth International Conference Advances in Civil, Structural and Mechanical Engineering – CSM 2017*, Zurich.

IPCC (2013) Climate change 2013 – The physical Science Basis.

SS3: Quality control procedures on the life-cycle management
of existing bridges
Organizers: J.C. Matos & J.R. Casas

Life-Cycle Analysis and Assessment in Civil Engineering: Towards an Integrated Vision – Caspeele, Taerwe & Frangopol (Eds)
© 2019 Taylor & Francis Group, London, ISBN 978-1-138-62633-1

COST Action TU1406 and main results on bridge lifecycle management

J.C. Matos
University of Minho, Guimarães, Portugal

J.R. Casas
UPCatalunya – BarcelonaTech, Barcelona, Spain

ABSTRACT

Life-cycle analyses are used in condition assessment of new and existing bridges, as well as for evaluation of maintenance strategies. Management systems, capturing different degradation processes, are very often used in relation to such life-cycle analyses methods in order to describe the actual and future condition. During the implementation of asset management strategies, maintenance actions are required to keep assets at desired performance levels. In case of roadway bridges, performance indicators are established for components, which alongside with the definition of standardized performance goals, allow to assess the accomplishment of quality control plans. In Europe there is a large disparity regarding the way performance indicators are quantified and goals specified. Therefore, a discussion at a European networking level, seeking to achieve a standardized approach in this subject, will bring significant benefits. COST Action TU1406 aims to achieve such goal by bringing together both research and practicing communities in order to establish a European guideline in this matter.

Each construction, during its life cycle, will face deterioration depending on several factors such as environmental condition, natural aging, material quality, execution of works and planned maintenance. Therefore, performance indicators (PIs) for the present and future structural conditions on deterministic and probabilistic level have to be defined and determined.

Structures like bridges are necessary for a functioning transport infrastructure network. Bridge performance goals (PGs) can be set in order to ensure bridge performance is in line with network level PGs. When defining bridge PIs, some difficulties may present themselves. First, the timescale for which network PGs are set is typically much shorter than the estimated service life of a bridge. Therefore bridge PGs should not only enable meeting the short term PGs, but also facilitate life cycle optimization. Furthermore, where

bridge management is traditionally focused on evaluating the condition of the bridge, the desired condition now needs to be expressed or translated into goals reflecting network performance.

As the focus on an efficient delivery of network performance increases, so does the interest in the relations between societal goals, performance indicators for both the road network and bridges or bridge elements. The implementation of asset management should increase the integration of network and bridge performance requirements. Network or even societal goals tend to be rather broad in their definition. Furthermore, there is often no exclusive relationship between performance indicators set at a lower level and goals at a higher level.

The objective of COST Action TU1406 is to investigate the way bridge PIs and key performance indicators (KPIs) are collected and quantified, how PGs are specified across Europe, and finally to produce guideline documents linking collection and quantification of PIs, KPIs, PGs, standards, and practices to decision making processes.

A methodology to collect PIs used across Europe is presented. This should be the basis of the future proposal for the definition of PIs to be adopted and used across Europe. The procedure is based on a deep analysis of existing bridge management policies and available documents for inspection and evaluation existing in European countries.

This paper shows an overview of PGs at different levels, from high-level strategic decisions to low level, system-specific requirements developed within the Action. The paper also explains how other performance aspects, like traffic safety, availability, economy, environmental and societal impacts could be quantified and used for the multi-objective bridge PGs assessment.

The aim is to establish good practices on the definition of quality control (QC) plans for roadway bridges.

Performance based design and assessment—levels of indicators

A. Strauss, L. Mold & K. Bergmeister
Institute of Structural Engineering, University of Natural Resources and Life Sciences, Vienna, Austria

A. Mandic
Department of Structural Engineering, Chair for Bridges, Faculty of Civil Engineering, University of Zagreb, Croatia

J.C. Matos
University of Minho, Guimarães, Portugal

J.R. Casas
UPC-Barcelona Tech, Barcelona, Spain

ABSTRACT

Road bridges represent functionally important and structurally complex components within the road network. In order to fulfil both the technical and social requirements of a road bridge, regular inspections are carried out over the entire service life. In addition, the cost of inspections and the maintenance of a bridge should be kept to a reasonable level. To balance all of these issues, quality control plans are created to prevent the risk of premature wear while minimizing costs and ensuring quality. Within Europe, many countries have developed their own management system. These systems generally differ in the type and the method of maintenance management. Therefore the quality of road bridges varies widely in Europe. In order to meet the European social and economic requirements for the maintenance of road bridges, the aim of the COST Action TU1406 (Matos, 2016) is to develop a standardized approach for these maintenance processes.

The concept of performance indicators (PI) is used to characterize the condition of a structure using quantitative and/or qualitative terms. Performance indicators are collected through inspections, non-destructive and destructive testing, monitoring and numerical and experimental modelling. Subsequently, these performance indicators are compared with defined performance goals (PGs) to determine the condition and the damage processes. Currently, there are major differences in Europe in the approach of identifying these performance indicators and defining their goals. Therefore, the aim of the COST Action TU1406 is to develop, with the help of scientists and practitioners, a standardized European guideline. Through Europewide surveys and homogenization processes, the key performance indicators for the condition assessment of bridges were defined. During its lifecycle, every building experiences abrasion-, aging- and damage processes, which are influenced by environmental conditions, material properties and traffic impacts as well as by the conditions and qualities of implementation. The performance indicators describe, in a holistic sense, the main influencing factors to guarantee the carrying capacity, the serviceability, the durability and thus the functionality over the planned technical lifetime. Therefore they are important variables for design, conservation strategies and maintenance management, as well as for the administration and the organization of the life cycle (see Fig. 1).

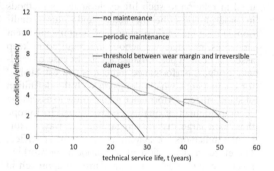

Figure 1. Conservation strategies for the life cycle of engineering structures.

REFERENCE

Matos, J.C. (2016) An overview of COST Action TU1406, Quality Specifications for Roadway Bridges (BridgeSpec). In: *Proceedings of the 5th International Symposium on Lifecycle Civil Engineering*, Delft, The Netherlands.

Life-Cycle Analysis and Assessment in Civil Engineering: Towards an Integrated Vision – Caspeele, Taerwe & Frangopol (Eds)
© 2019 Taylor & Francis Group, London, ISBN 978-1-138-62633-1

The indicator readiness level for the classification of research performance indicators for road bridges

M.P. Limongelli
Politecnico di Milano, Italy

A. Orcesi
IFSTTAR, Materials and Structures Department, University Paris-Est, France

A. Vidovic
University of Natural Resources and Life Sciences, Vienna, Austria

ABSTRACT

Road bridges are crucial elements in safety and functionality of the whole traffic infrastructure. Monitoring of performance and condition through an effective inspection and assessment regime is therefore an important part of an overall asset management strategy. Performance Indicators (PIs) are therein found as valuable quantities that can characterize the present and future structural conditions while accounting the Performance Goals (PGs) specified by the codes, owners and operators. PIs are based on the available knowledge from the inspection procedures that may be clustered into four main groups: visual inspections, destructive testing, non-destructive testing and structural health monitoring techniques. This knowledge is represented by so-called Performance Parameters (PPs) that are considered as a corresponding input for the estimation of Performance Indicators.

In Europe, large disparity has been recognized, regarding the way these Indicators are quantified and the goals specified. For that reason, COST Action TU 1406 aims at providing quality specifications for roadway bridges standardized at the European level. Either qualitative or quantitative PIs have been collected, PGs established and Quality Control (QC) plans defined. An additional goal within the scope of the Action is to investigate the practical implementation of innovative condition assessment, i.e. of the Research Performance Indicators that are not yet utilized for maintenance purposed, but are being investigated by research groups.

Proposal of a metric scale that is able to rate maturity level of the Indicators is thence presented. Indicator Readiness Level (IRL) framework ranks the maturity

and correspondingly checks the eligibility of the PIs for QC and related decision-making, as well as it underpins the additional needs that are required to bring the indicators to the level of full applicability within the QC. The IRL scales are to be included into the research database that not only verifies the indicators potential use in decision-making, but also has a potential to improve existing performance assessment methods within bridge management systems in Europe. This paper particularly focuses on the definition of the PIs and the IRL scale framework, and it presents the first results of the application of the maturity levels to both operational and research-based Performance Indicators related to corrosion mechanism.

REFERENCES

COST TU1406 Action (2016) *Quality Specifications for Roadway Bridges, Standardization at a European Level: eBook for the 3rd Workshop Meeting, Editors: Irina Stipanovic Oslakovic, Giel Klanker, Jose Matos, Joan Casas, Rade Hajdin, 20–21 October 2016* University of Twente, Faculty of Engineering Technology, Delft, The Netherlands

COST TU1406 (2016) WG1 Technical Report, Performance Indicators for Roadway Bridges of COST Action TU1406.

Limongelli, M. & Orcesi, A (2017) A proposal for classification of key performance indicators for road bridges. In: *39th IABSE Symposium-Engineering the Future, 21–23 September 2017, Vancouver*

Effects of multivariate data reduction of condition assessment of bridge networks on the Value of Information

L. Quirk
School of Engineering, University College Cork, Ireland

C.M. Hanley
Malachy Walsh and Partners, Cork, Ireland

J.C. Matos
ISISE, Civil Engineering Department, Minho University, Guimaraes, Portugal

V. Pakrashi
Dynamical Systems and Risk Laboratory, School of Mechanical and Materials Engineering, University College Dublin, Ireland
SFI Centre for Marine and Renewable Energy Ireland, University College Dublin, Ireland

ABSTRACT

For infrastructure stakeholders, the use of bridge management systems to manage existing bridge stocks has led to the accumulation of vast databases of condition ratings, primarily gathered through routine visual inspection protocols. Visual inspection is the principal condition assessment method; whereby defined components in the overall bridge structure are inspected and assigned a condition rating from a prescribed scale indicative of the damage state, based on the judgements of a trained inspector. Component level condition rating data can be manipulated in multiple ways to formulate a bridge condition index. The bridge condition index serves as a basis for decision making in relation to implementing maintenance and structural intervention actions in bridge network lifecycle management. The value of visual inspection is contingent on its ability to guide towards optimal decision making in bridge network management and thus is dependent on the accuracy of the bridge condition index used. The worst-conditioned component approach is used to assign an overall condition rating to the structure approximated to the rating of the component in the worst condition, which as a result significantly reduces the overall condition state of a bridge to that of a single component. Multivariate data reduction has been shown to use component-level condition rating data on a large scale to help define information-based, structure-specific weighting factors. This approach estimates the condition rating of the whole structure by combining condition ratings of important bridge components weighted by their significance to the structural integrity of the bridge. This paper considers the effects of multivariate data reduction of condition rating databases on the value of visual inspection. The value of information principle of decision theory is used to establish the difference in cost of a bridge management system decision process

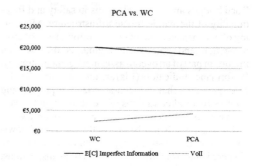

Figure 1. Results of integrating VoI analysis with PCA.

Table 1. Results of VoI analysis (in €).

	E[C] NI	E[C] PI	E[C] II	VoPI	VoII
Worst Case	−22518	−12034	−20097	10483	2420
PCA	−22559	−8728	−18370	13830	4188

implementing both the worst-case condition rating versus the multivariate data reduction technique.

REFERENCES

Hanley C., Matos J., Kelliher, D. & Pakrashi, V. (2017) Integrating multivariate techniques in bridge management systems. *Journal of Structural Integrity and Maintenance*, 2(3), 143–151.

Quirk, L., Matos, J., Murphy, J. & Pakrashi, V. (2017) Visual inspection and bridge management. *Journal of Structure and Infrastructure Engineering*, DOI: 10.1080/15732479.2017.1352000.

Znidaric, A., Pakrashi, V., O'Connor, A. & O'Brien, E. (2011) A review of road structure data in six European countries. *Proceedings of the ICE, Journal of Urban Design and Planning*, 164(4), 225–232.

Life-Cycle Analysis and Assessment in Civil Engineering: Towards an Integrated Vision – Caspeele, Taerwe & Frangopol (Eds)
© 2019 Taylor & Francis Group, London, ISBN 978-1-138-62633-1

Regular bridge inspection data improvement using non-destructive testing

M. Kušar
Faculty of Civil and Geodetic Engineering, University of Ljubljana, Ljubljana, Slovenia

N. Galvão
Civil Engineering Department, Minho University, Guimarães, Portugal

S. Sein
Department of Civil Engineering and Architecture, Tallinn University of Technology, Tallinn, Estonia

ABSTRACT

Comparison of different bridge inspections shows that they vary greatly. Their implementation is dependent on the amount and type of data that has to be obtained for each individual bridge. Detailed inspection provides reliable data but it is costly, time consuming and not suitable for large scale use. Regular bridge inspection, on the other hand, is cost-effective, quick and, in general, the most suitable for conducting large numbers of annual inspections. However, data gathered is in most cases based on visual observation only, and as such, it is less reliable. In order to improve the quality of acquired data, certain non-destructive tests (NDTs) could be integrated into regular bridge inspection practice, while maintaining the above-mentioned advantages of visual inspection.

Evaluation of NDT methods commonly used for bridge inspection based on several relevant criteria is therefore performed. Their relative importance is determined by experts from COST Action TU1406 by using Analytical Hierarchy Process. Results' reliability and test duration are considered to be the most important criteria, followed by results' interpretation complexity, cost and usability. Test standardization is considered to be the least important criterion.

For each of the NDT methods under consideration, values for each criteria are assigned and the utility function ($U_i = \{1 \dots 3\}$) is defined. The NDTs with the same applicability level are compared, and their possibilities for use in regular bridge inspection are assessed. The results show (Table 1) that all NDTs measuring material properties have high utility rating. Despite some of the methods being semi-destructive (phenolphthalein, probe penetration and pull off test) and rebound hammer having poor reliability, these tests are fast, inexpensive and undemanding

Table 1. NDT final classification.

Application	U_i	NDT
Material properties	2,71	Cover measurement
	2,55	Phenolphthalein test
	2,43	Probe penetration test
	2,42	Pull-off test
	2,22	Rebound hammer
Damage and defects	2,22	Impact echo
	1,86	Thermography
	1,83	Acoustic emission
	1,80	Ground penetrating radar
	1,63	Ultrasonic pulse echo
Corrosion	1,89	Half-cell potential
	1,82	Galvanostatic pulse
	1,82	Electrical resistivity
	1,65	Linear polarization resistance

to perform, making them suitable for complementary use to visual observation during regular bridge inspections.

For damage and defects assessment, the NDTs considered are less suitable for use during regular bridge inspections. Their biggest weakness is high complexity of the results' interpretation, followed by the on-site test duration. As large number of bridges need to be inspected daily, time consumption is of utmost importance giving these tests lower utility. Impact echo investigation is rated higher than other methods measuring damage and defects, and is at least conditionally suitable for use during regular bridge inspections.

Based on literature review, all non-destructive methods dealing with corrosion detection and assessment exhibit similar characteristics. Their use in regular bridge inspection should be limited to bridges exhibiting high degree of corrosion only.

*Life-Cycle Analysis and Assessment in Civil Engineering: Towards an
Integrated Vision – Caspeele, Taerwe & Frangopol (Eds)
© 2019 Taylor & Francis Group, London, ISBN 978-1-138-62633-1*

First results from a benchmarking of Quality Control Frameworks

A. Kedar
KEDMOR Engineers Ltd., Ramat Gan, Israel

S. Sein
Tallinn University of Technology, Tallinn, Estonia

ABSTRACT

During the benchmarking process of COST Action TU1406 a set of common roadway bridges from different European countries have been identified, selected and compared with an objective to validate the outcomes. Since the benchmarking was based on previously reported work, the bridge typologies, performance indicators and performance goals were selected according to the results of former working groups of this COST action. The first task of the case studies working group was to select suitable bridges from different countries, with sufficient amount of information and data. For each country, three different bridge prototypes where considered: girder bridge, arch bridge and frame bridge. For each prototype of bridge the bridges were ranked in priority order based on different criteria such as compatibility with the prototype defined, type and quality of existing data etc. At second stage, for each prototype a single bridge was selected as representative case study in order to identify the gaps between theoretical and practical approach. For each bridge, relevant Performance Indicators (PI) and Key Performance indicators (KPI) were identified and overall quality control plan was implemented by following the framework set by other WG of TU1406 action. The obtained PI's and KPI's were compared with goals through performance values in element or bridge level. The basis for benchmarking data was received from the bridge owners and operators, who gave the input for the existing situation and from research field who presented the input for possible calculation methods. At future third stage, it is intended to expand the work for each country case studies and compare the results in order to identify the main existing dispersion among the different Quality Control plans.

REFERENCES

AASHTO (2010) AASHTO Bridge Element Inspection Guide Manual.
DIN 1076 (1999) Engineering structures in connection with roads: inspection and tests [Ingenieurbauwerke im Zuge von Straßen und Wegen: Überwachung und Prüfung].

DSPAG (2012) Report on alternative measure for Glattfelden bridge.
European Cooperation in Science and Technology – COST (2014) *Memorandum of Understanding for COST Action TU1406.*
Frangopol, D. M., Strauss, A. & Kim, S. (2008). Bridge reliability assessment based on monitoring, *Journal of Bridge Engineering*, 13(3), 258–270.
Ishizaka, A. & Labib, A. (2011) Review of the main developments in the analytic hierarchy process. *Expert systems with Applications*, 38(11), 14336–14345.
Kanton Z. (2015) *Detail Assessment of Glattfelden Bridge.*
Kedar, A. (2017) *Work Group 4 – Implementation in a Case Study.* In: *Overview for the COST Action TU1406 Meeting.* COST Action TU1406 Riga Workshop – unpublished.
Kifokeris, D., Xenidis, Y., Panetsos, P., Matos, J. & Bragança, L. (2017) Bridge quality appraisal methodology: Application in the Strimonas bridge case study. In: *COST Action TU1406 Riga Workshop* – unpublished.
Mateus, R. & Bragança, L. (2011) Sustainability assessment and rating of buildings: Developing the methodology SBToolPT–H, *Building and Environment*, 46(10), 1962–1971.
NaRil (2011) Directive for calculation of existing road bridges [Richtlinie zur Nachrechnung von Straßenbrücken imBestand].
Panetsos, P. (2017) Work Group 4 – Sub Group B1: Detailed description of the existing data of the "Strimonas River Bridge", Greece and the suggested PI and KPI to be used and their values. *Research Report for the COST Action TU1406 WG4 Meeting*, 12 May 2017, Institut Français des Sciences et Technologies des Transports, de l'Aménagement et des Réseaux (IFSTTAR) 2017, Marne-la-Vallée, France, 104 p.
Ryjáèek, P. & Petrik, M. (2017) Work Group 4 – Sub Group B2: Case study – road concrete arch bridge Nerestce Research Report for the COST Action TU1406 WG4 Meeting. COST Action TU1406 Riga Workshop – unpublished.
Schellenberg, K., Vogel, T., Chevre, M. & Alvarez, M. (2013) Assessment of Bridges on Swiss National Roads. *Structural Engineering International, 4.*
Strauss, A. & Ivanković, A.M. (eds.) (2016) *WG1 Technical Report: Performance Indicators for Roadway Bridges of Cost Action TU1406*. Boutik, Braga 38 p.
Strauss, A., Ivankovic, A.M., Matos, J.C. & Casas, J.R. (2016) WG1 Technical Report: Performance Indicators for Roadway Bridges of COST Action 1406.

Life-Cycle Analysis and Assessment in Civil Engineering: Towards an
Integrated Vision – Caspeele, Taerwe & Frangopol (Eds)
© 2019 Taylor & Francis Group, London, ISBN 978-1-138-62633-1

Standardizing the quality control of existing bridges

V. Pakrashi
Dynamical Systems and Risk Laboratory, School of Mechanical and Materials Engineering, Ireland
Centre for Marine and Renewable Energy Ireland (MaREI), University College Dublin, Dublin, Ireland

H. Wenzel
Vienna Consulting Engineers, Austria

ABSTRACT

Bridges remain a key asset for the infrastructure in EU and elsewhere in the world and their quality are significantly important for commercial and societal well-being of different countries and communities. While the need of quality control of existing bridges is acknowledged by all owners, managers or other stakeholders, methods of such control can vary significantly based on ownership or managerial preference of a region or a country. Under such circumstances, standardizing the quality control of existing bridges is a major challenge, although the benefits seem to be significant. EU COST Action TU1406, Working Group 5 investigates various facets of this challenge by bringing the experience of several EU countries and experts together, while working closely with other experts around the world. Working Group 5 integrates experiences and information from scientists, engineers and stakeholders of bridge stock of various countries and identified challenges and lessons learnt from previous and existing experiences around the topic. With infrastructure maintenance management evolving over time and becoming more data-driven, it is of timely and strategic importance to engage in a dialogue around this topic. This paper presents the context around standardization of existing road bridges in relation to COST TU1406 and highlights the strategies of quality control based on existing knowledge. Some of the key challenges are highlighted and the focus and evolution of WG5 are detailed. Initial experience and understanding based on interaction with members contributing to WG5 is also presented in this paper. Finally, future visions of the working group are presented. This work will be of relevance for engineers and policy makers who would be involved in or influenced by a standardized framework for controlling the quality of existing bridges despite the disparate methods and levels of information, along with cultures of assessment or evaluation that exist in different bridge stocks.

REFERENCES

Hanley, C., Matos, J. & Pakrashi, V. (2016) Principal component analysis as a comparative technique for bridge management systems. *3rd Workshop Meeting COST Action TU1406*, Delft.

Hanley, C., Frangopol, D.M., Kelliher, D. & Pakrashi, V. (2016) Effects of increasing design traffic load on performance and life-cycle cost of bridges. In: *8th International Conference onBridge Maintenance, Safety and Management*, Foz do Iguaçu, Brazil.

Hanley, C., Matos, J., Kelliher, D. & Pakrashi, V. (2017) Integrating multivariate techniques in bridge management systems. *Journal of Structural Integrity and Maintenance*, 2(3), 143–151.

Hanley, C., Frangopol, D.M., Kelliher, D. & Pakrashi, V. (2017) Variations in Reliability with Bridge Live Load Definitions. In: *Proceedings of the ICE, Bridge Engineering*, 2(3), 143–151.

Hanley, C. (2017) *Effects of Disparate Information Levels on Bridge Management and Safety*. PhD Thesis. University College Cork, Ireland.

O'Connor, A., Pakrashi, V. & Salta, M. (2012) *Assessment and Maintenance Planning for Infrastructure Networks*. Transportation Research Board Annual Meeting, 2012, Washington DC, USA.

O'Connor, A. & Pakrashi, V. (2011) Recent Advances in Reliability based Assessment of Highway Bridges. Proceedings of ICASP11. In: *11th International Conference on Applications of Statistics and Probability in Civil Engineering*, Zurich, pp. 363–368.

Pakrashi, V., Kelly, J. & Ghosh, B. (2011) Sustainable prioritisation of bridge rehabilitation comparing road user cost. *Transportation Research Board Annual Meeting*, 2011.

Veit-Egerer, R., Widmann, M., Furtner, P. & Lima, R. (2013) Integrated asset management tool for highway infrastructure. In: *IABSE Symposium Report*, Vol. 99, No. 32, pp. 68–75. International Association for Bridge and Structural Engineering, May 2013.

Weninger-Vycudil, A., Hanley, C., Deix, S., O'Connor, A. & Pakrashi, V. (2015) Cross-asset management for road infrastructure networks. In: *Proceedings of the Institution of Engineers – Transport*, 168(5), 442–456.

Znidaric, A., Pakrashi, V., O'Connor, A. & O'Brien, E. (2011) A review of road structure data in six European countries. In: *Proceedings of the ICE, Journal of Urban Design and Planning*, 164(4), 225–232.

The case study of Chile—how quality control could improve better lifecycle management of bridges

M.A. Valenzuela
Pontificia Universidad Católica de Valparaíso, Valparaíso, Chile

ABSTRACT

Chile has a great road heritage in bridges, so it is necessary for the Road Administration to have a Management System for the maintenance of bridges, being at the level of developed countries, considering high risk condition due to earthquake and scouring process.

Although there have been some regional initiatives in this area, in the vast majority of cases maintenance plans are considered based on professional experience, evaluating the performance of the bridges either through the Department of Bridges or Regional Road Directions. These plans mainly consider the maintenance of traditional bridges (made of concrete – steel beams, concrete slab, concrete pillars, pillars and direct or deep foundations).

For that reason, following the "European Quality specifications for roadway bridges, standardization at a European level (BridgeSpec)", the Chilean technical team has started developing a proposal of new performance indicators and quality control procedures for the Life-Cycle management of bridges, in order to improve the current management programs, considering the importance of carrying out a diagnosis of their bridges and to implement, at the national level, a suitable management maintenance system of road bridges.

This paper presents a state of the art of the performance indicators, goals and quality control procedures applied via several start-up inspections in Chile, highlighting the advantage of this implementation in the structure lifecycle . The results provide a verification of the use of this performance indicators and quality control procedure under the framework of the Chilean Maintenance Program that will be applied in the national code of maintenance: Manual de Carreteras (Highway Manual).

REFERENCES

Casas, J.R. & Campos, J. (2016) Quality specifications for highway bridges: standardization and homogenization at the European level (COST TU-1406). In: *Proceedings of IABSE Conference*. Guangzhou (China) Zürich.

Márquez, M., Valenzuela, Arias, G. & Sepulveda, C. (2018) Management systems for inspection and maintenance of Chilean road bridges. In: *IABMAS Conference 2018 Melbourne*, Australia.

Stipanovic, I. & Klanker, G. (2016) Performance goals for roadway bridges. In: Proceedings 8th International Conference on Bridge Maintenance, Safety and Management, Foz do Iguaçu (Brazil). Taylor and Francis, London.

Strauss, A., Vidovic, A., Zambon, I., Dengg, F., Tanasic, N. & Matos, J.C. (2016) Performance indicators for roadway bridges. In: *Proceedings 8th International Conference on Bridge Maintenance, Safety and Management, Foz do Iguaçu (Brazil)*. Taylor and Francis, London.

Valenzuela, M.A., Valenzuela, N. & Romo, R. (2017) Management system for natural risk disaster on infrastructure: regional approach (GRDR). In: *11th International Bridge and Structure Management Conference. TRB Proceedings Mesa*, EEUU, Arizona.

Valenzuela, M.A., Peña-Fritz, A., Contreras, C., Valenzuela, N., Pineda, F. & Romo, R. (2018) Management of risk disasters at local level: Proposal of hyper-heuristic approach on start-up in Valparaíso. In: *Proceedings ICVRAM ISUMA UNCERTAINTIES*, Florianópolis, SC, Brazil.

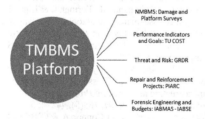

Figure 1. Diagram of the concepts included in the infrastructure management program.

SS4: Modeling time-dependent behavior and deterioration of concrete
Organizers: R. Wan-Wendner, M. Alnagger, G. Di Luzio & G. Cusatis

Life-Cycle Analysis and Assessment in Civil Engineering: Towards an Integrated Vision – Caspeele, Taerwe & Frangopol (Eds)
© 2019 Taylor & Francis Group, London, ISBN 978-1-138-62633-1

Size and shape effect in shrinkage based on chemo-mechanical simulations

L. Czernuschka, I. Boumakis, J. Vorel & R. Wan-Wendner
Christian Doppler Laboratory LiCRoFast, Institute of Structural Engineering,
University of Natural Resources and Life Sciences, Vienna, Austria

ABSTRACT

Due to safety demands and the growing requirement of sustainability, sound knowledge concerning the long-time performance of infrastructures becomes more relevant. The structure's long-term performance depends on different processes such as aging, shrinkage, as well as creep independently and in interaction. As widely known, the time-dependent behavior of concrete is influenced by the mix design and the environmental boundary conditions, which are ambient temperature and humidity. These determine the hydration reactions, the resulting temperature rise, and the drop in internal available water which in turn drive thermal strains, hygral shrinkage, creep and influence the spatial distribution of material properties. Objective of this contribution is the evaluation of code suggestions for the analysis of shrinkage in structural members compared to full 3D numerical multi-physics simulations. First the effects of size on the shrinkage evolution are investigated, followed by a study of cross-sections with substantially different effective thicknesses.

The numerical framework used for this investigation consists of a multi-physics model coupled with a mechanical model. The used multi-physics model is the well established hygro-thermo chemical model developed by Di Luzio et. al (2009a/b). This model captures the drying processes and the evolution of temperature, as well as the chemical reactions of concrete during hydration at early age and beyond. The outcomes of the multi-physics simulations, temperature, humidity and reaction degree fields, are coupled with the mechanical analysis by introducing thermal and hygral eigenstrain rates $\dot{\epsilon}_{sh} = \alpha_{sh}\dot{h}$ and $\dot{\epsilon}_T = \alpha_T\dot{T}$.

The presented cross-sections are discretized in space in a fully coupled chemo-mechanical model, which considers the influences of spatial gradients of physical entities, i.e. temperature, humidity, and hydration degree. Since the models provided by the

codes are typically beam models, the results obtained by the numerical simulations need to be comparable. As widely known, beam theory postulates, that the cross-sections need to stay plane. Therefore, 1 m long elements of the investigated cross-section are modeled, choosing the boundary conditions such, that plane cross-sections are enforced. This approach makes it possible to obtain effective shrinkage strains taking into account all relevant 3D effects yet stay comparable to the code based predictions.

For the evaluation of the effect of size on the shrinkage behavior, three substantially different sizes of an infinite slab are modeled. The chosen sizes are slabs with a nominal size of $h = 300$ mm, $h = 600$ mm and $h = 1200$ mm. The results show, that the *fib* model code prediction assumes the time function for the shrinkage more accurate than the Eurocode 2. Eurocode 2 on the one hand predicts a faster shrinkage evolution and on the other hand a lower asymptotic shrinkage value. Furthermore, with increasing size of the cross-section the tensile stresses increase while the compressive stresses in the core of the cross-section decrease.

In the second step the influence of the structure's shape is investigated. Therefore, a common T-beam is analyzed. It is observed that different shrinkage rates within the cross section are occurring, due to the faster drying of the narrow construction components. On top of axial shrinkage strains, also a curvature of the cross-section is observed.

REFERENCES

Di Luzio, G. & Cusatis, G. (2009a) Hygro-thermo-chemical modeling of high performance concrete. I: Theory. *Cement and Concrete Composites*, 31(5), 301–308.
Di Luzio, G. & Cusatis, G. (2009b) Hygro-thermo-chemical modeling of high performance concrete. I: Numerical implementation, calibration, and validation. *Cement and Concrete Composites*, 31(5), 309–324.

Life-Cycle Analysis and Assessment in Civil Engineering: Towards an Integrated Vision – Caspeele, Taerwe & Frangopol (Eds)
© 2019 Taylor & Francis Group, London, ISBN 978-1-138-62633-1

Three types of errors in the international norms for the design of concrete and reinforced concrete

R.S. Sanjarovskiy
L.N. Gumilyov Eurasian National University, Astana, Kazakhstan

T.N. Ter-Emmanuilyan
Kazakh-British Technical University, Almaty, Kazakhstan

M.M. Manchenko
Krylov State Research Center, St. Petersburg, Russia

ABSTRACT

Eurocode is a system which includes scientific developments and experience of outstanding scientists from various countries, motivated formulation of the main Principles and Rules, the classical mechanics and general theory of computing of elastoplastic systems, detailed and numerous experimental data. Nonlinearity of deformational properties of reinforced concrete at short and long term loadings is the basis of standards of *Eurocode* 2. Dependence "strain – deformation" of concrete has a descending interval and limited extension (*Whitney, Sargin, Hognestad, Emperger*); creep deformations are nonlinear from the very low levels of strain. In the article, three types of serious errors in the international norms of concrete and reinforced concrete caused by the violation of the *Eurocode* rules were identified and investigated. Two types of errors are associated with loss of instant non-linearity of concrete. This is an erroneous replacement of elastoplastic deformation by deformations of linear creep of concrete and substitution of the process of continuous loading of the structure by a jump from the elastic stage to the plastic hinge, bypassing the elastoplastic stage. The authors identify and analyze errors in the creep region, where, according to the managers and developers of the standards, there is an established consensus.

REFERENCES

Chiorino, M.A. (2014) Analysis of structural effects of time – dependent behavior of concrete: an internationally harmonized format. *Concrete and Reinforced concrete – Glance at Future, III All Russian (International) Conference on Concrete and Reinforced Concrete*, Moscow, 12–16 May 2014. vol. 7, plenary papers, pp. 338–350.

EN 1992-2 (2004) Eurocode 2: Design of the structures.

Pars, L.A. (1965) *A treatise on Analytical Dynamics*. Heinemann, London.

Sanjarovskiy, R.S. et al. (2015) Creep of concrete and its instant nonlinear deformation in the calculation of structures. *CONCREEP 10: Mechanics and Physics of Creep, Shrinkage, and Durability of Concrete and Concrete Structures; International Conference, Vienna, 21–23 September 2015*, pp. 238–247.

Life-Cycle Analysis and Assessment in Civil Engineering: Towards an Integrated Vision – Caspeele, Taerwe & Frangopol (Eds)
© 2019 Taylor & Francis Group, London, ISBN 978-1-138-62633-1

Deflections of reinforced concrete beams made with recycled and waste materials under sustained load: Experiment and *fib* Model Code 2010 predictions

N. Tošić, S. Marinković & I. Ignjatović
Faculty of Civil Engineering, University of Belgrade, Belgrade, Serbia

A. de la Fuente
Civil and Environmental Engineering Department, Universitat Politécnica de Catalunya, Barcelona, Spain

ABSTRACT

This study presents experimental results of long-term behavior of reinforced beams made from NAC, RAC, and HVFAC. Two 3.2 m span simply supported beams were cast from each concrete type and loaded after 7 and 28 days for a period of 450 days. The beams had a 160/200 mm cross-section with a 0.58% reinforcement ratio and were loaded in four-point bending in the thirds of the span. A fixed stress-to-strength at loading age ratio ($\sigma_c/f_{cm}(t_0)$) was selected for beams loaded at the same age. So as to simulate a realistic scenario for members loaded at an early age, the selected $\sigma_c/f_{cm}(t_0)$ ratios were high: 0.45 for beams loaded after 7 days and 0.60 for beams loaded after 28 days. This choice meant that all of the beams would exhibit non-linear creep behavior and this had to be taken into account in the analysis.

Deflections, cracking, and strains were measured on the beams. Mechanical properties, shrinkage, and creep were measured on accompanying concrete specimens.

Results showed similar mechanical properties of NAC and RAC (within 10%) and lower values for HVFAC (due to the effect of only one day of wet curing). After 450 days RAC had the largest shrinkage strain of −0.782‰ followed by NAC with −0.645‰ and HVFAC with −0.597‰. The shrinkage of RAC was 21% greater than that of NAC while the shrinkage of HVFAC was 7% smaller than that of NAC. The creep coefficient was only measured on RAC and HVFAC, and after 450 days it was 2.40 and 2.48 for RAC7 and RAC28, respectively, and 1.63 and 1.55 for HVFAC7 and HVFAC28, respectively.

Results of mid-span deflection measurements are given in Table 1 where $a(t_0)$ represents the initial de-flection measured 5 min after loading and $a(450)$ represents the deflection measured 450 days after loading. The final column in Table 5 represents the

Table 1. Mid-span deflection measurements.

Beam	$a(t_0)$ (mm)	$a(450)$ (mm)	$a(450)/a(t_0)$
NAC7	9.17	18.79	2.07
NAC28	8.11	16.51	2.04
RAC7	10.89	22.47	2.06
RAC28	6.23	14.69	2.36
HVFAC7	6.13	12.45	2.03
HVFAC28	4.04	8.72	2.16

ratio of 'final'-to-initial deflection $a(450)/a(t_0)$. For all six beams, this value lies in a very narrow range of 2.03–2.36.

Measured deflections were compared with prediction calculated using the *fib* Model Code 2010 model (FIB, 2013) and its rigorous procedure of numerical integration of curvatures as well as using all input variables calculated from code expressions based on the measured compressive strength. Results for NAC and RAC beam showed a good agreement between measured and calculated values. Whereas initial deflections were overestimated for all of the beams, the final deflections were overestimated by 12.8% for beam NAC7 and exactly predicted for beam NAC28, and overestimated by 2.4% and 8.8% for beams RAC7 and RAC28, respectively. For HVFAC, the results pointed to a very large overestimation of measured deflections. Final deflections were overestimated by 44.2% and 71.3% for beams HVFAC7 and HVFAC28, respectively, even though qualitatively.

REFERENCE

FIB (2013) *fib Model Code for Concrete Structures 2010*. International Federation for Structural Concrete (fib), Laussanne. doi: 10.1002/9783433604090.

Life-Cycle Analysis and Assessment in Civil Engineering: Towards an Integrated Vision – Caspeele, Taerwe & Frangopol (Eds)
© 2019 Taylor & Francis Group, London, ISBN 978-1-138-62633-1

An investigation into influential factors affecting the time to concrete cover cracking in reinforced concrete structures

F. Chen, H. Baji & C.-Q. Li
RMIT University, Melbourne, VIC, Australia

ABSTRACT

Steel corrosion has been identified as one of the most predominant deterioration mechanisms in reinforced concrete structures worldwide, which threatens the durability, reliability and serviceability of the structures. Corrosion of steel in reinforced concrete leads to concrete cover damages in the form of cracking, rust staining, spalling and delamination; however, cracking is suggested as the mainly useful indicator for structure assessment and raises most concerns to asset owners. Some empirical models are produced based on the experimental results to produce the relationship between corrosion and cracking in concrete cover. Analytical models, based on the fracture mechanics, have also been proposed (Li et al., 2006). Furthermore, many numerical models for simulation of corrosion-induced crack process can be found in the literature. Despite these comprehensive efforts, due to the nature of the corrosion and cracking process, which depends on various mechanical and environmental factors, some considerable discrepancies between the proposed predictive models and laboratory and field data have been reported. There is a general agreement that the corrosion rate, the concrete cover thickness and the concrete cover-to-bar diameter ratio are the most critical factors in prediction of time to crack initiation.

After extensive literature review, four analytical models are chosen in the present paper to investigate the most influential factors in the process of corrosion-induced cracking: Liu et al.'s model (Liu and Weyers, 1998); K. Bhargava et al.'s model (Bhargava et al., 2006); Chunhua Lu et al.'s model (Lu et al., 2011); El Maaddawy and Soudki's model (El Maaddawy and Soudki, 2007). The major concerns in most time to cover cracking models, such as porous zone, linear or nonlinear growth of corrosion products, corrosion products penetration in cracks and the residual elastic modulus are all included in these four selected models. Thus, more comprehensive comparisons and conclusions can be drawn based on the analysis of selected models.

Three sensitivity analysis methods are employed to investigate the influence of each parameter in the selected models on the time to cover cracking: 1) nominal range sensitivity analysis, 2) differential analysis, and 3) sensitivity index. Based on the results of sensitivity analysis, three factors related to the corrosion process, namely the corrosion rate, the expansion rate of corrosion products and the size of porous zone at the steel-concrete interface, are found to be the most influential factors affecting the time to cover cracking. As the lack of the available information on the size of porous zone and the uncertain nature of corrosion product properties, more research is needed for quantifying these variables than for developing more sophisticated models.

REFERENCES

Bhargava, K., Ghosh, A., Mori, Y. & Ramanujam, S. (2006) Analytical model for time to cover cracking in RC structures due to rebar corrosion. *Nuclear Engineering and Design*, 236, 1123–1139.

EL Maaddawy, T. & Soudki, K. (2007) A model for prediction of time from corrosion initiation to corrosion cracking. *Cement and Concrete Composites*, 29, 168–175.

Li, C.-Q., Melchers, R.E. & Zheng, J.-J. (2006) Analytical model for corrosion-induced crack width in reinforced concrete structures. *ACI Structural Journal*, 103.

Liu, Y. & Weyers, R.E. (1998) Modeling the time-to-corrosion cracking in chloride contaminated reinforced concrete structures. *ACI Materials Journal*, 95, 675–681.

Lu, C., Jin, W. & Liu, R. (2011) Reinforcement corrosion-induced cover cracking and its time prediction for reinforced concrete structures. *Corrosion Science*, 53, 1337–1347.

Development of fatigue life prediction for RC slabs under traveling wheel-type loading

K. Takeda & Y. Sato
Waseda University, Tokyo, Japan

ABSTRACT

Hamada and author have developed a fatigue life prediction method for RC slabs hereafter called as Hamada's equation (Takeda et al. 2017). It was developed based on the shear resisting mechanism and failure mechanism and can reasonably evaluate a lot of experimental results obtained by other researchers. However, it has limitation of applicable range and cannot apply to all of experimental results. In this study, authors focused on effects of loading speed, support condition and loading pattern on fatigue life of RC slabs and modify Hamada's equation to extend the applicability range.

It seems that fatigue life becomes short when the loading speed is extremely low. The reason might be because resisting area could be reduced caused by decrease in tensile strength of concrete which would contribute to evolution of higher crack density. Therefore, authors considered that effect of loading speed by extending a term of width of the beam-formed member in Hamada's equation.

In the case of slab with simply supported along longitudinal direction and can be treated as free ends along transverse direction, punching shear capacity may become smaller than that with normal support condition containing simple support along longitudinal direction and elastic support along transverse direction. In the case of slab with simple supports on all sides, on the other hand, punching shear capacity may become greater than that with the normal support condition because membrane action can be developed. The influences on different support conditions can be rationally considered in the proposed equation using a coefficient.

Figure 1 shows the S-N diagram of experimental results of traveling wheel-type loading tests by applying a proposed equation.

Fatigue life was defined as a load when shear failure would take place due to shear strength reduction under fatigue cyclic loading. Under this understanding of fatigue failure, the method to consider fatigue damage accumulated at a preceding loading block in stepped loading test was proposed. Figure 2 shows experimental observed fatigue life and predicted fatigue life by

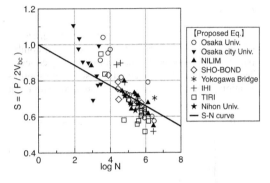

Figure 1. S-N diagram calculated by proposed equation.

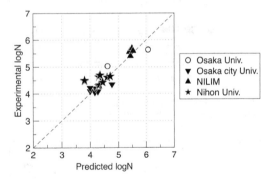

Figure 2. Observed and predicted fatigue life under stepped loading.

the proposed method. The proposed method updates punching shear strength at ever preceding loading block. The good agreement with experimental results is founded.

REFERENCE

Takeda, K., Hamada, N. & Sato, Y. (2017b) A study on fatigue life prediction of RC slabs under traveling wheel-type loads. In: *Proceeding of 25th Symposium on Developments in Prestressed Concrete*, pp. 129–134. (in Japanese).

Early damage detection of fastening systems in concrete under dynamic loading—model details and health monitoring framework

M. Hoepfner & P. Spyridis
Technical University Dortmund, Dortmund, Germany

ABSTRACT

This paper presents a method for the characterisation of the fatigue failure of fastening systems in combination with a coordinated health monitoring concept. The required hysteresis method is based on an approximation of the hysteresis loops using measurement data and hysteresis loop forms conformance criteria. The developed predictive model uses four adjustment parameters (determined through regression analysis), was applied to experimental fatigue experiments of anchor channels in concrete. Through this exercise, conventional characteristics of the hysteresis measurement method, deformation, stiffness, and dissipation energy, could also be determined for hysteresis loops with local non-conformities. Additional investigations of the adjustment parameters indicated the potential to identify an indicator for forecasting the experimental data at an early stage, in relation to specific changes in the shape, gradient, and area of hysteresis loops (see Figure 1).

Furthermore, the determination of the loop centroid – being a sensitive characteristic of the loop geometry, is achieved through numerical methods on the basis of measurement data (see Figure 2).

The evaluation of the horizontal and vertical positions indicates a continuously asymmetrical loop form and an imbalance between the loading and unloading phase. In a subsequent step, the above mentioned properties and alterations throughout the loading history are applied to a health monitoring concept with appropriate measurement technology and discussed.

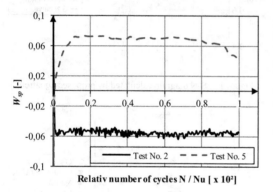

Figure 1. Value W_{sp} for test with final fatigue failure and test on level of fatigue limit capacity.

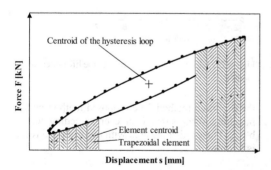

Figure 2. Trapezoidal elements definition and the hysteresis loop area.

REFERENCES

Bergmeister, K. & Wendner, R. (2010) Monitoring und Strukturidentifikation von Betonbrücken.
Block, K. & Dreier, F. (2003) Das Ermüdungsverhalten von Dübelbefestigungen.
Dratschmidt, F. & Ehrenstein, G.W. (1997) Threaded joints in glass fiber reinforced polyamide. *Polymer Engineering and Science*.
Lazan, B. J. (1968) Damping of materials and members in structural mechanics.
Strauss, A., Wan-Wendner, R., Vidovic, A., Zambon, I., Yu, Q., Frangopol, D.M. & Bergmeister, K. (2017) Gamma prediction models for long-term creep deformations of prestressed concrete bridges.

Study on the time variant alteration of chloride profiles for prediction purpose

F. Binder
ASFiNAG Service GmbH, Vienna, Austria

S.L. Burtscher
Burtscher Consulting GmbH, Vienna, Austria

A. Limbeck
Institute of Chemical Technologies and Analytics, Vienna University of Technology, Austria

The ASFiNAG road network contains more than 5000 bridges. Most of them are made of reinforced or pre-stressed concrete. In Austria strong winters are obligatory. Therefore de-icing measures are used for thawing. Additionally, more than half of the structures were built in the 1970's or 1980's. They have been exposed for many years to this chemical attack and accurate assessment and prognosis of the concrete degradation is vital for the economic preservation of the concrete structures. The LA-ICP-MS (Laser Ablation Inductively Coupled Plasma Mass Spectrometry) method presented in this paper is a fast and reliable analysis method for determination of the chloride content in the cement paste of existing concrete structures. The high depth resolution, the accuracy and the distinction of the concrete phases delivers meaningful data for assessment and forecast. For this purpose Chloride-profiles were determined in short intervals (Weekly). The paper focuses on the interactions between environmental loads and the chloride profile extracted from an affected RC-Component. Finally, a model for prediction of the chloride profile using environmental and monitoring data will be evaluated with regard to their input parameters.

Environmental action of the exposed concrete was assessed by the de-icing salt records, the in-situ monitoring of the humidity and Temperature, and additional weather data from other sources. Mean values for every month during almost 1 year were used in order to derive a relationship between the environ-mental actions, de-icing salt applications and the chloride profile. The variation of the surface concentration C_s was crucial, because if the model doesn't pick it correctly, the deviations can be very high. The results demonstrate that the contribution of each parameter in the diffusion dominated chloride transport was notably changing within one year (see Fig. 1 and 2). This indicates that the long-term chloride transport from a single measurement is not sufficient with the used model. Hence, long-term field data is essential to well understand the effect of all influential variables on chloride penetration into concrete.

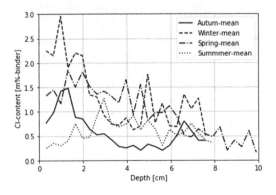

Figure 1. Chloride profile condensed to seasons at height $H2 = 1.25$ m.

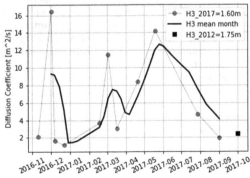

Figure 2. Development of chloride diffusion coefficient over time at height $H3 = 1.60$ m.

A rapid numerical simulation method of chloride ingress in concrete material

Y. Li, X. Ruan & Z.R. Jin
Department of Bridge Engineering, Tongji University, Shanghai, China

ABSTRACT

Concrete material has been widespread used in the construction of buildings, bridges and marine facilities, for its high strength and low cost. With the promotion of large scale construction to the extreme environment, the problem about material durability is more and more serious. At the recent decades, with the increase of research demand and the development of basic theory, the accuracy and the acceptability of the numerical simulation for the chloride ingress in concrete material has been gradually improved.

However, the problem of low efficiency in the precise simulation is still serious, and one of the important reasons is that the traditional methods rely on the complete solution at all time nodes. The huge size of nearly repetitive result leads to the difficulty in the model solving and the data storing, which also restrict the promotion of numerical simulation in full structural section.

In this paper, a rapid numerical simulation method is proposed, with the calculation of diffusion path caused by the aggregate particles and the solving of a small cement mortar model. As shown in Figure 1, the precise result of chloride ingress at all critical time nodes can be obtained from the conversion of the macroscale cement mortar model, based on the different length of diffusion path.

By analyzing the result of numerical experiments, the proposed method can greatly improve the efficiency of model solving and reduce the amount of data storing. These advantages will be more significant in the concrete simulation of long-term durability degradation.

At the simulation of local chloride diffusion, the proposed method can avoid the ion accumulation phenomenon at the back-end effectively, which usually happened in the traditional models, and describe the concentration field of chloride ion more accurately.

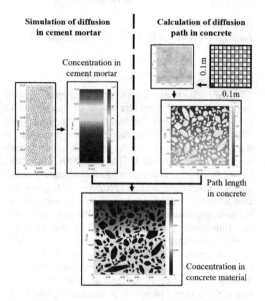

Figure 1. Flow of conversion with diffusion path.

REFERENCES

Angst, U. M. & Polder, R. (2014) Spatial variability of chloride in concrete within homogeneously exposed areas. *Cement and Concrete Research*, 56, 40–51.
Han, S.-H. (2007) Influence of diffusion coefficient on chloride ion penetration of concrete structure. *Construction and Building Materials*, 21, 370–378.
Jin, L., Zhang, R. & Du, X. (2017) Computational homogenization for thermal conduction in heterogeneous concrete after mechanical stress. *Construction and Building Materials*, 141, 222–234.
Pan, Z., Ruan, X. & Chen, A. (2014) Chloride diffusivity of concrete: probabilistic characteristics at meso-scale. *Computers and Concrete*, 13, 187–207.
Walraven, J. C. (1980) Aggregate Interlock: A Theoretical and Experimental Analysis. Delft University Press.

Life-Cycle Analysis and Assessment in Civil Engineering: Towards an Integrated Vision – Caspeele, Taerwe & Frangopol (Eds)
© 2019 Taylor & Francis Group, London, ISBN 978-1-138-62633-1

Concrete cover cracking under chloride-induced time-varying non-uniform steel corrosion

J. Zhang & P. Wang
Department of Civil Engineering, The University of Shanghai for Science and Technology, Shanghai, China

Z. Guan
State Key Laboratory for Disaster Reduction in Civil Engineering, Tongji University, Shanghai, China

ABSTRACT

Chloride-induced steel corrosion is the main influential factor of the safety and durability of reinforced concrete (RC) structures. Recently, the issue of chloride-induced steel corrosion for RC structures has attracted more and more attentions. However, with the progression of the research, some faults on the fundamental assumptions and mechanism have also been found. For example, the distribution of corrosion products in natural environment is more prone to be non-uniform rather than to be uniform, and the actual rust distribution of corrosion product is not time-invariant but time-varying. Therefore, in this study, a 2D chloride diffusion analysis for the case of concrete with a single rebar was first investigated with Fick's second law. Based on this, corrosion initiation time at each position around the steel cross section can be obtained, and time-varying non-uniform corrosion depth around the steel cross section at any time was calculated by multiplying the time-varying corrosion rate and corrosion time. Then, a numerical model was developed to study the concrete cover cracking under the time-varying non-uniform reinforcement corrosion. Four crack propagation paths were predefined in the model, including vertical and horizontal direction. And cohesive elements were introduced to simulate the crack evolution along these two directions. The numerical model was verified with the experimental results. The influence of environmental parameters, concrete cover depth, and rebar diameter on the evolution of concrete cover cracking was also investigated. The present research will be helpful to the life prediction of RC structures.

Life-Cycle Analysis and Assessment in Civil Engineering: Towards an Integrated Vision – Caspeele, Taerwe & Frangopol (Eds)
© 2019 Taylor & Francis Group, London, ISBN 978-1-138-62633-1

Analysis of coupled exposures considering the rapid chloride migration test and the accelerated carbonation test

M. Vogel, S. Schmiedel & H.S. Müller
Karlsruhe Institute of Technology (KIT), Karlsruhe, Germany

ABSTRACT

For a performance-based assessment of carbonation or chloride induced corrosion the models of the *fib* Model Code for Service Life (*fib* bulletin 34, 2006) are widely used. The relevant material parameters of these models are the inverse effective carbonation resistance $R_{ACC,0}^{-1}$ and the coefficient of chloride migration $D_{RCM,0}$ which are determined by performance testing. The measurement of these coefficients is realized by means of the accelerated carbonation test (ACC-test) and the rapid chloride migration test (RCM test). Finally, the risk of carbonation or chloride induced corrosion can be calculated.

But in reality concrete structures are exposed to combined actions. Therefore, a realistic performance-based durability assessment on concrete structures demands realistic performance testing methods. Against this background a new performance testing procedure will be presented that takes into account the coupled exposure of carbonation and chloride ingress.

For the experimental investigations nine different concrete mixtures (M1-M9) have been used. These concrete mixtures vary by three different binder types (OPC, OPC with fly ash and blast furnace cement) and three different water/binder-ratios (0.4, 0.5 and 0.6).

Strength tests as well as ACC-tests and RCM tests on cubical, cylindrical and beam specimens were realized. These experimental studies include 13 different tests which can be divided into

- strength tests (tests No. 1 and 2),
- the single ACC-tests (tests No. 3–6),
- the single RCM tests (test No. 7) and
- the coupled ACC-/RCM tests (tests No. 8–13)

Although a similar strength within the nine concrete mixtures was achieved, the inverse effective carbonation resistance and the coefficient of chloride migration vary widely. That means the concrete strength itself is no single indicator for a comprehensive description of the carbonation and chloride penetration resistance of concrete.

Therefore, the type of cement/additive has a significant influence on the durability parameters $R_{ACC,0}^{-1}$ and $D_{RCM,0}$. Concrete made of OPC obtains the lowest values of the inverse effective carbonation resistance, whereas blast furnace cement concretes show the highest values. OPC concretes with fly-ash are in the middle range. In relation to the chloride migration coefficient concrete made with blast furnace cement has the highest resistance against chloride diffusion. OPC concretes and OPC concretes with fly-ash have a significantly lower resistance against chloride ingress.

In the course of the test combination "first the accelerated carbonation test and second the chloride migration test" it was ascertained that the carbonation of the concrete leads to a deceleration of chloride ingress in the case of OPC concretes and OPC concretes with fly ash. On the other hand carbonation of blast furnace cement concretes shows an acceleration of chloride ingress.

The results of the second coupled test procedure "first the chloride migration test and second the accelerated carbonation test" include that chloride rich concretes are responsible for slowing down the carbonation process.

REFERENCE

fib bulletin 34 (2006) *Model Code for Service Life Design.* Fédération Internationale du Béton (fib), Lausanne.

Modelling of corrosion induced cracking in reinforced concrete

I. Lau, G. Fu, C.-Q. Li & S. De Silva
RMIT University, Victoria, Australia

ABSTRACT

Practical experience and observations suggest that corrosion affected reinforced concrete structures are more prone to cracking than other forms of structural deterioration. As such, corrosion is one of the main causes for premature deterioration in reinforced concrete. Once corrosion initiates, corrosion products are formed at the steel and concrete interface resulting in an expansive pressure exerted on the surrounding concrete. Eventually, a crack will form and propagate to a critical depth. At this point, the crack becomes unstable and suddenly propagates through the concrete cover.

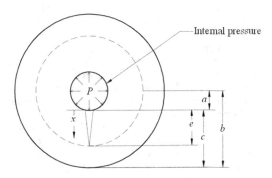

Figure 1. Schematic of a thick walled cylinder.

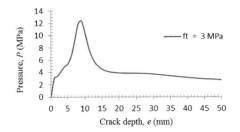

Figure 2. Pressure load (P) vs. crack depth (e) for $c = 50$ mm and $D = 16$ mm.

To determine this critical point, the stress intensity factor approach is adopted whereby the corrosion of reinforcing steel in concrete is modeled as a thick walled cylinder as shown in Figure 1. The analytical solution to stress intensity factor has been derived using the weight function method and is a function of two dimensionless parameters b/a and e/c.

The fracture criterion was then used to analyse corrosion induced crack propagation and an algorithm was developed in Matlab to carry out all computations (Wu et al. 2014, Lau et al. 2017). From figure 2, it is found that for a concrete cover of 50 mm and rebar diameter of 16 mm, the critical crack depth occurs at 9 mm from the rebar at a maximum pressure of 12.4 MPa.

A parametric study has been undertaken to study the effects of tensile strength and concrete geometry. It has been found that increasing the tensile strength increases the maximum pressure required to cause concrete cover cracking but the critical crack depth does not change. It has also been found that increasing the cover would increase the maximum pressure and critical crack depth required to cause concrete cover cracking. It has also been found that by increasing the rebar diameter, the maximum pressure decreases due to the larger surface area for the corrosion expansive pressure to act on. It can be concluded that the method derived in the paper can determine the critical crack depth for corrosion-induced cracking in reinforced concrete with reasonable accuracy.

REFERENCES

Lau, I., Fu, G.Y., Li, C.Q., Desilva, S. & Guo, Y.X. 2017. Critical crack depth in corrosion-induced concrete cracking. *ACI Structures Journal* (Accepted).

Wu, Z.M., Wu, X., Zheng, J.J., Wu, Y.F. & Dong, W. (2014) An analytical method for determining the crack extension resistance curve of concrete. *Magazine of Concrete Research*, 66(14), 719–728.

Life-Cycle Analysis and Assessment in Civil Engineering: Towards an
Integrated Vision – Caspeele, Taerwe & Frangopol (Eds)
© 2019 Taylor & Francis Group, London, ISBN 978-1-138-62633-1

Optimization of service life design of concrete infrastructures in corrosive environments under a changing climate

Z. Lounis
National Research Council, Ottawa, ON, Canada

ABSTRACT

The risk of failure of concrete infrastructures built in corrosive environments is increasing due to use of deicing salts, increased loads, inadequate maintenance and increased rate of deterioration due to climate change. Climate change leads to an increase in temperatures, which in turn leads to an increase in chloride diffusivity that yields an increase in probability of corrosion of reinforcing steel and a shortening of service life of concrete structures. The impact of temperature rise due to climate change on diffusivity was modeled using the Arrhenius relationship. Uncertainties in the parameters governing the service life, such as concrete cover depth, chloride threshold, chloride diffusion, surface chloride content are considered by modeling them as random variables. The optimum service life of concrete structures can be defined as the time at which the probability of corrosion reaches an acceptable value for different types of concrete, reinforcing steel and concrete cover depths. The time-dependent probability of corrosion of reinforcing steel embedded in concrete structures is formulated as a nonlinear optimization problem that is solved by the projected Lagrangian algorithm. The optimum service life of concrete structures can be defined as the time at which the probability of corrosion reaches an acceptable value for the different types of concrete,

reinforcing steel and concrete cover depths. The example of a concrete bridge deck illustrated the fact that the time-dependent probability of corrosion increases with temperature. After 40 years, the probability of corrosion increases by 37% and 77% for life cycle temperature increases of 3°C and 6°C, respectively compared to the reference case at 23°C. To reduce this probability of corrosion, corrosion-resistant steel reinforcement, high performance or/and higher concrete cover depth can be used. For the case of bridge decks reinforced with corrosion-resistant steel, the corrosion probability is reduced by half after 40 years compared to that associated with black steel for the climate scenario with 6°C life cycle temperature rise. The final selection of an optimal service life design should take into account the life cycle costs incurred during the service or a given life cycle. For the case of bridge decks, normal performance concrete and high performance concrete bridge deck alternatives were compared in terms of life cycle owner's costs and life cycle users' costs. In terms of initial costs, the HPC deck is about 10% more expensive than the NPC deck. However, in terms of life cycle owner's costs, the HPC deck is 24% less expensive than the NPC deck. Regarding the life cycle users' costs, the HPC deck users' costs are less than 1/3 of those of the NPC deck. Hence, the HPC deck is the optimal solution.

Simulation of crack propagation owing to deformed bar corrosion

S. Okazaki, C. Okuma & H. Yoshida
Faculty of Engineering, Kagawa University, Takamatsu, Japan

M. Kurumatani
Faculty of Engineering, Ibaragi University, Hitachi, Japan

ABSTRACT

Reinforced concrete structures located in coastal areas are strongly affected by corrosion of steel bars due to salt attack. Evaluating the residual performance of these structures is important in carrying out maintenance works and management. The structural and durability performance of such structures depends on the corrosion of the reinforcing bars and corrosion reduction amount.

In this research, by adopting a damage model based on fracture mechanics, we simulated crack propagation of concrete by corrosion expansion of a deformed rebar using FEM. This can reproduce fine crack propagation of concrete, and the effect of corrosion expansion on crack width can be investigated.

FEA using an isotropy damage model based on the fracture mechanics of concrete was adopted for the numerical simulation. In the corrosion expansion model in the research, only the displacement information is given in the analysis.

Distribution of the damage index of the concrete model equivalent to the crack distribution in the model using the deformed rebar is shown in the figure 1. The figure 2 shows the relationship between the number of steps and the crack width at the bottom of the concrete at the position right under the deformed rebar. Because of the difference in equivalent stress, the crack width in the same step is approximately 1.5 times larger in the round step than in the deformed reinforcing bars. It was suggested that it is necessary to grasp the shape of the rebar before-hand in estimating the corrosion amount of the reinforcement by the crack width.

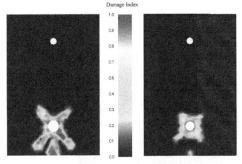

Displacement around steel bar, 0.01 mm

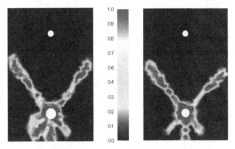

Displacement around steel bar, 0.20 mm

Figure 1. Crack distribution (left: round bar, right: deformed bar).

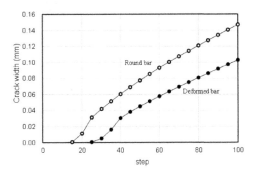

Figure 2. Crack width of bottom of concrete model.

REFERENCES

de Vree, J.H.P., Brekelmans, W.A.M. & van Gils, M.A.J. (1995) Comparison of nonlocal approaches in continuum damage mechanics. *Computers and Structures*, 55, 581–588, 1995.

Kurumatani, M., et al. (2013) Isotropic damage model based on fracture mechanics for concrete and its evaluation. *Transactions of JSCES Paper*, 2013, 20130015, 2013.8.

SS6: TRUSS ITN—reducing uncertainty in structural safety
Organizer: A. González

Life-Cycle Analysis and Assessment in Civil Engineering: Towards an Integrated Vision – Caspeele, Taerwe & Frangopol (Eds)
© 2019 Taylor & Francis Group, London, ISBN 978-1-138-62633-1

Mechanical characterisation of braided BFRP rebars for internal concrete reinforcement

S. Antonopoulou & C. McNally
School of Civil Engineering, University College Dublin, Dublin, Ireland

G. Byrne
Burgmann Packings Ltd, Dublin, Ireland

ABSTRACT

This study investigates the tensile behaviour of basalt fibre reinforced polymer (BFRP) composites that were developed using braiding as a manufacturing technique. Those materials will be introduced in concrete reinforcement applications. Three BFRP rebar sizes with a circular constant cross section and different braided configurations are developed and characterised with respect to their internal architecture. The braid angle on each layer of the rebar, varying from 10° to 45°, is an important parameter that has a direct impact on its performance characteristics. The effective longitudinal in-plane modulus (E_x^{FRP}) of of each braided sample is calculated numerically using the classical laminate theory (CLT) approach and then, tensile tests are performed according to the relevant standard.

Comparisons between analytical and experimental data demonstrate a significant influence of braiding parameters, like angle, number of layers, and fibre volume fraction on the mechanical properties of BFRP

Table 1. Test results for braided BFRP rebars.

Sample no	BFRP 1	BFRP 2	BFRP 3
OD (mm)	5	8	10
Fibre Volume Fraction (%)*	57.76	51.63	54.59
	Aver./ CoV	Aver./ CoV	Aver./ CoV
Maximum Load (kN)	5.46/0.05	17.84/0.01	30.60/0.02
Ultimate Tensile Strength (MPa)	277.84/0.05	354.99/0.01	389.64/0.02
Maximum Displacement (mm)	7.49/0.03	10.09/0.05	21.58/0.07
Ultimate Strain (%)	2.98/0.03	2.59/0.04	3.73/0.09
Elastic Modulus (GPa)	10.65/0.03	14.76/0.02	12.39/0.04

*Note: Numerically calculated using CLT approach.

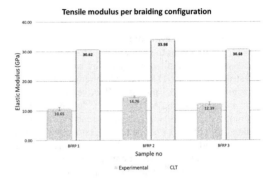

Figure 1. Tensile modulus of elasticity (E_x^{FRP}) – Experimental vs CLT results.

rebars. Their response is mainly dominated by textile architecture, which significantly affects localized properties, crack propagation and load redistribution in the material. The maximum tensile strength obtained, is comparable to the one of steel, although pultruded rebars can reach higher values, but with lower strain rate. In addition, it is noteworthy that all predicted moduli determined with CLT numerical approach are found to be higher than the test results and overestimate rebar's stiffness, most probably due to the degree of undulation from braiding process.

REFERENCES

ACI Committee 440 (2004) *Guide Test Methods for Fiber-Reinforced Polymers (FRPs) for Reinforcing or Strengthening Concrete Structures*. American Concrete Institute: ACI 440.3R-04.

Antonopoulou, S., McNally, C. & Byrne, G. (2016) Development of braided basalt FRP rebars for reinforcement of concrete structures. Fibre-Reinforced Polymer (FRP) Composites in Civil Engineering; In: *Proceedings of 8th International Conference (CICE2016)*, Hong Kong, China, 14–16 December 2016.

Antonopoulou, S. & McNally, C. (2017) Reliability assessment of braided BFRP reinforcement for concrete structures. *Proceedings of 27th European Safety and Reliability Conference (ESREL 2017)*, Portorož, Slovenia, 18–22 June 2017.

Use of post-installed screws in the compressive strength assessment of in-situ concrete

M.S.N.A. Sourav & S. Al-Sabah
Arup, Dublin, Ireland

C. McNally
University College Dublin, Ireland

ABSTRACT

Concrete strength evaluation of existing buildings is a major challenge in the field of structural engineering to assess the overall capacity of the structures. Non-destructive tests (NDTs) provide interesting approach for the easy, quick, and cost effective assessment of the concrete strength. The direct determination of compressive strength requires that concrete specimens taken from the structure be tested destructively. NDTs cannot yield absolute values of these properties as they measure other properties of concrete from which estimates of compressive strength can be inferred. Low cost and easy applicability contribute to the popularity of NDTs, though the low reliability of most NDTs in the assessment of compressive strength of concrete limits the use of NDTs in the practical field.

A new simplified NDT for the assessment of in-situ concrete strength, "Post-installed Screw Pullout (PSP) test", is presented in this paper. The theoretical consideration can be traced to the bonding action of deformed rebar in concrete. When failure pattern in bond test is dictated by complete pullout failure involving the crushing of concrete under the ribs of the rebar, compressive strength can be estimated from bond strength (Lorrain & Barbosa 2011). In the study of PSP test, a complete pullout failure mode is replicated with the use of screw installed in already hardened concrete. PSP test was investigated in mortar and concrete to study the different factors; compressive strength, presence, and the types of aggregates. Limestone, brick chips, and pumice as lightweight aggregate were used for the study. A preliminary study was conducted to attain a robust test set-up for PSP test.

Based on the experimental study of the PSP test, following conclusions can be stated on the basis of the analysis presented in the paper:

1. Load carrying behaviour of the screw used for the PSP test is affected by the presence of aggregates and the type of aggregates in the concrete.

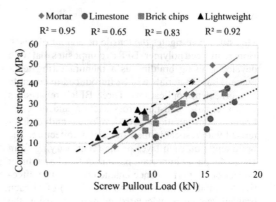

Figure 1. Compressive strength (MPa) vs screw pullout load (kN) for concrete considering the aggregates types.

2. The PSP test provides better correlation of compressive strength when the concretes with different aggregates are analysed separately (Figure 1).
3. In the strength assessment, PSP test shows low variability in terms of R-squared value, standard deviation, coefficient of variation, mean residual and RMSE in case of mortar and concrete with brick chips and lightweight aggregates as compared to the concrete having limestone aggregates.

The study confirms that the PSP test method is a viable test method with the potential to be reliable and reasonably accurate, yet cost effective and to contribute to the reduction of the uncertainty in the assessment of compressive strength of in-situ concrete.

REFERENCE

Lorrain, M.S. & Barbosa, M.P. (2011) Estimation of compressive strength based on Pull-Out bond test results for on-site concrete quality control. *Ibracon Structures and Materials Journal*, 4(4), 582–591.

Impact of input variables on the seismic response of free-standing spent fuel racks

A. Gonzalez Merino
Equipos Nucleares S.A., Maliaño, Spain
University College Dublin, Dublin, Ireland

L. Costas de la Peña
Equipos Nucleares S.A., Maliaño, Spain

A. González
University College Dublin, Dublin, Ireland

ABSTRACT

Free-standing racks are steel structures designed to store the spent fuel assemblies removed from the nuclear power reactor. Rack units rest in free-standing conditions submerged in the depths of the spent fuel pool. Figure 1 shows their typical arrangement spaced by only a few centimeters.

During a seismic event, free-standing racks undergo large displacements subjected to inertial effects and water-coupling forces through the surrounding fluid volume. Rack units may therefore rock, turn and twist around their vertical axis while sliding over the pool liner. The boundaries of these motions are provided by physical contacts appearing between racks and pool, fuel and racks, and also between rack units themselves. An accurate estimation of the transient response is therefore essential to achieve a safe pool arrangement and a reliable structural design. It deals with a transient dynamic response, a highly nonlinear behavior and a fluid-structure interaction problem. An ad-hoc analysis methodology based on finite element methods implements the hydrodynamic mass concept and takes advantage of dynamic contact elements. The accuracy of these simulations and the robustness of the computed outputs are highly sensitive to the uncertainties inherent to the input data.

Stochastic input data brings aleatoric and epistemic uncertainty to the rack seismic analysis. From the synthetic acceleration-time history of the earthquake to the heterogeneous features of the rack system, several sources of uncertainty exist. The manufacturing process itself may produce slight deviations in the dynamic properties and mass distribution of the rack units. Moreover, each unit is loaded with a different number of fuel elements according to the operation needs of the plant. Even the exact clearance spaces between units are hardly inspectable due to radioactive ambiance. Hence, all of these uncertainties propagate across the nonlinear transient analysis and affect the accuracy and robustness of the numerical outputs.

This paper carries out a 'one-factor-at-a-time' parametric analysis of five key input variables: acceleration time-history, rack mass, fuel loading, rack eigenfrequencies and hydrodynamic masses. This technique systematically varies an analysis parameter while keeping the others at their nominal value. It examines the impact on the transient outputs of most interest which include the maximal, minimal and final sliding displacements, the maximal force on support and the computational time. Quantification of these outputs is essential to ensure no collisions between units and pool walls and to check the structural integrity of both.

Numerical results are provided for a simple two-rack system as a source of insight into the uncertain seismic response of a real rack system. It is highlighted that the dispersion is much higher for the sliding displacements (CoVs up to 0.8) than for the maximal forces on support (CoVs under 0.2). However, no clear dependency is observed between the CoV and the range of variation of the input variables.

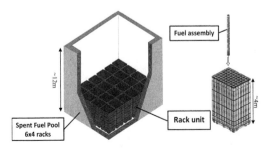

Figure 1. Spent fuel pool equipped with free-standing racks.

Life-Cycle Analysis and Assessment in Civil Engineering: Towards an Integrated Vision – Caspeele, Taerwe & Frangopol (Eds)
© 2019 Taylor & Francis Group, London, ISBN 978-1-138-62633-1

Surrogate infill criteria for operational fatigue reliability analysis

R. Teixeira, A. O'Connor & M. Nogal
Department of Civil, Structural and Environmental Engineering, Trinity College Dublin, Dublin, Ireland

ABSTRACT

Reliability analysis of Offshore Wind Turbine (OWT) fatigue damage is an intense, resource demanding task. While the current methodologies to design OWT to fatigue are quite limited in the way and amount of uncertainty they can account, they still represent a relevant share of the total effort needed in the OWT design process. The robustness achieved in the design process is usually limited.

To enable OWT to be more robust an innovative methodology that tackles the current limitations using a balanced amount of designing effort was developed. It consists in generating a short-term fatigue damage (D_{SH}) using a Kriging surrogate model that accurately accounts for uncertainty using an adaptive approach.

While accurate to account for uncertainty in (D_{SH}), depending on the number of variables considered in the analysis and the convergence intended for the fitting, the computational time needed to perform the full lifetime analysis, despite relatively low, can still be significant.

The current paper discusses then convergence of the Kriging surrogate model applied to (D_{SH}) of OWT towers. Different scenarios of convergence are exploited by using different criteria. The computational costs of using different convergence scenarios for the Kriging surface are presented and compared. The main goal is therefore to give the designer a rationale on how cost can be decreased and accuracy reasonably sacrificed using the mentioned approach to design robust OWT towers.

Results show that on a design basis two levels of approach may be efficient. In the first, if a very high computational cost is expected, a trade-off between accuracy and computational time must be considered and then, if the intention is to check how robust is the current design, then a full convergence of the surface should be pursued.

Life-Cycle Analysis and Assessment in Civil Engineering: Towards an
Integrated Vision – Caspeele, Taerwe & Frangopol (Eds)
© 2019 Taylor & Francis Group, London, ISBN 978-1-138-62633-1

Probabilistic decision basis and objectives for inspection planning and optimization

G. Zou & K. Banisoleiman
Lloyd's Register Group Limited, Southampton, UK

A. González
University College Dublin, Dublin, Ireland

ABSTRACT

Marine and offshore engineering has long been challenged with the problem of structural integrity management (SIM) for assets such as ships and offshore platforms due to the harsh marine environments, where cyclic loading and corrosion are persistent threats to structural integrity. SIM for such assets is further complicated by the very large number of welded plates and joints, for which condition surveys by inspections and structural health monitoring become a difficult and expensive task. Structural integrity of such assets is also influenced by uncertainties associated with materials, loading characteristics, fatigue degradation model and inspection method, which have to be accounted for. Therefore, managing these uncertainties and optimizing inspection and repair activities are relevant to improvements of SIM. This paper addresses probabilistic inspection planning and optimization by comparative analysis for a typical fatigue-prone structural detail based on reliability, life cycle costs (LCC) and value of inspection information (VoI). With the objective of clarifying the differences between the theoretical basis and objectives for probabilistic inspection optimization, three maintenance strategies are proposed in Table 1 and studied. Repair time for case 2 and inspection time for case 3 are optimized with the objectives of reliability maximization, LCC minimization and VoI maximization respectively. It can be seen from Table 2 that the optimal inspection times are different depending on the objective being considered. Figure 1 shows that planned inspection and repair in case 3 can help to achieve higher reliability with fewer repairs than repair without inspection in case 2. If the cost of unit inspection and repair is not negligible compared with failure consequence, as shown in Figure 2, expected maintenance costs would have the same order of magnitude as failure risk. Under the circumstance, it is suggested to use the optimization objective of life cycle cost minimization, which considers the costs of SIM. The paper proposes a simple approach for quantifying the VoI based on life cycle cost analysis for the three maintenance strategies in Table 1. It is concluded that the VoI is relevant to both the optimal maintenance decision with and without inspection.

Table 1. Maintenance strategies.

ID	Maintenance strategy
Case 1	Do nothing
Case 2	Time-based repair
Case 3	Inspection before repair

Table 2. Optimal inspection time.

Optimization objective	Optimal inspection time (year)
Reliability maximization	9th
LCC minimization	8th
VoI maximization	7th

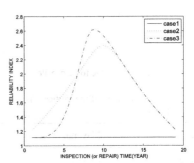

Figure 1. Fatigue reliability against inspection or repair time.

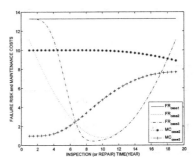

Figure 2. Failure risk (FR) and maintenance costs (MC) against inspection or repair time.

Characterization of hoisting operations on the dynamic response of the lifting boom of a ship unloader

G. Milana
Lloyd's Register Global Technology Centre, Southampton, UK
UCD, Belfield, Dublin, Ireland

K. Banisoleiman
Lloyd's Register Global Technology Centre, Southampton, UK

A. González
University College Dublin (UCD), Belfield, Dublin, Ireland

ABSTRACT

In the current paper, the impact of the hoisting operations, on the dynamic response of the lifting boom of a ship unloader, are taken into consideration. The lifting boom is used to carry out transient dynamic analysis, since it was recognized to be the single most representative element for studying the dynamic response of these structures. The response of the lifting boom was the result of dynamic analysis comprising two components: the structure (representing the waterside portion of the boom), and the applied load (expressed as different hoisting force profiles, as shown in Figure 1). A comparison between the different force profiles was carried out, in order to identify the parameters that mostly influence the dynamic behavior of the structure during a loading cycle.

Furthermore, a baseline case based on pseudo-static analysis (as standard recommendation FEM 1987) was introduced and comparisons carried out in terms of vertical displacement and bending moments, as shown in Table 1 and 2, respectively.

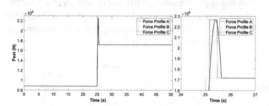

Figure 1. Comparison of the force profiles.

Table 1. Maximum value of the vertical displacement for each force profile.

Force Profile	Maximum vertical displacement
Baseline	0.496 m
Force Profile A	0.667 m
Force Profile B	0.543 m
Force Profile C	0.655 m

Table 2. Maximum value of the vertical bending moment for each force profile.

Force Profile	Maximum vertical bending moment
Baseline	1.79 10^7 Nm
Force Profile A	2.34 10^7 Nm
Force Profile B	1.90 10^7 Nm
Force Profile C	2.30 10^7 Nm

ACKNOWLEDGMENTS

This project has received funding from the European Union's Horizon 2020 research and innovation programme under the Marie Skłodowska-Curie grant agreement No. 642453 (http://trussitn.eu).

REFERENCE

Federation Europeenne de la Manutention (FEM). Revised 1987.1.001. *Rules for the Design of Hoisting Appliances*. 3rd Edition.

Monitoring crack movement on a masonry type abutment using optical camera system—a case study

F. Huseynov & E. O'Brien
University College Dublin, Dublin, Ireland

J. Brownjohn, K. Faulkner & Y. Xu
The University of Exeter, Exeter, UK

D. Hester
Queens University Belfast, Belfast, Northern Ireland

ABSTRACT

While the bridge stock around the world is aging, the demand put on transport infrastructure is continuously increasing making them to be kept in service much longer than they are originally designed for.

Figure 1. Test structure and vertical crack running along its full height.

Figure 2. Real view of the test layout. The *inset* in the caption are the L-shaped targets attached on the wing wall on both sides of the crack.

In a recent survey 1400 aging bridge structures are assessed through principal inspection in Cork, Ireland (Dromey et al., 2016). Among surveyed bridges, 28% are rated with at least significant damage and 81% with some damage. As a results of this study worst performing bridge components and the most frequent damage types are identified. The worst performing structural components found to be abutments and the most frequent type of damage, in particular for masonry type abutments which is the common type of design for historical bridges, found to be cracking and the loss of pointing.

This paper presents a case study that reveals useful information to a stakeholder in verifying the condition of its bridge structure. It is demonstrated through a field testing carried out on an abutment of a historical railway bridge that exhibits a vertical crack on its wing wall running through its full height. Although this does not pose danger to normal traffic it was observed that the crack moves under live load and obtaining real behaviour of the crack movement was crucial for the stakeholder to assess the condition of the structure, which is the main focus of this study. Figure 1 shows the vertical crack running along the full height of the test structure.

This is tackled by monitoring the crack movement on the abutment wing wall using an optical camera system pointing to optical targets attached on the structure. The real view of test layout is shown in Figure 2. As a result of this study, three-dimensional crack movement under train loading is obtained and the response of the test structure to a train loading is further analysed within the scope of this study.

REFERENCE

Dromey, L., Murphy, J.J., Rourke, B.O., et al. (2016) An analysis of a data set of 1,400 bridge inspections in County Cork. *Civil Engineering Research in Ireland*, 1, 85–90.

Life-Cycle Analysis and Assessment in Civil Engineering: Towards an Integrated Vision – Caspeele, Taerwe & Frangopol (Eds)
© 2019 Taylor & Francis Group, London, ISBN 978-1-138-62633-1

Outlier detection of point clouds generating from low-cost UAVs for bridge inspection

S. Chen, L.C. Truong-Hong & E. O'Keeffe
University College Dublin, Dublin, Ireland

D.F. Laefer
University College Dublin, Dublin, Ireland
New York University, New York, USA

E. Mangina
University College Dublin, Dublin, Ireland

ABSTRACT

Bridges are essential components of rod network to connect different services at daily basic. Increasing freight and environmental impact cause deficiencies of bridge's structures, which may cause catastrophic collapse. To prevent any sudden close of the bridge or extreme events, the bridges are often inspected in period time, for example, general inspection is carried on every two years. Although many methods have been developed to support to bridge inspection, the visual inspection with physic inspectors on the site is dominant. However, the visual inspection has many shortcomings: (1) subjective results; (2) requirement of heavy and/or special equipment; (3) traffic closure; (4) requirement of high-skill trained inspectors; (5) high risk for inspector; and (6) time consuming and expensive.

An alternative method can be using remote sensors, such as camera (Nishimura et al. 2012) for the inspection. However, these methods also have drawbacks: difficulty in acquiring details of an entire structure because of restriction of fixed view angles or associated occlusions problem. Recently, with development of robotics and computer vision, low cost UAVs have been introduced to the market, and using UAVs for bridge inspection became a competitive method with many benefits such as non-contact measurement, no requirement of traffic close or any heavy/special equipment, no need trained engineers. Additionally, UAVs provide better data coverage, especially in hard to rich area like the bottom side of the deck or higher part of the bridge's pylon.

Recent state-of-the-art of computer vision-based methods allow to generate accurate, high dense point cloud from UAVs-images with a single digital camera, which is often provided by laser scanning system. That can show UAVs' ability in capture 3D topographic data of structures. This has accelerated the application of using UAVs for infrastructure inspections, such as building modeling (Byrne et al. 2017), dam inspection (Hallermann et al. 2015), and road surface evaluation (Distresses & Zhang 2012).

Focusing on bridge inspection, this study investigates the utility of the Structure from Motion (SfM) approach to generate a point cloud from low-altitude aerial imageries collected by Unmanned Aerial Vehicle (UAV). Next, the paper also present a workflow based point density to remove redundant points known as outlier data points mainly caused by water surface or terrain. To find an effectively solution to resolve the outlier noise problem, two commonly used noise reduction methods, statistical based method and surface fitting method have been tested with various of parameter settings. It shows that, the SOR filter can efficiently remove most of the outlier noise in this situation. By searching 400 neighbors at each point, the FPR can achieve 2.54%, which means 97.46% of noise have been removed. But the accuracy and processing time can be further improved.

REFERENCES

Byrne, J. et al., (2017) 3D reconstructions using unstabilized video footage from an unmanned aerial vehicle 3D reconstructions using unstabilized video footage from an unmanned aerial vehicle. *Journal of Imaging*, 3, 15.

Distresses, S. & Zhang, C. (2012) An unmanned aerial vehicle-based imaging system for 3D measurement of unpaved road. *Computer-Aided Civil and Infrastructure Engineering*, 27, 118–129.

Hallermann, N., Morgenthal, G. & Rodehorst, V. (2015) Unmanned aerial systems (UAS) – Case studies of vision based monitoring of ageing structures. *International Symposium Non-Destructive*, 15–17 September 2015, Berlin, Germany. p. 15–17.

Nishimura, S. et al. (2012) Development of a hybrid camera system for bridge inspection. In: *Proceedings of the 6th International IABMAS Conference*, CRC Press, Stresa, Lake Maggiore pp. 2197–2203. Available from: http://www.crcnetbase.com/doi/abs/10.1201/b12352-328.

Using step-by-step Bayesian updating to better estimate the reinforcement loss due to corrosion in reinforced concrete structures

F. Schoefs
Research Institute of Civil Engineering and Mechanics, Université de Nantes, Nantes, France

B. Heitner, T. Yalamas & G. Causse
Phimeca Engineering, Clermont-Ferrand, France

E.J. O'Brien
School of Civil Engineering, University College Dublin, Dublin, Ireland

ABSTRACT

Probabilistic assessment of ageing structures has become an important research area as it interests not only researchers but investors, municipalities and governments. The most commonly used material for many important structures and infrastructure is reinforced concrete. Various degradations of such structures are manifest in the form of direct loss of reinforcement area. In this study a methodology is presented that is based on the time dependent stochastic modelling of the reinforcement loss (in [%]) due to corrosion and Bayesian updating. This latter can be applied in multiple steps during the lifetime of the structure in order to improve the estimate on the reinforcement loss.

An example application is studied where the corrosion propagation is modeled based on well-known formulaes, also used by Vu and Stewart (2000). In this case LogNormal (LN) distribution shows the best fit to the simulated reinforcement loss data at a given time (see Fig. 1). This LN distribution involves three parameters, which are to be updated: μ_{log}, σ_{log} and γ. The updating is performed at $year = 20$ and 40 using simulated data and Markov Chain Monte Carlo method.

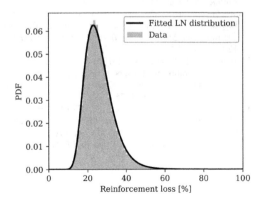

Figure 1. Simulated reinforcement loss at $year = 20$ and the fitted LN distribution.

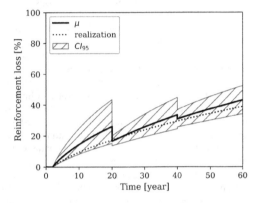

Figure 2. Results of Bayesian updating at $year = 20$ and 40.

Nine different cases are considered depending on the number (10, 20 or 50) and the quality (Coefficient of variation, $CoV = 0.05, 0.1, 0.2$) of the measurement data.

An example of the results of the simulated application can be seen in Figure 2. The main attributes of the reinforcement loss at every time step, i.e. the mean (μ), the $\mu +/-$ one standard deviation (σ) and the lower and upper bounds of the 95% confidence interval (CI_{95}) are computed. The realization that served as a basis for the updating (the simulation assumed that this realization represents the reality) is marked with dashed line.

In all the studied cases it has been possible to improve the estimation of the reinforcement loss. Comparing the results of the different cases, it is also found that collecting less data multiple times leads to higher rate of improvement than collecting more data at a single point in time.

REFERENCE

Vu, K.A.T. & Stewart, M.G. (2000) Structural reliability of concrete bridges including improved chloride-induced corrosion models. *Structural Safety*, 22, 313–333.

Life-Cycle Analysis and Assessment in Civil Engineering: Towards an
Integrated Vision – Caspeele, Taerwe & Frangopol (Eds)
© 2019 Taylor & Francis Group, London, ISBN 978-1-138-62633-1

Structural health monitoring of bridges: A Bayesian network approach

M. Vagnoli, R. Remenyte-Prescott & J. Andrews
Resilience Engineering Research Group, The University of Nottingham,
Nottingham, UK

ABSTRACT

The health state of the critical infrastructure of transportation networks, such as bridges, tunnels, roads and
railways, is influenced strongly by continuous effects
of external factors, such as wind, storm, heavy traffic, etc. Therefore, catastrophic failures of the critical
infrastructure, such as a bridge collapse, can occur
unexpectedly due to the degradation of the infrastructure, especially if the levels of degradation are
unknown. For example, between February 2016 and
April 2017 seven bridges around the world collapsed
suddenly, by causing fatalities and significant economic losses. Structural Health Monitoring (SHM)
techniques can assess the health state of an infrastructure by analysing its behaviour, and point out degraded
elements that require to be maintained. As a consequence, SHM methods can improve the safety of the
transportation network, and reduce the life cycle costs
of the infrastructure, by optimizing the maintenance
schedule. In this paper, a Bayesian Belief Network
(BBN) approach is proposed to assess the health state
of a beam-and-slab bridge, by relying on the analysis of its vertical acceleration. The BBN can monitor
the health state of a bridge, by taking account of the
health state of each bridge elements at the same time.
The BBN-based approach can also provide an assessment of the bridge health state every time when new
evidence of the bridge behaviour becomes available,
and consequently the health state of the bridge can be
monitored continuously. Moreover, the BBN can treat
both sources of uncertainty, i.e. epistemic and aleatory
uncertainties, which are not accounted for usually. The
BBN allows to merge the systematic knowledge of
the engineers, with the analysis of the measurement
of the bridge behaviour. Indeed, the qualitative part
of the BBN, i.e. the BBN structure, is developed by
considering the knowledge of bridge engineers. At
the same time, bridge engineers are also interviewed
with the aim of defining the Conditional Probability
Tables (CPTs) of the BBN. The CPTs, which represent the quantitative part of the BBN, are used to

investigate relationship between different bridge elements. The judgments of the experts are translated
into probabilities by using a Fuzzy Analytical Hierarchy Process (FAHP) (Kabir et al., 2016). The CPTs
are then updated by analysing the behaviour of the
bridge under changing health states, i.e. the acceleration of the bridge is analysed when the bridge
is healthy and damaged. A Damage Index (DI) is
assessed in order to update the CPTs by the means of its
probability density function (pdf) (Moughty & Casas,
2017). The damage index pdf is reconstructed by using
a Finite Mixture Model (FMM) approach, which is
chosen due to its ability in reliably reconstructing
the pdf with a low number of samples (Hoseyni et
al., 2015) (Vagnoli et al., 2017). The performance
of the BBN method is demonstrated by analysing
an FEM of a beam-and-slab bridge, where the damage is modelled as loss of stiffness at selected beam
elements.

The proposed BBN approach has demonstrated to
provide accurate results in detecting degraded health
state of the bridge, as well as good accuracy in
diagnosing the degraded elements of the bridge.

REFERENCES

Hoseyni, S.M., Di Maio, F., Vagnoli, M., Zio, E. & Pourgol-
Mohammad, M. (2015) A Bayesian ensem-ble of sensi-
tivity measures for severe accident mod-elling. *Nuclear
Engineering and Design*, 295, 182–191.
Kabir, G., Sadiq, R. & Tesfamariam, S. (2016) A fuzzy
Bayesian belief network for safety assessment of oil and
gas pipelines. *Structure and Infrastructure Engineering*,
12 (8), 874–889.
Moughty, J.J. & Casas, J.R. (2017b) Performance assessment of vibration parameters as damage in-dicators for
bridge structures under ambient excitation. *Procedia
Engineering*, 199, 1970–1975.
Vagnoli, M., Di Maio, F. & Zio, E. (2017c) Ensem-bles of
climate change models for risk assessment of nuclear
power plants. *Proceedings of the Institution of Mechanical
Engineers, Part O: Journal of Risk and Reliability*, DOI
1748006X17734946.

Noninvasive empirical methods of damage identification of bridge structures using vibration data

J.J. Moughty & J.R. Casas
Department of Civil and Environmental Engineering, Technical University of Catalonia (BarcelonaTech), Catalonia, Spain

ABSTRACT

Once identified, damage in bridges should be repaired promptly, as maintenance costs dramatically increase when damage is left unattended; therefore, it is vital that bridge owners conduct necessary inspections and structural testing on a regular basis, however, due to the large quantity of bridges, this is impractical. Instead, sensors can be used to indirectly infer the presence of damage using theoretical relationships, such as that of modal frequency and stiffness, or some other modal parameter, some of which are discussed in Moughty & Casas, (2017), however, the use of modal parameters as damage sensitive features have found that their performance may suffer from their sensitivity to environmental and operational conditions. Additionally, the non-stationarity of some vibration signals creates a problem when using standard modal techniques that employ Fourier-based transforms, as linear stationarity is assumed.

The present paper details a number of empirical vibration parameters that have been designed to be suitably applicable to the specific vibration signals that are typical from either; long duration, low energy stationary ambient signals or short duration, high energy non-stationary vehicle induced excitation signals. Two field experiments of progressively damaged bridges subjected to ambient and vehicle induced excitation, respectively, are used as case studies for the damage identification assessment. The vibration parameters assessed are; Cumulative Absolute Velocity (CAV), Cumulative Absolute Displacement (CAD), Distributed Vibration Intensity (DVI), Mean Cumulative Vibration Intensity (MCVI),

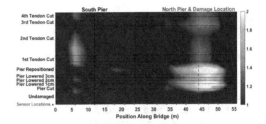

Figure 1. Evolution of CAV at each sensor (S101 Bridge).

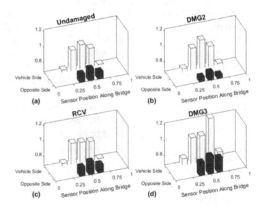

Figure 2. Normalized CAD values per sensor for damage scenarios; (a) Undamaged, (b) DMG2, (c) RCV & (d) DMG3.

Instantaneous Vibration Intensity (IVI). Table 1 is provided to give a breakdown of the applicability of each of the empirical vibration parameters to specific vibration properties that may influence their damage identification performance.

The results of the ambient induced excitation case found that both CAV & DVI exhibited detection and location capabilities (Figure 1). In the vehicle induced excitation case, all empirical parameters assessed identified the required damage events, with CAD & IVI providing the best results (Figure 2).

Table 1. Vibration parameter application classification.

Vib. Para. Property	CAV	CAD	DVI	MCVI	IVI
Fourier-Based	✗	✗	✓	✓	✗
Non-Stationary Signal Applicability	✓	✓	✗	✗	✓
Long Duration Signal Applicability	✓	✗	✓	✓	✗
Suitability to Ambient Induced Excitation	✓	✗	✓	✓	✗
Suitability to Vehicle Induced Excitation	✓	✓	✗	✗	✓

REFERENCE

Moughty, J.J. & Casas, J.R. (2017) A state of the art review of modal-based damage detection in bridges: development, challenges, and solutions. *Applied Sciences*, 7(5), 510.

Bridge condition evaluation using LDVs installed on a vehicle

A.D. Martínez Otero, A. Malekjafarian & E.J. O'Brien
University College Dublin, Dublin, Ireland

1 INTRODUCTION

A recently proposed method is to use an instrumented vehicle passing overhead to monitor the bridge condition. This is known as indirect or drive-by monitoring. Large bridge networks can be assessed using drive-by monitoring strategies, which has the potential to greatly reduce the cost of assessment per bridge. Moreover, the drawbacks related to visual inspection and direct instrumentation are avoided.

The drive-by version of standard curvature is known as Instantaneous Curvature (IC), which can be calculated from deflections measured from the vehicle. LDVs are able to measure the relative velocity between a vehicle and the pavement below as it travels at full road speed. In this paper, a structural health monitoring approach is proposed involving Laser Doppler Vibrometers (LDVs) installed on a vehicle. Relative velocities are measured to obtain the Rate of Instantaneous Curvature of the velocity (RIC). Standard deflection curvature is shown to be sensitive to local damage. RIC is obtained using the first derivative of IC with respect to time. A damage indicator obtained from RIC, the Difference Ratio (DR), is tested both in noise-free and noisy conditions.

2 RESULTS

The damage situation considered in this paper is illustrated in Figure 1 with a 50% loss of height crack at 5m distance from left support.

There is a noise associated with the measurements of the LDV installed in the vehicle that is considered in the simulation. Figure 2 shows RIC_{av} calculated using the noisy LDV data. Greater differences are noticeable around damaged location.

Figure 1. Damage scenario.

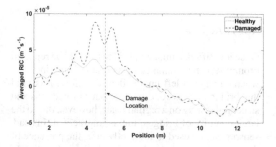

Figure 2. RIC_{av} for healthy and damaged cases under noisy conditions.

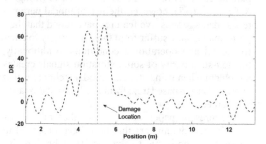

Figure 3. Difference Ratio (DR) of damage scenario in Figure 2 in presence of noise.

DR filtered results in noisy conditions are displayed in Figure 3. Crack location can be identified despite noise addition.

3 CONCLUSION

In this paper, a new damage location method is proposed using the relative velocity measurements obtained from a vehicle. The Rate of Instantaneous Curvature (RIC) is calculated from these measurements and is sensitive to damage. Difference Ratio (DR) is introduced for damage location purposes. Further research is needed to evaluate both the influence of an average highway road profile and the influence of lower damage applied to a bridge.

Life-Cycle Analysis and Assessment in Civil Engineering: Towards an Integrated Vision – Caspeele, Taerwe & Frangopol (Eds)
© *2019 Taylor & Francis Group, London, ISBN 978-1-138-62633-1*

On the bonding performance of Distributed Optical Fiber Sensors (DOFS) in structural concrete

A. Barrias & J.R. Casas
UPC, Department of Civil and Environmental Engineering, Technical University of Catalonia, Barcelona, Spain

S. Villalba
UPC, Department of Engineering and Construction Projects, Technical University of Catalonia, Terrassa (Barcelona), Spain

ABSTRACT

The use of fiber optic sensors (FOS) in Structural Health Monitoring applications has been studied and practiced with success for more than two decades. However, this has been mostly focused on discrete fiber optic sensors. Distributed optic fiber sensors (DOFS) offer the same advantages of point FOS with the advantage of enabling the monitoring of great extents of the structural element with the use of up to a single sensor, decreasing the implementation costs and facilitating a more global behavior of the structure with a simpler monitoring system. Within these, the Optical Backscatter Reflectometer (OBR) based DOFS are the ones which offer a better spatial resolution which is required for the detection of local damage. This technology has been studied and experimented in different applications (Villalba and Casas, 2013; Rodríguez, Casas and Villaba, 2015).

Nonetheless, there is still some uncertainty related to the performance of different bonding adhesives in the application of this technology to concrete structures. Due to the irregularities on the surface of this material and the heterogeneity of its aggregates, the study of the performance of the optimal bonding mechanism in these applications is of great interest and relevance. Moreover, the use of different spatial resolution and sampling acquisition inputs is also a field that requires further research.

In this way, the authors present a laboratory experiment where a reinforced concrete beam was instrumented with a 5-meter-long polyimide DOFS in a way that four equal segments were bonded to the bottom surface of the beam using for each segment a different type of adhesive: neutral cure silicone, polyester, epoxy and cyanoacrylate. Moreover, three strain gauges were also used to compare the results. The beam was then loaded, allowing for a direct comparison between all segments. Additionally to the comparison with the other instrumented sensors, it is also important the consideration and analysis of the associated Spectral Shift Quality (SSQ) values of the

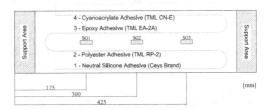

Figure 1. Instrumented sensors at beam bottom surface.

DOFS measurements to decide on the accuracy of the measurements.

From this experiment, it was possible to conclude a slightly better performance of the neutral cure silicone when adhered to the concrete in the elastic range of the loading. After cracking, all bonded segments are able to detect and locate the damage, but further analysis is required to overcome the drop of the associated SSQ values.

REFERENCES

Rodríguez, G., Casas, J. R. & Villaba, S. (2015) Cracking assessment in concrete structures by distributed optical fiber. *Smart Materials and Structures*. IOP Publishing, 24(3), 35005.

Villalba, S. & Casas, J. R. (2013) Application of optical fiber distributed sensing to health monitoring of concrete structures. *Mechanical Systems and Signal Processing*. Elsevier, 39(1), 441–451.

A machine learning approach for the estimation of fuel consumption related to road pavement rolling resistance for large fleets of trucks

F. Perrotta, T. Parry & L.C. Neves
University of Nottingham, Nottingham, UK

M. Mesgarpour
Microlise Ltd, Nottingham, UK

ABSTRACT

There remains a level of uncertainty concerning the methodological assumptions and parameters to consider in the estimation of road vehicle fuel consumption due to the condition of road pavements. In fact, recent studies highlighted how existing models can lead to very different results and that because of this, they are not fully ready to be implemented as standard in the life-cycle assessment (LCA) framework (Trupia et al. 2016). This study presents an innovative approach, based on the application of the Boruta algorithm (Kursa & Rudnicki 2010) (BA) and neural networks (NN), for the estimation of the fuel consumption of a large fleet of truck, which can be used to estimate the use phase emissions of road pavements.

The available dataset comes anonymized from the databases of truck fleet managers that collect and analyse data from standard sensors (SAE International 2016) installed on modern trucks to monitor and optimize the operational costs of their fleets. This is enriched with information from the HAPMS (the Highways Agency Pavement Management System) the management system of Highways England. In particular, the average gradient, radius of curvature, and measurements of roughness and macrotexture are fused with to the data of truck fleet managers based on GPS coordinates.

Based on the results of the BA, the gross vehicle weight, road gradient, vehicle speed, acceleration, the torque and revolutions of the engine, the gear used, three different wavelengths of pavement roughness (LPV03m, LPV10m and LPV30m) and the macrotexture have been identified to significantly impact the fuel consumption of the considered fleet of trucks. These are the variables that have been included in the developed model. This partially confirms the findings of previous studies and gives more confidence in this approach.

Results of the study show that NNs are suitable to analyse the large quantities of data being able to take into account the impact of rolling resistance-related parameters (road pavement roughness and macrotexture) on truck fuel consumption.

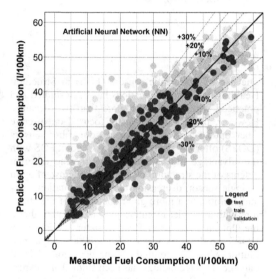

Figure 1. Fit of the developed NN.

Regarding possible improvements and future work, analysis of a wider ranged vehicle types (i.e. light trucks, vans and cars) vehicle speeds, road conditions representative of all the road network in UK and effect of weather (e.g. air temperature, wind speed, etc.) may help in obtaining more generally applicable results that would help in reducing uncertainties in the estimation of costs and environmental impact of the use phase of road pavements.

REFERENCES

Kursa, M.B. & Rudnicki, W.R. (2010) Feature selection with the boruta package. *Journal of Statistical Software*, 36(11), 1–13. Available from: http://www.jstatsoft.org/v36/i11/-paper.

SAE International (2016) SAE J1939-71, Vehicle application layer – surface vehicle recommended practice. *SAE International Standards*. Available from: http://standards.sae.org-/j1939/71_201610/.

Trupia, L. et al. (2016) Rolling resistance contribution to a road pavement life cycle carbon footprint analysis. *International Journal of Life Cycle Assessment*, 972–985.

Fuzzy-random approach to debris model for riverbed scour depth investigation at bridge piers

L. Sgambi
Faculty of Architecture, Architectural Engineering and Urbanism, Université catholique de Louvain, Louvain-la-Neuve, Belgium

N. Basso
Department of Architecture, School of Creative Science and Engineering, Waseda University, Tokyo, Japan

E. Garavaglia
Department of Civil and Environmental Engineering, Politecnico di Milano, Milan, Italy

ABSTRACT

The removal of sediment around bridge abutments and piers due to the erosive action of flowing water (i.e. scouring) is of the greatest concern to society. Currently it has been estimated that scour produced by rivers and streams causes about 60% of the total amount of bridge failures (Hamill, 1999). Underestimating this natural process can seriously threaten the overall safety of the infrastructure.

Several factors may affect the scour depth at bridge piers: flow intensity and sediment grading, flow depth, nature and occurrence of floods, side wall effects, sediment size, geometry and inclination of piers, etc. The depth of the scour hole in the sand adjacent to the bridge foundations can be estimated using theoretical models with hydraulic parameters. However, the uncertainty associated with the parameters involved in the evaluation (e.g. flow characteristics, debris, structural and geotechnical factors, etc.) makes it almost impossible to adopt a deterministic approach for the reliability analysis (Dordoni et al., 2010, Malerba et al., 2011).

Therefore, in order to properly assess the scour depth, both aleatory variability (i.e. due to randomness) and epistemic uncertainty (i.e. due to limited data and knowledge) must be considered. A fuzzy-probabilistic approach can take some of those uncertainties into account. This paper proposes an original method for modelling the debris action in river bridges. Based on fuzzy-random theory, both the aleatory variability related to the particle accumulation size and the epistemic uncertainty characterising fluvial hydraulics equations can be successfully modelled. As a consequence, the method here presented can describe in a more general, proper way the uncertainty of scour phenomenon.

The approach is applied to a case study (Parks Highway Bridge over the Tanana River, Alaska, Figure 1) in order to compare its results with the ones achieved by the U.S. Department of Transportation using a different procedure (equivalent pier depth method) and

Figure 1. Picture of the bridge structure and the pier in the riverbed (photo credit: Don Moe, http://donmoe.com/blog/2016/01/15/fairbanks-ak-to-denali-highway/).

prove its efficacy, thanks to the proximity of the values obtained.

As well as for the equivalent pier depth method, at the present, there are no data in scientific literature that can validate the results achieved with this original approach. Therefore, this work can be considered as a methodology paper. However, when more experimental data are available, the aleatory variables will be easily identified.

REFERENCES

Arneson, L.A., Zevenbergen, L.W., Lagasse, P.F. & Clopper, P.E. (2012) Evaluating scour at bridges. Report No. FHWA-HIF-12-003, U.S. Department of Transportation.

Dordoni, S., Malerba, P.G., Sgambi, L. & Manenti, S. (2010) Fuzzy reliability assessment of bridge piers in presence of scouring. *Fifth International Conference on Bridge Maintenance, Safety and Management*, Philadelphia, PA, 11–15 July 2010.

Hamill, L. (1999) *Bridge Hydraulics*. E&FN Spon, New York.

Malerba, P.G., Garavaglia, E. & Sgambi, L. (2011) Fuzzy-Monte Carlo simulation for the safety assessment of bridge piers in presence of scouring. In: Chang-Koon Choi (Ed), *2011 World Congress on Advances in Structural Engineering and Mechanics*. Seoul, Korea, 18–22 September 2011, Techno Press. pp. 558–606.

SS7: Application of probabilistic methods in fire safety engineering
Organizers: D. Rush, L. Bisby & R. Van Coile

Life-Cycle Analysis and Assessment in Civil Engineering: Towards an Integrated Vision – Caspeele, Taerwe & Frangopol (Eds)
© 2019 Taylor & Francis Group, London, ISBN 978-1-138-62633-1

Effect of modelling on failure probabilities in structural fire design

M. Shrivastava, A.K. Abu, R.P. Dhakal & P.J. Moss
University of Canterbury, Christchurch, New Zealand

ABSTRACT

Probabilistic structural fire design enables the assessment of loss of life or property that may occur from fire events. The effects of fire hazards on buildings are interpreted in terms of levels of damage in order to predict probabilities of failure or losses. As there is limited data on monitored response of buildings to real structural fires, current failure predictions are based on numerical simulations of structural behaviour. However, numerical simulations are model dependent, which implies that the same structure could have different responses if modelled under different configurations. These differences could be significant and may affect failure estimations, and hence the probability of failure. Uncertainty in structural modelling is classified as epistemic uncertainty. The response of a structural element is different if modelled as an isolated member (2D) or as part of a structural system (3D). A few researchers have compared structural response obtained from different structural modelling (Quiel and Garlock, 2010, Flint et al., 2006). It is observed that no work has been done so far in quantifying the risk derived from the analysis of different structural configurations. For the probabilistic assessment of a given fire hazard, the variation in structural response due to different styles of modelling (2D or 3D) produces different annual probabilities of structural response. The quantification of this epistemic uncertainty is important because it provides information about the risk involved in the design solution due to structural modelling. A noticeable variation in the probability of exceedance of structural response from different structural configuration may significantly affect the probability of failure which invariably governs the cost to reduce the failure risk. Therefore, this paper evaluates and compares the probability of exceedance of structural response given multiple fire scenarios derived using different structural models, 3D and 2D.

This procedure has been demonstrated with the help of a case study of a composite steel beam, modelled both as an isolated element and as a part of a 3D structure, exposed to a suite of fire profiles. The annual probabilities of failure for both structural types are evaluated and compared to highlight the effects of structural modelling on failure probability. It is observed that 3D modelling produces less deflections and comparatively less probability of exceedance of a level of displacement. It is also important to consider the computation time when selecting the structural configuration for modelling. 3D analysis may produce more accurate results but could be time consuming. It can be concluded that if a member is designed isolated then it will produce conservative results.

REFERENCES

Flint, G., Usmani, A., Lamont, S., Lane, B. & Torero, J.L. (2006) Fire induced collapse of tall buildings. *4th International workshop for Structures in Fire (SiF'06)*.

Quiel, S. & Garlock, M. (2010) Parameters for modeling a high-rise steel building frame subject to fire. *Journal of Structural Fire Engineering*, 1, 115–134.

Numerical analysis on the fire behavior of a steel truss structure

L.M. Lu
School of Mechanical and Civil Engineering, China University of Mining & Technology, Jiangsu, China
JiangSu Collaborative Innovation Center for Building Energy Saving and Construction Technology, Jiangsu, China

G.L. Yuan, Q.J. Shu & Q.T. Li
School of Mechanical and Civil Engineering, China University of Mining & Technology, Jiangsu, China

ABSTRACT

Due to the vigorous development of large-space multi-functional architectures, large-span steel structures have been widely used in recent years. At the same time, the fire-resistance design of these structures has got more and more attention. Since the traditional methods which apply the ISO-834 temperature curve to predict the fire performance of structures cannot be easily used for large space structures (Du, 2012). The researches on the structural response of large space steel truss structures under real fire scenarios are limited, especially considering whole heating-cooling process (Xue, et al. 2013, Fan, et al. 2014). Therefore, further researches are still necessary.

This paper conducted a comprehensive case study on fire behavior of a large space exhibition centre in Shanxi province China, as shown by Fig. 1.

The fire scenarios were simulated by FDS finite element package (Kevin, et al. 2010). Four fire scenarios were designed, as listed in Table 1. The fires were supposed to be fast fires and the fire fighting system were out of work.

The temperature field of the steel members for the exhibition centre was calculated using the simplified calculation method proposed in the Chinese Code: "Technical specification for fire protection of steel structure of buildings" (CECS 200–2006). It has been written into a MATLAB program. The air temperature around the steel truss member was determined from the temperature-time curves predicted by FDS modeling.

The structural performance of the steel trusses was simulated by ANSYS. The displacements and the equivalent stresses of steel members before and after heating in the four fire scenarios were compared and analyzed. The suggestions for fire resistance design are proposed: (a) to increase the cross-section of the structural member near the supports; (b) to reduce the outrigger dimension of the structure; (c) to avoid piling combustible goods near the corner of supports during service period of the structure, and control the area of fire source and arrange high exhibition platform in the high space zone.

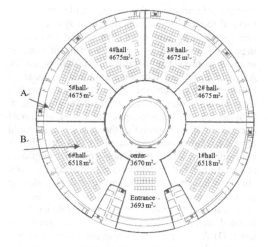

Figure 1. Layout of the exhibition halls.

Table 1. Fire scenarios.

No.	Fire source location	Fire source	Fire source area (m×m)
1	Near wall (A)	Exhibition items	18×18
2	Mid-span (B)	Exhibition items	18×18
3	Near wall (A)	Exhibition items	9×9
4	Mid-span (B)	Exhibition items	9×9

REFERENCES

Du, Y. & Li, G.Q. (2012) A new temperature–time curve for fire-resistance analysis of structures. *Fire Safety Journal*, 54(1), 113–120.

Fan, S., Shu, G.P., She G.J. & Liew, J.Y.R. (2014) Computational method and numerical simulation of temperature field for large-space steel structures in fire. *Advanced Steel Construction*, 10(2), 151–178.

Kevin M., Klein B., Hostikka S. & Jason, F. (2010) Fire Dynamics Simulator (Version 5) Technical Reference Guide. NIST special publication 1018-5, USA.

Xue, S.D., Xiong, J.L. & Li, Y. (2013) Empirical formula for air temperature in large space structure under fire. *Journal of Beijing University of Technology*, 39(2), 203–207.

The application of an LQI reliability based methodology to determine the fire resistance requirements for two Mumbai residential towers

D. Hopkin, S. Lay & A. Henderson
Olsson Fire & Risk, UK

ABSTRACT

This paper presents a case study concerning a probabilistic performance based structural fire engineering analysis which served to inform the fire resistance requirements for two adjacent high-rise residential concrete structures in Mumbai, India (Figure 1).

The analyses sought to define the probability of structural element failure in the event fire. Fire resistance solutions, i.e. 60, 90, 120 and 240 minutes, were trialed with the intent of ensuring that the proposed fire resistance solution results in an appropriately low probability of failure.

The acceptably low probability of failure, forming the acceptance criteria, was derived based upon a review of International structural reliability literature, including the Natural Fire Safety Concept Valorization Project (Schleich & Cajot 2011), EN 1990 (BSI 2002), ISO 2394 (ISO 2005) and the Joint Committee on Structural Safety (Faber & Vrouwenvelder 2001). Based upon the recommendations therein, an appropriately low probability of structural element failure in the event of fire was adopted in consideration of the relevant Society's Capacity to Commit Resources (SCCR). Considering the numerous factors featuring uncertainty / variability, e.g. fire load density, ventilation conditions, sprinkler reliability, construction tolerance, material properties,

Figure 1. Island City Centre, Mumbai – Buildings 1 and 2.

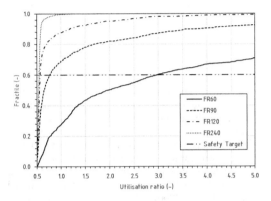

Figure 2. Utilization vs. cumulative probability for a 140 m^2 apartment – Evaluation of different slab fire resistance solutions.

etc., it was determined that a fire resistance period of 120 minutes was appropriate for the two towers (Figure 2). This was derived after a series of Latin Hypercube Sampling studies, which: (i) generated design fire conditions, (ii) undertook heat transfer analyses to a concrete slab element, and (iii) assessed the slab utilization throughout the full duration of the fire.

Sensitivity studies were also undertaken considering the impact of variance in core inputs, such as fire ignition rate, number of fatalities, and sprinkler intervention rates.

REFERENCES

BSI (2002) *BS EN 1990:2002 Basis of Structural Design*. British Standards Institution, London.
Faber, M. & Vrouwenvelder, T. (2001) *Probabilistic Model Code*. Copenhagen: Joint Committee on Structural Safety.
International Organisation for Standardisation (2015) *ISO 2394:2015: General Principles on Reliability for Structures*. ISO, Geneva.
Schleich, J. & Cajot, L.-G. (2001) *Valorisation Project – Natural Fire Safety Concept*. Profil Arbed, Luxembourg.

Target safety levels for insulated steel beams exposed to fire, based on Lifetime Cost Optimisation

R. Van Coile
Department of Structural Engineering, Ghent University, Belgium

D.J. Hopkin
Olsson Fire & Risk, UK
Fire Research Group, The University of Sheffield, UK

ABSTRACT

The absence of clear target safety levels for structural fire engineering severely hampers probabilistic structural fire design (Hopkin et al., 2017). In support of a generalized definition of target safety levels for structural fire safety engineering, optimum target safety levels for insulated steel beams are determined as a function of the fire characteristics by applying lifetime cost optimization (LCO) techniques, based on (Rackwitz, 2000). Where the fire development characteristics support the prospect of flashover, the Eurocode parametric fire curve is considered, otherwise fires are assumed to roam in search of fuel, leading to spatial variations in temperature, with thermal exposure to structural elements described via travelling fire methods. Fragility curves are derived as a function of, amongst others, the insulation thickness and fire load density, and applied in the LCO evaluations. The LCO results in an assessment of the optimum investment level as a function of the fire, damage and investment cost parameters characterizing the building.

Optimum reliability indices $\beta_{fi,opt}$ obtained for a representative UK office building floorplate are given in Figure 1, considering different nominal fire load densities $q_{F,nom}$. In Figure 1, DII is the damage-to-investment-indicator denoting the cost of fire-induced failure relative to the (rate of the) fire protection investment cost. The secondary X-axis indicates the expected yearly damage cost relative to the base construction cost, considering representative UK values for steel fire protection and discounting. Background to the applied parameters and their interpretation is given in the full paper.

It is intended that the current contribution can be a stepping stone towards rational and validated reliability targets for PBD in structural fire safety engineering. A final proposal of target reliability levels for structural fire design should be made by a code-making committee, informed by detailed studies such as that presented here. While further evaluations

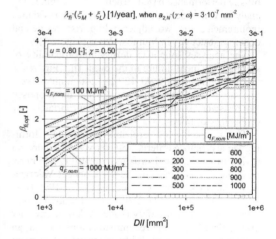

Figure 1. Optimum reliability index given a fully developed fire, $\beta_{fi,opt}$, for an office floorplate, with $q_{F,nom}$ the nominal fire load density.

are necessary to generalize the results, the following conclusions have been derived:

- The optimum level of structural fire safety is highly dependent on both the severity of fire-induced damage and the cost of improving fire resistance;
- The logarithm of the optimum failure probability is linearly related to the logarithm of the damage-to-investment ratio;
- Optimum fire resistance ratings are not sensitive to precise evaluations of input parameters. It is expected that safety targets can be based on order-of-magnitude assessments for input parameters.

REFERENCES

Hopkin, D., Van Coile, R. & Lange, D. (2017) Certain uncertainty-demonstrating safety in fire engineering design and the need for safety targets. *SFPE Europe*.

Rackwitz, R. (2000) Optimization—the basis of code-making and reliability verification. *Structural Safety*. 22, 27–60.

SS8: Bespoke models for marine structural management
Organizer: M. Collette

Integrated Computational Materials Engineering (ICME) techniques to enable a materials-informed digital twin prototype for marine structures

C.R. Fisher, K. Nahshon, M.F. Sinfield & D. Kihl
Naval Surface Warfare Center, Carderock Division (NSWCCD), West Bethesda, MD, USA

ABSTRACT

Residual stress from material processing and fabrication can severely degrade structural performance over a ship's lifecycle. However, the evolution of the residual stress distribution throughout the shipbuilding process is not well understood. Integrated Computational Materials Engineering (ICME) techniques enable linking disparate software codes across multiple length scales, thereby facilitating simulation of the entire material lifecycle – from material processing to fabrication to structural performance. Figure 1 shows a schematic of these linkages. This project is pairing computational simulation with physical measurement for verification and validation (V&V) of the linked finite-element analysis (FEA) tools.

The linked FEA tools enable simulation of representative marine structures. Concurrently, the residual stress in an analogous physical structure will be measured through each step of the fabrication process: incoming plate, welded assembly, and post-fatigue testing. The effort will follow specific areas within the component to understand the effects of fabrication and testing on residual stress magnitude and distribution. This fabrication-cycle material information is essential to understand the initial state of marine structures as the industry moves towards a digital twin prototype standard.

The first phase of this multi-year program is linking computational FEA codes is across the fabrication and performance phases. In this way, welding-induced residual stresses and distortions from welding are linked to fatigue and buckling performance requirements for marine structures. For this example, the

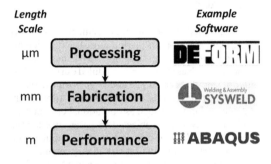

Figure 1. Graphical representation of the ICME-based linkage of FEA tools across multiple length scales.

SYSWELD weld simulation software was linked with *Abaqus FEA* tools. The translation of these results is accomplished using a newly developed Python script. Full details of extracting the results from *SYSWELD*, translating these results using the newly developed Python script, and including the results in *Abaqus FEA* are provided through an illustrative example analysis of an aluminum weldment.

Future developments will seek to link the material processing phase to the fabrication phase. This linkage is being investigated on the *DEFORM* software, which will be linked to *SYSWELD*. The goal is complete linkage across *DEFORM* → *SYSWELD* → *Abaqus FEA*. The development of this capability is motivated by the shipbuilding industries increased use of lightweight aluminum structures, in which the inclusion of the effects of welding in structural components can be critical to understanding structural response over that structure's lifecycle.

Adapting life-cycle management of ship structures under fatigue considering uncertain operation conditions

Y. Liu & D.M. Frangopol
Department of Civil and Environmental Engineering, ATLSS Engineering Research Center, Lehigh University, Bethlehem, PA, USA

ABSTRACT

Risk-informed life-cycle management has drawn increasing attention for marine infrastructure, including ship and offshore structures. Risk management measures such as repairs are scheduled during the lifetime of ships to keep fatigue and/or corrosion damages below prescribed levels. However, factors affecting the fatigue performance such as operational environment and stress levels are subjected to large uncertainties (Collette 2017). These uncertainties complicate the decision making process for risk management.

Marine structural management against fatigue damages has been a central focus within life-cycle research studies. Life-cycle management plans can be scheduled to achieve optimal performance of fatigue-sensitive structures during their service life. However, in the fatigue modeling used in the existing studies, the stress ranges are often based on assumptions from idealized operational conditions. Though probabilistic distributions are adopted to deal with the uncertainties in loadings, they are limited when ships are facing radically changed operation conditions. Changes of operational loading, routing decision, and climate condition affect the actual loading that the fatigue detail experienced (see examples of long-term stress range distributions in Figure 1). Therefore, the original plan of life-cycle interventions on ship structures can be not adequate when operation conditions are changed.

Decision making under uncertainties of operation scenarios of ships is a complex process. Multiple possible scenarios of operation conditions could lead to various consequences. Similar challenge of dealing with multiple climate change scenarios was recently investigated in the bridge maintenance process (Mondoro, Frangopol, & Liu 2017). An adaptable life-cycle management plan for future uncertain operational profiles based on *regret* function is proposed in this paper. *Regret* was introduced to the decision making process to counteract this epistemic type uncertainty effect (Bell 1982). The regret function is specifically prepared for trade-off analysis among alternative

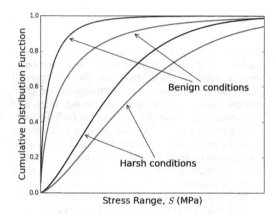

Figure 1. Cumulative distribution function of stress range.

consequences, thus is considered relevant for dealing with the problem herein.

The proposed life-cycle maintenance plan is illustrated using a fatigue detail of a ship structure. Both benign and harsh conditions are assumed to change the operational environment during the lifetime service of a ship under different stress ranges. The proposed adaptable lifetime maintenance plan is associated with the minimal regret facing the operation condition uncertainty. This approach can be used for planning lifetime maintenance interventions in the early stage for the risk management of ship structures under uncertain operation scenarios.

REFERENCES

Bell, D.E. (1982) Regret in decision making under uncertainty. *Operations Research*, 30(5), 961–981.
Collette, M. (2017) Uncertainty approaches in ship structural performance. In: *Handbook of Uncertainty Quantification*. Springer, pp. 1567–1588.
Mondoro, A., Frangopol, D.M. & Liu, L. (2017) Bridge adaptation and management under climate change uncertainties: A review. *Natural Hazards Review*, 19(1), 04017023.

Probabilistic service life management of fatigue sensitive ship hull structures considering various sea loads

S. Kim
Wonkwang University, Iksan, South Korea

D.M. Frangopol
Lehigh University, Bethlehem, PA, USA

ABSTRACT

The optimization process for service life management of fatigue-sensitive structures (e.g., ships and bridges) should consider the time-dependent structural degradation mechanisms under uncertainty (Frangopol & Soliman 2016, Kim et al. 2013). For this reason, significant efforts to assess and predict accurately the fatigue performance have been made (Kwon & Frangopol 2012). Ship structures subjected to repeated sea loadings experience structural deterioration over their lifetime due to fatigue (Soliman et al. 2016). The estimation of fatigue life is necessary for ensuring the structural safety, and managing the service life of ship structures. However, the accurate fatigue life prediction may be challenging due to the various uncertainties associated with loading and resistance effects. Climate change effects may result in more frequent abnormal sea loading, and more uncertainty in the fatigue life prediction of ship structures. For this reason, probabilistic fatigue life assessment considering various sea loading conditions, and the effect of abnormal sea loadings on the optimum service life management of deteriorating ship structures need to be investigated.

This paper presents the optimum service life management of fatigue sensitive ship hull structures considering various sea loading conditions. The effects of the mild, moderate and severe sea loading conditions on (a) the fatigue initiation and propagation, (b) the single-objective, and (c) multi-objective optimum service life management are investigated. The optimum service life management in this study is based on six objectives: maximizing the probability of fatigue crack damage detection, minimizing the expected fatigue crack damage detection delay, minimizing the expected repair delay, minimizing the damage detection time-based probability of failure, maximizing the expected extended service life, and minimizing the expected life-cycle cost. The design variables of the optimization are the inspection times. As a result, the relation among the loading conditions,

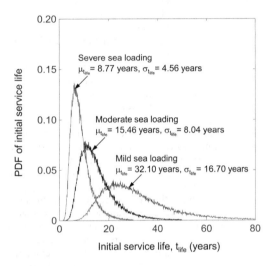

Figure 1. Initial service life of a fatigue sensitive ship hull structure under various sea loading conditions.

objective values and inspection times for service life management can be found.

REFERENCES

Frangopol, D.M. & Soliman, M. (2016) Life-cycle of structural systems: recent achievements and future directions. *Structure and Infrastructure Engineering*, Taylor & Francis, 12(1), 1–20.

Kim, S., Frangopol, D.M. & Soliman, M. (2013) Generalized probabilistic framework for optimum inspection and maintenance planning. *Journal of Structural Engineering*, ASCE, 139(3), 435–447.

Kwon, K. & Frangopol, D.M. (2012) System reliability of ship hull structures under corrosion and fatigue. *Journal of Ship Research*, SNAME, 56(4), 234–251.

Soliman, M., Frangopol, D.M. & Mondoro, A. (2016) A probabilistic approach for optimizing inspection, monitoring, and maintenance actions against fatigue of critical ship details. *Structural Safety*, Elsevier, 60, 91–101.

Ship motion and fatigue damage estimation via a digital twin

M. Schirmann, M. Collette & J. Gose
Department of Naval Architecture and Marine Engineering, University of Michigan, Ann Arbor, Michigan, USA

ABSTRACT

A digital twin is a dynamic virtual representation of a physical system that provides operational insight and support for life-cycle management decisions. Fundamental exploration of algorithms and approaches to link, compare, and fuse numerical models used in digital twins with real-world sensor data is still needed for ship performance prediction, condition assessment, and ultimately, significant implementation of digital twin technology in the marine industry. A preliminary digital twin framework for surface ships has been developed to yield time-and-place specific predictions of vessel motions and structural responses given weather forecast or hindcast data for a selected route. The weather model used in this work was NOAA WAVEWATCH III, and the framework code used a frequency-domain, two-dimensional strip theory program to estimate the vessel's response given input weather data. However, the framework code was written to be adaptable to other motion prediction models in both the time-domain and frequency-domain.

In this work, cumulative fatigue damage predictions for four routes in the Pacific Ocean were compared. A single frame of the simulation's animated significant wave height map is shown in Figure 1. Figure 1 also includes the four analyzed routes and each vessel's location at the current simulation time step. The fatigue damage predictions at the vessel's midbody stations for each of the four routes are compared in Figure 2. The vessel motion predictions for the Guam-Hawaii route were then analyzed to determine potential causes for the increased damage accumulation rate. This increase in cumulative damage for a single route stresses the importance of the ability to track and balance fatigue damage among ships in a fleet.

This work yielded a fatigue damage tracking and comparison method that could be performed using weather forecast or hindcast data rather than relying on structural surveys or installation of monitoring equipment. The ability to provide operators with response predictions would furnish a greater understanding of a weather forecast's implications, and thereby provide in-mission decision support. Furthermore, the predicted structural responses of the ship could be used to estimate the fatigue damage accumulated

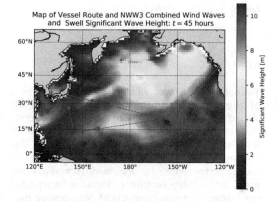

Figure 1. A single frame of the animated Pacific Ocean significant wave height map including the four routes analyzed for fatigue damage comparison and the location of each vessel at the current simulation time step.

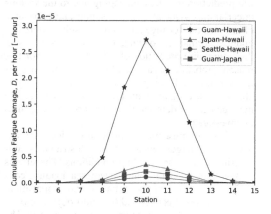

Figure 2. Comparison of cumulative fatigue damage, D, per hour at stations 5 through 15 of the vessel. Station 10 is the midship station, which is typically where the largest vertical bending moments occur.

during a voyage. The ability to track and compare cumulative fatigue damage of similar vessels based on encountered seas would support educated maintenance and deployment decisions to ensure fatigue damage equality among ships in a fleet.

SS9: Novel materials and systems for life-cycle structural health monitoring
   Organizers: V. Pakrashi, P. Cahill & P. Michalis

Life-Cycle Analysis and Assessment in Civil Engineering: Towards an Integrated Vision – Caspeele, Taerwe & Frangopol (Eds)
© 2019 Taylor & Francis Group, London, ISBN 978-1-138-62633-1

Innovative soft-material sensor, wireless network and assessment software for bridge life-cycle assessment

K. Loupos, Y. Damigos, A. Tsertou, A. Amditis, S. Lenas, C. Chatziandreoglou, C. Malliou & V. Tsaoussidis
Institute of Communication and Computer Systems, Athens, Greece

R. Gerhard, D. Rychkov & W. Wirges
University of Potsdam, Potsdam, Germany

B. Frankenstein
Teletronic Rossendorf GmbH, Rossendorf, Germany

S. Camarinopoulos
RISA Sicherheitsanalysen GmbH, Berlin, Germany

V. Kalidromitis & C. Sanna
TECNIC techniche E consulenzenell L'Ingegneria Civile Consulting Engineers, Rome, Italy

S. Maier & A. Gordt
University of Stuttgart, Stuttgart, Germany

P. Panetsos
Egnatia Odos A.E, Thessaloniki, Greece

ABSTRACT

Nowadays, structural health monitoring of critical infrastructures is considered as of primal importance especially for managing transport infrastructure however most current SHM methodologies are based on point-sensors that show various limitations relating to their spatial positioning capabilities, cost of development and measurement range. This publication describes the progress in the SENSKIN EC co-funded research project that is developing a dielectric-elastomer sensor, formed from a large highly extensible capacitance sensing membrane and is supported by an advanced micro-electronic circuitry, for monitoring transport infrastructure bridges. The sensor under development provides spatial measurements of strain in excess of 10%, while the sensing system is being designed to be easy to install, require low power in operation concepts, require simple signal processing, and have the ability to self-monitor

and report. An appropriate wireless sensor network is also being designed and developed supported by local gateways for the required data collection and exploitation. SENSKIN also develops a Decision-Support-System (DSS) for proactive condition-based structural interventions under normal operating conditions and reactive emergency intervention following an extreme event. The latter is supported by a life-cycle-costing (LCC) and life-cycle-assessment (LCA) module responsible for the total internal and external costs for the identified bridge rehabilitation, analysis of options, yielding figures for the assessment of the economic implications of the bridge rehabilitation work and the environmental impacts of the bridge rehabilitation options and of the associated secondary effects respectively. The overall monitoring system will be evaluated and benchmarked on actual bridges of Egnatia Highway (Greece) and Bosporus Bridge (Turkey).

The potential of energy harvesting for monitoring corroding metal pipes

F. Okosun & V. Pakrashi
Dynamical Systems and Risk Laboratory, School of Mechanical and Materials Engineering, Ireland
Centre for Marine and Renewable Energy Ireland (MaREI), University College Dublin, Dublin, Ireland

ABSTRACT

Prevention and monitoring of corrosion related degradation of pipelines is an important and established topic. Various corrosion monitoring techniques exist, and battery powered wireless sensors are now being employed for these techniques. Due to the limited lifetime of batteries, there has been a recent and growing interest in powering these sensors with energy harvested from the ambient environment, leading to the potential of ener-gy harvesting monitors for such purposes. This paper attempts to outline the potential of energy harvesting based corrosion monitoring, and provides a context around existing methods and technologies. An overview of the conventional corrosion monitoring techniques is presented, followed by a consideration of the energy harvesting technologies with a focus on vibration energy harvesting. Finally, existing research conducted around harvesting based monitoring of pipes are considered to highlight the potential evolution of this technology for Structural Health Monitoring (SHM) purposes.

REFERENCES

Cahill, P., Hazra, B., Karoumi, R., Mathewson, A. & Pakrashi, V. (2018) Vibration energy harvesting based monitoring of an operational bridge undergoing forced vibration and train passage. *Mechanical Systems and Signal Processing*, 106, 265–283.

Cahill, P., Jaksic, V., Keane, J., O'Sullivan, A., Mathewson, A., Ali, S.F. & Pakrashi, V. (2016) Effect of road surface, vehicle and device characteristics on energy harvesting from bridge-vehicle interactions. *Computer Aided Civil and Infrastructure Engineering* 31(12), 921–935.

Cahill, P., Ni Nuallain, N.A., Mathewson A., Karoumi, R. & Pakrashi V. (2014a) Energy harvesting from train induced response in bridges. *ASCE Journal of Bridge Engineering* 19(9), 04014034-1-11.

Cahill, P., O'Keeffe, R., Jackson, N., Mathewson, A. & Pakrashi, V. (2014b) Structural health monitoring of reinforced concrete beam using piezoelectric energy harvesting system. In: *Proceedings of EWSHM-7th European Workshop on Structural Health Monitoring*.

Kim, S., Lee, M. & Chou, P. H. (2015) Energy harvesting from anti-corrosion power sources. *Proceedings of the ISLPED'14 International Symposium on Low Power Electronics and Design*. pp. 363–368.

Metje, N., Chapman, D. N., Cheneler, D., Ward, M. & Thomas, A. M. (2011) Smart pipes-instrumented water pipes, can this be made a reality? *Sensors*, 11(8), 7455–7475.

Qiao, G., Hong, Y., Sun, G. & Yang, O. (2013) Corrosion energy: a novel source to power the wireless sensor. *IEEE Sensors Journal*, 13(4), 1141–1142.

Quirk, L., Matos, J., Murphy, J. & Pakrashi, V. (2018) Visual inspection and bridge management. *Journal of Structure and Infrastructure Engineering*, 14(3), 320–332.

Qureshi, F. U., Muhtaroglu, A. & Tuncay, K. (2015) A method to integrate energy harvesters into wireless sensor nodes for embedded in-pipe monitoring applications. *5th International Conference on Energy Aware Computing Systems and Applications, ICEAC 2015*.

Sun, G., Qiao, G. & Xu, B. (2011) Corrosion monitoring sensor networks with energy harvesting. *IEEE Sensors Journal*, 11(6), 1476–1477.

Yazdekhasti, S., Piratla, K.R., Atamturktur, S. & Khan, A. (2018) Experimental evaluation of a vibration-based leak detection technique for water pipelines. *Structure and Infrastructure Engineering*, 14(1), 46–65.

Ye, G. & Soga, K. (2012) Energy harvesting from water distribution systems. *Journal of Energy Engineering*, 138 (1), 7–17.

Yu, H., Zhou, J., Deng, L. & Wen, Z. (2014) A vibration-based MEMS piezoelectric energy harvester and power conditioning circuit. *Sensors*, 14, 3323–3341.

Yu, L., Giurgiutiu V. & Pollock, P. (2008) A multi-mode sensing system for corrosion detection using piezoelectric wafer active sensors. *Sensors and Smart Structures Technologies for Civil, Mechanical, and Aerospace Systems, Proceedings SPIE 6932*.

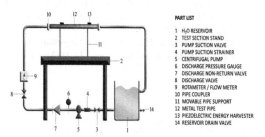

Figure 1. Experimental test-rig design for pipe corrosion assessment through vibration energy harvesting.

SS10: Value of structural health monitoring information for the life-cycle management of civil structures
Organizers: S. Thöns, G. Lombaert & M.P. Limongelli

Life-Cycle Analysis and Assessment in Civil Engineering: Towards an Integrated Vision – Caspeele, Taerwe & Frangopol (Eds)
© 2019 Taylor & Francis Group, London, ISBN 978-1-138-62633-1

Cost-based optimization of the performance of a damage detection system

A.C. Neves, J. Leander & R. Karoumi
KTH-Royal Institute of Technology, Stockholm, Sweden

I. González
Sweco AB, Stockholm, Sweden

ABSTRACT

From time to time we get to hear from the news that a bridge has collapsed, for the most part during the construction phase. A direct consequence of such event is material and structural damage, as well as the possible injuries and fatalities. Other consequences turn up more gradually in time after the incident, such as loss of reputation, loss of functionality, environmental issues, costs with compensations, *et cetera*. This type of situations can be avoided if Structural Health Monitoring (SHM) systems can detect early potential failures and timely withdraw the structure from service ahead of a likely disaster. Structural safety is the primary reason for the implementation of SHM but not less important is the significant cost reduction associated with routine maintenance and inspection procedures. In this sense, one of the remaining obstacles to fully implement SHM systems on our infrastructure deals with justifying the economic advantage of these systems, which is quite challenging to put into numbers.

This paper proposes a rational framework for the use of SHM in the decision-making process regarding the maintenance and monitoring of civil engineering structures, based on the optimal setup of the damage detection system that yields the minimum associated cost.

The proposed method is based on the Bayes Theorem, with which new information on the form of measurements performed in the structure is continually fed into and evaluated by the damage detection system. Acceleration measurements are obtained from the simulations of a numerical model of the bridge in healthy and damaged conditions. The data obtained in healthy condition is then used to train Artificial Neural Networks that are able to predict the healthy behavior of the structure, i.e. future acceleration. By looking at new data measurements, predictions errors are

computed and statistically characterized by means of a Gaussian Process and, based on the difference between predictions and effective measurements, damage can be identified.

As any system is prone to errors, it is not guaranteed that the damage detection system as proposed reveals always perfect results. It may happen that the system warns for damage when the structure is actually healthy or that it misses to detect existing damage. These are regarded as false diagnosis and are to be avoided or controlled as far as possible. In order to do so, the results obtained from the application of the Gaussian Processes on the Neural Networks are further analyzed using concepts such as damage index and detection threshold. A detection threshold is set for the damage index, where the former works as a boundary that separates indexes that are attributed to a healthy structure from the indexes that are attributed to a damaged structure, with some associated probability of wrong classification of the condition. There is thus a trade-off to be considered: if one wants a system on the safe side and decreases the threshold, one expects more false positives; on the other hand, if one wants a less strict system and increases the threshold, one expects more false negatives. The detection threshold that is selected for the system defines the equilibrium between these two errors. In this work, the optimal threshold is admitted to be the one that yields the lowest cost associated with false diagnosis. A simple method to estimate the expected total cost depending on the system performance is suggested.

The method is demonstrated with a case study of a fictitious railway bridge and within a context of made-up cost scenarios related with false diagnosis. The results presented in this paper give good indication and encourage further research work in regards of a more comprehensive analysis of the economic benefits following the implementation of SHM systems.

Life-Cycle Analysis and Assessment in Civil Engineering: Towards an Integrated Vision – Caspeele, Taerwe & Frangopol (Eds)
© 2019 Taylor & Francis Group, London, ISBN 978-1-138-62633-1

The integration of bridge life cycle cost analysis and the value of structural health monitoring information

G. Du
Danish Building Research Institute, Aalborg University, Copenhagen, Denmark

J. Qin
Danish Center for Risk and Safety Management, Department of Civil Engineering,
Aalborg University, Esbjerg, Denmark
Shanghai Institute of Disaster Prevention and Relief, Tongji University, Shanghai, China

ABSTRACT

Life cycle cost (LCC) is a vital economic evaluation tool for effective asset management towards sustainability. One obstacle in the current bridge LCC implementation is the inherent uncertainties for the input parameters, *i.e.* without knowing the structural performance, most of the existing LCC models are only established based on the assumptions of various life-cycle parameters. For instance, the remaining service life (RSL) of structural components, a key input in LCC, is highly related to the bridge deterioration process but often assumed deterministically in bridge LCC. With the advent of advanced structural health monitoring (SHM) techniques, the structural performance can be observed from the real-time monitoring data, whereas the structural RSL can also be estimated. Here, we estimate RSL taking account of the structural performance, which is expressed based on a time dependent ultimate limit state function, as elaborated in Qin et al. (2015):

$$RSL_{t_0} = \max\left\{ t \middle| R_0\theta_D\left(1-D(t)\right) - z\theta_S \mathbf{S}_t \geq 0 \quad t \geq t_0 \right\} - t_0$$

where $D(t)$ is a generic bridge deterioration function, as an accumulation process with time. An annual increment $\Delta_{D,i}$ has the distribution with uncertain expected value M_{μ_D} and the constant standard deviation σ_{Δ_D}. By applying the Bayes' rule, the updated RSL_t'' can be obtained from SHM and integrated into the existing LCC framework.

Another recognized issue is that, the trade-off between the benefit and the cost of using SHM system is rarely considered in the existing bridge LCC practice, i.e. the value of information (VoI). Therefore, we explore the application of VoI in the context of bridge SHM in civil engineering. Based on the

work presented by Raiffa & Schlaifer (1961), Benjamin & Cornell (1970), the value of the information on the SHM strategy e, is calculated as the cost difference with/without the additional information z_k from e, which is formulated as:

$$VoI = \arg\min_m E[C(a_m \mid z)] - \arg\min_m E[C(a_m)]$$

$$= \arg\min_m \sum_{k=0}^n C(a_m \mid z_k) \sum_{l=0}^n P(z_k \mid \theta_l) P'(\theta_l) -$$

$$\arg\min_m \sum_{l=0}^n C(a_m \mid \theta_l) P'(\theta_l)$$

This study proposes an improved LCC framework, integrating the real-time SHM information for estimating the life cycle parameter RSL. In addition, the trade-off between the SHM cost and the VoI on SHM is also illustrated using the pre-posterior Bayesian approach. A conceptual case study is presented for the illustrative purpose. This extended LCC framework provides a rational basis to assist optimal decisions for the asset management towards the maximum utility, considering the safety, cost and the environmental impact, all in a holistic manner.

REFERENCES

Benjamin, J.R. & Cornell, C.A. (1970) *Probability, Statistics, and Decision for Civil Engineers*. New York, McGraw Hill.

Qin, J., Thöns, S. & Faber, M.H. (2015) On the value of SHM in the context of service life integrity management. In: *12th International Conference on Applications of Statistics and Probability in Civil Engineering (ICASP12)*. pp. 1–8.

Schlaifer, R. & Raiffa, H. (1961) *Applied Statistical Decision Theory*.

The value of monitoring the service life prediction of a critical steel bridge

J. Leander & R. Karoumi
Division of Structural Engineering and Bridges, KTH Royal Institute of Technology, Sweden

ABSTRACT

If a bridge is judged unfit for service in a preliminary assessment, the bridge manager has to decide on interventions to secure the safety of the structure. Possible interventions are, e.g., repair, maintenance of critical components, upgrading, and demolition. Another strategy is to engage more accurate assessment methods with the purpose of extending the theoretical service life. The knowledge on the behaviour of a bridge can be improved by using in-service monitoring which reduces uncertainties in the loading conditions and the structural behaviour. It is, however, rarely used in practice for bridges due to an apprehension of costly implementations.

In the current paper, an evaluation of the value of information (VoI) is presented for a steel bridge in Sweden, the Old Lidingö Bridge, which is heavily afflicted with corrosion. A photo of the bridge is shown in Fig. 1. A specific decision scenarios has been considered with an aim of extending the service life of the bridge, using a monitoring system and a detailed deterioration model considering the combined effect of corrosion and fatigue. These two deterioration phenomena are the main causes for limiting the service lives of steel bridges. The corrosion process typically causes an increase in stress due to reduced cross-section area and develops local stress risers. Both effects are detrimental for the fatigue resistance. The aim with the presented investigation was to develop an assessment method where a sophisticated degradation model is combined with a reliability-based verification format and data from in-service monitoring.

A fatigue life assessment of bridges following the governing regulations is typically performed using linear damage accumulation and fatigue endurances described by S–N curves. The outcome is a damage index which is verified against an acceptable threshold value, see e.g. the Eurocode EN 1993-1-9. This damage index has no physical meaning and cannot be measured in practice. To enable a reliability updating considering results from inspections, a prediction model based on linear elastic fracture mechanics (LEFM) has been implemented.

In the current investigation, the corrosion was considered as a reduction of the cross-section area. No stress rise due to local discontinuities was modelled. The time dependent penetration depth was modelled as an exponential function

$$D(t) = r_{\text{corr}} \, t^{b_{\text{corr}}} \qquad (1)$$

where t is the exposure time expressed in years, r_{corr} is the corrosion rate experienced in the first year expressed in micrometres per year, and b_{corr} is the metal-environment-specific time exponent. A reduced cross-section area due to corrosion causes higher stresses in the remaining material. This time dependent effect alters the stress range which makes the stress intensity range dependent on the time and crack depth both.

The suggested reliability-based method for the fatigue assessment of steel bridges, allows a consideration of measured response and updating based on results from inspections. The paper demonstrates how the concept of VoI can be utilized, together with the suggested assessment method, as a tool for decision makers to secure the safety of critical bridges.

With tentative values in the model for VoI analysis, a procedure for quantifying the value of structural health monitoring (SHM) has been demonstrated. The concept has been validated using measured response from the Old Lidingö Bridge. For the case study, the monitoring is shown to greatly improve the estimated safety of the bridge.

Figure 1. A photo of the Old Lidingö Bridge in Stockholm, Sweden.

Structural monitoring and inspection modeling for structural system updating

A. Agusta & S. Thöns
Center for Oil and Gas -DTU, Technical University of Denmark, Kgs. Lyngby, Denmark

ABSTRACT

Structural monitoring (e.g. structural measurement and damage detection) and inspections have been widely applied as a part of structural maintenance. However, the potential of using the obtained information is not yet fully exploited to update the structural reliability and to support the maintenance and operation strategies. This paper describes the foundations, limitations, and characteristics of structural measurement, inspection (see e.g. Gandossi and Annis (2010), and damage detection (see e.g. Döhler and Thöns (2016)) modeling approaches and their utilization for the structural integrity management.

A new structural measurement modeling approach is developed by connecting the damage model uncertainty realizations thresholds $\hat{M}_D^{th}$ with two target damage probabilities:

$$\hat{M}_{D,1}^{th} : P(D_C|\hat{M}_{D,1}^{th}) = P(D_{C,1}^T) \quad (1)$$

$$\hat{M}_{D,2}^{th} : P(D_C|\hat{M}_{D,2}^{th}) = P(D_{C,2}^T) \quad (2)$$

where $P(D_{C,1}^T) < P(D_{C,2}^T)$. $P(D_{C,1}^T)$ and $P(D_{C,2}^T)$ are the target probabilities which may be derived for different consequences and safety costs classes according to e.g. ISO2394 (2015). With this method, the measurement outcomes have three distinct indication events that can be related to the system performance (see Figure 1). The developed approach distinguishes the prediction of the measurement outcomes (i.e. pre-posterior) and the situation when the outcome has already been obtained (i.e. posterior). The structural reliability updating procedures for both cases are presented in the paper.

Inspection and damage detection modeling approaches based on the signal/Damage Indicator Value distributions are introduced and used in conjunction with the Bayesian updating.

A structural system is considered to demonstrate the application of each modeling approach on the system level. The effects of the damage model uncertainty parameters (e.g. the expected value and the standard deviation) and the signal/Damage Indicator Value thresholds in updating the structural system

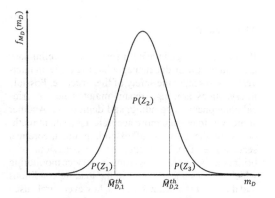

Figure 1. The example of damage model uncertainty distribution with two thresholds defined.

reliability are investigated and discussed. The results show that the proposed structural measurement modeling approach is able to update system damage probability depending on the system performance and provides indications to possible maintenance actions (e.g. necessity to repair or no repair). A further uncertainty reduction is achieved when the measurement outcomes is obtained (posterior). In terms of computational costs, the proposed structural measurement modeling approach is the lightest because it only requires a redefinition of the prior limit state equation to account for $\hat{M}_D$ in the defined intervals (see Figure 1) instead of Bayesian updating.

REFERENCES

Döhler, M. & Thöns, S. (2016) Efficient structural system reliability updating with subspace-based damage detection information. In: *Proceedings of the 8th European Workshop on Structural Health Monitoring (EWSHM 2016)*.

Gandossi, L. & Annis, C. (2010) *Probability of Detection Curves: Statistical Best-Practices*. European Commission, Joint Research Centre, Institute for Energy.

ISO2394 (2015) *General Principles on Reliability for Structures*. Geneva, Switzerland, Standard, International Organization for Standardization.

Life-Cycle Analysis and Assessment in Civil Engineering: Towards an Integrated Vision – Caspeele, Taerwe & Frangopol (Eds)
© 2019 Taylor & Francis Group, London, ISBN 978-1-138-62633-1

The effects of deterioration models on the value of damage detection information

L. Long
Department 7: Safety of Structures, BAM Federal Institute for Materials Research and Testing, Berlin, Germany

S. Thöns
Associate Professor, Department of Civil Engineering, Technical University of Denmark, Lyngby, Denmark

M. Döhler
Inria, I4S / IFSTTAR, COSYS, SII, Rennes, France

ABSTRACT

This paper addresses the effects of structural performance, deterioration, SHM system and damage detection algorithm parameters on the value of damage detection information. The quantification of the value of monitoring information for deteriorated structures is based on Bayesian pre-posterior decision analysis, comprising decision rules, structural system performance models, probabilistic reliability models, consequences analysis as well as benefit and costs analysis associated with monitoring results over their life cycle. Building upon these models, the value of damage detection information takes basis of the relevance and precision of damage detection information to ensure the structural integrity and reduce the potential structural system risks and expected costs throughout the service life before implementing damage detection system. The structural system performance is described with different deterioration models under extreme loading accounting for resistance, loading and model uncertainties. Monte Carlo simulation is used to find the cumulative probability of system failure throughout the life cycle. The updated probability of system failure given damage detection information is computed utilizing Bayesian updating theorem. The value of damage detection information is quantified by calculating the difference of the service life benefits with and without damage detection information. With the developed approach, the value of damage detection information for a statically determinate Pratt truss bridge girder subjected to different deterioration models is calculated. The structural system performances of the Pratt truss bridge girder are building upon series systems and are coupled with time-variant damage models describing continuously the deterioration process and structural resistance degradation throughout the service life. The analysis shows the impact of various deterioration models corresponding to different types of damages on the cumulative probability of bridge failure over time, which results in different service life benefits and values of damage detection information. The results can be used to develop optimal lifetime maintenance strategies before implementation of the damage detection system for bridges under different deterioration processes.

The value of visual inspections for emergency management of bridges under seismic hazard

M.P. Limongelli
Politecnico di Milano, Milano, Italy

S. Miraglia
Aalborg University, Aalborg, Denmark

A. Fathi
Amirkabir University of Technology, Teheran, Iran

ABSTRACT

In this paper, the value of information from visual inspections for the seismic emergency management of bridges is investigated. One of the major problems in the aftermath of an earthquake concerns the decision about possible traffic restrictions to issue for infrastructures like bridges. The bridge can be closed to traffic, it can be kept open, or traffic can be restricted in volume or velocity or limited to emergency vehicles. Each of these choices implies different consequences and corresponding risks. The decision regarding traffic restrictions can be made based on the so-called prior knowledge, i.e. without any information collected on the state of the bridge after the seismic event or exploiting information from a visual inspection performed after the earthquake. Collecting information has a cost, therefore it would be important to know, before the inspection is performed, if its cost is balanced by the benefit it brings in terms of risk reduction. A procedure based on pre-posterior Bayesian analysis (Schaifer & Raiffa, 1961, Faber & Thöns, 2013) is proposed to quantify the value of information from a visual inspection, before it is performed. The procedure has been applied to the case study of a two span reinforced concrete bridge located in a region at high seismic hazard (see Fig. 1). Given a specific earthquake scenario, the value of information from visual inspections in the phase of the emergency is computed using the Bayesian pre-posterior approach to decision making. The prior knowledge about the structural condition is modeled using the fragility curves computed for 5 different damage states. The traffic restrictions are computed fixing the minimum value of reliability index that the decision maker is willing to accept for the worst damage state. For this target value the new parameters of the distributions of the traffic load are computed and the values of the probability of failure in the different damage states computed. Both direct and indirect consequences of the traffic restrictions have been considered including cost of the bridge, of fatalities, environmental consequences due to increased pollution induced detours and economic costs related to increased fuel consumption and additional travel time.

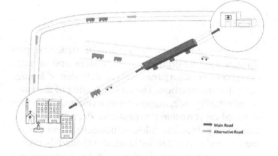

Figure 1. The hypothetical configuration for the case study.

Optimal decisions whether to perform an inspection before imposing traffic restrictions on a potentially damaged structure or take the decision based on the prior knowledge on the bridge, provided by the fragility curves, is identified by minimizing the expected total consequences. For the specific scenario considered herein, traffic restriction that excludes heavy trucks, but allows emergency vehicles is the best solution. The results are of course valid in the limited context of the hypothetical case study with the declared hypothesis. A sensitivity analyses with respect to the considered cost related to direct and direct consequences will be the object of future research efforts.

REFERENCES

Faber, M.H. & Thons, S. (2013) On the value of structural health monitoring. In: Steenbergen, R.D.J.M., van Gelder, P.H.A.J.M., Miraglia, S. & TonVrouwenvelder, A.C.W.M. (eds.) *Safety, Reliability and Risk Analysis: Beyond the Horizon*. CRC Press LLC. pp. 2535–2544.

Schlaifer, R. & Raiffa, H. (1961) Applied statistical decision theory.

A Bayesian network based approach for integration of condition-based maintenance in strategic offshore wind farm O&M simulation models

J.S. Nielsen & J.D. Sørensen
Department of Civil Engineering, Aalborg University, Aalborg, Denmark

I.B. Sperstad & T.M. Welte
Department of Energy Systems, SINTEF Energy Research, Trondheim, Norway

ABSTRACT

In the overall decision problem regarding optimization of operation and maintenance (O&M) for offshore wind farms, there are many decision parameters and many approaches for solving parts of the decision problem. Often, the parts of the problem closely related to the considered decision parameters are modelled accurately while other parts of the problem are modelled in a simplified manner, although the parts might interact. Simulation-based strategy models typically used for optimizing maintenance logistics strategies accurately capture system effects related to logistics. However, for condition-based maintenance (CBM) of deteriorating components, these models usually do not consider directly the influence of inspection, condition monitoring, and repair strategy on the failure rate. These effects are instead considered indirectly through high-level performance data, which does not capture the correct distribution of events in time (Welte et al. 2017).

The influence of the CBM strategy can be directly considered using a risk-based approach based on the Bayesian pre-posterior decision analysis (Nielsen and Sørensen 2017). An efficient approach to solve the decision problem is to apply stationary decision rules and discrete Bayesian networks for modelling of deterioration and effect of inspection and repair strategies. These approaches usually include only simple representations of the influence of weather, vessels, and system effects in relation to vessel utilization through the cost inputs.

In this paper, a novel approach developed within the LEANWIND.eu project based on Bayesian networks is presented for accurate integration of CBM in simulation-based strategy models. Using Bayesian networks, the probability distribution for the time of failure given no CBM is first estimated, and in the strategic O&M simulation model the time of failure is drawn from this distribution. Then, the conditional probability distribution for the time of the decision to make a preventive repair given the time of failure is

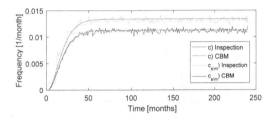

Figure 1. Comparison of distribution of inspections and CBM tasks in time for the risk-based model and the NOWIcob simulation-based strategy model (sim) for strategy c).

estimated, which accounts for the CBM strategy. This conditional distribution is then used by the simulation-based strategy model to generate the time when the planning of the preventive repair can start.

The approach has been implemented in the simulation-based strategy model NOWIcob (Hofmann and Sperstad 2013), and is illustrated in an example considering four CBM strategies for wind turbine blades. As shown in Figure 1, the approach approximately results in the correct distribution of inspections and CBM tasks in time. The costs found using the integration approach were generally higher than the estimates obtained by the risk-based model, as more effects relevant for estimating cost contributions were modelled in NOWIcob.

REFERENCES

Hofmann, M. & Sperstad, I.B. (2013) NOWIcob – A tool for reducing the maintenance costs of offshore wind farms. *Energy Procedia*, 35, 177–186.

Nielsen, J.S. & Sørensen, J.D. (2017) Computational framework for risk-based planning of inspections, maintenance and condition monitoring using discrete Bayesian networks. *Structure and Infrastructure Engineering*, 1–13.

Welte, T.M., Sperstad, I.B., Sørum, E.H. & Kolstad, M.L. (2017) Integration of degradation processes in a strategic offshorewind farm O&M simulation model. *Energies*, 10 (7), 925.

Life-Cycle Analysis and Assessment in Civil Engineering: Towards an Integrated Vision – Caspeele, Taerwe & Frangopol (Eds)
© 2019 Taylor & Francis Group, London, ISBN 978-1-138-62633-1

Structural integrity management with unmanned aerial vehicles: State-of-the-art review and outlook

M. Kapoor, E. Katsanos & S. Thöns
Department of Civil Engineering, Technical University of Denmark, Kgs. Lyngby, Denmark

L. Nalpantidis
Department of Mechanical & Manufacturing Engineering, Aalborg University, Aalborg, Denmark

J. Winkler
ATKINS – SNC LAVALIN, Copenhagen, Denmark

ABSTRACT

Over the last decade, Unmanned Aerial Vehicles (UAVs) have been used frequently for monitoring the construction and operation of civil infrastructure systems such as buildings, bridges, road networks, as well as industrial facilities and power plants. The unmanned aerial platform, being equipped with advanced equipment and sensors (e.g. high-resolution cameras and image stabilization), can collect data to be integrated into preliminary design workflows, robustly survey construction sites, monitor and document work-in-progress as well as inspect and monitor the condition of structures located in areas with limited accessibility. Given their operational simplicity along with the rather obvious time-and-cost-related benefits, there is a strong indication that UAVs can be effectively employed for Structural Integrity Management (SIM) and Structural Health Monitoring (SHM). However, the field of UAV research and application currently lacks a targeted employment of UAVs for the purpose. Along these lines, this paper provides an overview of the developments in UAV technologies, recent breakthroughs in sensor technologies, SHM and Value of Information analysis, the latter being oriented to facilitate an efficiency assessment of precision and cost dependent information. A state-of-the-art review of relevant scientific literature, research projects, and industrial studies, that attempt to integrate UAVs and SHM are shortly described and critically assessed herein. The direct implementation of UAVs for SHM through image-based methodologies such as image processing, automated crack detection, DIC etc., is discussed (e.g. Ham et al. 2016, Sankarasrinivasan et al. 2015). Additional UAV-based SHM techniques utilizing multi-sensor systems, contact based measurement strategies and hybrid systems are assessed (e.g. Na & Baek 2016, Khan et al. 2015). Efficiency assessment of UAV based SHM is presented taking basis in Value of Information analysis building upon a recent

studies by Kapoor (2017), Thöns (2018) and within the COST Action TU1402 (www.tu-1402.eu). The paper highlights the potentials of UAV-assisted SHM as an accurate, cost effective technique aimed towards optimizing performance and efficiency of the structural integrity management. Whereas the cost-related benefits of the image-based applications of the unmanned aerial platform are rather obvious, additional prospects hold potential for a methodical approach for the efficiency assessment of monitoring strategies and related technologies. Therein, the Value of Information analysis approach facilitates the assessment of the benefits and/or the optimization of UAV-based SHM strategies before their actual implementation.

REFERENCES

Ham, Y., Han, K.K., Lin, J.J. & Golparvar-Fard, M. (2016) Visual monitoring of civil infrastructure systems via camera-equipped Unmanned Aerial Vehicles (UAVs): a review of related works. Visualization in Engineering. 4 (1), 1. Available from: https://doi.org/10.1186/s 40327-015-0029-z.

Kapoor, M. (2017) *UAV Based SHM Value of Information Modeling.* MSc Thesis. Kongens Lyngby, Denmark, Denmark Technical University.

Khan, F., Ellenberg, A., Mazzotti, M., Kontsos, A., Moon, F., Pradhan, A. & Bartoli, I. (2015) Investigation on bridge assessment using unmanned aerial systems, 404–413.

Na, W. & Baek, J. (2016) Impedance-based non-destructive testing method combined with unmanned aerial vehicle for structural health monitoring of civil infrastructures. *Applied Sciences,* 7 (1), 15.

Sankarasrinivasan, S., Balasubramanian, E., Karthik, K., Chandrasekar U. & Gupta, R. (2015) Health monitoring of civil structures with integrated UAV and image processing system. *Procedia Computer Science,* 54, 508–515.

Thöns, S. (2018) On the value of monitoring information for the structural integrity and risk management. *Computer-Aided Civil and Infrastructure Engineering,* 33, 79–94.

Metamodeling strategies for value of information computation

M.S. Khan
IITB-Monash Research Academy, Indian Institute of Technology Bombay, Mumbai, India

S. Ghosh & J. Ghosh
Department of Civil Engineering, Indian Institute of Technology Bombay, Mumbai, India

C. Caprani
Department of Civil Engineering, Monash University, Melbourne, Australia

ABSTRACT

Structural health monitoring (SHM) involves various activities and techniques that assist in establishing the state of a structure. The information obtained through SHM in the form of structural performance inspection, load and hazard estimations etc. are used to update the prior structural model predictions. Consequently, SHM assists the stakeholders in adopting the most optimal actions for the management of a structure.

The Value of information (VoI) is a framework that attempts to formally quantify SHM benefit. However the application of VoI framework especially to more practical problems is restricted primarily due to its high computational demands (Konakli et al. 2016). For VoI calculation, the underlying reliability model has to be repeatedly called for various inspection values that could be measured.

This paper investigates two such metamodeling techniques for the normalized expected VoI $\left(\widehat{EVI} = \frac{EVI}{EVI_{PI}}\right)$ evaluation – PCE and kriging, where EVI is the expected VoI in case of imperfect information (measurements simulated with error) and EVI_{PI} is the expected VoI in case of perfect information (measurements simulated without error). A simple VoI problem with a simple capacity minus demand based reliability model is adopted with two possible decision actions – do nothing and immediate replacement of the component. The $\widehat{EVI}$ is estimated for the 5th year using metamodels and is compared wit that obtained using crude MCS with 10^4 samples.

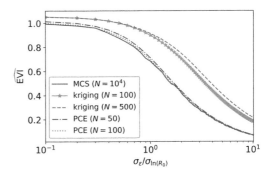

Figure 1. $\widehat{EVI}$ as a function of normalized error.

It is found that a PCE of order 3 with 10^2 samples provides very good approximation to the MCS results. However the results from kriging are poor due to the poor fit of the type of variogram models explored and the non-stationarity of the random process. Further work is required to develop a working kriging metamodel for the VoI estimation. The reduction in computational cost through metamodeling can enable more practical applications of the VoI framework.

REFERENCE

Konakli, K., Sudret, B. & Faber, M.H. (2016) Numerical investigations into the value of information in lifecycle analysis of structural systems. *ASCE-ASME Journal of Risk and Uncertainty in Engineering Systems, Part A: Civil Engineering 2* (3), B4015007.

SS11: Design for robustness of steel and steel-concrete composite structures
Organizers: J.-F. Demonceau & J.-P. Jaspart

Life-Cycle Analysis and Assessment in Civil Engineering: Towards an Integrated Vision – Caspeele, Taerwe & Frangopol (Eds)
© 2019 Taylor & Francis Group, London, ISBN 978-1-138-62633-1

Robustness of steel structures subjected to a column loss scenario

J.-F. Demonceau, M. D'Antimo & J.-P. Jaspart
Urban and Environmental Engineering Research Unit, University of Liege, Belgium

ABSTRACT

Recent events such as natural catastrophes or terrorism attacks have highlighted the necessity to ensure the structural integrity of buildings under an exceptional event. According to the Eurocodes and some different other national design codes, the structural integrity of civil engineering structures should be ensured through appropriate measures but, in most cases, no precise practical guidelines on how to achieve this goal are provided.

The Eurocode dealing with the structural integrity of structures is Eurocode 1, Part 1-7. For the above mentioned reason, this Eurocode is presently under revision with the objective of providing improved general recommendations to ensure an appropriate structural robustness in case of an exceptional event, knowing that the detailed application rules related to specific structural solutions would be reflected in the associated specific Eurocodes, i.e. in Eurocode 2 for concrete structures, in Eurocode 3 for steel structures…

In this framework, the robustness of steel and composite steel-concrete building frames is investigated at the University of Liege for many years following the so-called "alternative load path method" (Demonceau, 2008, Demonceau & Jaspart, 2010, Demonceau et al, 2011, Comeliau et al, 2012, Huvelle et al, 2013, Huvelle et al, 2015, Demonceau et al, 2017) with the final objective to propose design requirements useful for practitioners to mitigate the risk of progressive collapse considering the conventional scenario "loss of a column" further to an unspecified event.

The present paper gives a global overview of the ongoing researches in the field of robustness at the University of Liège and of the adopted strategy aiming at deriving design requirements to be proposed for future implementation in the Eurocodes. In the following sections, the main outcome of recent researches with references to recently published papers will be reflected.

Firstly, the investigations conducted to study the quasi-static response of 2D steel and steel-concrete composite frames subjected to column loss scenarios are introduced in Section 2. Secondly, based on the knowledge gained from the studies presented in Section 2, the behaviour of 3D structures has been investigated and, in Section 3, it is demonstrated how the developments achieved for 2D frames can be easily extended to predict the behaviour of 3D structures. Finally, the dynamic effects which may be associated to a column loss are considered in Section 4. In particular, it is explained how the dynamic response of steel and steel-concrete structures can be derived from the quasi-static response.

REFERENCES

Comeliau, L., Rossi, B. & Demonceau, J.-F. (2012) Robustness of steel and composite buildings suffering the dynamic loss of a column. *Structural Engineering International: Journal of the International Association for Bridge and Structural Engineering*, 22/3, 323–329.

Demonceau, J.-F. (2008) *Steel and Composite Building Frames: Sway Response Under Conventional Loading and Development of Membrane Effects in Beams Further to an Exceptional Action*. PhD Thesis. University of Liege (freely downloadable at: http://hdl.handle.net/2268/2740).

Demonceau, J.-F., Comeliau, L., Hoang, V. L. & Jaspart, J.-P. (2017) How can a steel structure survive to impact loading? *The Open Civil Engineering Journal*, 12, 434–452.

Demonceau, J.-F., Comeliau, L. & Jaspart, J.-P. (2011) Robustness of building structures – recent developments and adopted strategy, *Steel Construction – Design and Research* Ernst & Sohn (Wiley Company). Vol. 4/3. pp. 166–170.

Demonceau, J.-F. & Jaspart, J.-P. (2010) Experimental test simulation a column loss in a composite frame. *Advanced Steel Construction*, 6, pp. 891–913.

Huvelle, C., Hoang, V. L., Jaspart, J.-P. & Demonceau, J.-F. (2015) Complete analytical procedure to assess the response of a frame submitted to a column loss. *Engineering Structures*, 86, 33–42.

Investigation of the column loss scenario of one composite steel and concrete frame

G. Roverso, N. Baldassino & R. Zandonini
Department of Civil, Environmental and Mechanical Engineering, University of Trento, Trento, Italy

ABSTRACT

Accidental events, such as impact loading or explosions, are rare events with a very low probability of occurrence. A structure, affected by accidental events, can be strongly damaged causing very high human losses and economic consequences.

Column removal in multi-storey buildings is an effective and popular damage scenario recommended for progressive collapse investigations. The collapse of an extended portion of the structure after a local damage can be avoided if the damaged part is able to redistribute loads to the undamaged parts, so that a new stable equilibrium configuration is achieved. A structure fulfilling this requirement is 'robust'. Redundancy, ductility and alternate load paths provide deformation capacity and collapse resistance ensuring a robust structural response.

In this study, the response of composite steel-concrete frames under column removal is considered. These models were validated against experimental data obtained from two full-scale tests conducted at the University of Trento within the 'RobustImpact' research project (European Commission 2012). The tests on two 3D composite steel and concrete frames subjected to a central column loss allowed investigating the role of the beam-to-column connections and of the concrete slab for the forces redistribution. The specimens (Fig. 1) were ground floor sub-frames 'extracted' from two reference buildings designed in accordance to the Eurocodes. The frames had the same overall dimensions, but a different, symmetric and asymmetric, configuration of the column layout. The tested frames were capable of developing important rotation in the joints and large deformations of the concrete slab associated with membrane forces.

3D finite element models representing full-scale frames (Fig. 2) were set up by means of the software Abaqus (Simulia 2014). Beam and shell elements were used to simulate the structural elements and both mechanical and geometrical non-linearities were implemented. The beam-to-column flush endplate connections were modelled accurately to consider

Figure 1. The first tested frame.

Figure 2. The finite element model.

the mechanical components of the joints according to EN1993-1-8 (2005). This paper focuses on the numerical model of the first symmetric frame and reports its validation against the experimental results.

REFERENCES

EN1993-1-8 (2005) *Eurocode 3: Design of Steel Structures: Part 1-8: Design of Joints. Brussels*, European Committee for Standardization.

European Commission (2012) Robust impact design of steel and composite building structures – 'RobustImpact'. Research Programme of the Research Fund for Coal and Steel, Grant Agreement number RFSR-CT-2012-00029.

Simulia Dassault Systems (2014) *ABAQUS: Analysis Users Manual Version 6.14*.

Life-Cycle Analysis and Assessment in Civil Engineering: Towards an Integrated Vision – Caspeele, Taerwe & Frangopol (Eds)
© 2019 Taylor & Francis Group, London, ISBN 978-1-138-62633-1

Design of steel and composite structures for robustness

N. Hoffmann, U. Kuhlmann & G. Skarmoutsos
Institute of Structural Design, University of Stuttgart, Germany

ABSTRACT

Nowadays the demand for a robust design of structures is increasingly important and the consideration of the robustness of a building should not be neglected. In the sense of robustness it is not necessarily meant to only mitigate a local damage but also a progressive collapse of the entire structure due to this local damage.

One possibility, given in the current Eurocodes, is to accept the local damage, for example a loss of a column, and to activate alternate load paths in order to redistribute the available and additional loadings into the intact part of the structure. Applying this alternate load path method especially the additional requirements on joints have to be respected. Therefore, referring to joints not only the load bearing capacity is important but also the available rotation capacity, so that sufficient ductility in this crucial structural part is provided. It is important that the joints are not only exposed to hogging moment loading but due to the column loss also to sagging moment loading. Furthermore, the capability of the joints to redistribute the moment loading into a normal force has to be provided.

Within experimental investigations the behavior of composite joints concerning their possibility to fulfil these requirements was investigated. Thereby focus is given to the redistribution of the moment loading into a normal force loading and especially to the rotational behavior of the composite joints. Additionally, two experimental tests on composite frames under column loss were performed in order to evaluate the behaviour of the joints in a complete frame. Due to further numerical studies the observations were confirmed and the range of parameters was expanded.

At the moment only deemed-to satisfy rules are available in the Eurocodes, no explicit verification of the rotation capacity. One possibility to perform such verification is given through the so-called beamline method. The aforementioned method compares the available moment-rotation behaviour of a joint with the required moment-rotation behaviour resulting out of the relevant structure. Investigations were executed how to make use of this method to verify the joints demands resulting from the loss of a column.

Life-Cycle Analysis and Assessment in Civil Engineering: Towards an Integrated Vision – Caspeele, Taerwe & Frangopol (Eds)
© 2019 Taylor & Francis Group, London, ISBN 978-1-138-62633-1

Behaviour of an innovative joint solution under impulsive loading

M. D'Antimo, J.-F. Demonceau & J.-P. Jaspart
Urban and Environmental Engineering Research Unit, University of Liege, Belgium

ABSTRACT

The FREEDAM project aims at developing an innovative joint solution able to dissipate significant seismic energy through the activation of friction components at the joint level; these components are customized to keep their mechanical properties when submitted to cyclic actions (D'Antimo 2018). The idea is to develop a solution able to undergo huge rotational demands without exhibiting high plasticity level in the joint components.

Within this project, the University of Liège is involved in the robustness assessment for the proposed innovative joint solution. Since several years, different contribution have been achieved in the robustness field; in particular, the behaviour of structures subjected to the exceptional event "loss of a column" is under investigation, using the alternative load path method, with the objective of proposing practical guidelines and formulation allowing ensuring an appropriate robustness to the structures (Demonceau, 2008, Demonceau & Jaspart, 2010, Demonceau et al, 2011, Comeliau et al, 2012, Huvelle et al, 2013, Huvelle et al, 2015, Demonceau et al, 2017, Colomer et al, 2017).

Having as final goal the robustness assessment of structures equipped with FREEDAM joints, a preliminary experimental campaign on FREEDAM joints subjected to severe impulsive loading have been performed. Double-sided beam-to-column connections were tested by means of a dropping mass falling from a fixed height. The weight of the mass and the dropping height were designed in order to dissipate a chosen amount of energy. To track the deformation of the tested specimens during the test, Digital Image Correlation (DIC) have been used. The main aim of the experimental campaign is the characterization of the joint behaviour at different level of strain rates. The main objective is to estimate the DIF (Dynamic Increase Factor) of the joint to be use in the structural analysis of frames equipped with the proposed solution.

To this scope, a detailed numerical modelling, validated against the experimental evidence, is currently under development. The numerical tool is a valid aid to perform parametrical analysis extending the experimental results and to derive the complete behaviour of the joint under impulsive loading protocol. The paper covers the first results of the experimental campaign and the ongoing activities in the framework of the robustness assessments of the FREEDAM joints.

REFERENCES

Colomer Segura, C., Hamra, L., D'Antimo, M., Demonceau, J.-F. & Feldmann, M. (2017) Determination of loading scenarios on buildings due to column damage. *Structures,* 12, 1–12.

Comeliau, L., Rossi, B. & Demonceau, J.-F. (2012) Robustness of steel and composite buildings suffering the dynamic loss of a column. *Structural Engineering International: Journal of the International Association for Bridge and Structural Engineering,* 22/3, 323–329.

D'Antimo, M., Demonceau, J.-F., Jaspart, J.-P., Latour, M. & Rizzano, G. (2018) Preliminary study on beam- to-column joint under impact loading. *Open Construction and Building Technology Journal* (accepted for publication).

Demonceau, J.-F. (2008) *Steel and Composite Building Frames: Sway Response Under Conventional Loading and Development of Membrane Effects in Beams Further to an Exceptional Action.* PhD Thesis. University of Liege (freely downloadable at: http://hdl.handle.net/2268/2740).

Demonceau, J.-F., Comeliau, L., Hoang, V. L. & Jaspart, J.-P. (2017) How can a steel structure survive to impact loading? *The Open Civil Engineering Journal,* 12, 434–452.

Demonceau, J.-F., Comeliau, L. & Jaspart, J.-P. (2011) Robustness of building structures – recent developments and adopted strategy, *Steel Construction – Design and Research,* Ernst & Sohn (Wiley Company). Vol. 4/3, pp. 166–170.

Demonceau, J.-F. & Jaspart, J.-P. (2010) Experimental test simulation a column loss in a composite frame. *Advanced Steel Construction,* 6, 891–913.

Huvelle, C., Hoang, V. L., Jaspart, J.-P. & Demonceau, J.-F. (2015) Complete analytical procedure to assess the response of a frame submitted to a column loss. *Engineering Structures,* 86, 33–42.

Influence analysis of group studs stiffness in accelerated construction steel-concrete composite small box girder bridges

Y. Xiang
College of Civil Engineering and Architecture, Zhejiang University, Hangzhou, China
Cyrus Tang Center for Sensor Materials and Applications, Zhejiang University, Hangzhou, China

S. Guo
College of Civil Engineering and Architecture, Zhejiang University, Hangzhou, China

ABSTRACT

The steel–concrete composite beam (SCCB) combines the steel beam with the concrete panel together by the connectors to jointly bear load and make full play of the two materials' strength. The steel–concrete composite structure can be quickly connected and constructed by using group studs-shear connectors.

Although some of Chinese and abroad scholars have done a number of numerical simulation analysis and experimental researches on the deflection, slip and natural frequency of steel–concrete composite beams with different group stud shear connectors under static, dynamic and fatigue load etc., the studies before this focused mainly on the problem of single beams. For the practical engineering, the faced structure is a whole bridge with SCCBs or the rapid construction SCCB bridges, in which the concrete plates will be assembled by transverse beams, cross-braces or prestressing. The behavior of the bridges not only is subjected to vehicle load, but also subjected to composite action and influence of the pre-tension in the concrete plates. As a result, the behavior of the whole bridge is more complex than that of single beam.

Based on the comparison of numerical analysis results and measured values of a single steel–concrete composite small box girder with group steel nail connector, this paper presented a simplified finite element modeling of SCCB or bridges by using shell elements plus spring elements in elastic work condition for service state without consideration of local stress analysis.

A accelerated construction composite bridges, composed of five small box girders, with 40 m span, and 16 m width (as shown in Figure 1), is analyzed by the proposed method, and its behavior under load and the influence of the different spring element stiffness on mechanical performance were explored.

The results show that the bridge with group studs according to design of completely shear stiffness, can realize rapid construction of the medium-short

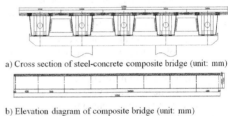

a) Cross section of steel-concrete composite bridge (unit: mm)

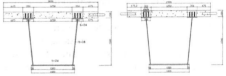

b) Elevation diagram of composite bridge (unit: mm)

c) Cross section of side girder and internal girder

Figure 1. Steel-concrete composite girder bridge (unit: mm).

span SCCB bridges The proposed method has enough precision, and is simple, convenient and fast.

ACKNOWLEDGMENTS

This work is financially supported by the Fundamental Research Funds for the Central Universities of China (2018 Zhejiang University) and the Cyrus Tang foundation of China.

REFERENCES

Guo, S. (2017) *Static Behavior Analysis and Experimental Investigation of Accelerated Construction High Performance Steel-Concrete Composite Small Multi-Box Girder Bridges*. Hangzhou, Zhejiang University.

Xiang, Y. & Guo, S. (2017) Parameter analysis of push-out specimens with different group studs in accelerated steel and concrete composite beams construction under complicated stress condition. *China Journal of Highway*, 30 (3), 246–254 (in Chinese).

Life-Cycle Analysis and Assessment in Civil Engineering: Towards an Integrated Vision – Caspeele, Taerwe & Frangopol (Eds)
© 2019 Taylor & Francis Group, London, ISBN 978-1-138-62633-1

Development of a design-oriented structural robustness index for progressive collapse

C. Praxedes & X.-X. Yuan
Department of Civil Engineering, Ryerson University, Toronto, Canada

ABSTRACT

Performance-based structural design has gained attention in the past decades. Structural robustness is an important system performance indicator that evaluates whether a system is susceptible to local or total failure and identifies critical elements of the system. In a progressive collapse analysis, the robustness of a structure describes its capacity to sustain loads and confine damage after an external, often sudden and shocking, event. Usually, the initiating event is of localized nature relative to the structure dimensions.

Many authors have proposed indices in order to quantify structural robustness. The approaches can be divided into two categories, namely, deterministic and probabilistic. The deterministic approach makes use of structural properties determined from the intact and damaged systems. In this approach the definitions of robustness can be further divided into two subgroups: stiffness- and response-based measures. The probabilistic approach, by definition, accounts for the uncertainty of system properties and external loadings. This approach can also be divided into two subgroups: probability of failure- and risk-based.

This study aims to provide a practical metric for structural robustness. An idealized Daniels system is used to assess the practicality of the proposed index. The analysis consists of determining whether the index is consistently affected by several structural influencing factors that are usually considered in design to prevent progressive collapse. Existing indices are also assessed for a comparative assessment of structural robustness. Five existing robustness indices were selected for the comparative study. The selected indices were (1) the stiffness-based index by Lu et al. (1999); (2) the response-based indices by Brett and Lu (2013) and Frangopol and Nakib (1991); (3) the probability-based index, by Lind (1995); and (4) the risk-based index by Baker et al. (2008).

The quantification of robustness for progressive collapse analysis must indicate the susceptibility that a damaged structure collapse after an initial damage. In the present study, a conditional survival function, $S(n_{pr} \mid d)$, is proposed to be used to describe the probability of system survival when the damage progresses to n_{pr} elements given the initial damage, d.

Considering the total probability of damage progression to n_{pr} elements given the initial damage, $P(n_{pr} \mid d)$, it is proposed that the robustness index be defined as:

$$\rho = \frac{S(n_{pr} \mid d)}{P(n_{pr} \mid d)} = \frac{S_{pr,d}}{P_{pr,d}} \tag{1}$$

The defined index assess the system capacity to confine damage after an initial failure event. The relationship between a cascading failure and the failure of the whole system is associated with the concepts of progressive collapse, thus it is considered that the proposed index is more practically relevant to progressive collapse design.

The comparative analysis assessed the following influencing factors, material behaviour, structural redundancy, compartmentalization and dynamic effects. The proposed robustness index is of simple calculation and also provided consistent results irrespective of the factor analysed. Previously proposed indices, on the other hand, either responded counter-intuitively or were not affected by some of the properties evaluated. It is expected that the use of the proposed robustness indicator will assist in the performance-based design structural systems for robustness in order to prevent progressive collapse.

REFERENCES

Baker, J.W., Schubert, M. & Faber, M.H. (2008) On the assessment of robustness. *Structural Safety*, 30 (3), 253–267.

Brett, C. & Lu, Y. (2013) Assessment of robustness of structures: Current state of research. *Frontiers of Structural and Civil Engineering*, 7, 356–368.

Frangopol, D.M. & Nakib, R. (1991) Redundancy in highway bridges. *Engineering Journal–American Institute of Steel Construction Inc.* 28 (1), 4550.

Lind, N.C. (1995) A measure of vulnerability and damage tolerance. *Reliability Engineering & System Safety*, 48 (1), 1–6.

Lu, Z., Yu, Y., Woodman, N.J. & Blockley, D. (1999) A theory of structural vulnerability. *The Structural Engineer*, 77 (18), 17–24.

Life-Cycle Analysis and Assessment in Civil Engineering: Towards an Integrated Vision – Caspeele, Taerwe & Frangopol (Eds)
© 2019 Taylor & Francis Group, London, ISBN 978-1-138-62633-1

Performance metrics for seismic-resilient steel braced frame buildings

O. Serban & L. Tirca
Concordia University, Montreal, Canada

ABSTRACT

Disaster resilience of a building consists in its capacity to restore full functionality after natural hazards (Bruneau & Reinhorn 2004). Recently, several studies have raised concerns about the vulnerability of existing urban infrastructure including the building stock. In order to assess the potential economic losses and functionality disruption, the resilience-based design approach is recommended. However, the current code designs method is force-based and not resilience-based. Moreover, using the current code seismic design method, the human lives are protected, the potential damage of a building is controlled through drift limits, the structural collapse is mitigated but the level of building functionality in the aftermath of a hazardous event is still unknown. Nowadays, there are no deterministic design procedures developed to assess the building structure's response to multi-hazards loading nor strategies to mitigate the building nonlinear response under such loads. Moreover, predicting factors that affect the duration of business interruption, business reallocation, and recovery time is not a straightforward process.

Designing low-damage structural systems is a key component of seismic-resilient buildings. Once the acceptable level of damage is decided by owners or other decision makers, a measurement system with performance objectives can be used.

The main focus of this study is to develop system-level resilience performance metrics that includes the performance of structural and non-structural components, loss assessment and recovery time prediction after an earthquake event. This methodology can also be extended to assess the building functionality in the aftermath of other natural hazards.

The proposed framework for resilience-based design is similar with that for performance-based design. For example, an office building is evaluated to the Basic Safety Objective which is intended to be generally consistent with the safety level expected in ASCE/SEI 41 (2013) standard. In other words, the building should remain standing under extremely rare events (2%/50 years hazards) and any other damage or loss is acceptable. On the other hand, for more frequent events (10%/50 years hazards), the building should remain stable with significant reserve capacity while any other damage to the non-structural components should be controlled. Buildings designed according to the current code need to satisfy the collapse safety criteria (FEMA P695, 2009), in which the adjusted collapse margin ratio, should be greater to or equal to the minimum permissible adjusted collapse margin ratio corresponding to 10% probability of collapse.

According to the proposed framework, damage levels are defined on the IDA curves (Vamvatsikos & Cornell, 2002) computed through a rigorous nonlinear time history analysis. Then, based on fragility curves, losses and recovery time are estimated in the aim of calculating the seismic resilience. Hence, to assess the building structure's performance under an extreme event, an accurate and computational efficient numerical model that is able to capture the nonlinear response from yielding to failure is required.

The proposed framework for seismic-resilient buildings is intended to be a tool for decision makers.

REFERENCES

American Society of Civil Engineers (ASCE 41). (2013) *Seismic Rehabilitation of Existing Buildings.* Reston, Virginia.

Bruneau, M. & Reinhorn, A.M. (2004) Seismic resilience of communities conceptualization and operationalization. In: Fajfar, P. & Krawinkler, H. (eds.) *Performance-Based Seismic Design Concepts and Implementation*; *Proceedings of the International Workshop, 28 June–1 July 2004,* Bled, Slovenia.

Federal Emergency Management Agency (FEMA). (2009) *Quantification of Building Seismic Performance Factors (FEMA P695).* Washington, D.C, FEMA.

Vamvatsikos, D. & Cornell, C.A. (2002) Incremental dynamic analysis. In *Earthquake Engineering and Structural Dynamics* 31 (3), 491–514.

Post-failure torsion capacity and robustness of encased tubular arch spring connections

Ph. Van Bogaert, K. Schotte & H. De Backer
Civil Engineering Department, Ghent University, Belgium

ABSTRACT

A particular reaction of steel tubular arches fixed to a concrete abutment is the torsion moment. Since the torsional capacity and stiffness of circular sections are rather high, large torsion clamping may be expected. However, whether connectors like steel strips or headed studs perform well for torsion is not entirely clear. Three tests are reported, that were intended as a first approximation and mainly to detect what are the failure mechanisms in this type of connection.

The setup consists of a truncated parallelogram concrete slab of 1.2 by 1.2 m and 0.15 m thickness with a minimum of reinforcing rebars. To reduce the torque magnitude and to use limited auxiliary equipment, the tests are a scaled situation of a real situation by a factor of approximately 10. Each sample included a vertical steel tube 50/8 S 235 welded to a horizontal UPN 80 S 235 profile. The latter allows its connection by load cells to the vertical UPN profiles and thus the application of 2 equal horizontal forces, which act as a torsion moment at the base of the vertical tube. The first test was intended as a trial and also aimed to detect whether any natural bond would exist between the encased part of the steel tube and the concrete slab. During the test it became evident that natural bond was inexistent.

In the 2nd and 3rd test the steel tubes were equipped with respectively strip- and stud connectors. The second test has demonstrated the high degree of robustness of strip connection for torsion, as well as the fact that from the first loading a composite action of steel and concrete must exist. The first modification of the resisting system is due to exceeding compression of the surrounding concrete, followed by a second modification, because high of pressure in the contact area between both materials. The strip connection eventually fails through yielding of the steel strips as reinforcement of the composite cross-section. In

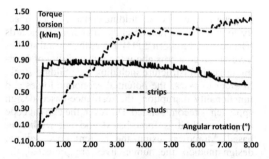

Figure 1. Diagrams for strips and studs combined.

addition, the strip connection is rather flexible for torsion and does not really act as perfect clamping, since it appeared to be 70-times weaker than the torsion stiffness of the steel tube.

The third test showed that headed studs need to be supplemented by rebars, since failure occurred prematurely, due to concrete splitting, starting at the contact area of the stud shaft. This is easily avoided by normal reinforcement. In addition, the connection, subjected to torsion does not show any robustness, since after first modification of the strength mechanism, no further increase of the torsion moment was possible. However, the torsion stiffness of the connection is distinctly higher compared to the previous type. Still, it appeared to be 13-times weaker than the torsion stiffness of the steel tube. The rotation-torsion moment diagrams of both types of connectors are shown in Fig. 1.

Further experimental research would be welcomed to improve knowledge about this torsion connection and might perhaps result in establishing reliable data about the torsion stiffness, the combined action of various types of connections and the degree of robustness of torsion clamping of steel bridge superstructure in concrete abutments.

SS12: Repair and self-repair of concrete
Organizers: N. De Belie, K. Van Tittelboom, D. Snoeck & E. Gruyaert

Life-Cycle Analysis and Assessment in Civil Engineering: Towards an Integrated Vision – Caspeele, Taerwe & Frangopol (Eds)
© 2019 Taylor & Francis Group, London, ISBN 978-1-138-62633-1

Optimizing nutrient content of microbial self-healing concrete

Y.Ç. Erşan & Y. Akın
Department of Civil Engineering, Abdullah Gul University, Kocasinan, Kayseri, Turkey

ABSTRACT

Cracks in microbial self-healing concrete are autonomously sealed by the specifically selected bacteria that induce calcium carbonate precipitation. One of the governing parameters of this process is the production of dissolved inorganic carbon (i.e. CO_2) through different microbial pathways such as aerobic respiration, ureolysis and nitrate reduction. So far, studies mostly focused on bacteria selection and their bench-scale applications to explore the limits of the concept and meanwhile the role of nutrients were overlooked. Crack repair in microbial self-healing concrete depends on the amount and the rate of calcium carbonate precipitation and the reactions are limited by the availability of the reactants. In some studies, nutrients were provided in capsules while in others they were supplied as admixtures. When added as admixtures bioavailability of the nutrients becomes important for crack healing as they disperse in the mixture and a significant portion stays far from an individual crack. Therefore, it is necessary to optimize the nutrient content of the microbial self-healing concrete. This study focuses on previously suggested nitrate reduction pathway for microbial self-healing concrete and its corresponding nutrients Ca-nitrate (CN) and Ca-formate (CF) which are added as admixtures. We initially tested the coupled effect of the nutrients on mortar properties and kept the CF:CN ratio constant at 2.50:1.00. The CN doses in the mixtures were increased by 0.25% (wt/wt cement) increments up to 3.00% and the CF doses were set relatively. All the mixtures were investigated in terms of workability, setting time and strength development. Results showed that CN-CF dosage should not be higher than 2.00–5.00% (wt/wt cement) to maintain the workability of the mortar in the acceptable region. The admixtures, mostly CF, improved the strength development of the mortar, yet no proportionality between the improvement and the dosages could be achieved. The highest compressive strength achieved in different mixes was 65 MPa. These initial results defined the limit values for further experiments planned on the nutrient leaching from mortar specimens. Variation in mortar properties and nutrient bioavailability was recorded and the optimum nutrient content range was defined as 3.5% to 7% which should be chosen depending on the expectations *in situ*. In this range, the NO_3-N leaching rates varied between $0.71\,mg \cdot L^{-1} \cdot d^{-1}$ and $1.65\,mg \cdot L^{-1} \cdot d^{-1}$. For the same range, $HCOO^-$ leaching rates were varied between $17.10\,mg \cdot L^{-1} \cdot d^{-1}$ and $35.80\,mg \cdot L^{-1} \cdot d^{-1}$. Further research is necessary to more accurately suggest nutrient contents for different environmental conditions. Moreover, the $CaCO_3$ precipitation performance of the bacteria should be tested under these leaching conditions in order to define a specific nutrient concentration. Overall, current results create a base for further research on cost-benefit optimization of microbial self-healing concrete.

Life-Cycle Analysis and Assessment in Civil Engineering: Towards an Integrated Vision – Caspeele, Taerwe & Frangopol (Eds)
© 2019 Taylor & Francis Group, London, ISBN 978-1-138-62633-1

Microencapsulated spores and growth media for self-healing mortars

K. Paine, I. Horne, L. Tan, T. Sharma, A. Heath & R. Cooper
University of Bath, UK

J. Virgoe, D. Palmer & A. Kerr
Lambson, UK

ABSTRACT

This paper reports on a study to use synthesised gelatin/acacia gum microcapsules as a carrier for spores and growth media (nutrients and precursor) for self-healing mortar. The microcapsules were designed to transition between hydrated and dried conditions, converting from a rubbery soft state to a glassy stiff state; such that they would survive the wet mixing process but rupture successfully upon crack formation.

The encapsulations carried out for this project were of two types: encapsulation of dormant bacterial spores; and encapsulation of the growth media and water.

Spores of *Bacillus pseudofirmus* DSM 8715 were used as they had been shown in earlier work to have potential for crack repair in concrete (Sharma et al. 2017). The encapsulation of spores proved to be relatively straightforward.

The growth media used consisted of calcium acetate as the precursor and glucose and yeast extract as the source of nutrients for spore germination and cell growth. Due to the particularly disruptive nature of the calcium acetate, encapsulation took a significant amount of refining to identify the maximum loading achievable in the water phase and then balance this with the highest ratio of oil phase:water phase. Development settled on a water:gelatine: growth media

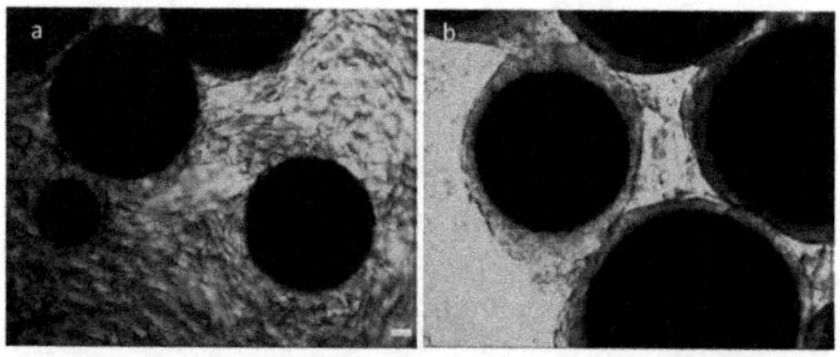

Figure 1. Encapsulation of growth media (a) addition of the emulsion to the external phase; and (b) after 2 hours wall formation was largely complete.

mixture ratio of 80:10:15. This allowed a oil:water ratio of around 46% by weight which led to good wall formation and capsule stability (Figure 1).

The microcapsules were added to cement mortars and pastes to investigate their effect on early-age properties and their self-healing capability. Isothermal conduction calorimetry showed that the microcapsules had few effects on the hydration of cement and that they survived the mixing process as designed.

Mortars made with microcapsules were cracked after 28 days and subject to cyclic wetting and drying for up to six months. Clear differences in crack closure and precipitation of calcite were observed. Indeed, the control mortars (without microcapsules) showed no healing at all whilst those in which microcapsules were added showed clear signs of precipitation within the crack. The implication is that the constituents of the microcapsules released upon cracking and played a role in calcite formation.

The mortar with bacteria and growth media performed best. The tests to determine the self-healing capability by means of capillary absorption. Calcite precipitation and crack closure was observed in the mortar that contained both encapsulated spores and growth media.

The research has shown that it is possible to encapsulate both spores and growth media in microcapsules and that these microcapsules survive the mixing process and release their cargo on crack formation. The work has shown that there is potential for synthesised gelatin/acacia gum microcapsules as a carrier of spores and growth media for self-healing concrete.

REFERENCE

Sharma, T.K., Alazhari, M., Heath, A., Paine, K. & Cooper, R.M. (2017) Alkaliphilic *Bacillus* species show potential application in concrete crack repair by virtue of rapid spore production and germination then extracellular calcite formation. *Journal of Applied Microbiology*, 122 (5), 1233–1244.

*Life-Cycle Analysis and Assessment in Civil Engineering: Towards an
Integrated Vision – Caspeele, Taerwe & Frangopol (Eds)
© 2019 Taylor & Francis Group, London, ISBN 978-1-138-62633-1*

Use of fiber-reinforced self-healing cementitious materials with superabsorbent polymers to absorb impact energy

D. Snoeck, T. De Schryver, P. Criel & N. De Belie
Magnel Laboratory for Concrete Research, Ghent University, Ghent, Belgium

ABSTRACT

One of the major flaws with concrete remains the susceptibility to cracking. The behavior and self-healing of cementitious materials with synthetic microfibers and superabsorbent polymers (SAPs) during and after a static four-point-bending test have been proven successful. It was however not known how this material reacts when subjected to impact loading.

In this study, a reference mixture and a mixture containing 1 m% of SAP were tested at an age of 28 days and stored in different healing conditions (Wet/Dry cycles and at $95 \pm 5\%$ RH) (Snoeck et al., 2014, Snoeck and De Belie, 2015b, Snoeck and De Belie, 2015a, Snoeck et al., 2016). After 28 days of healing the specimens are impacted and healed again. The applied impact test is a Drop-Weight-Test (DWT) and the healing is microscopically monitored and studied by means of natural frequency analysis. Specimens containing SAPs show a more ductile behavior during impact loading compared to reference samples. This ductile behavior enhances multiple cracking of the strain-hardening cementitious materials causing a possible higher amount of autogenous healing, which was confirmed by natural frequency analysis. The evolution of the natural frequencies shows a superior healing caused by SAPs in both Wet/Dry-cycling and storage at $95 \pm 5\%$ RH.

A strain-hardening cementitious composite with SAPs is able to withstand impact loading due to multiple crack formation and high ductility. Autoge-nous healing of the formed multiple cracks leads to an effective regain in natural frequency of the stud-ied plate material.

Superabsorbent polymers are able to provide stress initiators to increase the amount of multiple cracking and thus the ductility to withstand impact loading.

Overall, the specimens containing SAPs had a superior ductile behavior during impact loading compared to the REF specimen. The same can therefore also be con-cluded concerning the impact resistance.

The Wet/Dry healing condition was far more superior in terms of healing due to the availability of water. Some of the natural frequency properties were able to be regained.

More healing and regain in natural frequency properties were observed in specimens containing superabsorbent polymers. They are able to promote autogenous healing. Even in a relative humidity con-dition, samples with SAPs showed healing due to the uptake of moisture by the SAPs.

Overall, this strain-hardening material with su-perabsorbent polymers is very interesting to use when impact loading is expected and healing is needed.

REFERENCES

Snoeck, D. & De Belie, N. (2015a) From straw in bricks to modern use of microfibres in cementitious composites for improved autogenous healing – a review. *Construction and Building Materials*, 95, 774–787.

Snoeck, D. & De Belie, N. (2015b) Repeated autogenous healing in strain-hardening cementitious composites by using superabsorbent polymers. *Journal of Materials in Civil Engineering*, 04015086, 1–11.

Snoeck, D., Dewanckele, J., Cnudde, V. & De Belie, N. (2016) X-ray computed microtomography to study autogenous healing of cementitious materials promoted by superabsorbent polymers. *Cement and Concrete Composites*, 65, 83–93.

Snoeck, D., Van Tittelboom, K., Steuperaert, S., Dubruel, P. & De Belie, N. (2014) Self-healing cementitious materials by the combination of microfibres and superabsorbent polymers. *Journal of Intelligent Material Systems and Structures*, 25, 13–24.

The role of silicate salts in self-healing properties of cement pastes

M. Stefanidou, V. Kotrotsiou & F. Kesikidou

Laboratory of Building Materials, Civil Engineering Department, Aristotle University of Thessaloniki, Thessaloniki, Greece

ABSTRACT

The study investigates the influence of silicate salts (Na and K) and nano-particles and the influence of the curing environment on the mechanical and physical properties of cement pastes. 8%, of the cement content, solution of potassium and sodium silicate were added to fresh matrix. In a second series of pastes containing the salts, 1,5% nanoSiO$_2$ and nano-CaO were also added. The pastes were subjected to a stable force in their fresh state in order to deliberate crack them. Curing variables included moist curing and cycles of moist-drying were tested. The final specimens were tested for compression strength, porosity and microstructure at the age of 28 and 90 days. Scanning electron microscope (SEM) imaging analysis assisted in verifying the results.

For pastes with silicate salts, the test results showed greater improvement in the mechanical properties of samples subjected to moist curing than those subjected to cyclic curing. The compressive strength of the specimens in no case exceeded those of the control group but the samples with silicate salts were the ones closest to the reference. The role of nanoparticles needs investigation as they seem to form dense structure and managed to fill empty spaces. SEM imaging revealed a number of needle-like crystals in the paste which were likely the primary reason for the mechanical properties.

Life-Cycle Analysis and Assessment in Civil Engineering: Towards an
Integrated Vision – Caspeele, Taerwe & Frangopol (Eds)
© 2019 Taylor & Francis Group, London, ISBN 978-1-138-62633-1

Life cycle assessment of Self-Healing Engineered Cementitious Composite (SH-ECC) used for the rehabilitation of bridges

P. Van den Heede & N. De Belie
Magnel Laboratory for Concrete Research, Department of Structural Engineering,
Faculty of Engineering and Architecture, Ghent University, Ghent, Belgium

F. Pittau & G. Habert
Chair of Sustainable Construction, Institute of Construction and Infrastructure Management,
Swiss Federal Institute of Technology Zurich (ETH Zurich), Zurich, Switzerland

A. Mignon
Polymer Chemistry and Biomaterials Group, Department of Organic and Macromolecular Chemistry,
Ghent University, Ghent, Belgium

ABSTRACT

In Europe, more than 167 million tons of concrete were produced in 2015, of which roughly 30% was used in the infrastructure sector. A huge amount of concrete is expected to be used in the next decades for the rehabilitation of bridges. Typically, the major damages in concrete bridge structures are related to freeze-thaw and exposure to chlorides, which may cause cracks and rebar corrosion. In this paper the potential of using self-healing engineered cementitious composite (SH-ECC) for rehabilitation of bridges is analyzed with incorporation of 1 m% of two in-house developed superabsorbent polymers (SAPs) by Mignon (2016) and 2 v% of oil-coated synthetic polyvinyl alcohol (PVA) microfibre. The two types of SAP were a synthetic acrylic acid + acrylamide (AA+AM) based SAP and a semi-synthetic acrylic acid + modified sodium alginate (AA+AlgMOD) based SAP. The overall mixture proportions of the SH-ECC were in agreement with Snoeck & De Belie (2016). A life cycle assessment (LCA) methodology cf. the ISO 14040 standards (Finkbeiner et al. 2006) was adopted to compare the global warming potential (GWP) of this novel material and technique for bridge repair with other repair solutions, such as rehabilitation involving traditional ordinary Portland cement (OPC) concrete and repair with Ultra-High Performance Fibre Reinforced Concrete (UHPFRC). In this study, the chosen functional unit (FU) is 1 m^2 of repaired bridge. The bridge Log Čezsoški, nearby Bovec, Slovenia, described by Habert et al. (2013), was used as reference. This bridge has a surface of 292.5 m^2. Life cycle inventories were compiled using mainly input from the Ecoinvent 3.4 database. The LCA calculations were performed in SimaPro 8.

The calculation output revealed that the impact due to the production of materials is responsible for the major contribution in carbon emissions, regardless of the rehabilitation system used. The proposed SH-ECC solutions have a significantly higher impact than the traditional OPC concrete solution. This goes for both the AA+AM and AA+AlgMOD based SAP options. Nevertheless, if the thickness of replaced concrete could be limited to 30 mm instead of the full thickness of 80 mm and if the waterproofing membrane would no longer be needed, the SH-ECC option becomes a lot more feasible. Then, a reduction in global warming potential of up to 50% in comparison with the OPC concrete solution would be possible. UHPFRC is also a valid solution for bridge repair. If used as partial replacement of the original concrete layer, it is able to save approximately 30% of the carbon emissions.

When the aspect of service life, as well as the uncertainties inherent to maintenance and durability are taken into account, both the SH-ECC and UHPFRC solutions would be pronounced carbon saving alternatives. When assuming a standard triangular error distribution, the OPC concrete solution has the highest GWP value, with an emission between 265 and 619 kg CO_2 eq per m^2 of bridge repair, while the alternative solutions, SH-ECC and UHPFRC, are able to save between 55–70% and 59–74%, respectively.

REFERENCES

Finkbeiner, M., Inaba, A., Tan, R., Christiansen, K. & Klüppel, H.J. (2006) The new international standards for life cycle assessment: ISO 14040 and ISO 14044. *International Journal of Life Cycle Assessment*, 11 (2), 80–85.

Habert, G., Denarié, E., Šajna, A. & Rossi, P. (2013) Lowering the global warming impact of bridge rehabilitations by using ultra high performance fibre reinforced concretes. *Cement and Concrete Composites*, 38, 1–11.

Mignon, A. (2016) *Effect of pH-Responsive Superabsorbent Polymers on the Self-Sealing and Self-Healing of Cracks in Concrete.* PhD Thesis. Ghent, Ghent University.

Snoeck, D. & De Belie, N. (2016) Repeated autogenous healing in strain-hardening cementitious composites by using superabsorbent polymers. *Journal of Materials in Civil Engineering*, 28 (1), 04015086–1–04015086–11.

Self-healing concrete vs. conventional waterproofing systems in underground structures: A cradle to gate LCA comparison with reference to a case study

S. Rigamonti, E. Cuenca, A. Arrigoni, G. Dotelli & L. Ferrara
Politecnico di Milano, Milano, Italy

ABSTRACT

The use of self-healing concrete in real applications is a challenging opportunity due to its potential to extend the service life of structures by minimizing or even eliminating maintenance and conservation tasks. In this paper the environmental sustainability of self-healing concrete is analyzed. The particular case study deals with the waterproofing and protection of a 24000 m^2 underground parking garage of a new shopping center in Novara (Italy). For the construction of the parking a novel waterproofing system was used, consisting of a concrete treated with crystalline admixtures which allows to achieve the aforementioned self-healing capacity by resealing the cracks. In order to evaluate and quantify the environmental benefits and burdens involved in this self-healing ability the Life Cycle Assessment (LCA) methodology has been used (Guinée et al. 2011) together with service life prediction models (fib 2006).

The study assumes a cradle-to-gate with options scenario, which includes the stages of the life cycle from extraction of virgin materials to construction phase, corresponding to A1-A5 phases of the EN 15804 standard. The environmental performances of the analyzed structure are then compared to those of a similar construction made of reinforced concrete without the use of crystalline admixtures, but with alternative waterproofing systems. Specifically, the environmental impact of three different waterproofing techniques for the parking have been assessed. First, the use of the crystalline admixture (CA) as a waterproofing agent has been considered. Afterwards, the results have been compared with those obtained from two classic bitumen-based waterproofing systems which differ in their composition and with those obtained using a more modern flexible polyolefin (FPO) membrane. The latter has the advantage of being fully bonded to the protected concrete element, thus avoiding water underflow or migration between the structure and the membrane.

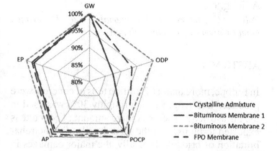

Figure 1. Comparison of the results obtained from the LCA.

The following impact categories were considered: Global warming (GW), Acidification (AP), Ozone depletion (ODP), Photochemical oxidation (POCP), Eutrophication (EP). The environmental impacts resulting from the adoption of the crystalline admixture and the FPO membrane solution are approximately the same. Conversely, although the variability of the result is in the range of a few percentage points, the bitumen-based materials generates higher impacts (Figure 1).

The results have shown that the adoption of a waterproofing crystalline additive is a solution that, from the environmental point of view, can compete with the most recent waterproofing membranes. It has also to be pointed out that, for structures with a very large surface like a parking, the influence of the waterproofing on the total environmental foot-print resulted to be limited to maximum 10%, due to the high emissions associated to the production of concrete and steel.

REFERENCES

fib (2006) fib Bulletin 34, Model code for service life design.
Guinée, J., Heijungs, R., et al. (2011) Life cycle assessment: Past, present, and future. *Environmental Science & Technology*, 45, 90–96.

Efficiency of manual and autonomous healing to mitigate chloride ingress in cracked concrete

K. Van Tittelboom
*Magnel Laboratory for Concrete Research, Department of Structural Engineering,
Faculty of Engineering and Architecture, Ghent University, Belgium*

B. Van Belleghem
*Magnel Laboratory for Concrete Research, Department of Structural Engineering,
Faculty of Engineering and Architecture, Ghent University, Belgium
Strategic Initiative Materials (SIM), Belgium*

R. Callens
*Magnel Laboratory for Concrete Research, Department of Structural Engineering,
Faculty of Engineering and Architecture, Ghent University, Belgium*

P. Van den Heede
*Magnel Laboratory for Concrete Research, Department of Structural Engineering,
Faculty of Engineering and Architecture, Ghent University, Belgium
Strategic Initiative Materials (SIM), Belgium*

N. De Belie
*Magnel Laboratory for Concrete Research, Department of Structural Engineering,
Faculty of Engineering and Architecture, Ghent University, Belgium*

ABSTRACT

Cracks are inevitably present in concrete structures and create preferential paths for the penetration of corrosion-inducing substances such as chlorides. As soon as a critical amount of chlorides has reached the location of the steel reinforcement, chloride-induced corrosion may occur, which is one of the main deterioration mechanisms of reinforced concrete structures. If left untreated, cracks lead to a reduction in service life and repair of cracks is thus absolutely inevitable to make concrete structures more sustainable (Otieno et al., 2016).

However, repair works impose high direct and indirect costs. Moreover, current practices only result in crack repair after being detected during inspection and when the budget for repair has become available. Due to this, aggressive agents, may already have entered the concrete matrix before manual crack repair. Therefore, it could be more efficient to trigger crack repair autonomously at the moment of crack appearance. Through embedding encapsulated healing agents, crack formation will coincide with capsule breakage and release of the healing agent.

As the autonomous healing mechanism occurs in a less controlled way, the effectiveness of autonomous and manual crack repair is compared in this study. Moreover, a comparison is made with manual crack healing after cracks have been exposed to a chloride rich environment for a certain time. Two type of healing agents were considered, a commercial polyurethane based agent and a commercially available water repellent agent.

Cylindrical concrete samples with and without self-healing properties were prepared and exposed to an accelerated chloride diffusion test. Samples with self-healing properties, containing encapsulated healing agent, were healed at the moment the crack was created. For other samples, cracks were manually healed through injection of the healing agent into the crack. For some series crack injection was performed before starting the chloride diffusion test, for other series, this

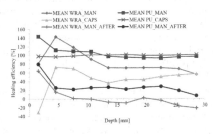

Figure 1. Mean healing efficiency as a function of the crack depth for all test series under investigation.

was done after the samples had been exposed for 3 weeks to the chloride solution. After the test, the chloride content in the vicinity of the crack was determined and the crack healing efficiency was calculated at different depths below the exposed surface (Figure 1).

When comparing the different test series among each other, it can be observed that series of which cracks were repaired after chloride exposure, show low (to even negative) healing efficiencies. It can thus be concluded that applying a healing agent after exposure to chlorides is less efficient. The beneficial effect of autonomous or manual crack healing before chloride exposure can be noticed from the healing efficiencies of the other test series. The best healing efficiencies were obtained by manual or autonomous repair with polyurethane before exposure.

REFERENCE

Otieno, M., Beushausen, H. & Alexander, M. (2016) Chloride-induced corrosion of steel in cracked concrete – Part I: Experimental studies under accelerated and natural marine environments. *Cement and Concrete Research*, 79, 373–385.

Establishment of spraying repair technology for concrete structures using drone

T. Iyoda
Department of Civil Engineering, Shibaura Institute of Technology, Tokyo, Japan

K. Nimura
Seibu construction Co. Ltd, Saitama, Japan

T. Hasegawa
Department of Electrical Engineering, Shibaura Institute of Technology, Tokyo, Japan

ABSTRACT

In recent years, infrastructure deterioration is remarkable in Japan. Measures against the deterioration of the infrastructure are serious issues. Inspection and management of bridges, tunnels, etc. are carried out periodically by management organizations. This work is often done at high altitudes. In addition, there are lack of personnel involved in checks and work. In recent years, it is expected that utilization of universal availability unmanned aerial vehicles (drone) is possible. In this research, we proposed spraying repair materials to concrete structures using drone in Figure 1. Specifically, repair materials were sprayed on wall structures and tunnels using drone. As a result, it turned out that it is difficult to make the drones fly stably. It also turned out that repaired material sprayed was "unevenness". Therefore, we devised further measures for stable flight. We report the experimental results with the signature of 2D sensor and LED which measures the distance between the drone and the structure by the equipment installed in the drone and the possibility of the application in Figure 2.

Figure 2. Improvement for our drone.

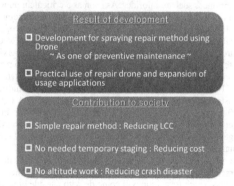

Figure 3. Result for development of repairing drone and contribution to society.

CONCLUSION

Fig. 3 shows the results and contribution of this research. Ease of use: Improvement of operability of the spray drone was confirmed by utilizing the area sensor. By visually confirming by LED, the hit rate to the target is also easily transversal. Also, the burden on the pilot could be reduced.

Spraying unevenness: By operating the nozzle, it was possible to eject perpendicularly with the object at all times. From this fact, remarkable reduction in spraying unevenness can be expected.

The results of this research may be able to reduce dangerous work by working drones on behalf of humans in the near future.

Figure 1. Repairing drone.

Polymer Flexible Joint as a structural repair method for reducing stress concentrations in cracked concrete structures

Ł. Zdanowicz, M. Tekieli, B. Zając & A. Kwiecień
Cracow University of Technology, Cracow, Poland

ABSTRACT

Local stress concentrations in concrete due to imposed deformations such as concrete shrinkage or temperature changes are one of the main reasons of damage of concrete members such as concrete floor on ground and concrete pavements. Many traditional approaches of repair (i.e. epoxy resin) can cause new damages because the bonded material are usually too stiff. An alternative can be the Polymer Flexible Joint (PFJ) method. PFJ is a method for repairing damaged structures. This approach allows to carry loads and bear relatively large deformations.

The aim of this paper is to describe the influence of application of PFJ on concrete beams, examined in four-point bending (4PB) tests before and after repair. The research program included 63 4PB tests in order to analyze the influence of polymer on the repair effectiveness. The effectiveness was defined as a change of stress capacity and strains capacity between original and repaired elements. The strain field on side surfaces of the specimens was measured with digital image correlation (DIC) method.

The example of stress response in 4PB tests for original and repaired element is presented in Figure 1. Due to the polymer layer, bending stiffness of repaired element was lower than of the original one and therefore higher values of CMOD were noticed for repaired elements. Comparing strain capacities, significantly higher deformability is observed in repaired specimens than in original ones. The effectiveness of repair in terms of load capacity is usually lower than 100% (AV = 87–95%) and it depends on polymer thickness and specimen's geometry. Contrary to the stress capacity of repaired specimens, their strain capacity is much higher than original one (for repaired specimens AV = 106–230%). The thicker polymer joint, the higher the strain capacity.

Results obtained from DIC method show clearly the mechanism of crack propagation under applied load. Based on these analyses, a hypothesis was made

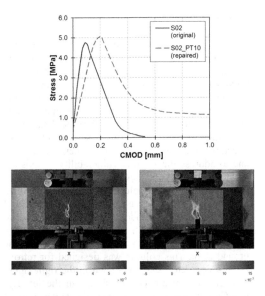

Figure 1. Example of load response of the notched beam in 4PB (top) and strain field obtained using DIC for original (bottom left) and repaired specimen (bottom right).

that polymer flexible joint allows to redistribute strains in joint and does not cause high stress concentration in this area. However, to prove this assumption more research is needed.

REFERENCES

Kwiecień, A. (2012) *Polymer Flexible Joints in Masonry and Concrete Structure*, Monograph 414, Series: Civil Engineering. Cracow, Poland, Wydawnictwo Politechniki Krakowskiej (in Polish).

Tekieli, M., De Santis, S., de Felice, G., Kwiecień, A. & Roscini, F. (2017) Application of digital image correlation to composite reinforcements testing. *Composite Structures*, 160, 670–688.

Structural column retrofitting of school building using Ferrocement Composites in Vigan, Ilocos Sur, Philippines

J.M.C. Ongpeng, V. Pilien, A. Del Rosario, A.M. Dizon, K.B. Aviso & R.R. Tan
De La Salle University, Manila, Philippines

ABSTRACT

The Philippine lies along the Pacific ring of fire which is the most active earthquake zone. Several earthquakes and volcanic eruptions occur due to a continuous series of oceanic trenches, plate movements, volcanic arcs and belts. Currently, civil engineering community including private and government sectors are joining forces to prepare before the "Big One". One strategy is to investigate and assess the integrity of structures to strengthen old and weakened structures and to address the seismic vulnerability of existing buildings. In this paper, a case study on an expansion of a two-storey 46-year old school building in Vigan, Ilocos Sur, Philippines is investigated. The school expansion includes an addition of 3rd floor level to the existing structure. Structural analysis and design was done in identifying the most critical column in the ground floor. This column was designed for retrofit using – concrete jacketing shown in the right figure and ferrocement composites shown in the left figure. Concrete jacketing increases the dimensions of the existing columns using fastened deformed bars as additional longitudinal bars and steel ties with concrete cover. On the other hand, Ferrocement composites uses wire mesh reinforcement fastened to an existing column with lightweight concrete. Comparative life-cycle assessment of the two methods using SimaPro was

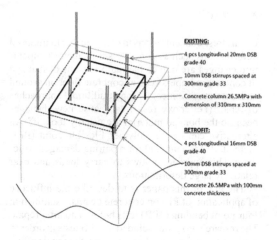

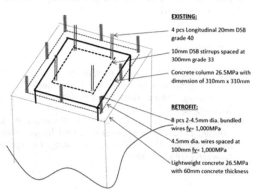

done for the critical column on a cradle-to-gate condition. Results showed that Ferrocement composites was effective in replacing concrete jacketing in terms of strength and was more environmentally friendly.

REFERENCES

Kaish, A.B.M.A., Jamil, M., Raman, S.N., Zain, M.F.M. & Nahar, L. (2018) Ferrocement composites for strengthening of concrete columns: A review. *Construction and Building Materials*, 160, 326–340.

Kan, A. & Demirboğa, R. (2007) Effect of cement and EPS beads ratios on compressive strength and density of lightweight concrete. *Indian Journal of Engineering and Materials Sciences*, 14 (2), 158–162.

Madandoust, R., Ranjbar, R.M. & Mousavi, S.Y. (2011). An investigation on the fresh properties of self compacted lightweight concrete containing expanded polystyrene. *Construction and Building Materials*, 25 (9), 3721–3731.

National Ready Mixed Concrete Association Fleet Benchmarking and Costs Survey (2013) National Ready Mixed Concrete Association (NRMCA).

Oreta, A.W.C. & Ongpeng, J.M.C. (2011) Modeling the confined compressive strength of hybrid circular concrete columns using neural network. *Computers and Concrete*, 8 (5), 597–616.

An optimum strategy for FRP-strengthening of corrosion-affected reinforced concrete columns

H. Baji, C.Q. Li & F. Chen
RMIT University, Melbourne, Australia

W. Yang
Victoria University, Melbourne, Australia

ABSTRACT

Due to their excellent mechanical and chemical properties, ease and efficiency of application and relatively low cost, strengthening using Fibre Reinforced Polymers (FRP) is emerging as an alternative to the traditional methods is strengthening of reinforced concrete columns (Binici, 2008). Although considerable research on strengthening of RC structures using Fiber Reinforced Polymers (FRP) composites has been undertaken, prediction of optimum strengthening time has not adequately been studied (Baji et al., 2017). This paper presents a methodology, conceptually illustrated in Figures 1, for optimal strengthening of corrosion-affected short RC columns with low eccentricity.

An optimization problem based on minimization of total expected cost is formulated, from which optimum strengthening time and number of required FRP layers can be obtained. The optimum repair time, t_r, is found in a way that before and after strengthening using FRP, the reliability index, β, is less than the acceptable probability of failure, β_a. The optimum strengthening problem is mathematically formulated as follows,

Minimize: $E(C_T) = E(C_R) + E(C_F)$ (1a)

Subject to: $\beta(t_r) \leq \beta_a ; \beta(t_L) \leq \beta_a$ (1b)

where, E denotes the expected value, t_r = strengthening time, t_L = predefined lifetime of the structure, C_T = total cost of maintenance, C_F = cost of failure and C_R = cost of repair and strengthening actions. For derivation of the reliability index, an ultimate limit state involving resistance and dead and live loads is defined. The renewal theory is used for finding the total expected cost.

Application of the proposed methodology is presented in a worked example for a corrosion-affected reinforced concrete column strengthened with Fibre Reinforced Polymers (FRP) sheets. The results from

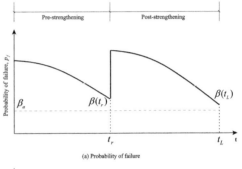

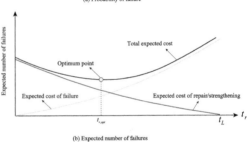

Figure 1. The optimal strengthening strategy.

a worked example show that an optimum solution to the formulated strengthening problem exists, and it is sensitive to the corrosion rate and cost of failure to cost of strengthening ratio. The proposed approach is useful in the development of an effective strengthening schedule for existing corrosion-affected RC columns.

REFERENCES

Baji, H., Yang, W. & Li, C.Q. (2017) An optimum strengthening strategy for corrosion-affected reinforced concrete structures. *ACI Structural Journal*, 114, 1591–1602.

Binici, B. (2008) Design of FRPs in circular bridge column retrofits for ductility enhancement. *Engineering Structures*, 30, 766–776.

SS13: Life-cycle of slope and river bank protection system considering soil bioengineering as well as conventional structures
Organizers: G. Kalny, H.P. Rauch & A. Strauss

Life-Cycle Analysis and Assessment in Civil Engineering: Towards an Integrated Vision – Caspeele, Taerwe & Frangopol (Eds)
© 2019 Taylor & Francis Group, London, ISBN 978-1-138-62633-1

Degradation processes of wooden logs in soil bioengineering structures

G. Kalny, K. Rados, B. Berntatz, B. Winkler & H.P. Rauch
Institute of Soil Bioengineering and Landscape Construction, University of Natural Resources and Life Sciences, Vienna, Austria

ABSTRACT

Taking into consideration the reliability of a construction during their life cycle, it is important to have knowledge regarding the durability and degradation processes of the materials being used. In terms of soil bioengineering works, wooden logs can be considered as important auxiliary building materials. Laboratory and field investigations were undertaken for the purpose of quantifying the degradation processes of a range of 0 to 20-year-old wooden logs, and of analyzing the most relevant impact parameters. The results showed that the changing over a period of time is a less important factor than the constitution of the wood, such as the inclusion of branches and cracks, and the distance of the annual rings. Field investigations showed that the back 5 cm of the logs had the highest decay ratios. The proportion of lower strength and higher decay ratios increased in older age classes.

Development of a concept for a holistic LCA model for soil bioengineering structures

M. von der Thannen, S. Hoerbinger & H.P. Rauch
Institute for Soil Bioengineering and Landscape Construction, BOKU, Vienna, Austria

R. Paratscha, R. Smutny & A. Strauss
Institute of Structural Engineering, BOKU, Vienna, Austria

T. Lampalzer
Austrian Service for Torrent and Avalanche Control, Regional Headquarters Vienna, Burgenland, Austria

ABSTRACT

The consequences of climate change constitute, at the same time, a challenge and an opportunity for our future society. Climate change adaption and mitigation, as well as a reduced availability of resources can only be met with substantial advances in technology and science. In view of these challenges it can be stated that climate change is having comprehensive effects on the entire field of civil engineering. To assess the use of energy, and detect potential environmental burdens, the concept of Life Cycle Assessment (LCA) has been developed (EN ISO 14040, 2009). Different models of LCA have already been in use for specific products and processes. However, a model for application in the field of soil bioengineering has still not been developed. Although, soil bioengineering is of increasing importance as there is a high demand for engineering solutions which take into consideration not only technical issues but also ecological and socio-economic values (Rauch, Sutili, & Hoerbinger, 2014).

In the frame of a research project at the University of Natural Resources and Life Sciences, Vienna, Austria funded by the ACRP, a conceptual approach for an LCA model for the field of soil bioengineering has been developed. The focus, so far, has been on the product and the construction phase (von der Thannen et al., 2017), but in this paper, the model will be extended for the purpose of simulating and analyzing the use phase of soil bioengineering constructions. Therefore, the maintenance work has to be considered, as well as the service life time of the construction materials. Figure 1 shows the system diagram developed for soil bioengineering structures. Incorporated into the diagram are three maintenance concepts that are generally described in soil bioengineering literature: follow-up maintenance, development maintenance and conservation maintenance.

Depending on the type of construction and the pursued objective a maintenance scenario has to be

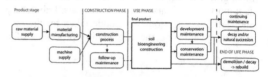

Figure 1. System diagram for soil bioengineering structures including maintenance.

evolved individually. Furthermore, the structure type, the pursued objective and the maintenance concept are influencing the end of life scenario. To simplify the procedure soil bioengineering constructions are classified into three classes: plant based structures, combined structures and nature based structures. Finally, three diagrams that illustrate the development for each class are presented by analyzing the use phase and the end of life scenario. The results will provide key elements to install a holistic LCA Model for soil bioengineering structures.

REFERENCES

EN ISO 14040. (2009) *Umweltmanagement – Ökobilanz – Grundsätze und Rahmenbedingungen; Environmental Management – Life Cycle Assessment – Principles and Framework*. (Stand: November 2009 ed.). Berlin, Beuth.

Rauch, H.P., Sutili, F. & Hoerbinger, S. (2014) Installation of a riparian forest by means of soil bio engineering techniques–monitoring results from a river restoration work in Southern Brazil. *Open Journal of Forestry*, 4, 161–169.

von der Thannen, M., Hoerbinger, S., Paratscha, R., Smutny, R., Lampalzer, T., Strauss, A. & Rauch, H.P. (2017) Development of an environmental life cycle assessment model for soil bioengineering constructions. *European Journal of Environmental and Civil Engineering*, 1–15. doi: 10.1080/19648189.2017.1369460.

Life-Cycle Analysis and Assessment in Civil Engineering: Towards an Integrated Vision – Caspeele, Taerwe & Frangopol (Eds)
© 2019 Taylor & Francis Group, London, ISBN 978-1-138-62633-1

Soil bioengineering: Requirements, materials, applications

H.P. Rauch, M. von der Thannen & C. Weissteiner
Institute of Soil Bioengineering and Landscape Construction, University of Natural Resources and Life Sciences, Vienna, Austria

ABSTRACT

Soil bioengineering is a construction technique that uses biological components for hydraulic and civil engineering solutions. In Europe the origin has historical roots from when the society started protecting their infrastructure against natural disasters. Nowadays, these techniques are used as "soft" engineering solutions considering not only technical aspects but also ecological and aesthetic values. The procedures of designing and implementation are similar compared to conventional civil engineering structures. The most distinctive difference to conventional civil engineering structures is the use of different construction materials. Soil bioengineering solutions are based on the application of living plants and other local available additives. These different construction materials are most relevant to benefit in the context of ecosystem services.

Another soil bioengineering key factor is the fact that the services of such "living" engineering structures correlate with the dynamic development and capacity of the plants and properties of the auxiliary materials. Furthermore, it is depending on the field of application and the specific required soil bioengineering function. A state of the art soil bioengineering application takes environmental as well as engineering framework conditions into account, considering beside the process planning, design and construction also monitoring and maintenance of soil bioengineering systems.

The paper gives an overview of the most relevant requirements of soil bioengineering systems in different fields of applications. The functional capability of materials will be analysed and discussed by the means of several soil bioengineering examples.

Service life planning for Austrian river bank protection structures

R. Paratscha, A. Strauss & R. Smutny
Institute of Structural Engineering, BOKU, Vienna, Austria

M. von der Thannen & H.P. Rauch
Institute for Soil Bioengineering and Landscape Construction, BOKU, Vienna, Austria

T. Lampalzer
Austrian Service for Torrent and Avalanche Control, Regional Headquarters Vienna, Burgenland, Austria

ABSTRACT

Currently the used assessment methods for Austrian protective structures are referencing technical and economic aspects. In order to guarantee a sustainable planning of these structures, it is necessary to introduce environmental impact indicators and to compare them with technical and economic considerations.

By applying the service life planning design approach, it is possible to calculate economic and environmental indicators. The International ISO standard series 15686 provides a good framework for such a calculation. According to ISO 15686-1, service life planning design is the process of preparing the brief and the design for the structure and its parts to achieve the design life (ISO, 2011).

In this study environmental and economic indicators for six different construction types of riverbank protection structures are calculated and compared. Conventional structures (concrete, stone, wood) as well as soil bioengineering structures are analysed. The methodology is based on the framework for environmental life cycle assessment (LCA) and life cycle costing (LCC) of buildings and civil engineering works. Since there is no normative regulation regarding sustainable planning in the field of protective structures, a specific conceptual approach was developed, covering all relevant life cycle stages (cradle to grave).

The aim of the study is to demonstrate a methodical approach showing how to implement sustainable life cycle thinking in the early planning stage of riverbank protection systems as a strategic decision making tool.

In this study, the calculation of the two environmental indicators of Global Warming Potential (GWP) for the presentation of emissions and Cumulative Energy Demands (CED) is presented in order to show the consumption of energy resources. The life cycle costs (Equation 1) were calculated using the present value method. The LCC consists in construction costs (CC), the present value of the maintenance costs (MC) and the present value of reconstruction costs (RC).

$$LCC = CC + MC + RC \qquad (1)$$

To create a comparative life cycle analysis, it is very important for all structures to adopt the same framework conditions. These framework conditions exist in methodology, data sources, system boundary and functional unit. In addition, such a comparison may only be made on the basis of technical feasibility and compliance with the safety guidelines. Safety should be the top priority. The lifecycle considerations that have been performed in this study can only be related to the inventories used. By changing the inventories, the result can vary significantly.

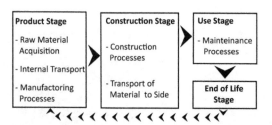

Figure 1. System flow diagram according to EN ISO 15978(CEN, 2011).

REFERENCES

CEN (2011) EN 15978: 2011 11 – Sustainability of construction works – Assessment of environmental performance of buildings – Calculation method. *Planning & Implementation of Construction Projects, Buildings, Other Buildings*. Brussels, European Committee for Standardisation (CEN).

ISO (2011) Buildings and constructed assets – Service life planning – Part 1: General principles and framework. *ISO 15686-1: 2011 05 15*. Geneva, International Organization for Standardization (ISO).

Life-Cycle Analysis and Assessment in Civil Engineering: Towards an Integrated Vision – Caspeele, Taerwe & Frangopol (Eds)
© 2019 Taylor & Francis Group, London, ISBN 978-1-138-62633-1

Specialisation for the ecoengineering sector in the Mediterranean environment ECOMED

P. Sangalli
Sangalli Coronel y Asociados SL and EFIB (European Federation of Soil and Water Bioengineering), San Sebastian Gipuzkoa, Spain

G. Tardío
Technical University of Madrid, Madrid, España

G. Zaimes
Technological Educational Institute of Kavala, TEIKAV, Kavala, Greece

ABSTRACT

Soil l degradation and are also useful to protect natural riparian areas affected by fast environmental changes and they ensure a significant contribution to the long-term protection and mitigation against all forms of soil loss and erosion.

Ecologically based approaches currently represent a very small percentage of the stabilisation works undertaken due largely to gap in awareness and skills amongst practitioners (Stokes et al, 2014). In order to avoid this gap in the Mediterranean region, seven countries are involved in the Ecomed project, an Erasmus plus project. As part of the ECOMED project the skills gap within the industry were researched in order to develop solutions and training aimed at specialisation within the Mediterranean Bioengineering sector.

The aim of this project is to generate a sector-specific theoretical and practical syllabus essential for the specialization process of the Mediterranean Ecoengineering sector.

Bioengineering works have a clear dynamic response: at first the initial rigidity is offered by the inert materials but, as time progresses, the indigenous vegetation plays an increasing role in the stabilization of the site.

The general scheme of bioengineering works spins around the following axis:

- The use of indigenous species potential for the short- and long-term degraded land restoration.
- The use of the nearby materials for the project design. Bioengineering works have low-carbon project features.
- The improvement of the ecosystem resilience by fostering and triggering the site's natural ecosystem evolution.

In order to establish the best design process as well as the formation needed in order to project and execute soil Bioengineering techniques, fourteen works done in seven countries will be selected and analysed, in order to establish different parameters like the undreined and effective sher strength and root strength, the assessment of the deterioration of wooden elements, the characterisation of the vegetation, the ecological evolution of the vegetation, …

In the communication will present the results of this analysis and the conclusions in order to establish the possibility of the application of Soil bioengineering techniques in Mediterranean area

REFERENCES

EFIB (2015) European Guidelines for Soil and Water Bioengineering, European Federation of Soil and Water Bioengineering, Tardio, G., Mickovski, S.B., Stokes, A. & Devkota, S. (2017) Bamboo structures as a resilient erosion control measure. *Proceedings of the Institution of Civil Engineers – Forensic Engineers*, 170 (2), 72–83.

Tardio, G. & Mickovski, S.B. (2016) Implementation of eco-engineering design into existing slope stability design practices. *Ecological Engineering*, 92, 138–147.

Zaimes, G.N, Lee, K.-H., Tufeckioglu, M., Long, L.A., Schultz, R.C. & Isenhart, T.M. (2012) The effectiveness of riparian conservation practices in reducing sediment in Iowa streams. In: Hendriks, B.P. (ed.) *Agricultural Research Updates*. Volume 2. Hauppauge, NY, Nova Science Publishers. pp. 117–166.

Zaimes, G.N., Ioannou, K., Iakovoglou, V., Kosmadakis, I., Koutalakis, P., Ranis, G., Emmanouloudis, D. & Schultz, R.C. (2016) Improving soil erosion prevention in Greece with new tools. *Journal of Engineering Science and Technology Review*, 9, 66–71.

Life-Cycle Analysis and Assessment in Civil Engineering: Towards an Integrated Vision – Caspeele, Taerwe & Frangopol (Eds)
© 2019 Taylor & Francis Group, London, ISBN 978-1-138-62633-1

Development and challenges of soil bioengineering applications to vegetated riprap

P. Raymond
Terra Erosion Control, Nelson, British Columbia, Canada

S. Tron
ÖGUT, Austrian Society for Environment and Technology, Vienna, Austria

I. Larocque
MacDonald Hydrology Consultants Ltd., Kimberley, British Columbia, Canada

ABSTRACT

This study demonstrates how vegetated riprap, which is the incorporation of live cuttings and soil amendment within conventional riprap along watercourses, is an alternative to conventional riprap alone. Vegetated riprap provides erosion protection of streambanks, softens the rock appearance, and enhances fish and wildlife habitat along shorelines.

This innovative solution has been designed and implemented by Terra Erosion Control Ltd. for the past 11 years. Large and medium scale trial experiments were carried out over a six year period with successful results. In 2012 the design was included within the City of Calgary Design Guidelines for Erosion and Flood Control Projects for Streambank and Riparian Stability Restoration (AMEC, 2012), and vegetated riprap has been used extensively along medium and large size rivers in Alberta Canada.

The initial vegetated riprap design was first installed along the Columbia River (Trail, Canada). Two key applications compose the vegetated riprap: brush layers and vegetated pockets. In 2012 and 2013, field excavation studies revealed root development restriction from the use of protective wooden panel boards used to protect the live cuttings from the riprap. This observation as well as the desire to use more biodegradable material lead to the development of the revised design.

The initial and revised design of brush layers and vegetated pockets achieved very good survival and growth rate. The brush layers also showed full canopy closure and propagation beyond the planting area. These conditions improve fish habitat by creating shade, cover and input of small organics (Slaney & Zaldokas 1997) as well as reducing water temperatures for fish habitat (Harris 2005) and decreasing the velocity of water flow (Darby 1999).

In 2012 and 2013, excavation studies were conducted on a 6 and 7 year growth *Salix bebbiana* and *Populus balsamifera*. Root growth observed was good but limited to below the protective wooden panel boards. There was no sign of riprap displacement, roots were observed to envelop the rocks, create a network binding the rocks together, and grow back to their initial diameter.

In 2013, a cutting of *Salix bebbiana* was excavated. It showed that root development within the riprap layer along the cutting has a mean diameter significantly lower and higher tensile strength than within the substrate below the riprap layer. The results also show that roots increase the cohesion below the riprap structure. This additional cohesion significantly decreases with the radial distance from the cutting. However, as the cuttings were placed almost adjacent to one another, the total cohesion was higher due to the overlap among the different root systems.

Overall, this study shows that vegetated riprap enhances fish and wildlife habitat along watercourses as well as successful root growth within and below the riprap, which further strengthens the riprap and increases overall streambank erosion protection.

REFERENCES

AMEC. (2012) City of Calgary Design Guidelines for Erosion and Flood Control Projects for Streambank and Riparian Stability Restoration.

Darby, S.E. (1999) Effect of riparian vegetation on flow resistance and flood potential. *Journal of Hydraulic Engineering*, 125 (5).

Harris, R.R. (2005) Monitoring the effectiveness of riparian vegetation restoration. Final Report. Prepared by Center for Forestry, University of California, Berkeley. Prepared for California Department of Fish and Game, Salmon and Steelhead Trout Restoration Account Agreement No. P0210566.

Slaney, P.A. & Zaldokas D. (1997) Fish habitat rehabilitation procedures. BC watershed restoration project, 0-7726-3320-7.

Tron, S. & Raymond, P. (2014) Analysis of root reinforcement of vegetated riprap. *Geophysical Research Abstracts*, 16, EGU2014-4928.

Life-Cycle Analysis and Assessment in Civil Engineering: Towards an Integrated Vision – Caspeele, Taerwe & Frangopol (Eds)
© 2019 Taylor & Francis Group, London, ISBN 978-1-138-62633-1

The limits of mechanical resistance in bioengineering for riverbank protection

A. Evette & D. Jaymond
Irstea, UR LESSEM, University of Grenoble Alpes, Saint-Martin-d'Hères, France

A. Recking & G. Piton
Irstea, UR ETGR, University of Grenoble Alpes, Saint-Martin-d'Hères, France

H.P. Rauch
Department of Civil Engineering and Natural Hazards, University of Natural Resources and Life Sciences, Vienna, Austria
Institute of Soil Bioengineering and Landscape Construction, Vienna, Austria

P.-A. Frossard
Hepia, Filière Gestion de la Nature, Jussy, Switzerland

ABSTRACT

Soil bioengineering techniques for riverbank protection are nature-based engineering solutions fulfilling both erosion control and ecological functions. In general dimensioning and impact assessment of potential soil bioengineering interventions are the basis and precondition to establishing soil bioengineering as standardized hydraulic and civil engineering measures. Nevertheless, very few dimensioning methods exist, which is the main obstacle standing in the way of mainstreaming bioengineering techniques. In this study 18 bioengineering constructions that have resisted alpine flood events were documented and analyzed empirically. They were classified according to stream slope, shear stress and age since completion. The maximum river longitudinal slope was 2.5%. We also considered six experimental works built in the Geni'Alp project, over these limits of use, with river longitudinal slope in the 4–10% range as well as the shear stress resistance values of riverbank bioengineering works from the literature. These 21 new sites from the literature, comprising a wide variety of bioengineering techniques, and the Geni'Alp experimental works were analyzed. Finally, 21 shear values for river reaches implementing 51 bioengineering techniques were calculated for the maximum past flood, taking into account the most relevant local hydraulic conditions.

SS14: Advanced NDT for visualization and quantification of concrete deterioration and repair effects
Organizers: T. Shiotani, E. Verstrynge, D.G. Aggelis & P. Pahlavan

Case study on determination of remaining bearing capacity of cantilevered balconies of high rise buildings

B. Craeye
EMIB Research Group, Faculty of Applied Engineering, University of Antwerp, Antwerp, Belgium
DuBiT Research Group, Department of Industrial Sciences & Technoloy, Odisee University College, Aalst, Belgium
Department of Structural Engineering, Ghent University, Zwijnaarde/Ghent, Belgium

W. Gijbels
Betonadvies Gijko, Concrete Consultancy Company, Sint-Niklaas, Belgium

L. De Winter, M. Maes & T. Soetens
Sanacon, Concrete Consultancy Company, Zwijnaarde/Ghent, Belgium

D. Vanermen
Establis, Structural Engineering Agency, Antwerp, Belgium

ABSTRACT

In 2011 one of the cantilevered galleries of a high rise residential building in Leeuwarden, the Netherlands (constructed in 1965), fell down. In the late 60s and the mid-70s, a significant amount of reinforced concrete structures were built and nowadays suffer severely from varying concrete damaging mechanisms. In more than 50% of the cases, corrosion is the main deteriorating mechanism. Furthermore, the cause of the failure of the balcony lies in a combination of (i) inaccurate execution and inappropriate positioning of the reinforcement rebars, (ii) a higher load on the structural element and (iii) partial corrosion (pitting) of the concrete steel. In this present case study three reinforced concrete high rise buildings, situated near Brussels and whose design dates from 1963, are being investigated thoroughly by means of a protocol based on CUR Publication 248. These buildings consists of 17 floors, and have a different orientation (Figure 1). The prefabricated balconies of the building, who are the solely entrance path to the apartment units, have a length of 33 m per floor and represent a total area of approximately 2500 m². By means of the four-step inspection protocol, the structural safety of the balconies is being evaluated and the residual bearing bending capacity is determined according to the Eurocode and fib Model Code regulations: (i) file investigation and visual inspection of the concrete elements, (ii) non-destructive determination of the reinforcement and strength characteristics, (iii) destructive testing (core drilling, potential mapping and chloride profiling), and (iv) evaluation of the structural safety of the balconies. It has to be determined whether the structural integrity of these reinforced concrete elements is endangered. The influence of concrete strength class, steel quality and rebar position is being identified for this particular case by means of a sensitivity analysis. By means of this protocol the risk of failure of the cantilevered balconies is estimated and potential danger zones are identified, in order to come to an appropriate building repair strategy. Approximately 25% of the investigated balconies can be considered safe as the resistant bending moment exceeds the applied one in ULS. More than 60% of the balconies have a decreased safety level and 15% are in a critical state. The main reasons of the structural unsafety of these floors are: 1) incorrect positioning of the tension reinforcement, and 2) considerable pitting corrosion of the tension reinforcement underneath cracked areas. By means of a sensitivity analysis of the applied models, it was found that especially the effective depth and steel quality of the reinforcement have a decisive effect on the bearing capacity of the cantilever reinforced concrete balconies, working in pure bending. The strength class of the concrete has a negligible effect.

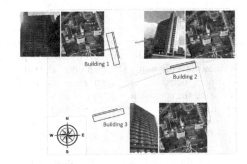

Figure 1. General view on the three inspected buildings.

REFERENCE

de Jonker, M., Mans, D.G. & Wijte, S.N.M. (2012) *Research on and Evaluation of Constructive Safety of Cantilevered Concrete Floors of Gallery Apartments* (in Dutch). The Netherlands, CUR Publication 248, CURNET.

Localization and characterization of damage modes in reinforced concrete by means of acoustic emission monitoring during accelerated corrosion and pull-out testing

C. Van Steen & E. Verstrynge
Department of Civil Engineering, KU Leuven, Leuven, Belgium

M. Wevers
Department of Materials Engineering, KU Leuven, Leuven, Belgium

ABSTRACT

Reinforcement corrosion is believed to be the most common and most expensive deterioration mechanism in existing reinforced concrete (RC) structures (Andrade, 2010). One of the important damage modes due to corrosion is the deterioration of bond between steel and concrete (Auyeung, 2000). In this paper, the Acoustic Emission (AE) technique is evaluated for damage detection, characterization and localization during the corrosion process and during pull-out tests. RC prisms with smooth or ribbed reinforcement were corroded up to different target corrosion levels: 0%, 1.5%, 5% and 10% mass loss. The corrosion process was accelerated in the lab by imposing a constant direct current while the specimens were partially immersed in a 5% sodium solution. One of the specimens of each rebar type and every corrosion level was monitored continuously with the AE technique. At target corrosion levels, pull-out tests were carried out to study the bond capacity. During these tests, AE monitoring was performed as well on every specimen. Results show that AE is able to detect damage due to corrosion and that the moment of cracking can be determined from cumulative AE energy curves (figure 1). During pull-out tests, debonding damage

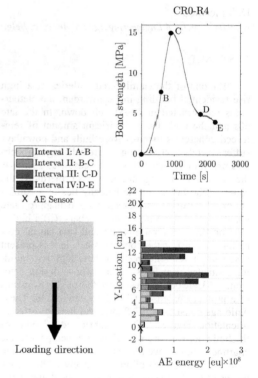

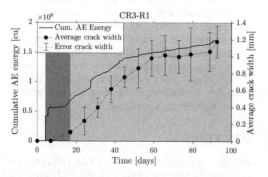

Figure 1. Average crack width and cumulative AE energy versus time for sample CR3-R1 during corrosion.

Figure 2. Cumulated AE energy of linear located AE events during pull-out testing.

was successfully detected, characterized and located from AE curves (figure 2).

REFERENCES

Andrade, C. & Macini, G. (2011) *Modelling of Corroding Concrete Structures*. RILEM Bookseries 5, Madrid, Spain.

Auyeung, Y. (2000) Bond behaviour of corroded reinforcement bars. *ACI Materials Journal*, 97 (2), 214–221.

Analysis of fatigue behaviour of single steel fibre pull-out in a concrete matrix with micro-CT and acoustic emission

M. De Smedt, C. Van Steen, K. De Wilder, L. Vandewalle & E. Verstrynge
Department of Civil Engineering, KU Leuven, Leuven, Belgium

ABSTRACT

This paper reports the first results of an experimental programme investigating the fatigue behaviour of steel fibre reinforced concrete (SFRC). As a first step, uniaxial monotonic and cyclic pull-out tests of individually embedded steel fibres in a concrete matrix are performed. An experimental setup is developed to perform these pull-out tests in combination with the advanced non-destructive measurement techniques of Acoustic Emission (AE) sensing (Grosse and Ohtsu, 2008) and X-ray microfocus Computed Tomography (micro-CT scans) (Landis and Keane, 2010). The varying parameters are the type of hooked-end fibre, the embedded length, the fibre orientation, and the loading pattern.

The different stages during monotonic fibre pull-out (Pompo et al., 1996) are detected by their corresponding acoustic emissions. The AE activity is higher for 5D hooked-end fibres compared to 3D hooked-end fibres. Furthermore, it is shown that applying 2000 cycles before the maximum pull-out force does not lead to a distinct fatigue behaviour.

The localisation of AE events, combined with micro-CT scans of the fibre position, visualise the damage during the different pull-out stages, as shown in Fig. 1. Furthermore, the type of occurring process is distinguished by the measured amplitude of the AE signal. Other damage patterns, such as concrete cracking of the specimen, are detected as well.

Lastly, outliers with respect to the bond-slip behaviour of the pull-out tests are visually inspected and confirmed based on the micro-CT scans. The position of air voids (example given in Fig. 2) influences the anchorage surface of the fibre and thus the overall bond-slip behaviour.

In conclusion, the combination of the AE technique and the micro-CT scans leads to a deeper insight and better understanding of the monotonic and cyclic pull-out behaviour of steel fibres embedded in concrete.

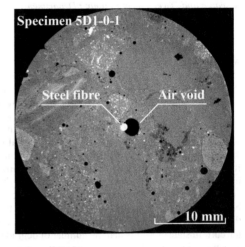

Figure 2. Horizontal slice of micro-CT scans.

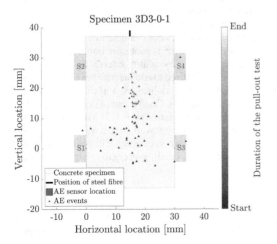

Figure 1. Localisation of AE sources in a vertical plane.

REFERENCES

Grosse, C. & Ohtsu, M. (2008) *Acoustic Emission Testing*. Springer.
Landis, E. & Keane, D. (2010) X-ray microtomography. *Materials Characterization*, 61, 1305–1316.
Pompo, A., et al. (1996) Analysis of steel fibre pull-out from a cement matrix using video photography. *Cement and Concrete Composites*, 18, 3–8.

Non-destructive inspection method for bending strength estimation of polymer concrete

C. Saito, M. Okutsu, M. Nakagawa & S. Yanagi
Civil Engineering Project, Access Network Service Laboratories, Nippon Telegraph and Telephone Corporation, Japan

H. Takahashi
Research and Development Planning Department, Nippon Telegraph and Telephone Corporation, Japan

ABSTRACT

Polymer concrete (PC) is a construction material with high compressive strength and bending strength. Moreover, PC provides both compressive strength and bending strength by using concrete alone, so it is used instead of reinforced concrete (RC) as a material of underground structures, such as PC-made manholes. It is difficult to visually grasp the signs of deterioration in PC strength before cracking occurs. At the stage that cracks can be detected by visual observation, since the bending strength is greatly reduced, it is a risk that the state of insufficient strength is overlooked. Under the current state of affairs that the number of degraded structures in use will increase from now onwards, repairing and reinforcing such structures at an appropriate timing necessitates a means for measuring the strength of PC by a simple operation.

However, at present, the only effective means of measuring PC strength is destructive inspection performed by cutting core specimen from structure. The problems with the means are twofold: it necessarily damages the shape and function of the structure under inspection, and it is expensive. Accordingly, to avoid those problems, it is desirable to perform inspection on a non-destructive manner. Therefore, by devising a new inspection method, namely, the first non-destructive inspection method for measuring strength of a PC structure, we are aiming to significantly improve inspection work in terms of time and cost.

In this study, bending-fracture tests and stress-strain measurements were carried out on ten age-deteriorated samples collected from PC manholes in use. Since bending strength F approximates to static elastic modulus E well according to a linear equation with accuracy of determination coefficient $R^2 = 0.80$, it is expected that F can be estimated by measuring E. We selected ultrasound as a method for acquiring elastic modulus and conducted field experiments. Fifty-two PC manholes were measured ultrasonically and destructively. Plotting the elastic wave velocity propagating on the surface and the bending strength showed the correlation of $R^2 = 0.81$ and $\sigma = 1.60$ (MPa).

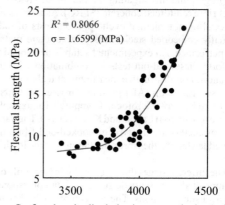

Figure 1. Correlation between bending strength and surface-sound-wave velocity.

In this way, it was possible to establish a method for non-destructively estimating bending strength of PC for the first time and verify the accuracy of that method. It will thus help to considerably improve maintenance and management of PC installations.

However, estimating bending strength by exploiting surface waves does not evaluate internal soundness at all; therefore, there is room for improvement in estimation accuracy by adding internal soundness to the evaluation. We tried to add the number of years elapsed to explanatory variables as a means of evaluating internal soundness. Coefficient of determination $R^2 = 0.88$, standard deviation ($\sigma = 1.31$ MPa), and accuracy are significantly improved. In this way, by considering the degree of progress of internal degradation, we were able to estimate bending strength at high accuracy.

REFERENCE

Bedi, R., Chandra, R. & Singh, S.P. (2013) Mechanical properties of polymer concrete. *Journal of Composite Materials*.

*Life-Cycle Analysis and Assessment in Civil Engineering: Towards an
Integrated Vision – Caspeele, Taerwe & Frangopol (Eds)
© 2019 Taylor & Francis Group, London, ISBN 978-1-138-62633-1*

Study of the applicability of the polarization resistance method in analysis and experimental measurement of electric conductivity

C. Okuma
Graduate School of Engineering, Kagawa University, Takamatsu, Japan

S. Okazaki & H. Yoshida
Faculty of Engineering, Kagawa University, Takamatsu, Japan

ABSTRACT

Although the service life of concrete structures is approximately 50 years, a considerable amount of them in Japan are past 50 years after construction. Demolish all these concrete structures presents financial challenges. Therefore, it is necessary to maintain them and prolong their life. Recently, the polarization resistance method is highly perceived as a nondestructive test method to evaluate the corrosion rate of rebar in concrete structures.

In the polarization resistance method, a weak current is passed through both the ends of a reinforcing bar, the potential change of the reinforcing bar from the natural state is measured, and the polarization resistance is measured based on the difference between the current and the potential change.

Apparent polarization resistance could be measured from polarization resistance method. To obtain polarization resistance, the measured value is multiplied by the area to be measured, the range where the current flows. Since the polarized area to be measured is influenced by the material heterogeneity such as rebar corrosion or altered concrete, it is not evaluated well using boundary element method. Therefore, specimens were prepared and the polarization resistance was measured.

Experiments were conducted in which apparent polarization resistance, true polarization resistance, and concrete resistance of reinforced concrete test pieces were measured. Conductivity analysis of the test piece was carried out to investigate influence of curing period on electrical dispersion characteristics.

Moreover, we conducted an electrical conduction analysis considering material heterogeneity such as rebar corrosion or altered concrete by using finite element method. Figure 1 shows current density distribution diagram normalized at center of reinforcing bars. From analysis result, it was confirmed that measurement area was affected by material heterogeneity such as rebar polarization resistance or altered concrete. Finally, the results were compared with measured polarization resistance value.

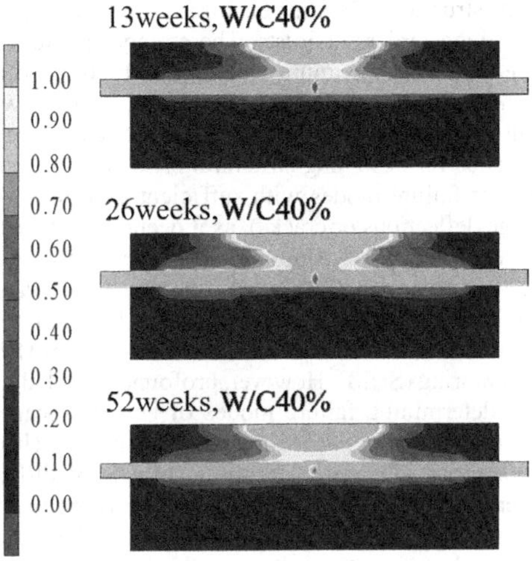

Figure 1. Normalized current density.

Comparing true polarization resistance of experiment and analysis, the analysis was higher. We believe that polarization area of experiment is constant.

In order to diagnose corrosion of reinforcing bars from polarization resistance, it is necessary to calculate the correct polarized area.

REFERENCES

Yoshida, H. et al. (2013) Numerical analysis study on polarization resistance evaluation considering current dispersion range. In: *Public Works Association Proceedings*, A2. Vol. 69, No. 2, pp. I_667–I_688.

Kobayashi, K. et al. (2001) Evaluation of corrosion rate of rebar with polarization resistance method. In: *Public Works Association Proceedings*. Vol. 50, No. 669, pp. 173–186.

Yokota, M. et al. (1991) Evaluation of corrosion of reinforcing bars in concrete by AC impedance method. In: *Nondestructive Evaluation Symposium on Civil Engineering Lecture Papers*. pp. 241–246.

Damage mechanisms analysis of reinforced concrete beams in bending using non-destructive testing

S. Pirskawetz, G. Hüsken, K.-P. Gründer & D. Kadoke
Bundesanstalt für Materialforschung und -prüfung (BAM), Berlin, Germany

ABSTRACT

Beams that are loaded in transverse direction are one of the main structural elements used in reinforced concrete structures. Bending beams are used, for example, in bridge girders or joists. The strength related failure modes at maximum loading can be divided into bending and shear failure that could occur with or without indication. Conventional design concepts used in practice for designing structural elements in bending aim at failure modes with sufficient indication (e.g. large deflections or cracks), as it occurs in the case of secondary flexure-compression-failure. These indicators can also be used to identify structural changes of civil infrastructure systems (e.g. bridges) using non-destructive testing methods for Structural Health Monitoring (SHM). However, profound knowledge on the determining failure modes of bending beams is required for the reliable application of structural health monitoring. Therefore, resulting effects of different thermal and mechanical loads on the response of a structure are investigated by means of a bridge demonstrator. In preparation of tests on this large-scale bridge demonstrator, application as well as data acquisition and synchronization of different measuring systems were tested on four-point bending beams on laboratory scale.

For this purpose, different non-destructive testing (NDT) methods have been used for analyzing the deformation behavior. The suitability of the different measuring techniques will be discussed by means of the obtained experimental results. The deformation behavior of a reinforced concrete beam having a span of 2.75 m was investigated to characterize the failure modes by NDT methods and compared with classical measuring techniques (e.g. deformation measurements by displacement transducers). The bending tests have been accompanied by Acoustic Emission (AE) analysis, optical deformation measurements using Stereophotogrammetry (SP) and Digital Image Correlation (DIC).

The results of the applied optical deformation measurements are in good agreement with deflections measured by means of classical displacement transducers. Figure 1 shows the displacement of the SP

Figure 1. Deflection of the beam at maximum load.

markers at maximum load. The three-dimensional detection of deformations and resulting strains with optical techniques is beneficial for analyzing the damage mechanism of loaded structures and related crack formation and propagation. Similar results can hardly be obtained by classical techniques or, if possible, extensive technical efforts are necessary.

However, the application of optical deformation measurements for continuous monitoring of structures is limited at present. Monitoring of crack formation and crack growth can be realized by acoustic emission testing in an adequate manner. The conducted tests have shown that the transition from the uncracked Phase 1 to the cracked Phase 2 can be reliably detected and combined with crack localization. On the contrary, the localization of further crack growth and formation of new cracks in Phase 2 is limited due to the interrupted sound paths which does not any longer allow the assumption of the material being isotropic. Consequently, the positioning of sensors on already cracked structures needs to be optimized considering existing cracks and other specific characteristics of the damaged structure. The influence of noise was not part of the tests discussed here, but is important and must be considered for bridge monitoring.

The practical findings obtained by the preliminary tests on laboratory scale have been incorporated in the design of the bridge demonstrator and selected measuring systems. The ongoing measurements conducted on the bridge demonstrator are the basis for the monitoring of structures in service.

A comparative study of acoustic emission tomography and digital image correlation measurement on a reinforced concrete beam

Y. Yang, F. Zhang & D.A. Hordijk
Delft University of Technology, Delft, The Netherlands

K. Hashimoto & T. Shiotani
Kyoto University, Kyoto, Japan

ABSTRACT

In the recent years, advanced measurement techniques for identifying the damage of existing Reinforced Concrete (RC) structures have been developing rapidly. Acoustic Emission tomography (AET) shows promising features in exploring the internal damages (Shiotani et al., 2014). In the algorithm the time delay of wave signals caused by the local damages of the structure is assumed to be uniformly distributed over the calculated distance. That results a nominal velocity distribution over the structure as an indication of the damage distribution. However, relationship between velocity from AET and damage of concrete is not clarified in RC structural members.

This paper presents a comparative study on the nominal wave velocity obtained from AET and the physical damage of a RC beam in terms of crack opening based on a beam test. The crack width of the specimen was determined by both Digital Image Correlation (DIC) and the conventional LVDT measurements.

The crack distribution obtained by DIC and the nominal wave speed from AET are compared in Figure 1. Despite that the local variations of the wave velocity are not always in line with the measured crack pattern, the nominal wave velocity from AET still can reflect the damage of the specimen at a larger scale.

A supplementary comparison between the LVDT measurements and AET shows a stronger relationship between crack width and wave velocity when the crack width is small (<0.1 mm), shown in Figure 2. It shows that the scatter of the AET can be improved by averaging the nominal wave velocity of the cells along the cracked section.

The paper shows that AET is a realistic NDT tool to estimate the damage distribution in terms of cracks. To be able to extend the upper bound of the detectable crack width (0.1 mm in this paper), a different excitation source in combination with sensors covering

Figure 1. Displacement measured by DIC and nominal wave speed derived from AET, at load $P = 80$ kN.

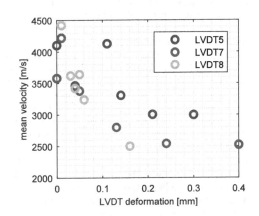

Figure 2. Relationship between average wave velocity and elongation of LVDTs.

different frequency range is recommended. In addition to the AET, conventional AE measurement can further improve the measurement of the crack development.

REFERENCE

Shiotani, T., Osawa, S., Kobayashi, Y. & Momoki, S. (2014) Application of 3D AE tomography for triaxial tests of rocky specimens. In: *31st Conference of the European Working Group on Acoustic Emission, Dresden.*

Life-Cycle Analysis and Assessment in Civil Engineering: Towards an Integrated Vision – Caspeele, Taerwe & Frangopol (Eds)
© 2019 Taylor & Francis Group, London, ISBN 978-1-138-62633-1

Ultrasound pulse velocity to measure repair efficiency of concrete containing a self-healing vascular network

E. Tsangouri, J. Lelon, P. Minnenbo, D.G. Aggelis & D. Van Hemelrijck
Department of Mechanics of Materials and Constructions (MeMC), Vrije Universiteit Brussel (VUB), Brussels, Belgium

ABSTRACT

Self-healing of concrete is not science-fiction anymore. In recent years, research studies were focusing on concrete that includes encapsulated repair agent and has the ability to autonomously detect cracks and trigger the reparation processes. The design of self-healing concrete is today optimized and considers 1D or 2D tubular vascular networks that are embedded into concrete during casting. The tubes are empty and sound and the agent is stored and embedded in concrete reservoirs till the moment that cracking occurs. Due to internal stresses, the tubes break at the crack plane and the repair agent is released into the tubes reaching and filling the crack.

In laboratory preliminary tests, vascular network is designed by combining a pair of tubular capsules attached to 3D printed reservoirs. The reservoir is made by polyamide, a material that does not corrode and is not reactive to concrete. The geometry is given in Figure 1a. The 3D printed rectangular reservoir box stands at the compressive zone of a beam and is connected to tubes that carry the agent at lower levels where the capsules stand. As shown in Figures 1b-c, the reservoir system in this case is built to cover the middle zone of a beam tested in three-point bending. At the top of the reservoir a tube is added and is used to fill up from exterior the reservoir in case that additional agent is required. The agent is a low-viscosity and expansive polyurethane-based polymer that is traditionally used to fill up crack with opening up to 0.5 mm.

Similar to small-scale beams, a slab is casted and a reservoir system is embedded (Figure 2a). The steel-reinforced slab is tested under four-point bending. At first testing cycle, cracks develop at bottom tensile zone and at second loading phase the crack sealing and reopening after healing are evaluated. This time, the tubular capsules are ar-ranged in plane (and in angle shown in Figure 2b) in a way to cover the tensile-loaded lower zone. To create the capsules network and due to capsules length limitations, three types of circular PVC nodes are fabricated (Figure 2c). The same as in small-scale tests, polyurethane-based agent is used. The agent is supplied into the capsules at four locations indicated in Figure 2b (marked A to D) by a flexible

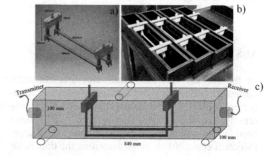

Figure 1. a) Vascular network design; b) Molds with attached tubes; c) Configuration and transducers position.

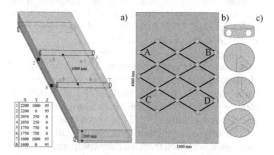

Figure 2. a) Design with transducers position; b) Interior view of the vascular network; c) Detail of connection node.

tube that reaches the top and extends at the exterior of the slab.

The structural performance is previously assessed using several monitoring techniques (acoustic emission, ultrasound transmission, digital image correlation, etc.) but repair efficiency was not quantified. In this study, ultrasound pulse velocity measurements are performed on small-scale concrete beams and on real-size slabs in order to measure regain in mechanical properties after healing. Up to eight piezoelectric PZT transducers are attached on concrete and receive the transmitted acoustic signals providing an integrated monitoring system. Significant wave velocity recovery is measured and associated to stiffness and (partial) strength recovery after healing.

Combining X-Ray imaging and acoustic emission to measure damage progression in ultra-high-performance-concrete

R. Kravchuk, D. Loshkov & E.N. Landis
University of Maine, Orono, Maine, USA

ABSTRACT

Adding sufficient quantities of ductile fibers into a brittle matrix such as concrete has long been known to transform a brittle material to a relatively ductile one. That transformation is made possible by a number of well-known toughening mechanisms including, fiber-matrix debonding and pull-out, additional matrix cracking, as well as fiber bending and fracture. In the work presented here, split cylinder fracture of fiber reinforced ultra-high-performance concrete (UHPC) was examined using two complementary techniques: x-ray CT and acoustic emission (AE). 50-mm diameter specimens of two different fiber types were scanned both before and after load testing.

From the CT images, fiber orientation was evaluated to establish optimum and pessimum specimen orientations, at which fibers would have maximum and minimum effect, respectively. As expected, fiber orientation affected both the peak load and the toughness of the specimen, with the optimum toughness being between 20 and 30% higher than the pessimum. Cumulative AE energy was also affected commensurately. Post-test CT scans of the specimen were used to measure internal damage. Damage was quantified in terms of internal energy dissipation due to both matrix cracking and fiber pullout using calibration measurements for each. Fiber pullout was the dominant energy dissipation mechanism, however, the sum of internal energy dissipation measured amounted to only 60% of the total energy dissipated by the specimens as measured by the net work of load. AE signals were classified using an artificial neural network. AE sources corresponding to fiber pullout produced more cumulative energy, but the difference was not as great as was measured using CT.

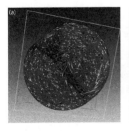 

Figure 1. 3D Renderings of damaged specimen.

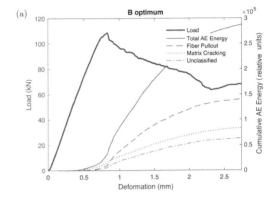

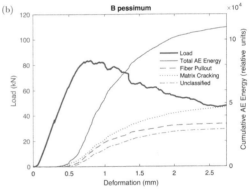

Figure 2. Energy dissipation measured by AE source type.

It is postulated that localized compaction of the UHPC matrix as well as internal friction between fractured fragments makes up the balance of internal energy dissipation. While the CT analysis provided damage information at the end of the test, pattern-based analysis of recorded AE waveforms provided information on the manifestation of different dissipation mechanisms as they occur during the test.

REFERENCE

Landis, E.N., Kravchuk, R. & Loshkov, D. (2018) Experimental investigations of internal energy dissipation during fracture of fiber reinforced ultra-high-performance concrete. *Frontiers of Structural and Civil Engineering*, (in press).

SS15: PROGRESS—Provisions for Greater Reuse of Steel Structures
Organizers: P. Kamrath & P. Hradil

Life-Cycle Analysis and Assessment in Civil Engineering: Towards an
Integrated Vision – Caspeele, Taerwe & Frangopol (Eds)
© 2019 Taylor & Francis Group, London, ISBN 978-1-138-62633-1

Environmental- and life cycle cost impact of reused steel structures: A case study

S. Vares, P. Hradil & S. Pulakka
VTT Technical research Centre of Finland Ltd, Finland

V. Ungureanu
Politehnica University of Timişoara, Romania

M. Sansom
Steel Construction Institute, UK

ABSTRACT

This theoretical study presents the Life Cycle model of a steel framed, single-story industrial hall with the floor area of 480 m². The paper shows greenhouse gas impacts (GWP) and life cycle costing (€) of the building for the first life cycle and for the case of steel frame and envelope reuse. The study pointing out benefits and loads and showing the results for new building and for the reused building with possible end of life scenarios. Calculations based on Life Cycle Assessment method (EN 15978) and Life Cycle Costs (ISO 15686-5).

According to result, steel reuse case show about 12% less GWP emission (in total 80.9 tCO2) and 10% less costs (17 k€) than in case when the steel structure is used in the first time (Fig. 1–2). The difference is not big because steel industry uses recycled steel in new steel production.

When the study takes into account also Module D (load and benefits beyond system boundaries) (Figure 3) the net benefit (GWP) in case of reused steel structures is in total −90.3 tCO$_2$ (−188 kgCO$_2$/m²). The result is positive when the future end-of-life (EoL)

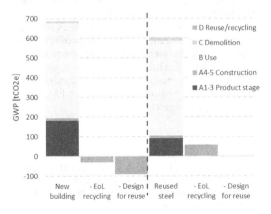

Figure 1. Total GWP (LCA method) of a new building (left) and reused building (right) including possible EoL scenarios.

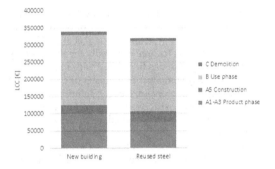

Figure 2. Total Life Cycle Costs (LCC) method.

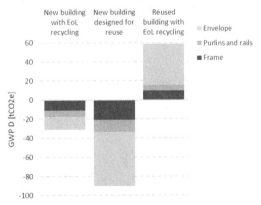

Figure 3. Loads and benefits beyond the system boundary (Module D).

scenario is less efficient (today's reuse vs. future recycling).

REFERENCES

EN 15978 (2011) Sustainability of construction works. Assessment of environmental performance of buildings. Calculation method.
ISO 15686-5 (2017) Buildings and constructed assets. Service life planning. Part 5: Life cycle costing.

Assessment of reusability of components from single-storey steel buildings

P. Hradil & L. Fülöp
VTT Technical Research Center of Finland Ltd., Espoo, Finland

V. Ungureanu
Politehnica University Timisoara, Timisoara, Romania

ABSTRACT

The paper presents the current development of the method for prediction of reusability of building components and whole structures (Hradil et al. 2017). The present method is focused especially for single-storey steel buildings, but can be applied with some modifications for any component (or cluster of components) reclaimed from a demolished or refurbished building. It enables classification of various building parts and products through a procedure to calculate their reusability index. These values can be further used to produce a single reusability indicator of the whole end-of-life scenario (e.g. the complete or partial building reuse).

The calculation was demonstrated on the comparative study of an industrial building with three different configurations of the load-bearing frame (see Figure 1) and three different end-of-life scenarios.

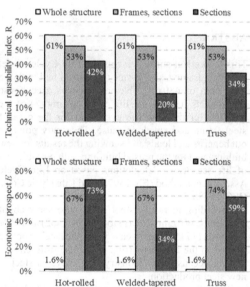

Figure 2. Results of the comparative study.

The scenarios included: (a) reuse/relocation of the whole structure, (b) reuse of the primary structure with the purlins and rails reconditioned and sold as general hot-rolled sections and (c) reduction of the whole steelwork to reusable sections. The aggregated results (see Figure 2) may be useful in planning of new building or assessment of existing ones.

The present research received funding from European Commission's RFCS project PROGRESS under the grant agreement No 747847.

REFERENCE

Hradil, P., Talja, A., Ungureanu, V., Koukkari, H. & Fülöp, L. (2017) Reusability indicator for steel-framed buildings and application for an industrial hall. In: *Eurosteel 2017*. Copenhagen, Ernst & Sohn.

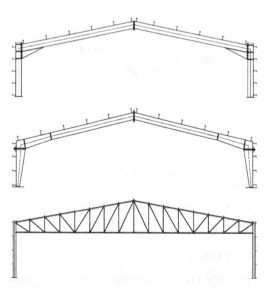

Figure 1. Three different types of the primary structure.

Calculating the climate impact of demolition

P. Kamrath
Paul Kamrath Ingenieurrückbau GmbH, Germany

ABSTRACT

Demolition techniques and the management of construction and demolition waste of buildings are key issues in the development of sustainable construction. Prevention, reuse and recycling are the basic approaches to waste management.

Sustainable construction does not finish at the end-of-life of a building. Since demolition waste is one of the biggest components of all waste, its impact on resources is non negligible. Thus landfilling should be avoided and use of the recycled material needs to be enhanced.

This paper takes data from different projects to calculate the climate impact of demolitionworks. With the data of machines and trucks, the used hours of excavators and the duration of each project the amount of energy and CO_2 produced during demolition work is calculated. The paper does not focus on one single project, but takes several projects into account. In the end all data is normalized to get the CO_2-amount produced for each cubicmeter of a building's volume. Mainly, only data for housing was used, so at the end there is a span differentiating between small, medium-sized and big houses.

Thus, the paper takes a first view to the climate impact of demolition and firstly gives data to analyze the end-of-lifetime of a building from a climate view.

To avoid waste during demolition it is important to plan the demolition carefully, and to be aware of the building's construction materials beforehand. In many cases the masonry is partly contaminated. If the chemical behavior of the different materials is not accounted for, the resulting mineral waste cannot be recycled. If contaminated materials such as rock wool is not identified (e.g. beneath the floors) separation will not be possible after demolition. The key is taking samples prior to the demolition of houses. The goal of the chemical analysis of such materials is to find hidden contaminants such as heavy metals. Smaller contaminations should also be considered as it may be

Figure 1. New materials need the invention of new tools. Deconstruction of insulation plates by an excavator with a newly invented sharp peeler tool.

possible to use such materials where sealing is possible after rebuilding (e.g. with parking lots). Planning from demolition up to the new building may thus help to avoid landfilling.

The rate of recycled steel and metal is high in general. Since the early times of deconstruction, steel construction parts were separated and sold to scrap metal merchants. Mainly, the economic aspect of metal as a resource are still driving the effort to separate any metal from other materials. In opposite, the main aspect of separating waste from mineral debris is to keep the costs per ton of waste removal little, since pollution of mineral debris tends to higher fees and a bad quality of recycled gravel.

ACKNOWLEDGEMENTS

The present research received funding from European Commission's RFCS project PROGRESS under the grant agreement No 747847.

Deconstruction, recycling and reuse of lightweight metal constructions

P. Kamrath
Paul Kamrath Ingenieurrückbau GmbH, Germany

M. Kuhnhenne, D. Pyschny & K. Janczyk
RWTH Aachen University, Germany

ABSTRACT

Demolition techniques and the management of construction and demolition waste of buildings are key issues in the development of sustainable construction. Therefore, it is necessary to define the different terms clearly and to differentiate them from each other. This paper gives an overview of various terms and definitions from the field of deconstruction. Furthermore, existing demolition methods are shown and an outlook on the use of robots on the construction site in the future is given. Finally, the planned works within the project PROGRESS regarding the reuse of metal building envelopes are described.

Any building has a limited lifetime. At the end of life, there exist typically three possibilities, the most appropriate option depending on costs, environmental conditions and other local issues such as preservation orders (Dorsthorst & Kowalczyk, 2002):

1. **Deconstruction**: A clearance could extend the lifetime and could be an alternative to demolition. During a building clearance, any non-load bearing parts of the building will be deconstructed. The rebuild process starts with the old skeleton.
2. **Reuse of Structure**: Deconstruction and reuse of the structure itself could be an alternative for some structures, especially those made of steel. This method helps to generate a second life for the load bearing structure e.g. for bridges or halls at another place.
3. **Demolition**: Complete demolition is the typical end-of-life scenario. To avoid waste and landfilling, reuse and recycling of materials should be taken into account.

If a building is being planned today, a lot of energy is done in reducing the needed heat-energy and to optimize the building process. The goal is to lower the producing costs and thus to use preproduced parts such as claddings. When it comes to the

Figure 1. Hybrid element consisting of steel, foam and cardboard.

end-of-lifetime all cladded parts are difficult to put in to recycling, because different materials were used, which are difficult to separate decades later.

Figure 1 shows as a typical example the separation from a three-material roof, which was build from corrugated iron, Styrofoam and cardboard.

For many buildings, it is the failure or deterioration of the envelope that precipitates its premature demolition. This can be aesthetic deterioration, changing architectural design or, as is often the case, the need to update the envelope to modern standards of thermal performance. Therefore, the deconstruction and management of construction and demolition waste of the existing building envelopes is a key issue in the development of sustainable construction.

REFERENCE

Dorsthorst, B.J.H. & Kowalczyk, T. (2002) Design for recycling. CIB Publication 272, Paper n.8. In: *Proceedings of the CIB TG39 Deconstruction Meeting, Design for Deconstruction and Materials Reuse, Karlsruhe, Germany*. Rotterdam, CIB.

Modelling and experimental testing of interlocking steel connection behaviour

P. Matis, T. Martin, P.J. McGetrick & D. Robinson
School of Natural and Built Environment, Queen's University Belfast, Belfast, UK

ABSTRACT

Today's urban environment and transportation networks rely heavily on the use of steel load carrying structures. For these structures, the two main connection methods (i.e. bolting and welding) are based on century old technologies. Both are expensive and time consuming and they cannot always facilitate easy disassembly for material reuse. While new manufacturing capabilities offer the potential to develop faster and most cost-effective connection methods, a major effort to this end has yet to be undertaken. It is clear that by improving and refining these features of multi-storey steel buildings, considerable savings in both weight and cost could be achieved. To achieve these savings, improved construction efficiency and heightened material reuse, a new class of interlocking steel connections using precise, computer-controlled, advanced manufacturing techniques in laser cutting has been developed that rely on neither bolting nor welding. This new joint method connects steel members that have precisely shaped ends in an interlocking approach.

This paper presents the experimental testing and numerical modelling of the flange of the new interlocking steel connection in direct tension. Tensile dogbone tests were carried out to establish the basic material stress-strain response. These tests were carried out in accordance with BS EN ISO 6892-1 (2016) with the use of an appropriate Digital Image Correlation (DIC) camera system. In addition, a series of experimental tests of the flange plate connections were performed to capture the behaviour and failure modes of the new steel connection under tension with the same DIC high resolution camera system. Full load-displacement curves were recorded for each of these tests including into the post-ultimate range and the failure mode of the flange connections was observed and verified. A simplified two-dimensional (2D) finite element (FE) model has been created using the ABAQUS software. The connection model accounts for material and geometric non-linearity, large deformation and contact behaviour. Fracture of the material was not incorporated into the model. Contact is critical to model the

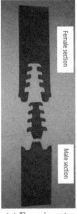 

(a) Experiment (b) Implicit Static solver (c) Explicit dynamic solver

Figure 1. Failure modes of the FE simulation (ABAQUS) and experiment of the flange connection in direct tension.

tensile behaviour of the joint and was modelled using a surface to surface contact interaction taking into consideration friction between the surfaces.

The predicted load-displacement curves closely agree with the experimental curve. The model also predicts well the failure mode of the connection which occurs due to fracture at the base of the lower male section as shown in Figure 1. The comparison with the experimental data shows that the 2D FE model has a very good level of accuracy while significant computational resources and memory usage can also be saved by using this 2D FE model, rather than a detailed 3D FE model.

REFERENCES

BS EN ISO 6892-1 (2016) *Metallic Materials—Tensile Testing Part 1: Method of Test at Room Temperature*. London, BSI Group.

Dassault Systems Simulia (2014) *ABAQUS Analysis User's Manual*. Version 6.14.

SS16: Life-cycle asset management for railway-structures (LeCIE)
Organizers: T. Petraschek, A. Hüngsberg, N. Friedl, U. Staindl, G. Lener & A. Strauss

Life cycle assessment for civil engineering structures of railway bridges made of steel

G. Lener & J. Schmid
Unit of Steel Construction and Mixed Building Technology, University of Innsbruck, Innsbruck, Austria

ABSTRACT

The contribution presents the WEB based prognosis tool "LeCIE" (Life Cycle Assessment for Civil Engineering of Railway Bridges). LeCIE aims for developing a wide ranging concept for a predictable life cycle management of engineering structures in the rail sector. In order to reach this scope, inspection results regarding damages and damage processes are combined with probabilistic degradation predictions, forecasting procedures, monitoring and assessment methods. This includes the development of a framework for the life cycle's management system considering singular as well as combined damages, structural components and the overall structure of concrete-steel bridges, steel bridges and composite ones. The elaboration of the whole task was performed by the University of Natural Resources and Life Science, Vienna (BOKU) and the University of Innsbruck (UIBK) in collaboration.

The most relevant basis for the development of life cycle management tools is the best possible knowledge about functional degradation processes of support structures. A sensitivity analysis of the frequent degradation processes known has been carried out by evaluating the inspection results. Based on the knowledge acquired and an extended data collection, it has been possible to realize a conceptual study for a new procedure in order to deduct degradation functions.

Figure 1. Bridge editor in LeCIE prognosis tool – photo: ÖBB.

The new developed WEB based prognosis tool "LeCIE", which will be demonstrated in this contribution, allows to create, edit and delete new and existing bridges in a free input format. The bridge objects are classified according to the priority of the railway line and can be identified on the parameters of line number and kilometer distance. The analysis of the above mentioned processes, degradation analysis as well as life cycle assessments, can be adopted for the whole bridge object, groups of structural elements or single defined elements. Figure 1 shows the staring section of the bridge editor in "LeCIE", which includes photos from the objects.

Life-Cycle Analysis and Assessment in Civil Engineering: Towards an Integrated Vision – Caspeele, Taerwe & Frangopol (Eds)
© 2019 Taylor & Francis Group, London, ISBN 978-1-138-62633-1

Approach on network-wide sustainable asset management focused on national funding

R. Liskounig
SCHIG mbH, Vienna, Austria

ABSTRACT

Life Cycle Civil Engineering focuses very often on merely technical parameters. When discussing public funding of infrastructure networks additional issues have to be taken into account.

SCHIG mbH is a subsidiary of Austria's Ministry of Transport, Innovation and Technology. In 2015 SCHIG mbH started to develop an indicator to measure the preservation of value of the Austrian railway infrastructure, based on a constant level of quality, from which the efficient allocation of resources for reinvestment and maintenance could be derived.

The normative framework for the Austrian railway infrastructure and its funding is defined on European and national level.

The Austrian Ministry of Transport has introduced social, socio-economic and environmental goals for railway infrastructure.

Mobility should be designed socially, traffic has to be safe, environmentally beneficial and organised efficiently. These general objectives can be observed throughout the various levels of strategy-building.

Starting with Austria's General Transport Strategy for all Transport modes the emphasis is shifted slightly towards strengthening of the market position of railways within the competing modes of transport.

In the strategy for the largest railway network in Austria (belonging to ÖBB-Infrastruktur AG) market position of the railway is considered an important issue. Particular importance is also given to safety.

The efficient organisation of transport and optimal use of financial resources is seen as of crucial importance at all strategic levels.

Funding agreements between the Ministry of Transport and ÖBB-Infrastruktur AG define indicators by which to measure the achievement of strategic objectives and the fulfilment of the contractual agreements.

ÖBB-Infrastruktur AG provides an annual Network Status Report indicating the condition of the most significant assets within the Network.

The focus of the approach is set to the strategic level. However, infrastructure managers are independent as to the methods by which the strategic goals are achieved operationally. As indicators and decisions have to be robust it must be ensured that the output of the operational – often bottom up – management of the network provides reliable and correct data as a basis for further strategic decision-making.

The current value of the network is the basis of all further research. This consideration is closely connected to qualitative (technical, logistic, operative, etc.) requirements at any given date.

Although liable to change in the technical, social and environmental fields quality requirements must be fulfilled to a constant degree in order to maintain the value of the network.

SCHIG mbH is working on a framework to support decision makers in the allocation of public funding.

In addition to the detailed information about the assets used by the asset manager, consideration will be given to a top-down approach. A set of indicators suitable for network-level decision making on issues of public funding is in preparation.

REFERENCES

Funding Contracts. https://www.bmvit.gv.at/verkehr/eisenbahn/finanzierungsvertrag_oebb.html.
General Transport Strategy. https://www.bmvit.gv.at/verkehr/gesamtverkehr/gvp/downloads/gvp_gesamt.pdf.
Network Status Report. https://presse.oebb.at/file_source/corporate/presse-site/Downloads/Publikationen/Publikationen%20aus%20Konzern/OEBB_Infra_Netzzustandsbericht2016.pdf.

Life-Cycle Analysis and Assessment in Civil Engineering: Towards an Integrated Vision – Caspeele, Taerwe & Frangopol (Eds)
© 2019 Taylor & Francis Group, London, ISBN 978-1-138-62633-1

Decision-making framework and optimized remediation for railway concrete bridges deteriorated by carbonation and chloride attack

A. Vidovic, I. Zambon & A. Strauss
Institute of Structural Engineering, Department of Civil Engineering and Natural Hazards,
University of Natural Resources and Life Sciences, Vienna, Austria

D.M. Frangopol
ATLSS Research Center, Lehigh University, Bethlehem, PA, USA

ABSTRACT

Considering the importance of bridge transportation infrastructure, it must be ensured that existing stock of bridges is being properly maintained. Web-based Bridge Management System (BMS) that serves as a tool for the assessment, evaluation and decision-making of existing reinforced concrete railway bridges has been recently developed. This paper presents the methodology behind the tool, which supports decisions whether a bridge should be repaired and/or replaced. Accordingly, particular focus is given to the forecast of the durability of reinforced concrete bridges and development of an effective maintenance and repair strategy in order to increase the service life at minimum cost.

The existing BMSs commonly base decisions on engineering judgement and visual evaluation of bridge condition, where the more complex systems contain algorithms that treat the data to produce the optimum maintenance strategy at both project and network level, i.e. related to individual bridges or entire bridge stock, respectively. Herein, the BMS is mainly being developed at the project level – for individual bridges

and/or components. Moreover, the decision-making takes basis on physical models of carbonation process and chloride ingress, as corrosion is being the most widespread type of damage of concrete structures.

The application of the procedure therefore starts with the assignment of norm specific degradation and prediction models to the discretized components of the considered bridge. Each component is furthermore being represented with suggested interventions for considered extent of damage, following with a selection of an optimized intervention strategy for the component that exhibits the worst behavior. The first objective is the minimization of the life-cycle cost function, whereas subsequent is the minimization of the deterioration model itself. After the verification of compatibility of optimized intervention with accompanying components is carried out, the optimized strategy is applied to other components and/or cluster of components, and finally brought as a solution to other deterioration processes, if applicable. The decision-making procedure is validated – from deterioration model being updated with monitoring information to verification of the durability and suitability of the interventions.

443

Life-Cycle Analysis and Assessment in Civil Engineering: Towards an Integrated Vision – Caspeele, Taerwe & Frangopol (Eds)
© 2019 Taylor & Francis Group, London, ISBN 978-1-138-62633-1

Short-, mid- and long-term LCM prognosis of heavy maintenance and replacement demand for bridge structures at 3 selected railway routes analysing different maintenance strategies

R. Veit-Egerer & G.J. Rajasingam
FCP Fritsch, Chiari & Partner ZT GmbH, Vienna, Austria

T. Petraschek, L. Rossbacher, N. Friedl & U. Staindl
Austrian Federal Railways ÖBB, Vienna, Austria

ABSTRACT

The subject of the presented investigation is the application of a LCM model developed by FCP in close collaboration with the Austrian Federal Railways (ÖBB) for bridge structures along three selected routes from the Austrian railway network:

– Nordbahn
– Pyhrnbahn
– Südbahn

The objective of the investigation is a prognosis of the heavy maintenance and replacement demand for bridges for different forecast periods (12/50 and 120 years). Three different maintenance strategies were simulated and applied for an in-depth analysis:

– Simplified LCC projection without considering structural condition ratings
– Planned preventive maintenance (considering structural condition ratings)
– Do-Minimum Strategy (considering structural condition ratings)

In the course of the elaboration the following methodical approach was chosen:

– Clustering of data records according to bridge types (depending on bridge length and con-struction material)
– Detailing with regard to technical service life and maintenance strategy
– Definition of the appropriate cost models for maintenance and replacement measures using database benchmarks
– Simplified cost approach for routine maintenance and structural inspection according to the Austrian

regulation RVS 13.05.11 "Lebenszykluskostenermittlung für Brücken"
– Derivation of interventions schedules (intervention type and the corresponding point of time) as a consequence of implementing the latest condition ratings into the bridge-type-dependent degradation models and prognosis of subsequent progression of condition over time
– Calculation of the cost schedules for each analysed maintenance strategy

The results are elaborated for each route and for each of the three listed maintenance strategies in terms of the following parameters:

– BoQ (bill of quantities) per year and per intervention type
– BoQ (bill of quantities) per year and per bridge material type
– Costs per year and per intervention type
– Costs per year and per bridge material type
– Impact of respective interventions on the progression of condition according to the underlying degradation models.

In the first instance the derived results were evaluated route by route. As these three routes differ in terms of historical and constructional characteristics the differences between the three routes were ana-lysed in a subsequent LCM comparison.

The presented prediction tool provides accurate prognoses of maintenance scheduling on network level to be expected and indicates reasonable LCC estimations. As only some major asset database parameters are considered the results do not reflect the entire knowledge from the local sites respectively (these must be incorporated separately in a subsequent, more refined LCC analysis).

444

Execution time estimation of recovery actions for a disrupted railway track inspection schedule

M.H. Osman & S. Kaewunruen
University of Birmingham, Birmingham, UK

ABSTRACT

Disruptions in railway track inspection schedules demand recovery action to avoid losing track possession while delivering an adequate level of track operational safety. Otherwise, the disrupted schedule may trigger consequences such as track component damage and train delay. To effectively respond to the disrupted situation, execution time of a deterministic recovery action must be presented to decision maker(s) (Osman, M.H. Bin et al, 2016). Integrated analyses of track reliability and profitable capacity are found sufficient to estimate the execution time when managing a disrupted track inspection.

An estimated profit on a train journey over a definite interval (t_o, t_m), $\tilde{P}$ with the assumption that a disruption occurs at time $t_d < t_r$ can be calculated from the following equation (Vaurio 1994; Khadem et al. 2010):

$$\tilde{P}(t_o, t_m) = \int_{t_o}^{t_m} (c_a x - c_b) g_s(x) dx \qquad (1)$$

where c_a is expected profit per unit time while the system which is n consecutive track segments of the journey, is operating, and c_b is the penalty cost incurred due to the system having to be operated at low traffic speed. Meanwhile, $g_s(x)$ is a probability density function of a random variable X; the time difference between the restoring point and the state of the system which has passed an alert limit. A sudden shift in a degradation path is defined as a (*soft*) failure. In this study, the system of interest is subject to on-board inspection and is maintained to function according to (n, k): F system configuration. The proposed method in (Shmueli 2003) was used for system reliability computation because it does not involve recursive computation.

Given a time dependent recovery cost function $c_r(t)$, no financial aid is needed to manage a disruption as long as the estimated profit is greater than the recovery cost at the execution time t_r. Intuitively, the shortest time length between t_d and t_r can be

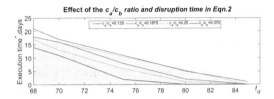

Figure 1. Downward trend in response time over t_d.

determined by solving the following optimization model for t_r.

$$\underset{t_d < t_r < t_{ins}}{\text{Min}} \left| c_a t_d \int_{t_o}^{t_d} g(x) dx + \int_{t_d}^{t_r} (c_a x - c_b) g(x) dx - (c_{ins} + c_r(t_r)) \right| \qquad (2)$$

Fig. 1 exhibits a downward trend in the earliest execution time over t_d and is justified at least for four values of ratio of c_a/c_b. Obviously, longer time is needed if a disruption occurs at early stage of inspection interval $t_{ins} = 85$. As the value of t_d approaches the inspection time, a maintenance team can execute the predetermined recovery action immediately, after the schedule is disrupted. This privilege: however, gradually disappeared when we decreased the c_a/c_b ratio.

Overall, the findings exhibit the advantages of determining an estimation of execution time for the case of the disrupted schedule.

REFERENCES

Khadem, M., et al. (2010) Profit-generating capacity for a freight railroad. *Transportation Research Record*, 7400 (August), 1–17.

Osman, M.H., Bin, et al. (2016) Disruption management of resource rchedule in transportation sector: Understanding the concept and strategy. *Procedia Engineering* 1295–1299.

Shmueli, G. (2003) Computing consecutive-type reliabilities. *IEEE Transactions on Reliability*, 52 (3), 367–372.

Vaurio, J.K. (1994) A note on optimal inspection intervals. *International Journal of Quality & Reliability Management*, 11 (6), 65–68.

Structural assessment and rehabilitation of old stone railway bridges

P.G. Malerba
Department of Civil and Environmental Engineering, Politecnico di Milano, Milan, Italy

D. Corti
Director of Società Subalpina Imprese Ferroviarie, Domodossola, Italy

ABSTRACT

This paper will present the structural assessment and the rehabilitation design of some stone arch bridges of the historical Domodossola-Locarno railway. The railway meanders through the Vigezzo Valley which stretches from Domodossola in norther Italy to Locarno, in the Swiss Canton of Ticino (Fig. 1).

Mainly famous for its breath-taking views of the Alps, the railway is both an essential feature in the tourist industry of the area and an important transportation infrastructure. Though the years, the line has been a dependable means of transport for the people of the Vigezzo Valley and also provided a connection between the Simplon and the Saint Gotthard lines.

The construction of the railway begun in 1913 and was completed in 1925 (Weder & Pfeiffer 1997). With a total length of 52 km (32 of which on the Italian side and 20 on the Swiss side), it has a gauge of 1000 mm and is powered by a continuous electrical current of 1440 V.

The morphology of the terrain entailed what, at the time of construction, were challenging infrastructural works. Among these are 35 tunnels, 35 viaducts, 25 bridges and 19 short-span decks (shorter than 5 m). 52 of these structures are single or multi-arch stone bridges, 18 are reinforced or pressed concrete, 6 are made of steel and 3 are mixed structures.

In 2013, SSIF (the company in charge of maintenance and management of the line) started implementing a plan that will end in 2021 and that will involve the detailed inspection and/or refurbishment of bridges and viaducts evaluated as critical. As part of this plan, detailed inspections and surveys were carried.

In 2016 the first repairing works were completed. With reference to a sample of three stone arch bridges, the first part of the paper will describe the most frequently recurring damages due to age and the wear of time (Malerba 2014). The second part, on the basis of the assessment about their actual serviceability, highlights the needs both for a full seismic rehabilitation and the strengthening of the connection between spandrel-walls/vault, exhibiting in many cases wide gaps as an explicit signal of detachment (Fig. 2). The

Figure 1. ABe 8/12 type train on a typical stone arch bridge.

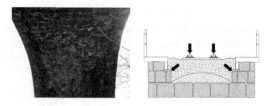

Figure 2. a) Disconnection of the edge ribs of the arches; b) mechanism of disconnection.

paper ends with the presentation of the main criteria followed in the rehabilitation design.

REFERENCES

Malerba, P.G. (2014) Inspecting and repairing old bridges: Experiences and lessons. *Structure and Infrastructure Engineering*, 10 (4), 443–470.

Weder, C. & Pfeiffer, P. (1997) *Centovalli Valle Vigezzo. La ferrovia, il paesaggio, la gente.* Zurich, AS Verlag & Buchkonzept.

SS17: INFRASTAR—fatigue reliability analysis of wind turbine and bridge structures
Organizers: E. Brühwiler, E. Niederleithinger & J.D. Sørensen

Fatigue reliability analysis of Cret De l'Anneau viaduct: A case study

A. Mankar, S. Rastayesh & J.D. Sørensen
Aalborg University, Aalborg, Denmark

ABSTRACT

Fatigue of reinforced concrete structures is often not considered for civil engineering structures due to the fact that dead loads of reinforced concrete structures are very high (for case of normal strength concrete) while live loads on these structures are relatively small which leads to very small stress variations during service duration of the structure. However, particularly for bridge structures with increased use of high strength concrete and increase in traffic loads this scenario is reversed and fatigue verification of these structures becomes much more important for the safety.

Most of the bridges in Switzerland built during the last 50 years are reinforced concrete bridges and they typically experience more than 100 million cycles of fatigue load during design lifetime. This is especially the case for reinforced concrete decks of such bridges exposed to traffic loads during their lifetime which are not designed for fatigue (Schläfli & Brühwiler, 1998).

Bridge engineers in the industry use PM rule of linear damage accumulation along with Wöhler curves from codes and standards, for new structures and existing structures for fatigue verification of existing bridges and often with the result to replace an existing bridge or at-least the deck of the bridge.

A best way forward could be to use reliability methods (a probabilistic approach) to obtain a more detailed assessment of the bridge and thereby a better basis for decision making.

This paper presents a reliability-based framework for reliability assessment with respect to fatigue failure of Crêt de l'Anneau viaduct as a case study, where the MCS department at EPFL, Lausanne, Switzerland has installed a long term monitoring system for estimating strains in the structure deck slab. As part of reliability-based framework stochastic modelling of fatigue strength of reinforcing bars along with stochastic modelling of fatigue loads will be presented. Calibration of fatigue safety factors will also be presented. The reliability value obtained will be compared with required reliability of structures as recommended by (SIA-269, 2016) Swiss Standard for Existing structures and (EN 1990, 2002).

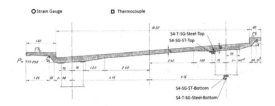

Figure 1. Crêt de l'Anneau viaduct cross section.

For details about monitoring system, reference is made to (MCS, 2017), Figure 1.

The current age of the bridge is 70 years, and it is investigated the bridge can be used for additional 50 years, i.e. a total of 120 years. The reliability is assessed for the reinforced concrete deck slab with respect to fatigue failure of the reinforcement, as this position is often the critical location.

The annual and cumulative reliability indices based on First Order Reliability method (FORM), (Madsen, et al., 2006) & (Sørensen, 2011) obtained can be compared with target reliability indices indicated in (SIA-269, 2016) and (EN 1990, 2002) respectively, to obtain a range of fatigue safety factors (FDF) required to obtain and maintain the annual and accumulated target reliability indices.

REFERENCES

EN 1990 (2002) *EN 1990 Basis of Structural Design*. Brussels, European Committee for Standardisation.

Madsen, H.O., Krenk, S. & Lind, N.C. (2006) *Methods of Structural Safety*. New York, Dover Publications.

MCS (2017) *Surveillance du Viaduc du Crêt de l'Anneau par un monitoring à longue durée*. Lausanne, MSC.

Schläfli, M. & Brühwiler, E. (1998) Fatigue of existing reinforced concrete bridge deck slabs. *Engineering Structures*, 20 (11).

SIA-269 (2016) *Existing Structures – Bases for Examination and Interventions*. Zurich, Swiss Society of Engineers and Architects.

Sørensen, J.D. (2011) *Notes in Structural Reliability Theory and Risk Analysis*. s.l., Aalborg University.

"Pocket-monitoring" for fatigue safety verification of a RC bridge deck slab

I. Bayane & E. Brühwiler
Laboratory of Maintenance and Safety of Structures (MCS), École Polytechnique Fédérale de Lausanne (EPFL), Lausanne, Switzerland

ABSTRACT

Road and rail bridges and in particular their reinforced concrete or orthotropic steel decks are subjected to significant fatigue loading. Fatigue safety verification requires information about the fatigue resistance, the critical details, and the fatigue loading action effect including past and future traffic as well as environmental conditions.

The last development of the monitoring techniques with the high capacity of acquisition and storage is providing the possibility to examine bridges in real time for long-term monitoring.

Consequently, there is a significant potential to use monitoring systems to investigate the current structural condition with the goal to verify the fatigue safety of existing bridges. Structural monitoring provides relevant data and reduces the uncertainties of various involved parameters.

Structural engineers have to show that there is no fatigue problem and that bridges can remain in service without intervention.

The verification of the fatigue safety must be based on direct measurements of the structural response at relevant locations since it is not economical nor practical to instrument the whole bridge structure. In fact, the instrumentation of the structure needs to be optimized by placing sensors at critical locations thereby reducing the number of sensors to the minimum such that the obtained information will be reliable and sufficient to perform the fatigue safety verification. In this context, the methodology of "pocket monitoring" is proposed to characterize both the structural response and the effects of fatigue action.

The objective of this paper is to present the methodology of fatigue safety verification using the "pocket monitoring" as applied to an existing bridge structure. The concept of the pocket monitoring will be presented, with the data processing algorithms.

The effect of temperature and its influence on the monitoring period and the reliability of the results will be discussed. The structural behavior and the fatigue safety will be examined.

Within this framework, a steel-concrete composite road viaduct in Switzerland, shown in Figure 1, was instrumented according to the "pocket monitoring" approach with the objective to verify the fatigue safety.

Figure 1. View of the investigated steel-concrete composite viaduct.

Fatigue safety verification of a steel railway bridge using short term monitoring data

B. Sawicki & E. Brühwiler
Laboratory of Maintenance and Safety of Structures (MCS), École Polytechnique Fédérale de Lausanne (EPFL), Lausanne, Switzerland

M. Nesterova
Laboratoire Sécurité et Durabilité des Ouvrages d'Art (SDOA), The French Institute of Science and Technology for Transport, Spatial Planning, Development and Networks (Ifsttar), Champs-sur-Marne, France

ABSTRACT

The case study of the fatigue safety examination of railway bridge is presented. The riveted steel arch structure is carrying a single railway track on two main girders. The bridge was built in 1897 in Switzerland.

The proof engineer conducted an overall assessment using updated load models from the Swiss Standard SIA 269 for existing structures. The results of these calculations indicated sufficient resistance at Ultimate Limit State, but the calculated fatigue safety was not sufficient and the replacement of the main girders was recommended.

To challenge this recommendation, detailed examination was performed using data obtained from short term monitoring. The monitoring system was installed on the main girders of one span using 14 strain gauges and two omega gauges. Strains were measured during the passage of the trains over a period of four weeks.

Analysis of measured data revealed that the load models given in the SIA 269 Standard were overly conservative and non-realistic. Additionally, the girders were designed as single span simply supported beams but the partial fixity due to the connection of the adjacent beams led to a continuous girder action with considerable hogging moments reducing thus fatigue stresses.

The histograms of fatigue stress ranges were constructed and it was shown that the measured fatigue stress ranges are smaller than the fatigue endurance limit of the given riveted detail of the beams, indicating a theoretically infinite fatigue life.

Figure 1. Railway bridge with the scaffolding for installation of monitoring.

No structural intervention is thus necessary. This case study shows that a simple short term monitoring is very useful for the fatigue examination of bridges allowing for reducing or avoiding (costly) interventions on structures.

ACKNOWLEDGEMENT

This project has received funding from the European Union's Horizon 2020 research and innovation program under the Marie Skłodowska-Curie grant agreement No 676139.

SS18: The impact of BIM and web technologies in the life-cycle of our built environment
Organizers: P. Pauwels, K. McGlinn & V. Malvar

Life-Cycle Analysis and Assessment in Civil Engineering: Towards an Integrated Vision – Caspeele, Taerwe & Frangopol (Eds)
© 2019 Taylor & Francis Group, London, ISBN 978-1-138-62633-1

Investigation of the lifetime extension of bridges, using three-dimensional CIM data

T. Yamamoto, K. Konuma & T. Yaguchi
Pacific Consultants Co., Ltd., Osaka, Japan

H. Furuta, H. Tsuruta & N. Ueda
Kansai University, Osaka, Japan

ABSTRACT

In recent years, as the age of infrastructure such as road bridges has progressed, maintenance and management have become major social issues in Japan.

Over time, preventive maintenance has followed the long-life repair plan based on periodic inspection. However, financial resources for executing maintenance, as well as the number of technical experts engaged in public and private sectors, will likely decrease.

Along with the rapid progress of ICT in recent years, the introduction of BIM/CIM is being promoted. MLIT is now promoting "i-Construction" in Japan.

The main application method is to shift the data format from conventional two-dimensional data to three-dimensional data, which will then be used to unitarily manage the design, construction, maintenance and investigation of bridges.

Meanwhile, CAE is well known from preceding the design method that consistently uses one piece of three dimensional data in the machine manufacturing field of automobiles and other vehicles.

The design method, databased system, and asset management system based on CAE make it possible to connect everything in the future. Therefore, in the field of civil engineering, we believe that the design method using CAE that will be suited to the IoT era.

Specifically, "maintenance small bridge" development by the materials to be used and structural details that we devised avoids the fatal deterioration damage during the service period and minimize maintenance and management. The ultimate goal is to reduce the burden of future maintenance and management.

In this research, we devised a flow for "maintenance small bridge" development, and decided to outline the utilization of 3-dimensional CIM data incorporating environmental conditions as a foothold. The contents of the investigation are as follows.

We focus on salt damage that is the main degradation phenomenon in the concrete bridge. Specifying the places where salt easily accumulates, based on the results of flying salt simulation using three-dimensional data, enables us to extend the service life of a structure and to minimize the LCC.

This paper describes the results of the study and future prospects.

Life-Cycle Analysis and Assessment in Civil Engineering: Towards an
Integrated Vision – Caspeele, Taerwe & Frangopol (Eds)
© 2019 Taylor & Francis Group, London, ISBN 978-1-138-62633-1

A generic model for the digitalization of structural damage

A. Hamdan & R.J. Scherer
Technische Universität Dresden, Dresden, Germany

ABSTRACT

Over the last decade, the possibilities of detecting structural damage have significantly increased, due to new techniques like identifying material cracks through scanners, drones or other advanced measuring instruments. Nevertheless, the methods of storing the measured data are still limited to graphical representations of the damage or datasets which have no context to the building. Due to this, the damage of structures must be evaluated and categorized manually by experts. The possibilities of a computer-aided analysis, in which tools for filtering or reasoning could be executed are restricted because of a missing data model for structural damage. Therefore, a model to digitalize structural damage is developed and discussed in this article. The data of the structure is stored in a BIM model which uses a standardized data format, the Industry Foundation Classes (IFC) and by using a multi model approach, both models are connected.

Life-Cycle Analysis and Assessment in Civil Engineering: Towards an Integrated Vision – Caspeele, Taerwe & Frangopol (Eds)
© 2019 Taylor & Francis Group, London, ISBN 978-1-138-62633-1

Modelling risk paths for BIM adoption in Singapore

X. Zhao
School of Engineering and Technology, Central Queensland University, Australia

ABSTRACT

Building information modelling (BIM) has been transforming the architecture, engineering, and construction (AEC) industry in many countries (Azhar, 2011, Zhao, 2017), including Singapore. In recent years, the government of Singapore, through the Building and Construction Authority (BCA), has highlighted productivity in the construction industry, and initiated a Construction Productivity Roadmap to transform the construction industry and raise its productivity (BCA, 2011). Because of the potential to enhance construction productivity, driving BIM adoption has been seen as one strategic thrust (Teo et al., 2015) of Singaporean construction and building sector. BIM adoption can offer numerous benefits to the users, but it should not be based on an oversimplified assumption that BIM implementation does not bring about risks. Indeed, there were diverse BIM-related risks, such as technological, contractual, and legal issues that have been well reported in the previous studies

The objective of this study is to model the paths of risks associated with BIM adoption in the context of Singapore. To achieve the objective, 16 risks categorized into nine groups were identified from a literature review, and a questionnaire survey was conducted with 42 professionals in Singapore. This study collected data relating to the likelihood of occurrence (LO) and magnitude of impact (MI) of the risks associated with BIM adoption in Singapore. The LO was rated using a five-point scale: 1=rarely (LO < 20%); 2 = somewhat likely (20% ≤ LO < 40%); 3 = likely (40% ≤ LO < 60%); 4 = very likely (60% ≤ LO < 80%); and 5 = almost definite (LO > 80%). The MI was evaluated using another five-point scale: 1 = very small; 2 = small; 3 = medium; 4 = large; and 5 = very large. The hypothetical risk paths were tested using partial least square-structural equation modelling.

The results confirmed the risk categorization and supported seven significant risk paths: "inadequate relevant knowledge and expertise" contributed to "poor information sharing and collaboration"; "poor information sharing and collaboration" significantly resulted in "low data quality"; "low data quality" further contributed to "liability for data input"; "liability for data input" significantly contributed to "cost overrun with BIM"; "data ownership issues" could contribute to "cultural resistance", "poor information sharing and collaboration" and "cost overruns with BIM". These seven risk paths formed four chains of risk paths. "Data ownership issue" and "inadequate relevant knowledge and expertise" were the root risks in two chains of risk paths, suggesting that they were the primary source of all the other risk categories and should be emphasized. "Data ownership issue" can either directly result in "cost overrun with BIM", or indirectly caused cost overrun through "poor information sharing and collaboration", "low data quality", and "liability for data input".

The findings of this study enable practitioners to understand the risks associated with BIM adoption, take measures to mitigate the root risks, and assure the potential benefits of BIM.

REFERENCES

Azhar, S. (2011) Building information modeling (BIM): Trends, benefits, risks, and challenges for the AEC industry. *Leadership and Management in Engineering*, 11, 241–252.

BCA (2011) *Construction Productivity Roadmap*. Singapore, Building and Construction Authority.

Teo, E.A.L., Ofori, G., Tjandra, I.K. & Kim, H. (2015) The potential of Building Information Modelling (BIM) for improving productivity in Singapore construction. In: Raidén, A.B. & Aboagye-Nimo, E. (eds.) *31st Annual ARCOM Conference*. Lincoln, UK, Association of Researchers in Construction Management.

Zhao, X. (2017) A scientometric review of global BIM research: Analysis and visualization. *Automation in Construction*, 80, 37–47.

Life-Cycle Analysis and Assessment in Civil Engineering: Towards an Integrated Vision – Caspeele, Taerwe & Frangopol (Eds)
© 2019 Taylor & Francis Group, London, ISBN 978-1-138-62633-1

Using semantic technologies to improve FM asset information management processes

J.W.B. Kibe
University College London – IEDE, London, UK

ABSTRACT

An increasing demand in social & professional connectivity fuelled by the pressures of rapidly evolving technological advances is placing great demands on current Facilities Management (FM) information technologies to deliver more interactive data and knowledge management to facilitate decision making processes. As a result, the industry is rapidly moving away from traditional silo mentality, introducing and adopting a plethora of intelligent/smart systems with the intention of interlinking and clarifying information across the various disciplines within the industry. This paper seeks to identify the key challenges of asset information management in context to FM and explore the potential for Semantic technologies to resolve some of these challenges. The literature review provides an overview of the concepts of semantic and linked technologies to establish the value of the proposed approach while a case study is introduced to give an insight into the challenges of managing asset data in an active FM centric environment. This paper hypothesises that the use of sematic web & linked data technologies can provide a centralised global context to the information stored in the Asset Information Model to effectively improve FM asset information management processes.

The management of information & data is a prevalent problem for FM industries and the vision is that semantic web will allow for documents, files and information to be transformed in-to machine readable datasets that can be used in a variety of web services. Semantic Web represents the next evolution of information technology and is primarily concerned with the meaning and not the structure of data. Unlike traditional relational databases where meanings and relationships between data items are inflexible, SW enables data linkage across various and diverse sources to be understood and read by computers. Numerous studies suggest that the employment of semantic technologies such as RDF and ontology to XML documents will allow for this knowledge to be extracted

and shared across a digital platform. RDF and ontologies provide a means to build the semantic bridges required for data exchange and application integration. Documents will continue to render a significant part of the knowledge base stored on local desktops particularly in the FM industries and as such there an urgent need for researchers to look into more into ways to reconciling documents and semantic technologies to describe, store, catalogue, and retrieve information in a systematic and efficient manner.

The scope of the case study was limited to the management of information pertaining to the maintenance and management of building systems in a specific hospital building in Oxford, UK. Applied methods for collect asset data/information within the hospital included manual data collection from over 50 Sub-contractors & 12 In-house technicians, 11 managers (financial, contracts, projects etc.), automated data collection from FM systems and smart gadgets and semi-automated data capture through the application of both manual and automated processes. Documents produced for asset management purposes typically include asset schedules, predictive preventive maintenance sheets, task sheets, service sheets, service manuals, risk assessment forms, log sheets, instruction sets, survey sheets, trackers etc.

The study found that on a strategic level, interviewee's attributed the challenge in managing this information to three key items i.e. the lack of; a single centralised and relational repository system, communication between the FM inter-departmental teams and software integration across FM asset management systems. As a result the FM team was struggling with various compliance issues including; the loss of documents and poor document control which inadvertently had a significant impact on their revenue. This paper considers how document data modelling approaches such as XML-based data modelling approach can be integrated with semantic technologies to effectively facilitate for a mechanism in which FMs can query digital documents to store, retrieve and share valuable information.

GENERAL SESSIONS
GS1: Probability theory and applied structural reliability methods

Vulnerability of critical slopes by using continuous Bayesian networks

D. De León, D. Delgado & E. Solorio
Universidad Autónoma del Estado de México, Mexico City, Mexico

L. Esteva
Institute of Engineering, Universidad Autónoma del Estado de México, Mexico City, Mexico

ABSTRACT

The paper deals with the vulnerability assessment of a critical slope in Mexico throughout the use of a Bayesian network with continuous variables that consider the slope geometry and random soil properties and random rainfall features. The soil properties change according to the random rainfall intensity, which is modeled as exponentially distributed, making the process highly dynamical. The influence of each parameter, especially the phreatic water level, on the safety level is explored through continuous functions. Practical mitigation recommendations may be devised by analyzing the BNet and modeling variations on these variables.

The proposed formulation is based on a four-steps sequence:

1. Calculation of the resisting and acting moments, for the changing conditions of the soil properties for each trial in a MC simulation process.
2. The depth of the water level is calculated as a random variable considering the rainfall effect on the soil conditions and, therefore, affecting the slope conditions.
3. Calculation of the slope failure probability, by fitting a lognormal distribution to the simulated safety factors and calculating:

$$P_f = P(SF < 1) \quad (1)$$

Comparison between an allowable failure probability, obtained by minimizing the expected total cost (Ang and De León, 2005) including the costs of the slope-failure and those of the mitigation actions adopted, with

4. The calculated failure probability. If the calculated value is larger than the allowable one, a new MC simulation under mitigation measures (considered by a reduction on the water level) leads to a new failure probability until its value is below the acceptable one. Table 1 shows the statistics of the soil properties.

Fig. 1 shows the modeling of the slope considered for the illustration.

Table 1. Statistics of soil properties.

Variable	Mean Value	Standard deviation	Distribution Type
Cohesion (kPa)	14	3	Normal
Friction Angle (°)	30°	6	Normal
Volumetric weight (kN/m^3)	18	4	Normal

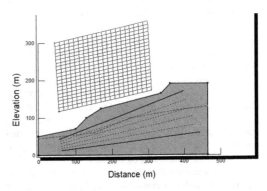

Figure 1. Slope geometry model (initial condition). Dimensions in m.

REFERENCE

Ang, A. & De Leon, D. (2005) Modeling and analysis of uncertainties for risk-informed decision in infrastructures engineering. *Journal of Structure and Infrastructure Engineering*, 1 (1), 19–31.

Life-Cycle Analysis and Assessment in Civil Engineering: Towards an Integrated Vision – Caspeele, Taerwe & Frangopol (Eds)
© 2019 Taylor & Francis Group, London, ISBN 978-1-138-62633-1

A fast and efficient approach to solution of structure system reliability

N. Xiao
College of Civil Engineering and Architecture, Zhejiang University, Hangzhou, China

Y. Chen
Foshan Rail Transit Design and Research Institute Co. Ltd., Foshan, China

F.Y. Lu
College of Civil Engineering and Architecture, Zhejiang University, Hangzhou, China

ABSTRACT

In recent decades, the research on the reliability of structural systems has been a hot topic. The traditional structural system reliability analysis method depends mainly on the branch and bound technology of the structure failure path, which is leading to difficulty in identification of the main failure mode and huge workload in calculation of the system failure probability (Dong & Xia, 1995; Dong & Li, 1995; Thoft-christensen & Murotsu, 2012).

Park & Choi *et al.* (2004) proposed an efficient algorithm for fast evaluation of the failure probability of complex structure, the main idea of the algorithm is addressed as following. The failure probability for component i in a structure system with n components is denoted α_i, ranking the failure probability of each element $\alpha_1' > \alpha_2' > \alpha_3' > \cdots > \alpha_n'$ according to their size, if the number of statically indeterminate structure system is s, the failure probability of the structure system is then defined as $P_{system} = \alpha_{s+1}'$. Although the algorithm has a high computational efficiency, the order of components failure in the failure path is not considered in their algorithm in which the ultimate bearing capacities of all components in the structure are assumed to be complete correlation random variables. This assumption adopted has great difference with the failure process in the practical structures.

In view of this, combined with the β-bound method and joint failure probability branch bound method, a new method for fast and efficient solution of structural

system reliability is put forward in this paper. Firstly, the safety margin functions of structural element and the corresponding failure probability are determined by application of structure analysis principle, the element with maximal value of failure probability is then removed and substituted by virtual force to form a new structure, and then reanalysis for the new structure is carried out. By following the steps above, until the mechanism is formed, the structure system reliability is finally determined to be the failure probability of the component in the last step. Case studies comparing with those existing methods show that the proposed method in this paper is simple, less calculating workload, high efficient calculation accuracy as well as failure path available, which is suitable for the system reliability calculation of large scale structure.

REFERENCES

Dong, C. & Li, Z.N. (1995) Accurate calculation theory of structural system reliability. *Structure & Environment Engineering*, 3, 46–50.

Dong, C. & Xia, R.W. (1995) Research progress of reliability evaluation theory of modern structural system. *Advances in Mechanics*, 25 (4), 537–548.

Park, S., Choi, S., Sikorsky, C. & Stubbs, N. (2004) Efficient method for calculation of system reliability of a complex structure. *International Journal of Solids and Structures*, 2004 (41), 5035–5050.

Thoft-christensen, P. & Murotsu, Y. (2012) Application of structural systems reliability theory. *Springer Science & Business Media*.

Failure probability of a designed nonlinear structure taking into account the uncertainty of Fourier phase

R. Huang, T. Sato, C. Wan, A. Ahamed & L. Zhao
Key Laboratory of Concrete and Pre-stressed Concrete Structure of Ministry of Education, Southeast University, Nanjing, Jiangsu, China
Jiangsu Advanced Institute of Seismic Resistant Technology for Mechanical and Electrical Equipment, Nanjing, China

ABSTRACT

We develop a procedure to calculate the failure probability of a structure designed by using the so called yield strength demand spectrum. The seismic performance of the designed structure is usually checked through a nonlinear dynamic analysis by using the response spectrum compatible earthquake motion (RSCEM). The RSCEMs are always simulated by assigning a proper Fourier phase and the nonlinear response of deigned structure is strongly affected by this assigned Fourier phase. Fourier phase of earthquake motion possess strong uncertainty due to the inherent feature of uncertainties of earthquake mechanism and propagating path of earthquake wave.

Earthquake motion is an uncertain process that can be simulated using a developed model by modifying the fractional Brownian motion process. So based on Monte Carlo method, we will simulate 100000 different RSCEMs and then analyze the structural failure probability considering the uncertainty of Fourier phase, as well the uncertainty of RSCEMs. The failure criterion of the designed structure is an exceedance ratio of nonlinear displacement beyond the assigned ductility to design the structure. The modified fractional Brownian model is also controlled by two parameters, namely Hurst index H and variance σ_0^2, leading to the change of simulated Fourier phase. Effects of these two parameters on failure probability, so, will as well be studied.

Considering a SDOF (single-degree of freedom) nonlinear structure through Clough model subjected to the generated RSCEMs and the calculated failure probability is 0.00081. The effects of Hurst index and variance on failure probability of this SDOF nonlinear structure are also achieved and illustrated as follows. When variance is constant and the failure probability increases as Hurst index increases. When Hurst index is constant and the failure probability decreases as variance increases.

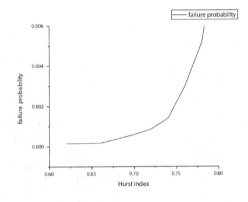

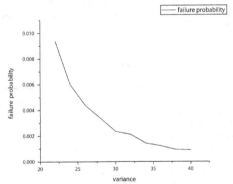

REFERENCES

Guo, X. & Cheng, G.D. (2010) Recent development in structural design and optimization. *Acta Mechanica Sinica*, 26 (6), 807–823.

Qiu, Z.P., et al. (2013) Structural reliability analysis and reliability-based design optimization: Recent advances. *Science China Physics, Mechanics and Astronomy*, 56 (9), 1611–1618.

Yu, B. (2011). *Structural Reliability Analysis Based on Ultimate Bearing Capacity and Seismic Ductility Demand.* Doctoral Dissertation, Guang Xi University (in Chinese).

Life-Cycle Analysis and Assessment in Civil Engineering: Towards an Integrated Vision – Caspeele, Taerwe & Frangopol (Eds)
© 2019 Taylor & Francis Group, London, ISBN 978-1-138-62633-1

Iterative point estimate method for probability moments of function

W. Fan, H. Guo, J. Wei & Z. Li
Key Laboratory of New Technology for Construction of Cities in Mountain Area (Chongqing University),
Ministry of Education, Chongqing, China
School of Civil Engineering, Chongqing University, Chongqing, China

P. Deng
Chongqing Jianzhu College, Chongqing, China

ABSTRACT

Point estimate method (PEM) is one of the simplest and most efficient approaches for the analysis of stochastic function or system. The precision of results and the efficiency of implement are two indexes to evaluate the performance of the point estimate algorithm, and the precision is more important. For most of PEMs, the precision depends on the number of nodes. However, the number of nodes is always determined subjectively and empirically. For example, both Zhao & Ono's method and Zhou & Nowak's method require n×m+1 nodes for function or system of n-dimension dependent variables, however, the number of nodes m used in the point estimates for functions of single random variables is always determined subjectively and empirically which is one of the main disadvantages of PEMs.

Table 1. The analysis times of general function for different methods.

	X_1	X_2	X_3	X_4	Total times
Zhao & Ono ($m=5$)	5				20
Zhao & Ono ($m=7$)	7				28
DIPEM	57	13	13	13	96
AIPEM	27	7	7	7	48

*X_1, X_2, X_3 and X_4 are the random variables of function.

Table 2. The precision of general function for different methods.

	μ_Y	$M_{Y,2}$	$M_{Y,3}$	$M_{Y,4}$
Zhao & Ono ($m=5$)	1.0000	0.9813	0.7326	0.2820
Zhao & Ono ($m=7$)	1.0000	0.9994	0.9619	0.6993
DIPEM	1.0003	1.0000	1.0000	1.0000
AIPEM	1.0003	1.0000	1.0000	1.0000

*the precision denotes the ratio of numerical values to exact values obtained by symbolic integration, and μ_Y, $M_{Y,2}$, $M_{Y,3}$ and $M_{Y,4}$ are the mean value, variance, third and fourth central moment, respectively.

In this work, two PEMs have been proposed to ensure the precision of results. One is the direct iterative point estimate method (DIPEM), in which all probability moments from different nodes are compared step by step until the results converge. The other is the adaptive iterative point estimate method (AIPEM), in which the degree of nonlinear for function is deduced via the difference of lower statistical moments from different nodes, then the number of necessary points is determined rationally and the corresponding moments are obtained.

Several numerical cases involving univariate polynomial function, multivariate polynomial function and general polynomial function are investigated to verify the precision and efficiency of proposed DIPEM and AIPEM. Table 1 and table 2 shows the results for the general polynomial function.

According to the analysis results, it can be found that the efficiency of existing PEMs is the best, but the results may be inaccurate, especially for higher statistical moments. The precision of DIPEM and AIPEM are always high enough, and AIPEM is more efficient relatively, naturally more practical. And the same conclusion can be obtained in other cases.

REFERENCES

Li, H.S., Lü, Z.Z. & Yuan, X.K. (2008) Nataf transformation based point estimate method. *Chinese Science Bulletin*, 53 (17), 2586–2592.

Miller III, A.C. & Rice, T.R. (1983) Discrete approximations of probability distributions. *Management Science*, 29 (3), 352–362.

Rahman, S. & Xu, H. (2004) A univariate dimension-reduction method for multi-dimensional integration in stochastic mechanics. *Probabilistic Engineering Mechanics*, 19 (4), 393–408.

Rosenblatt, M. (1952) Remarks on a multivariate transformation. *Annals of Mathematical Statistics*, 23 (3), 470–472.

Xu, H. & Rahman, S. (2004) A generalized dimension-reduction method for multidimensional integration in stochastic mechanics. *International Journal for Numerical Methods in Engineering*, 61 (12), 1992–2019.

Zhao, Y.G. & Ono, T. (2000) New point-estimates for probability moments. *Journal of Engineering Mechanics*, 126 (4), 433–436.

Life-Cycle Analysis and Assessment in Civil Engineering: Towards an
Integrated Vision – Caspeele, Taerwe & Frangopol (Eds)
© 2019 Taylor & Francis Group, London, ISBN 978-1-138-62633-1

An assessment of the inherent reliability of SANS 10162-2 for cold-formed steel columns using the Direct Strength Method

M.A. West-Russell, C. Viljoen & E. van der Klashorst
Department of Civil Engineering, Stellenbosch University, Stellenbosch, South Africa

ABSTRACT

Cold formed steel (CFS) columns may be designed using the direct strength method (DSM), a modern method recommended to replace the effective width method. DSM require as input the elastic buckling stability properties of the CFS section, which may be computed using a finite strip analysis.

Model factors that quantify the ability of DSM to predict experimental capacity for various section shapes are provided by Ganesan and Moen (2012).

The reliability of CFS columns is assessed in this contribution for a range of load ratios for self-weight, imposed- and wind loading. Load combinations are applied according to SANS 10160-1, an adaptation of EN 1990. A single pre-qualified C section is considered, with varied column lengths to allow investigation of both local- and global buckling dominated failures. A C section with a web-stiffener allowed investigation of distortional buckling failures. On the resistance side, only model uncertainty is accounted for probabilistically. The elastic buckling loads for local, distortional and global buckling are considered deterministic values. Self-weight and imposed loads are probabilistically described as per the recommendations of Holický (2009), while the South African probabilistic wind load model of Botha (2016) is used.

The global buckling mode yielded the lowest levels of reliability, with reliability indexes ranging from $\beta = 1.68$ for the STR-P load combination to $\beta = 2.78$ for the STR load combination. The low reliability is driven by a lack of reliability on the resistance side, due to fairly high model uncertainty. The sufficient safety margin of the loading side partially compensate for this when imposed or wind load components are significant in terms of the total loading. These low values should be cause for concern, especially where significant dead loads are supported. It is recommended that the capacity reduction factor for CFS compression members be revisited.

Life-Cycle Analysis and Assessment in Civil Engineering: Towards an Integrated Vision – Caspeele, Taerwe & Frangopol (Eds)
© 2019 Taylor & Francis Group, London, ISBN 978-1-138-62633-1

Failure probability estimation in high dimensional spaces

K. Breitung
ERA Group, Technical University of Munich, Munich, Germany

ABSTRACT

In high dimensional spaces standard methods for failure probability estimation as FORM/SORM run into problems.

First it is difficult to find all relevant design points, second the estimation methods using the curvatures are becoming inaccurate in such circumstances.

The subset simulation method—used for some years—is often described as very efficient for this task, but unfortunately it gives incorrect results for complex problems, since it cannot distinguish between local and global minimum distance points. But this is an essential prerequisite for algorithms calculating failure probabilities. So its use is problematic.

Here an idea for approximating failure probabilities in such circumstances is outlined. Instead of trying to find immediately estimates, it tries to keep control of the limit state function and its behavior.

This is done by minimizing it on centered hyperspheres. Since these are compact sets it is in general possible to locate the global minima there with reasonable numerical effort.

So in a nutshell, instead of finding design points by solving the equation system for the Lagrangian of the distance function under the condition that the limit state function is equl to zero.

Here now one solves the Lagrangian for the limit state function under the condition that the distance to the origin is constant.

Then further methods shall be applied for example dimension reduction and response surfaces around the found design points to find structural description of the limit state surface here.

Life-Cycle Analysis and Assessment in Civil Engineering: Towards an Integrated Vision – Caspeele, Taerwe & Frangopol (Eds)
© 2019 Taylor & Francis Group, London, ISBN 978-1-138-62633-1

Reliability based design of temporary structures

E. Vereecken, W. Botte & R. Caspeele
Ghent University, Ghent, Belgium

ABSTRACT

Currently, the design of scaffolds follows EN 12811 and several other 'codes of good practice'. It is, however, unclear whether these design guidelines follow a reliability-based approach similar as considered for the design of traditional long-term structures (see e.g. EN 1990). Therefore, the objective of this work is to derive partial factors for temporary structures, more specifically for façade scaffolds. These partial factors are to be based on appropriate safety levels, considering the short life time of these structures and the possible reuse of the elements. For this purpose, a probabilistic calculation method is set up, based on Latin Hypercube Sampling and FORM analyses. For these probabilistic calculations, a reference period of one year is adopted. The samples obtained from LHS are used as input for structural analyses performed in SCIA Engineer. The results from SCIA Engineer are used as input for the FORM analyses to determine the reliability index associated to failure due to yielding, the interaction of normal forces and bending moments or buckling.

In order to be able to evaluate the structural reliability levels obtained through the probabilistic calculations executed on different scaffolds, target reliability levels for the temporary structures under consideration were necessary. This target reliability can be based on a human safety criterion or on economic considerations. For human safety, three target reliability levels are proposed: 2.3, 2.9 and 3.1. When economic considerations are investigated, the minimum of the objective function as proposed by Rackwitz (2000) varies depending on the assumed value of the ratio of the failure cost to the initial cost, discount rate, scaffold class, among others. The most influencing parameter is the ratio of the failure costs to the initial costs. When varying this parameter, the reliability index corresponding to the absolute minimum of the objective function varies between 2.53 and 3.49. Nevertheless, in general, the objective function is quite constant around its minimum, thus slightly lower or higher values for β could also lead to a feasible economic optimal solution. If more detailed data on the parameters is available in

Table 1. Adjusted partial factors for the design of temporary structures based on an optimization procedure, the APFM and the suggested values.

β_t	Partial factor	Optimisation procedure	APFM	Suggested value
2.5	γ_G	1.35	1.34	1.35
	γ_Q	0.80	0.99	1.00
	γ_W	0.80	1.08	1.00
	γ_M	1.00	1.05	1.00
3.0	γ_G	1.0	1.41	1.35
	γ_Q	1.2	1.20	1.20
	γ_W	1.2	1.26	1.20
	γ_M	1.0	1.07	1.05
3.5	γ_G	1.35	1.49	1.50
	γ_Q	1.40	1.45	1.50
	γ_W	1.40	1.46	1.50
	γ_M	1.10	1.10	1.10

practice, a more detailed investigation of the target reliability can be performed. Nevertheless, the target value of β should not less than 2.3, which is the limit for human safety. For the determination of adjusted partial factors as executed in this contribution, three target reliabilities are considered: 2.5, 3.0 and 3.5.

Finally, two methods are applied to determine the partial factors: the Adjusted Partial Factor Method (APFM) (as proposed in *fib* Bulletin 80) and an optimisation procedure based on least square averaging. The results of both methods are compared in Table 1 to come to a final suggestion for the partial factors for the three safety levels. It should however be emphasized that further calculations are needed before coming to final recommendations.

REFERENCES

fib (2016) *fib* Bulletin 80: *Partial Factor Methods for Existing Concrete Structures*.
Rackwitz, R. (2000) Optimization—the basis of code-making and reliability verification. *Structural Safety*, 22 (1), 27–60.

Reliability-based analysis of tensile surface structures designed using partial factors

E. De Smedt, M. Mollaert & L. Pyl
Vrije Universiteit Brussel, Brussels, Belgium

R. Caspeele
Universiteit Gent, Ghent, Belgium

ABSTRACT

The design of tensile surface structures is based on expert judgement. A standardized design approach as for traditional buildings is not yet available for tensile surface structures. To achieve a partial factor framework similar to that for traditional buildings (i.e. within the frame of the Eurocodes), research into structural reliability calculations for tensile surface structures is needed. This paper studies a 6 m by 6 m hyperbolic paraboloid cable-net structure (fig. 1). The structure is designed using partial factors of which the partial factor for the material is investigated in more detail. Therefore, the cable-net is designed according to three material safety factors: 1.2, 1.5 and 2.0 and the reliability index is calculated for the catenary and arching cables for each structure subjected to either snow or wind load. As expected, an increase of the material safety factor results in an increase of the reliability index (Table 1).

If the structure is designed according to the Eurocode for a consequence class 2 and a 50 year reference period (material safety factor: 1.5, EN 1993-1-11) the reliability index of the cable-net proves to be 7.36 which is almost twice as high as the target reliability index 3.8 suggested by the Eurocode for the reliability based design (EN 1990). In this case no safety factor is taken into account for the pretension, if this factor would have been included in the calculations the section would increase and the reliability index would be even higher than 7.36. This implies that the design approach of the Eurocode for flexible structures should be further investigated.

The structural behavior of the cable-net under snow load, i.e. the increase of tension in the catenary cables and decrease of tension in the arching cables and the opposite under wind uplift, is also noticed in the distributions. Under snow load, the central safety factor of the arching cables 2.73 is significantly higher than for the catenary cables 1.36 and the opposite happens

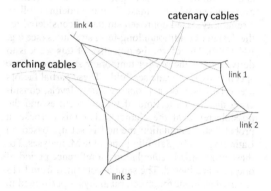

Figure 1. Perspective view of the cable-net. Designation of the cable-groups and turnbuckles (links 1 up to 4).

Table 1. Cross-section $A_{cc/ac}$ and reliability indices $\beta_{cc/ac}$, of the catenary cables (cc) and arching cables (ac) of the different cases.

				Snow		Wind	
Case	γ_m	A_{cc} mm²	A_{ac} mm²	β_{cc} (–)	β_{ac} (–)	β_{cc} (–)	β_{ac} (–)
1	1.2	8.33	10.44	4.00	33.47	16.98	5.51
2	1.5	10.51	13.14	7.36	40.39	23.18	8.67
3	2.0	14.20	17.64	12.15	41.89	31.32	13.48

under wind uplift, 1.65 for the arching cables and 2.17 for catenary cables.

REFERENCES

CEN (2001) EN 1990: Eurocode 0: Basis of structural design.
CEN (2006) EN 1993-1-11: Eurocode 3: Design of steel structures – Part 1–11: Design of structures with tension components.

GS2: Durability

The resistance to salt penetration of the high-strength fly-ash concrete used to make composite girder bridges

H. Ito
Toyama Prefectural University, Japan

K. Kubota, H. Kuriyama & T. Izumiya
Kawada Industrial, Inc., Japan

ABSTRACT

Concrete structures that are capable of performing safely for long periods are integral to the achievement of a sustainable society. The steel members inside composite girder bridges will often become corroded when they are exposed to salt for a long period. The purpose of this study is to investigate the salt penetration resistance of high-strength fly-ash (FA) concrete.

To achieve this purpose, a saltwater immersion test was conducted, using cylindrical specimens of FA concrete. In addition, a salinity penetration analysis was carried out determine the relationship between the depth of the concrete on composite girder bridges and salt damage.

A composite girder bridge is also called "pre-flexed beam bridge". The pre-flexed beams and concrete used to construct the bridge are pre-stressed. The concrete used to make bridges of this type must develop high strength quickly. It also needs to have a high level of workability, so it can be placed into narrow spaces.

For these reasons, the workability and mechanical strength of FA concrete were examined using the procedures before the salinity penetration test was performed. Then, specimens prepared according to several different formulations were immersed in saltwater for one year, to perform the saltwater immersion test (Figure 1). The diffusion coefficient of each type of specimen was calculated from the chloride ion concentration and distribution within the sample. The diffusion coefficients obtained were used to predict when embedded steel bars would start to corrode under varying salinity conditions, using a salinity penetration analysis program.

As the results of comparison of diffusion coefficients, changes in the diffusion coefficients of the specimens over time are shown in Figure 2. H-1, H-2 and N-2 are FA-free cases. HS-1, HF-1, NF-1, NF-2 and HF-2 are FA-added cases. From the results, it was confirmed that the diffusion coefficient of concrete with added FA becomes less as the immersion time increases. And, it was confirmed that the diffusion coefficients of concrete with added FA that was immersed for 9 months or 12 months were less than those of the FA-free concrete, indicating that the concrete with added FA has improved salinity penetration resistance.

As the summaries of salinity penetration analysis, the results demonstrated quantitatively that the addition of FA suppresses damage of concrete by salt. It was calculated that if concrete is placed 100 meters from the beach, with concrete covering steel bars to a depth of 70 mm, the start year for embedded steel bar corrosion would be delayed about 80 years, as compared with FA-free concrete.

Figure 1. Saltwater immersion situation.

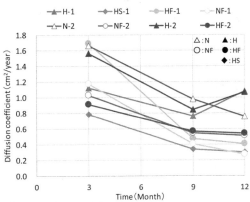

Figure 2. Relationship between the diffusion coefficient of chloride ions and the immersion period.

Influence of re-application of surface penetrant on progression of carbonation of concrete with surface penetrant

Y. Sakoi, M. Aba & Y. Tsukinaga
Department of Civil Engineering and Architecture, Hachinohe Institute of Technology, Japan

ABSTRACT

The durability of concrete structures is very important. To increase the durability of concrete structures, it is important to prevent penetration of some deterioration factors (CO_2, chloride ion, water and so on) from surrounding environment. In future, it is expected that the amount of structures that need maintenance will increase, and it is important to make maintenance planning and conducting with consideration of life-cycle-cost of structures.

In terms of the improvement of durability of concrete, the surface quality (denseness of concrete surface) of a concrete structure is most important to prevent the penetration of deterioration factors (water, ions, gases and so on) from the outside of the concrete structures. To improve the denseness of the surface layer of concrete, there are various factors, such as materials, construction method, curing method after casting of concrete and so on. In recent years, many surface penetrants have been examined as one of the methods to increase of durability of concrete structures, and these have been also applied to actual structures. There are many kinds of surface penetrants, and its modified effects and modified mechanism are different depending on its main constituents. However, any surface penetrant modifies a concrete surface and prevents penetration of some deterioration factors from the surrounding environment.

The main objective of this study is to examine the influence of timing of a painting of a surface penetrant on durability for painted concrete and to examine the effect of re-painting and re-deterioration for painted concrete. The investigation of the influence of timing of a painting and re-painting of penetrant is the objective of this study.

Some of the results are shown in Figure 1 and 2. Figure 1 shows the influence of the timing of applying a Na type surface penetrant for fresh concrete, and Figure 2 shows the influence of the timing of re-applying a penetrant.

The prevention effect of carbonation of concrete was different with the timing of painting, penetrant type, re-applying or not and so on. However, in any cases, it was clear that the effect of prevention of carbonation was higher due to applying a penetrant earlier.

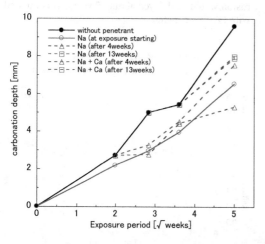

Figure 1. Influence of the timing of applying Na type penetrant.

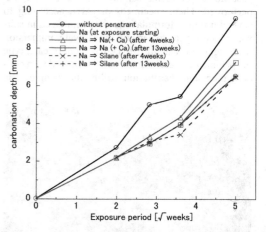

Figure 2. Influence of re-applying a penetrant (Na type + Na type, Na type + Silane type).

Formulation of the conditions of the destruction of the passivation film of steel bar by chloride ion in high pH environment

N. Hashimoto & Y. Kato
Tokyo University of Science, Noda, Japan

ABSTRACT

Steel bars in concrete structures usually maintain a passivation state due to high pH in the concrete pore solution. However, it is known that the passivation film of a steel bar in concrete is destroyed by chloride attack in a marine environment. Among past studies in the field of concrete engineering, there is little theoretical explanation of the mechanism of breakdown of the passivation film around a steel bar; therefore, a theoretical model of the mechanism needs to be developed. This paper derives such a theoretical model and verifies its validity based on experimental data.

The relationship between the critical chloride ion concentration for passivation film breakdown, the steel bar potential and pH can be expressed as:

$$a_{Cl^-} \geq A^{\frac{\chi}{2}pH} \exp(-BE) \qquad (1)$$

where a_{Cl^-}, E, A and B are the concentration of chloride ions concentration [mol/L], the steel bar potential [V] and constants, respectively. When the steel bar potential is constant, the relationship between pH and critical chloride ion concentration for passivation film breakdown is as shown in Figure 1. The gradients of the plotted lines are almost constant, not depending on the steel bar potential. When the steel bar potential is constant, the relationship between the steel bar potential and the critical chloride ion concentration for passivation film breakdown is as shown in Figure 2. The gradients of the plotted lines are constant, not depending on the pH.

As a result, it can be concluded that the relationship between the steel bar potential and the critical chloride ion concentration for passivation film breakdown is an exponential relation, and the relationship between the pH and the critical chloride ion concentration for passivation film breakdown is a linear relation.

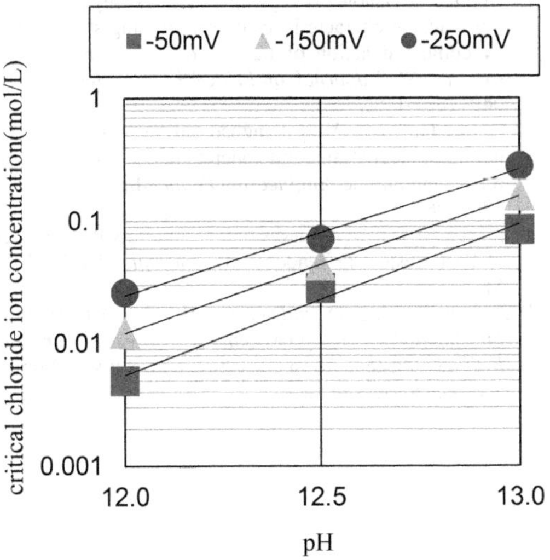

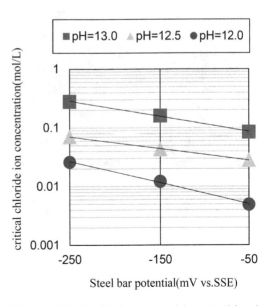

Figure 1. Relationship between pH and critical chloride ion concentration for passivation film breakdown.

Figure 2. Relationship between steel bar potential and critical chloride ion concentration for passivation film breakdown.

Simulating low-frequency and long-term fatigue loading for life-cycle structures

F. Li & J. Zhao
Jiangsu Key Laboratory of Environmental Impact and Structural Safety in Engineering, China University of Mining and Technology, Xuzhou, Jiangsu, China

ABSTRACT

Environmental action and the effect are the main issue of life-cycle structures, but an important reminding is necessary to be drawn that, the environmental action does not exist solely; in fact, it exists together with some mechanical actions, which may result in coupled effects of environment and mechanics, called corrosion fatigue damage/failure. To reveal the complex mechanism of the coupled effect, experimental simulation is necessary, in which the long-term combined action of environment and mechanics is needed to be applied for the specimens. So far, the main difficulty in the long-term simulation of the combined action is to conveniently apply the long-term fatigue load to investigate the issue of corrosion-fatigue or erosion-fatigue. To solve this difficulty, the authors have invented a mechanical fatigue machine which is suitable for long-term fatigue loading. The invented machine has the advantages of simple structure, easy processing, convenient and quick assembly, low cost and stable long-term performance. It is convenient to realize fatigue loading with multi-specimen and multi-amplitude at the same time and suitable for long-term fatigue loading tests. The machine has successfully applied in several experiments with long-term combined action of corrosion/erosion and fatigue. In this paper, the invented fatigue machine and the applications, i.e., a long-term fatigue loading for corrosion-fatigue issue of beam specimens, a long-term fatigue loading for corrosion-fatigue issue of bond specimens, and a long-term fatigue loading for erosion-fatigue issue of little column specimens, are introduced.

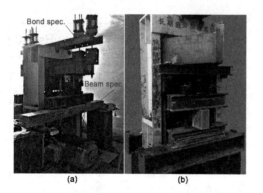

Figure 1. Long-term fatigue loading for: (a) four bond specimens and two beam specimens, and (b) two little column specimens.

REFERENCES

Li Fumin, Fan Li & Yuan Yingshu. A kind of fatigue loading machine. Chinese Patent: 201110110971.8. 2011.04.29. (in Chinese).
Li Fumin, Luo Xiaoya, Wang Kejin, Ji Yongsheng (2017) Pitting damage characteristics on prestressing steel strands by combined action of fatigue load and chloride corrosion. *ASCE Journal of Bridge Engineering*, 22 (7), 04017023-1-12.
Li Fumin, Qu Yaxu, Wang Jianghao (2017) Bond life degradation of steel strand and concrete under combined corrosion and fatigue. *Engineering Failure Analysis*, 80 (10), 186–196.
Schupack, M. (1978) A Survey of the Durability Performance of Post-tensioning Tendons. *ACI Journal*, 75 (10), 501–510.
Wang Jianghao (2017) *Study on the Chloride Penetration in Interfacial Zones of Epoxy Joints of the Precast Segmental Bridges*. Master's Thesis of China University of Mining and Technology, Xuzhou.

Study on the spatial distribution of the chloride ion supply in the superstructure of an open-type wharf

Y. Tanaka, Y. Kawabata & E. Kato
National Institute of Maritime, Port and Aviation, Port and Airport Research Institute, Kanagawa, Japan

ABSTRACT

Reinforced concrete structures in ports and harbors are exposed to severe environments in terms of chloride-induced deterioration. However, it is difficult to precisely predict the chloride ion penetration in concrete because such penetration is spatially varied not only in open-type wharfs, but also in structural members.

The aim of this study is to discuss the factors behind the spatial variability of penetration of chloride ions. In this study, the chloride ion content was measured at many points in two slabs that were extracted from the same 30-year-old open-type wharf. The factors that caused the variability in the chloride ion supply were then discussed with reference to the obtained data for chloride ion content and the estimated distribution of the surface chloride ion content.

Figures 1 and 2 show the estimated distribution of C in Slab-L and Slab-S, respectively. The white circles denote the investigated points. The values of C_0 differed significantly in two slabs. The distributions of C_0 were spatially varied and the distributions along the back-side of both slabs were relatively higher than those at both the middle and front-side of the slabs. In other words, regardless of the location of the slabs, the quantity of chloride ion supply at the back-side of a slab was relatively larger than that at the middle and front-side of the slabs.

The distribution of C_0 was spatially varied in a slab and the quantity of the chloride ion supply for a slab differed significantly between the investigated slabs. The factors of the variability of spatial distribution of C_0 are the location of the slab in a wharf, the environmental conditions such as sea levels and the other structural members surrounding the slab. It is important to take the factors of the variability of spatial distribution of C into consideration when administrators or investigators of a wharf decide the area for investigation and monitoring. This makes the maintenance work of a wharf more efficient.

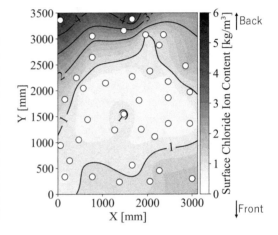

Figure 1. The estimated distribution of the surface chloride ion content of Slab-L.

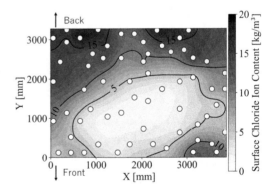

Figure 2. The estimated distribution of the surface chloride ion content of Slab-S.

GS3: Concrete structures

Application of 2D micro-scale image analysis on concrete surface for evaluating concrete durability under various environments

T. Chlayon, M. Iwanami & N. Chijiwa
Tokyo Institute of Technology, Tokyo, Japan

ABSTRACT

Many studies in marine concrete durability implied that concrete surface treatments immensely affected permeability resistance and the attached biomass amounts (Lv et al. 2015, Coombes et al. 2017). However, the parameters for describing the concrete skin appeared limited. Recently, numbers of research attempted to implement 2D-image analysis on a concrete surface (Ozen & Guler 2014, Han et al. 2016), yet the studies focusing on microscales remained limited. This paper introduced methods for obtaining concrete skin parameters by comparing micro-image data from four concrete surface conditions: normal, rough by bleeding, emulated abrasion, and acidic attack (Fig. 1).

The backscattered electron image (BEI) indicated that the four cases had vast changes in microcracks and cement-hydrate particles on the concrete skin. Thus, the parameters regarding microcracks and particle size distribution became eligible for identifying the differences. In addition, applying certain image filters on the captured images resulted in significant improvements in the image threshold output.

At ×100 magnification, there was dissimilar in microcrack length and thickness among the regular surface, surface with an abrasion effect, and the surface under acidic attack. Microcracks became wider and significantly increased in number when the samples were subjected to acid solutions. Also, the particle size distribution analysis indicated that the rough surface had the highest particle amounts, while the abrasive surface had more considerable particle length than the standard case.

Data acquired by the proposed method agreed well with the visual inspection of the BEIs. Future studies might include investigation at other magnification scales, and later on attempted utilizing these to establish relationships among the surface microcrack amounts, particle size distribution, and other durability factors such as water absorption rate, residual surface chloride or chloride diffusion coefficient. The proposed parameters could be widely used in many surface-related durability studies of concrete including chloride ingress, acidic attack, wetting-drying, biofouling, and scouring issues.

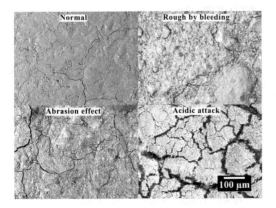

Figure 1. Concrete surfaces types; normal rough by bleeding, applying abrasive force, acidic attack.

REFERENCES

Coombes, M.A. Viles, H.A. Naylor, L.A. & La Marca, E.C. (2017) Cool barnacles: Do common biogenic structures enhance or retard rates of deterioration of intertidal rocks and concrete? *Science of the Total Environment*, 580, 1034–1045.

Han, J., Wang, K., Wang, X. & Monteiro, P.J.M. (2016) 2D image analysis method for evaluating coarse aggregate characteristic and distribution in concrete. *Construction and Building Materials*, 127, 30–42.

Lv, J.F., Mao, J.Z. & Ba, H.J. (2015) Influence of Crassostrea gigas on the permeability and microstructure of the surface layer of concrete exposed to the tidal zone of the Yellow Sea. *Biofouling*, 31 (1), 61–70.

Ozen, M. & Guler, M. (2014) Assessment of optimum threshold and particle shape parameter for the image analysis of aggregate size distribution of concrete sections. *Optics and Lasers in Engineering*, 53, 122–132.

Robustness of flat slabs against progressive collapse due to column loss

T. Molkens
Sweco Belgium nv, Hasselt, Belgium

ABSTRACT

Current design codes are requesting for a safe building in the ultimate limit state and in the accidental load case. The last one can be guaranteed by adding tying reinforcement, or by checking the structure against the consequences of column loss. It seems that flat slabs can develop even on an elastic way an alternative load path. By doing so, the floor slabs behave elastically which makes that no extra ordinary costs are needed to keep the adjacent structures in service and to recover the building to his original safety level. Several aspects were studied, results are shown and where possible easy handy rules for the design are presented which doesn't require much efforts from the designer

Besides a general introduction to the problem situation a first part of the article concerns the modeling of the structure by the aid of FEM. Mesh size, column discretization, reinforcement determination and material parameters are discussed. Several verifications or validations are done where the model is compared with analytical known solutions.

First important topic which is more investigated in depth is the so-called aspect ratio, which is equal to the ratio of the long span divided by the small one from a rectangular panel the links four columns. Out of the obtained deformations it can be concluded that the slab will react mainly elastically, so the simple bending theory can fit, and straight forward calculations can be used to investigate the problem of an internal column loss. Due to the torsional bending moments it comes out that finally the total amount of reinforcement which is parallel to the long direction of the panel A_y will be always sufficient and if the maximum reinforcement perpendicular to the previous one (A_x) is used over the whole panel also this one can cover positive bending and torsional moments in case of column loss. In Figure 1 this reinforcement is called "practical", also provisions out of EN 1991-1-7 are shown.

Influence of the column dimensions (or stiffness parameters) are also investigated on the bending

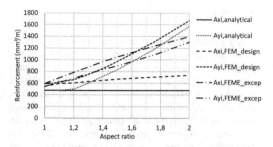

Figure 1. Total amount of lower reinforcement Ax in a panel.

behaviour and reaction forces but this seems to ne not important. However, the influence of this on the punching force, needed for this verification, is important and indicates that for non-slender ($k > 0{,}06$ = ratio column size/span) an extra punching verification is needed for such an exceptional load case. Besides this also the influence of load ratio, precast plank profiles, voiding elements and costs are briefly discussed.

With a very limited cost of about 1.5% of the structural costs, flat slabs can be provided for a reinforcement which guarantee that the slab react elastically with column loss.

REFERENCES

Blaauwendraad, J. (2010) *Plates and FEM*. Dordrecht, Springer.

Droogné, D., Caspeele, R., Taerwe, L. & Herraiz, B. (2017) Parametric study and reliability-based evaluation of alternate load path in reinforced concrete slabs. In: *39th IABSE Symposium, Vancouver, 21–23 September 2017*.

Girkmann, K. (1946) *Flächentragwerke*. Wien-New York, Springer Verlag.

Gulvanessian, H., Calgaro, J.-A. & Holicky, M. (2012) *Designers' Guide to Eurocode*. London, ICE Publishing.

Timoshenko, S.P. & Woinowsky-Krieger, S. (1970) *Theory of Plates and Shells*. Signapore, McGraw-Hill.

Life-Cycle Analysis and Assessment in Civil Engineering: Towards an Integrated Vision – Caspeele, Taerwe & Frangopol (Eds)
© 2019 Taylor & Francis Group, London, ISBN 978-1-138-62633-1

Prestressed concrete roof girders: Part I—deterministic and stochastic model

A. Strauss & B. Krug
University of Natural Resources and Life Sciences, Vienna, Austria

O. Slowik & D. Novák
Brno University of Technology, Brno, Czech Republic

ABSTRACT

Shear strength of concrete beams has been the subject of many controversies and debates and it has been extensively studied over the last five decades. The ultimate capacity of reinforced concrete beams subjected to combined shear and flexure is affected by many phenomena and uncertainties, such as existing multiaxial states of stresses, the anisotropy induced by the diagonal concrete cracking, the interaction between concrete and reinforcement (bond) and the brittleness of the failure mode. A large number of shear tests have been performed during the last decades in order to obtain valuable information about the shear transfer mechanisms.

But there is still a lack of clarity e.g. with respect to the effects of normal-shear force interaction and the material properties on the shear performance. Therefore, it was of high interest to shed more light on the normal-shear force interaction of wide-span prestressed reinforced concrete lightweight roof elements by means of innovative probabilistic numerical analyses and experimental testing. Extensive experimental studies on small specimens, and small and full scale beams have been performed in order to comply the required information for complex material laws as they are implemented in advanced probabilistic nonlinear numerical analyses. It was understood from the early beginning that to develop very good numerical model for comparison with experiment of real structure would be impossible without the proper knowledge of fracture-mechanical parameters. Therefore comprehensive experimental study of material parameters was performed. By means of an advanced identification approach based on artificial neural network modelling the following three fundamental parameters of concrete were identified: modulus of elasticity E_c, tensile strength f_{ct}, and specific fracture energy G_f. The compressive strength of concrete f_c was measured by means of standard cubic compression tests.

In addition to the probabilistic analysis of the shear normal-force interaction and the associated safety considerations, another objective of this paper is to present the complex approach for computational modeling of destructive tests based on data from fracture mechanical experiments. The framework developed in this context will be presented on the benchmark of scaled T shaped and full scale LDE7 TT roof girders, that were produced by Franz Oberndorfer GmbH & Co KG (Strauss et al. 2017). The shear destructive test under laboratory conditions were performed on ten scaled girders, whereas Figure 1 displays the whole testing, modeling and assessment procedure. Finally, the results of this research support in establishing robust deterministic and stochastic modeling techniques, see Section 5, which are also the basis for a reliability-based optimization of this type of girders and for advanced statistical methods.

In this paper – Part I, FEM computational modelling and basic statistical modelling is described. Computational models were used for sensitivity analysis described in the part II (Lehký et al. 2018) and analysis using semi-probabilistic as well as fully probabilistic approaches is described in the part III (Novák et al. 2018) of a three-paper series.

REFERENCES

Lehký, D., Novák, D., Novák, L. & Šomodíková, M. (2018) Prestressed concrete roof girders: Part II surrogate modeling and sensitivity analysis. In: *Proceedings of IAELCCE Conference*. Ghent, Belgium (in press).

Novák, D., Novák, L., Strauss, A. & Slowik, O. (2018) Prestressed concrete roof girders: Part III – Semi-probabilistic design. In: *Proceedings of the IAELCCE Conference*. Ghent, Belgium (in press).

Strauss, A., Krug, B., Slowik, O. & Novak, D. (2017) Combined shear and flexure performance of pre-stressing concrete T-shaped beams: Experiment and deterministic modeling. *Structural Concrete*, 1–20. https://doi.org/10.1002/suco.201700079.

Prestressed concrete roof girders: Part II—surrogate modeling and sensitivity analysis

D. Lehký, D. Novák, L. Novák & M. Šomodíková
Brno University of Technology, Brno, Czech Republic

ABSTRACT

The paper describes a particular part of complex stochastic modeling and design of a TT-shaped precast prestressed concrete girder failing in shear, namely surrogate modeling and sensitivity analysis. Both were used in order to reduce high computational effort related to utilization of a 3D nonlinear FEM model created in GID-ATENA Science software environment(Červenka et al. 2007).

Two types of surrogate models have been developed: (1) artificial neural network model (ANN, Lehký & Šomodíková 2017) and (2) polynomial chaos expansion model (PCe, Ghanem & Spanos 1991). The parameters of both models were set up using 30 nonlinear FEM simulations of girder with random samples generated by LHS method from uncorrelated stochastic model. For model validation, additional 30 simulations were used, in this case generated from correlated stochastic model. Both models were used for sensitivity analysis described in this paper (part II) and also for structural optimization and detailed statistical and reliability analysis using semi-probabilistic as well as fully probabilistic approaches described in the part III of a three-paper series.

In case of sensitivity analysis, three methods were utilized and compared: (i) Spearman non-parametric rank-order statistical correlation sensitivity, (ii) sensitivity analysis in terms of coefficient of variation (CoV), and (iii) sensitivity analysis in terms of Sobol sensitivity indices obtained by postprocessing of PCe. A comparison of the selected results for all of the studied sensitivity methods is depicted in Figure 1. Note that Spearman-U corresponds to uncorrelated data set. From the results, the following selected conclusions can be drawn:

- The most dominant group of parameters is concrete material as the dominant failure mode of this prestressed girder is shear.
- Statistical correlation certainly played a role in the detection of sensitivity.
- Comparing the three sensitivity methods, some small differences among individual methods

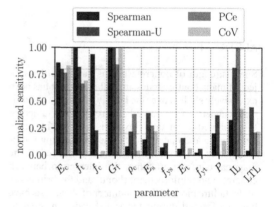

Figure 1. Comparison of sensitivity factors obtained using various methods.

occured due to their different principles. For example, Sobol indices reflect the interactions between parameters to the model output variance compared to CoV method.

The obtained information was used to set up a stochastic model and surrogate models in an optimum manner and was employed in the subsequent determination of selected uncertain design parameters followed by load-bearing capacity and reliability assessment using semi-probabilistic as well as fully probabilistic approaches.

REFERENCES

Červenka, V., Jendele, L. & Červenka, J. (2007) *ATENA Program Documentation – Part 1: Theory*. Prague, Czech Republic, Cervenka Consulting.

Ghanem, R.G. & Spanos, P.D. (1991) *Stochastic Finite Elements: A Spectral Approach*. Berlin, Springer.

Lehký, D. & Šomodíková, M. (2017) Reliability calculation of time-consuming problems using a small-sample artificial neural network-based response surface method. *Neural Computing and Applications*, 28, 1249–1263.

Prestressed concrete roof girders: Part III—semi-probabilistic design

D. Novák, L. Novák & O. Slowik
Institute of Structural Mechanics, Faculty of Civil Engineering, Brno University of Technology, Czech Republic

A. Strauss
Institute of Structural Engineering, University of Natural Resources and Life Sciences, Austria

ABSTRACT

The paper describes integration/application of the modelling of nonlinearity and uncertainty to predict shear failure behaviour of prestressed concrete girders. The whole approach is complex, going from fracture-mechanical laboratory experiments and advanced deterministic 3D computational modelling of girders to stochastic modelling, sensitivity and surrogate modelling, as described in Part I and Part II. The aim was to assess the variability of shear response and to present and verify alternative design procedures in comparison with fully probabilistic design.

The combination of non-linear finite element method and reliability analysis is strong tool for realistic modelling of structures. On the other hand, it is still highly time consuming to perform huge non-linear stochastic models with many stochastic input variables. Due to this fact, structural designers are interested in semi-probabilistic methods to determine design value of response, which are able to greatly reduce number of needed nonlinear calculations.

Beside normative methods Partial Safety Factors (PSF) and global safety factor according to EN 1992-2, it is shown how to determine design value of ultimate shear capacity by ECoV method (Červenka 2013), ECoV modified by Schlune et al. (2012) and by numerical quadrature – point-estimate method suggested by Rosenblueth (1975).

The most accurate, on the other hand the most time consuming approach, is fully probabilistic method based on Monte Carlo (MC) type simulation. In this case it is necessary to identify type of distribution function and moments of response, for this purpose we can use commonly known Latin Hypercube Sampling method (LHS). LHS was performed by FReET (Novák et al., 2014), multipurpose probabilistic software for statistical, sensitivity and reliability analysis of engineering problems.

The range of design values obtained is really large (from 122 kN to 210 kN). Normative approaches are naturally conservative – lowest 3 values. Advanced reliability approaches resulted in high design values (highest 3 values). Fully probabilistic method

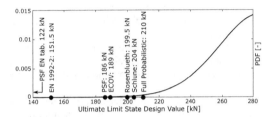

Figure 1. Comparison of ultimate limit state design values determined by safety formats and advanced reliability methods.

determined the highest R_d. The most efficient method in this application was ECoV modified by Schlune, but when full stochastic model was assumed the standard ECoV by Červenka would be much more suitable regarding computational demand of needed simulations and obtained accuracy of estimated R_d. Semi-probabilistic approaches represent generally useful tools to determine design value of resistance without time-consuming statistical simulation. They provide usually very good results and can be recommended as superior technique in comparison with normative approaches.

REFERENCES

Červenka, V. (2013) Reliability-based non-linear analysis according to fib Model Code 2010. *Structural Concrete Journal fib*, 14, 19–28.

Novák, D., Vořechovský, M. & Teplý, B. (2014) FReET: Software for the statistical and reliability analysis of engineering problems and FReET-D: Degradation module. *Advances in Engineering Software*, 72, 179192, ISSN: 0965–9978.

Rosenblueth, E. (1975) Point estimates for probability moments. In: *Proceedings of the National Academy of Sciences USA*, 72, 3812–4.

Schlune, H., Gylltoft, K. & Plos, M. (2012) Safety formats for non-linear analysis of concrete structures. *Magazine of Concrete Research*, 64, 563–574. London, The Geological Society.

Basalt fiber for strengthening of compressed structural elements in concrete and reinforced concrete: Finite element modeling

T. Zhelyazov
Technical University of Sofia, Sofia, Bulgaria

The Basalt Fiber Reinforce Polymer (BFRP) is a relatively new, promising material that can be used in line with the carbon or glass fiber reinforced polymers for construction.

The contribution is focused on the numerical simulation of the mechanical response of concrete specimens loaded in compression and externally strengthened by BFRP.

A hybrid constitutive model is used to model the mechanical response of the concrete: a multilinear isotropic hardening for compression and a damage-based model

$$D = 1 - \frac{\varepsilon_{\inf}(1 - A_t)}{\varepsilon_{eqv}} - \frac{A_t}{e^{B_t(\varepsilon_{eqv} - \varepsilon_{\inf})}} \quad (1)$$

$$\varepsilon_{eqv} = \sqrt{\varepsilon_1^2 + \varepsilon_2^2 + \varepsilon_3^2} \quad (2)$$

– for tension (Figure 1).

The unidirectional basalt fiber fabric is modeled as a material of transverse isotropy (i.e. the compliance matrix for BFRP has five independent constants).

After having defined the constitutive relations on the mesoscale (within the finite element), the response of the strengthened element on the macroscale is obtained by finite element analysis (Figure 2).

In the employed incremental procedure, for each value of the driving force parameter a nonlinear stating analysis based on the Newton – Raphson procedure is performed.

If the damage model is activated for a given finite element the damage variable is evaluated on the basis of the stress and strain distributions obtained for the current value of the applied load. If the upper damage threshold is reached, the finite element is deactivated (initiation of a mesocrack).

The constitutive models will be used subsequently in numerical simulation of other structural elements such as reinforced concrete columns of rectangular cross-section and concrete or reinforced concrete beams with internal and external reinforcement in composite material.

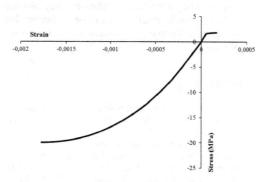

Figure 1. Concrete: a hybrid material model.

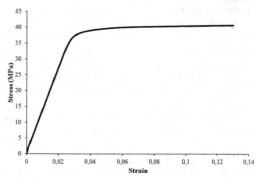

Figure 2. Stress – strain relationship obtained by using the damage – based constitutive model.

GS4: Nonlinear analysis and structural optimization

Optimized design and life cycle cost analysis of a duplex welded girder bridge

B. Karabulut & B. Rossi
KU Leuven, Sint-Katelijne-Waver, Belgium

G. Lombaert
KU Leuven, Leuven, Belgium

D. Debruyne
KU Leuven, Gent, Belgium

ABSTRACT

Stainless steel, as Chromium Nickel (Cr-Ni) alloys, has recently gained increased interest for construction owing to the combination of excellent corrosion resistance and mechanical strength. In particular, duplex grades, with a balanced austenite ferrite microstructure, show greater proof strength and ductility than that of standard austenitic stainless steel groups. The increased interest in stainless steel is also due to recent research demonstrating the good fatigue resistance of duplex welded components, raising an even higher interest in bridges made of these grades in corrosive environments. The EN 1.4162 and 1.4062 duplex grades are characterized by a lower Nickel and Molybdenum content which results in a more stable cost, but at the same time more prone to pitting corrosion. Additionally, reference research today exists proving that those grades have comparable corrosion resistance to the austenitic stainless steel grades EN 1.4307 and EN 1.4404 (Merello et al. 2003).

The ultimate objective of this study is to calculate the possible weight reduction in an existing carbon steel girder bridge, when lean duplex welded components are used. The considered reference case is a highway bridge, replacing the designs already performed by a civil engineering design company named "GRID International" for both S355 and S690 carbon steels (Pedro et al. 2017). The benefit from the greater mechanical properties of the latter are considered according to the latest version of EN 1993-1-4 (2015).

The fatigue design is made using the hot-spot stress method combined with finite element models (FEM) to assess the local stress distribution, considering eight critical details along the girder (Fig. 1). Computation of the hot spot stress is dependent on the FE model topology. Therefore, a sensitivity analysis was carried out beforehand, investigating several FEA parameters such as: mesh size; mesh topology; element type; integration method; detail length; extrapolation method and influence of actual modelling of the butt welds. The highest sensitivity is obtained in mesh size study.

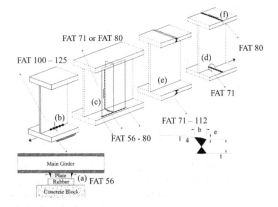

Figure 1. Scheme of the selected critical details.

The fatigue life of all the considered details were found to be satisfactory with a higher number of loading cycles than the design value proposed by EN 1993-1-9 (2005).

Based on previous published research of Rossi et al. (2017) on the life cycle cost assessment of painted and hot-dip galvanized steel bridges, the initial costs of the carbon and the stainless steel option are compared, as well as the total Net Present Value at the life horizon of the bridge.

REFERENCES

Merello, R., Botana, F.J., Botella, J., Matres, M.V. & Marcos, M. (2003) Influence of chemical composition on the pitting corrosion resistance of non-standard low-Ni high-Mn-N duplex stainless steels. *Corrosion Science*, 45 (5), 909–921.

Pedro J.O., Reis A. & Baptista C. (2017) *High Strength Steel (HSS) S690 in Highway Bridges: Comparative Design.* Copenhagen, Eurosteel 2017.

Rossi, B., Marquart, S. & Rossi, G. (2017) Comparative life cycle cost assessment of painted and hot-dip galvanized bridges. *Journal of Environmental Management*, 197, 41–49.

Decision criteria for life cycle based optimisation in early planning phases of buildings

C. Dotzler, P. Schneider-Marin, C. Röger & W. Lang
Institute of Energy Efficient and Sustainable Design and Building, Technical University of Munich, Munich, Bavaria, Germany

ABSTRACT

Overall, the building sector is responsible for 33% of the global final energy consumption and nearly 25% of the greenhouse gas emissions (Dean et al. 2016, p. 4). To restrict the global warming to 2°C, the CO_2-emissions have to be reduced by 60% in 2050 compared to 2012, for example (IEA 2015). Thus, the key EU targets until 2030 are the increase in energy efficiency of 27% and the coverage of the total energy consumption with renewable energy (European Commission 2017).

But what does this mean for planning new buildings and how can energy consumption and CO_2-emissions be reduced in practice? The reduction of energy demand for the use phase has a high impact. But with the optimisation of this phase to a nearly zero energy standard the embodied energy and environmental impacts of the building construction and services come to the fore. Thus, there must be a certain discussion about dependencies of different ecological and economic impacts as well as about the definition of main drivers.

The research project 'Design2Eco' evaluates data from recent office buildings to detect the crucial building components which influence the ecological and economic performance the most. The focus is on detecting optimisation potentials which can be initiated at early design stages or could still be influenced easily in early planning phases. This should avoid disadvantageous planning decisions which are difficult to reverse and which have a high impact on the ecological or economic quality of the building during the whole life cycle.

The study of Life Cycle Assessment (LCA) und Life Cycle Costing (LCC) shows, that there are certain dependencies on architectural design and there is a lack of ecological information on data of building services.

The building parts that can be identified as most influential on the LCC are floors and ceilings. They

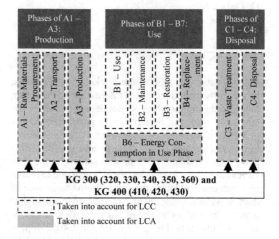

Figure 1. Considered life cycle phases and cost types at LCA and LCC referred to (DIN EN 15804, DIN 276-1).

also play a significant role for the LCA. Thus, they are a strategic parameter independent of the building geometry for two main reasons: their significant surface area and the high exchange rates of the surfaces.

REFERENCES

Dean, B., Dulac, J., Petrichenko, K. & Graham, P. (2016) Towards zero-emission efficient and resilient buildings. Global Status Report 2016. In: *21st Conference of Parties (COP21)*. Available from: https://wedocs.unep.org/bitstream/handle/20.50011822/10618/GABC-Report_Updated.pdf?sequence=1&isAllowed= [Accessed 11 December 2017].

European Commission. (07.01.2018). *EU Climate Action.* Available from: https://ec.europa.eu/clima/citizens/eu_en [Accessed 7 January 2018].

IEA (2015) *Energy Efficiency Market Report 2015.* Available from: https://www.iea.org/publications/freepublications/publication/MediumTermEnergyefficiencyMarketReport2015.pdf [Accessed 11 December 2017].

Nonlinear reliability analysis of RC columns designed according to Chinese codes

D.-G. Lu, J.-S. Wang & Z.-M. Chang
School of Civil Engineering, Harbin Institute of Technology, Harbin, Heilongjiang, China

ABSTRACT

In most of the previous studies, the reliability analysis of RC columns only subjected to the combined axial compression and uniaxial bending was considered (Ruiz & Aguilar 1994, Diniz & Frangopol 1997). It was assumed that the axial load and the bending moment are perfectly correlated, and the load eccentricity is a fixed deterministic quantity. These assumptions may lead to overestimation or underestimation of the reliability analysis results of RC columns. Frangopol et al. (1996) studied the change rule of reliability of RC columns under the random correlation of bending moment and axial force.

This paper presents an investigation on assessing the safety levels as well as main influencing factors in reinforced concrete (RC) columns designed according to the current Chinese design code of concrete structures (MOHURD 2010). For analyzing the nonlinear properties of RC columns subjected to both compression and bending, the fiber element model is used for generating the moment-curvature (M-φ) curves. The columns with four kinds of sections designed according to the current code provisions are analyzed by using the nonlinear section analysis program developed in HIT, and the M-φ curves are obtained for different axial compression ratio. The M-φ curves are then used to generate the load-deflection curves of these columns by way of incrementally increasing curvature. The ultimate moment-axial force (M-N) relation curves are furthermore sought to generate failure surfaces under sequential and proportional load paths.

To investigate the statistical properties of M-N relation, six basic random variables are considered, i.e. the yielding strength and strain of concrete, the limit strength and strain of concrete, the yielding strength and the elasticity modulus of steel. Six hundred columns are analyzed by using the Monte Carlo method. The M-N relation curve is shown in Figure 1. It is found that the peak stress and strain at the core concrete as well as the yielding strength of steel have the most effects on the variability of M-N interaction relation. The iso-reliability contours of RC columns with different load ratio and eccentric distance are generated using the Monte Carlo method. Finally, the effects of concrete strength, reinforcement ratio and stirrup ratio on the M-N relation are examined. Reliability

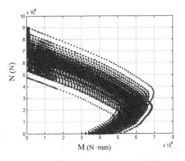

Figure 1. Random M-N curves.

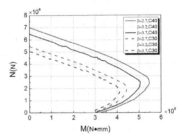

Figure 2. Reliability contour.

contour of different concrete strength is shown in Figure 2. The results of this study can be used to support the development of improved evaluation and design specifications for RC columns.

REFERENCES

Diniz, S. & Frangopol, D. (1997) Reliability bases for high-strength concrete columns. *Journal of Structural Engineering*, 123 (10), 1375–1381.

Frangopol, D., et al. (1996) A new look at reliability of reinforced concrete columns. *Structural Safety*, 18 (2–3), 123–150.

Ministry of Housing and Urban-Rural Development of the People's Republic of China (2010) GB50010-2010 Chinese design code of concrete structures. (in Chinese).

Ruiz, S. & Aguilar, J. (1994) Reliability of short and slender reinforced – concrete columns. *Journal of Structural Engineering*, 120 (6), 1850–1865.

Risk analysis for the impact on traffic sign bridges

T. Braml
HFR Ingenieure GmbH, Munich, Germany

M. Keuser
University of German Armed Forces, Munich, Germany

S. Petry
HFR Ingenieure GmbH, Munich, Germany

ABSTRACT

Traffic sign bridges are structures meant to carry traffic signals. They are often built as single-span structures, based on reinforced concrete foundations and are arranged directly next to the roadways. Because of their exposed location, traffic sign bridges represent a risk in regard to a vehicle crash. In the framework of a research project, impact events of recent years on German motorways are evaluated by a qualitative and quantitative risk analysis according to DIN EN 1991-1-7. For the qualitative risk analysis, a hazard matrix is developed taking into account different factors. Expectation values are developed and overall risk factors are determined. For the quantitative risk analysis, failure probabilities and sequences are evaluated and the risk of a vehicle impact on traffic sign failure is computed. By the use of dynamic transient computation methods, different crash scenarios are simulated on a reference traffic sign bridge, showing until which limits the reference structure can withstand the impact.

Figure 2. View of a retaining bloc.

Finally, the importance of constructive measures is discussed with regard to the previous mentioned risk of a vehicle impact.

At traffic sign bridges dynamic FE calculations with the program system Siemens NX as well as probabilistic calculation with Strurel were carried out. Based on the evaluation of impact events at the traffic sign bridge, the probability of occurrence of an impact event for the german motorway was calculated. It has been found that, the current load approaches for traffic signs lead to reliabilities that meet the requirements of DIN EN 1990. The dynamic calculations showed that the load capacity of the selected traffic sign bridge is 2045 kN in terms of impact loads for a chosen reference traffic sign bridge. In the next step the analysis of a static substitute load will be carried out. In this calculation, the effect of a vehicle restraint system will be included.

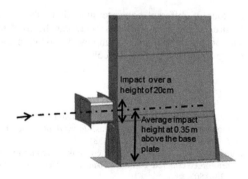

Figure 1. Initial condition for the crash analysis.

Life-Cycle Analysis and Assessment in Civil Engineering: Towards an
Integrated Vision – Caspeele, Taerwe & Frangopol (Eds)
© 2019 Taylor & Francis Group, London, ISBN 978-1-138-62633-1

Dependency of punching shear resistance and membrane action on boundary conditions of reinforced concrete continuous slabs

B. Belletti, S. Ravasini & F. Vecchi
Department of Engineering and Architecture, University of Parma, Parma, Italy

A. Muttoni
Ecole Polytechnique Fédérale de Lausanne, Lausanne, Switzerland

ABSTRACT

The design of reinforced concrete flat slabs can be governed at failure by punching shear close to concentrated loads or columns. Punching shear resistance formulations provided by codes are usually calibrated on results from tests on isolated specimens that simulate the slab zone within the points of contraflexure around a column. Indeed, the behavior of actual flat slabs can be different than isolated specimens due to the beneficial contributions of moment redistributions and membrane actions, Figure 1, that cannot take place in the conventional experiments. Geometrical features, reinforcement layouts and in-plane forces provided by external vertical elements, such as shear walls, can affect the membrane action and, consequently, the punching shear resistance. The shrinkage of concrete is also considered to study the difference of punching shear strength between actual and isolated slabs, and its effects on CMA.

This paper presents a study on reinforced concrete continuous flat slab whose lateral expansion is restrained by the presence of vertical elements considering also shrinkage effects. Nonlinear finite element

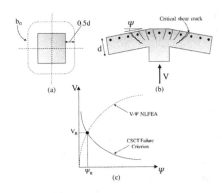

Figure 2. (a) Assumption of control perimeter at 0.5d from the edge of the column; (b) Punching shear strength correlated to the crack opening; (c) Punching shear resistance at the intersection between non-linear load-rotation curve and CSCT failure criterion.

analyses (NLFEA) are performed with a multi-layered shell approach with the crack model PARC_CL 2.0 implemented in Abaqus Code (Belletti et al.). The numerical results are then post-processed adopting the Critical Shear Crack Theory (CSCT) failure criterion to obtain the punching shear resistance, Figure 2.

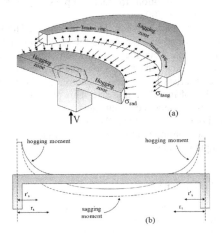

Figure 1. (a) Membrane Action – radial and tangential stresses (σ_{rad} and σ_{tang}) and effect of the tension ring on the hogging area; (b) Moment redistribution between hogging and sagging areas.

REFERENCES

Belletti, B., Scolari, M. & Vecchi, F. (2017) PARC_CL 2.0 crack model for NLFEA of reinforced concrete structures under cyclic loadings. *Computers and Structures*, 191, 165–179.

Cantone, R., Belletti, B., Manelli, L. & Muttoni, A. (2016) Compressive membrane action effects on punching strength of flat RC slabs. *Key Engineering Materials*, 711, 698–705.

fib – International Federation for Structural Concrete (2013) *fib Model Code for Concrete Structures 2010*. Berlin, Verlag Ernst & Sohn.

Muttoni, A., Fernández Ruiz, M. & Simões, J.T. (2017) The theoretical principles of the critical shear crack theory for punching shear failures and derivation of consistent closed-form design expressions. *Structural Concrete*, 1–17.

Kriging-based heuristic optimization of a continuous concrete box-girder pedestrian bridge

V. Penadés-Plà
Institute of Concrete Science and Technology (ICITECH), Universitat Politècnica de València, Valencia, Spain

T. García-Segura
Department of Construction Engineering and Civil Engineering Projects, Universitat Politècnica de València, Valencia, Spain

V. Yepes & J.V. Martí
Institute of Concrete Science and Technology (ICITECH), Universitat Politècnica de València, Valencia, Spain

ABSTRACT

The structural optimization aims to determine the best solutions for the project objectives while guaranteeing the structural constraints. The heuristic algorithms follow an intelligent process in which the design variables are modified for the purpose of optimizing the objective function and verifying the constraints. However, the structural optimization problems depend on a large number of design variables with various constrains. This causes that the computational cost remains excessive (Simpson et al. 2004). One effective solution to carry out the optimization with a low computational cost is the use of approximate response surfaces obtained by surrogate models or metamodels. One of the most encouraging metamodels used in the structural optimization is the kriging model (Cressie 1990).

In this study, a three-span continuous box-girder pedestrian bridge will be cost-optimized in two different ways: first using a conventional heuristic optimization, and then using a metamodel-based heuristic optimization based on kriging. Finally, a comparison between both methodologies will be shown and the convenience of kriging model for the robust design will be discussed. The comparison between conventional heuristic optimization and kriging-based heuristic optimization is carried out by means of the mean cost of nine solutions for each sample size (10, 20, 50, 100, 200, and 500).

Figures 1 and 2 show the main results. The horizontal dashed line represents the results obtained by conventional heuristic optimization while the solid line represents the results obtained by the kriging-based heuristic optimization according to the different number of initial sample size. This paper proves that the results obtained using metamodels to the structural optimization are really close to the conventional heuristic optimization, with a significant saving of computational cost.

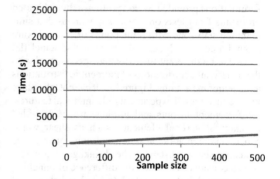

Figure 2. Comparison of the mean time spent.

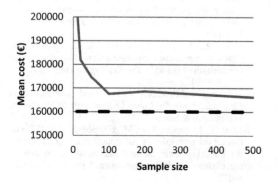

Figure 1. Comparison of the mean cost-optimized bridges.

REFERENCES

Cressie, N. (1990) The origins of kriging. *Mathematical Geology*, 22 (3), 239–252.
Simpson, T.W., et al. (2004) Approximation methods in multidisciplinary analysis and optimization: A panel discussion. *Structural and Multidisciplinary Optimization*, 27 (5), 302–313.

GS5: Earthquake engineering

A simple estimation method of the probability distribution of residual displacement and maximum bending moment for pile supported wharf by earthquake

T. Nagao
Research Center for Urban Safety and Security, Kobe University, Kobe, Japan

P. Lu
Graduate School of Engineering, Kobe University, Kobe, Japan

Technical standards for civil engineering works in Japan introduced two-stage design input earthquake ground motions after the 1995 Kobe earthquake. One of the two-stage design earthquake ground motions is called as the Level-one earthquake ground motions which presumably occur with certain degree of frequency during the design working period of infrastructure. Technical standards for port and harbor facilities applies reliability-based design for earthquake resistant design of wharves against the Level-one earthquake ground motions.

Earthquake ground motions can be expressed by the multiplication of source, path and site amplification characteristics in the frequency domain. Among the three characteristics, site amplification characteristic is known to greatly differ from site to site. Therefore, probability distribution of earthquake ground motion is thought to greatly differ from site to site. However, the technical standards for port facilities does not take the variation of input seismic motion at the site of interest into account for the reliability-based design of wharves.

It is necessary to evaluate both the residual displacement and the maximum bending moment generated in piles for the earthquake resistant design of pile supported wharves. As pile supported wharves are constructed on coastal soft ground, residual deformation occurs in soil layers by earthquake. It should be noted that both the residual displacement and the maximum bending moment of pile supported wharves are strongly affected by the deformation of the ground. Therefore, two-dimensional finite element earthquake response analysis considering both the non-linear characteristics and the effects of liquefactions of soil layers must be used for the evaluation of the residual displacement and the maximum bending moment for pile supported wharves. The problem of the application of that kind of analysis is the computational load. The computational load is particularly noteworthy in cases where probability distribution of those indices shall be evaluated because designers must conduct that time-consuming analyses many times in order to calculate the probability distribution.

This study aims at proposing a simple estimation method of probability distribution of residual displacement and the maximum bending moment of pile supported wharves considering the variation in earthquake ground motion. We conducted two-dimensional finite element earthquake response analyses by using 136 seismic input motions considering the variations in the site amplification factors and obtained the probability distribution of residual displacement and the maximum bending moment of pile supported wharves. A simple estimation method for probability distribution of residual displacement and maximum bending moment was proposed by using the results of three earthquake response analysis. It was shown that the proposed method enables to estimate probability distributions of residual displacement and maximum bending moment on the conservative side with enough accuracy and the method is highly applicable to the practical design.

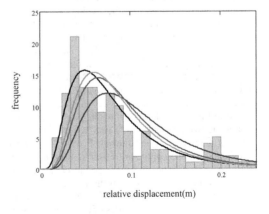

― log-normal distribution
― simple estimation (x=1.0)
― simple estimation (x=0.75)
― simple estimation (x=0.5)

Machine learning implementation for a rapid earthquake early warning system

F. Sihombing & M. Torbol
Ulsan National Institute of Science and Technology, Ulsan, South Korea

ABSTRACT

The occurrence of an earthquake is a random phenomenon, to which a rate can be attached, but the result are catastrophes. To reduce the losses, both social and economic, a comprehensive risk reduction program must be planned and applied. One important tool at the society disposal is an earthquake early warning system (EEWS) that provides information on the earthquake epicenter and other key parameters. Many issues arise in EEWS research, such as rapid estimation and accuracy of the earthquake source and parameters. This study presents a machine learning algorithm for EEWS, the algorithm is: neural network and deep learning. The algorithm works in a timely manner with various accuracy for the earthquake parameters.

In this study, we proposed an EEWS framework using the data-driven method and the deep learning techniques. We built a dataset using the earthquake events with magnitudes from $7 > M > 3.5$ and occurred from 2007 until 2017. We use it to estimate the underlying earthquake parameters (the distance, the depth, the local magnitude, the S-wave arrival time, etc.). We built three different datasets of earthquake event. The first dataset included information from the first five stations detecting the earthquake. The second dataset included information from the first three stations detecting the earthquake. The third dataset included information from the first station detecting the earthquake and the time difference between first and second station detecting P-wave. The neural network was trained using earthquake waveform features.

The estimations of the earthquake parameters using neural network in this study are better when more sensors detecting the earthquake P-wave. Using the third dataset, when only two sensor stations detecting P-wave, neural network estimates the parameters of interest of the earthquake with the uncertainty larger than first and second dataset. While for the first and second dataset, the uncertainties of the estimations are within the same magnitude.

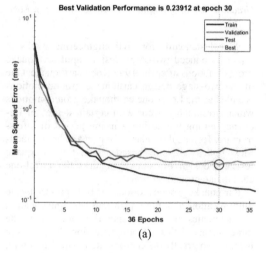

Figure 1. Earthquake magnitude estimation using neural network.

As more extensive earthquake-data are available and the advance of computational methods it opens new opportunity for data-driven EEWS research. Neural network is an example of the data-driven parameter estimation method. In this study we proposed an EEWS Deep Learning framework and using neural network as its estimation tools to understand the feasibility of the method. The results shows the trained neural network capable to estimate the earthquake parameter rapidly with limited information.

REFERENCES

Allen, R.M. (2007) The ElarmS earthquake early warning methodology and application across california. *Earthquake Early Warning Systems*, 21–43.

Bose, M., Wenzel, F. & Erdik, M. (2008) PreSEIS: A neural network-based approach to earthquake early warning, for finite faults. *Bulletin of the Seismological Society of America*, 98 (1), 366–382.

Experimental investigation of seismic behaviour of corroded RC bridge piers

X. Ge & N.A. Alexander
University of Bristol, Bristol, UK

M.M. Kashani
University of Southampton, Southampton, UK

ABSTRACT

Reinforcement concrete (RC) brides, especially those in the marine environment, are greatly threatened by corrosion. The seismic behaviour of corroded RC piers is extremely important, since piers are the most vulnerable component of a bridge. The investigation of corrosion effects can help people evaluate the safety of old RC structures and predict the remaining service life of RC structures in the extreme environment. In RC structures, corrosion can result in significant degradation of mechanical properties of reinforcement, debonding between concrete and steel bars and concrete cover cracking. With these effects, the seismic capacity of RC bridge piers is significantly reduced. The influence varies when the corrosion level is different. Previous researched showed that non-uniform pitting corrosion can change the buckling mechanism of reinforcing bars. Corrosion can cause ductility reduction of corroded steel reinforcement. The yield and ultimate loads of steel bars in tension decrease with increasing corrosion degree. The decreasing trend of strength capacity is more obvious than the trend of deformation in tensile tests of corroded steel bars. Ma et al. (2012), and Kashani et al. (2017) conducted monotonic or cyclic tests to study the nonlinear behaviour of corroded RC structures. They found that RC columns with higher corrosion degree exhibit poor hysteresis response, more stiffness loss, steep descending branch in the envelope curve and more energy dissipation. This phenomenon is more obvious to those RC columns with low horizontal reinforcement. Corrosion-damaged shear links can lead to premature buckling of vertical reinforcement. To corroded RC columns excited by the cyclic load, ultimate deformation capacity reduction is more obvious than ultimate load capacity reduction.

This study aims to investigate the non-linear behaviour of corroded RC columns under seismic excitation. To this end, a set of shaking table tests are conducted on an uncorroded, and two corroded medium-scale cantilever RC columns with different corrosion levels. The ground motion record of Manjil, Iran earthquake (1990) is chosen from FEMA P695 (2009) as the input signal. In order to reduce the specimen preparation period, admixed chloride method and external current accelerating method are

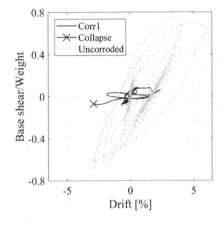

Figure 1. The hysteresis loops of the uncorroded column and Corr1 in extensive damage level (300%) test.

employed during corroding procedure. Two columns are tested with ground motion scale factors of 25%, 300% and 500% to examine the different performance of corroded RC columns in slight, extensive and complete damage levels. Aiming to detect some structural characteristics, such as natural frequency and damping ratio. Low-amplitude white-noise tests are carried out on the specimen before the first and between each round of tests. The test results, along with the uncorroded column, are analysed in terms of reinforcement corrosion rate, hysteresis behaviour, stiffness degradation and time-effective frequency response. An example hysteretic response of tested columns is shown in Figure 1.

REFERENCES

FEMA P695 (2009) *Quantification of Building Seismic Performance Factors*. Washington, DC, Federal Emergency Management Agency.

Kashani, M.M., Crewe, A.J. & Alexander, N.A. (2017b) Structural capacity assessment of corroded RC bridge piers. *Proceedings of the Institution of Civil Engineers–Bridge Engineering*, 170 (1), 28–41.

Meda, A., Mostosi, S., Rinaldi, Z. & Riva, P. (2014) Experimental evaluation of the corrosion influence on the cyclic behaviour of RC columns. *Engineering Structures*, 76, 112–123.

Seismic isolation design for Chaijiaxia Yellow River Bridge with steel triangular plate damper

L. Zhou, J. Peng, Y. Wu & Z. Yin
Shanghai Urban Construction, Design and Research Institute, China

ABSTRACT

Chaijiaxia Yellow River Bridge is a curved cable-stayed bridge with two uneven pylons, functioning as a major local connecting facility. The span arrangement of the bridge is (46.8 + 49.2 + 364.0 + 49.2 + 46.8 + 42.862)m. The curved segment of the bridge is located at the low pylon side with a radius of 600m and total length of 181m. Semi-floating system is adopted with vertical and transverse supports only. Elevation of the bridge is illustrated in Figure 1.

Chaijiaxia Yellow River Bridge is located in Lanzhou with high seismic activity. In addition, the bridge has significant spatial variability due to the curved girder and large span. Finite element analysis is employed to study its seismic performance. The results indicate that the bottom sections of the south pylon and north transition pier under transverse earthquake fail to satisfy the code requirements.

In order to improve its seismic performance, seismic isolation design is employed. A newly developed type of steel triangular plate dampers as illustrated in Figure 2 are adopted in the transverse direction. The yielding forces of the dampers are optimized to 2000 kN, 7000 kN, 8000 kN and 3000 kN for south transition piers, south pylon, north pylon and north transition piers, respectively.

The seismic response of the bridge with the designed seismic isolation system is recalculated, which indicates that the damper system is very effective in reducing the internal forces in the auxiliary and transition piers, whose bending moments decreased from 35% to 95%. The girder seismic responses in three boundary conditions, (1) no transverse constrain, (2) with damper system and (3) with constant transverse constrains are presented in Figure 3. The result suggests that the damper system can assist the bridge to achieved satisfying balance.

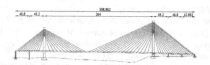

Figure 1. Elevation of Chaijiaxia Yellow River Bridge (unit: m).

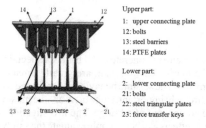

Figure 2. Steel triangular plate damper.

Upper part:
1: upper connecting plate
12: bolts
13: steel barriers
14: PTFE plates

Lower part:
2: lower connecting plate
21: bolts
22: steel triangular plates
23: force transfer keys

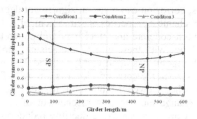

(a) Transverse girder displacement contrast

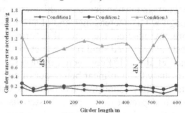

(b) Transverse girder acceleration contrast

Figure 3. Seismic responses in different boundary conditions.

REFERENCES

Astaneh-Asl, A. & Black, R.G. (2001) Seismic and structural engineering of a curved cable-stayed bridge. *Journal of Bridge Engineering*, 6 (6), 439–450.

Liu, T. Shen, X. & Ye, A. (2014) Hysteretic model parameter of the steel triangular plate damper in bridges. *Structural Engineers*, 30 (6), 54–60.

Siringoringo, D.M. & Fujino, Y. (2007) Dynamic characteristics of a curved cable-stayed bridge identified from strong motion records. *Engineering Structures*, 29 (8), 2001–2017.

Life-Cycle Analysis and Assessment in Civil Engineering: Towards an
Integrated Vision – Caspeele, Taerwe & Frangopol (Eds)
© 2019 Taylor & Francis Group, London, ISBN 978-1-138-62633-1

Assessing economic risk for businesses subject to seismic events

L. Hofer, M.A. Zanini, F. Faleschini & C. Pellegrino
Department of Civil, Environmental and Architectural Engineering, University of Padova, Padova, Italy

ABSTRACT

Recent seismic events have shown how industries and productive processes can be significantly vulnerable and suffer high financial losses in case of earthquake occurrence. Often for risk analysts and researchers it is difficult to define a priori the potential losses extension due to seismic damage (Webb et al. 2002). Recent experiences highlighted how companies may be extremely vulnerable to earthquakes and undergo huge financial losses, often characterized by high uncertainty in estimations and capable of leading to bankruptcy, due to the impossibility of resuming production in good time.

In the field of earthquake engineering, the Pacific Earthquake Engineering Research (PEER) Centre developed a probabilistic Performance-Based Earthquake Engineering (PBEE) framework to assess the mean annual frequency of a chosen variable (Porter 2003) based on the calculation of a triple integral equation. This PEER-PBEE methodology was developed with the aim to improve the decision-making procedures about the seismic performance of constructed facilities. The final decision variable has to represent the addressed problem and can be the reconstruction cost, the business interruption time, or any other variable of interest. However, in this field, few studies analyzing the impact of business interruption caused by earthquakes on industrial productive processes are present.

For all these reasons, this work aims at formulating a new framework for probabilistic seismic risk assessment for the production chains of enterprises. This framework is able to take into account both non-structural losses (NS) of components and the consequences of business interruption (BI). In particular, this work defines a procedure for assessing the Expected Annual Loss due to business interruption (EAL_{BI}), the Expected Annual Loss due to structural damage on non-structural element (EAL_{NS}), and for the Expected Annual Total Loss (EAL_T), representing total losses (non-structural on process components and direct business interruption) suffered by a company. Lastly, the above framework is applied to a

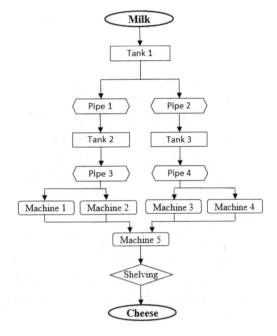

Figure 1. Production scheme of the cheese-producing facility.

case study represented by a typical Italian cheese-producing facility, whose production process layout is shown in Figure 1, and results are widely discussed.

REFERENCES

Porter, K.A. (2003) An overview of PEER's performance-based earthquake engineering methodology. In: *Proceedings of the Ninth International Conference on Applications of Probability and Statistics in Engineering.* San Francisco, CA.

Webb, G.R., Kathleen, J.T. & Dahlhamer, J.M. (2002) Predicting long-term business recovery from disaster: A comparison of the Loma Prieta earthquake and Hurricane Andrew. *Global Environmental Change Part B: Environmental Hazards,* 4 (2–3), 45–58.

Electricity supply reliability modelling for public sector facilities in view of seismic disaster risk

G. Shoji
University of Tsukuba, Tsukuba, Japan

I. Matsushima
Former Graduate Student, University of Tsukuba, Tsukuba, Japan

ABSTRACT

Uninterrupted supply of electricity to public sector facilities such as medical institutions, government buildings, fire stations, police stations and evacuation centers is crucial for the continuing functioning of the facilities during a rapidly unfolding situation due to a natural disaster such as an earthquake or serious accident (Rose et al. 1997, Chang et al. 2000, Shinozuka et al. 2002, Shinozuka et al. 2004). Therefore, electrical backup systems for the public sectors should be analyzed. We proposed models to estimate the system reliability of the electric power supply by the installation of backup systems for electric power users in the public sectors. A conceptual framework was first developed, and then numerical models to evaluate reliability based on a system optimization scheme were developed. A scenario was presented in which backup systems are installed and the effects on risk management for the facilities were evaluated.

Estimation of system reliability of electrical power supply after installation of backup systems is shown in Figure 1. Initially, target facilities should be selected and locations should be determined within a specified area. Next, seismic hazards in the area should be estimated by modelling seismic activity based on a probabilistic approach. Target facilities may then be classified and grouped by the chosen seismic hazard index. The decision-makers may then install backup systems according to the number of classifications of the derived seismic hazard index. Lastly, considering the seismic hazards, a system optimization scheme is adopted to estimate the system reliability of the electrical power supply by installing the backup systems. Iteration through all combinations of the number of seismic hazard classifications and the number of alternative systems is required. Through this iteration, the optimal combination of backup systems is derived to maximize the system reliability of the electrical power supply to target facilities within the constraints of life cycle costs and mechanical efficiency of the backup systems.

A scenario by the installation of backup systems to the public facilities is presented. In the scenario, four alternative backup systems for electrical power

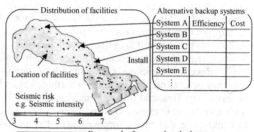

Figure 1. Framework of proposed method.

Table 1. Alternative power backup systems.

System	Relationship of power supply and demand over time	Alternative systems
I		Supplement power by service wires
II		Backup power generator
III		Power supply from off-network sources such as power supply vehicles
IV		Distributed generation systems

—: Supply, …: Demand,
-·-: Supplied by backup systems

supply from system I to system IV are proposed in Tables 1 & 2. System I is an idealized backup system in which power is supplemented by a secondary service wire, system II uses a backup power generator, and system III uses a power supply from, for example, power supply vehicles owned by electric companies. In addition, system IV is an idealized system using new distributed energy supplies such as renewable energy systems and fuel cell systems. Because whether or not an electrical power failure occurs in the network, minimum electrical power should be supplied to maintain the function of facilities, after a power failure, each backup system supplements the electrical power supplied through the normal networks.

In the scenario, the public facilities are assumed to be located in a city having a population of several hundred thousand, and it is assumed that there are 95 facilities in an area about 20 km wide from north to south and about 30 km wide from east to west by referring the facilities on Kawasaki City in Japan. This scenario has three cases for the parameters for system IV, as also shown in Table 2 because system IV is based on advanced state-of-the-art energy supply

technology so that the sensitivity of dominant parameters should be checked in the optimization scheme. Service periods of target facilities that the sectors manage and maintain are varied in 5-year increments from 5 to 30 years. Service periods up to 30 years are reasonable because innovations in energy supply technology regarding backup systems have become rapid in recent years.

Table 3 shows the optimal combinations of the backup systems and the constraint costs $\bar{C}$ in the service periods for the three cases shown in Table 2. From Table 3, considering all combinations of assumed service periods and seismic risks, among the combinations, system I is selected mostly for case 1, system IV mostly for case 3, and system I for 20, 25, and 30 years in case 2. In case 3, the mechanical efficiency η_j of system IV is larger and the initial cost c_j^I of system IV is lower than in the other cases, and hence the number of combinations for selecting system IV increases. This indicates that the mechanical efficiency η_j and the initial cost c_j^I of the backup systems are dominant and important parameters from the point of view of the sensitivity of the proposed system optimization model.

Table 2. Mechanical efficiency and costs including initial costs, running costs, and maintenance costs of the systems. Cases in efficiency and initial cost of system IV.

System	Efficiency η_j	Initial cost c_j^I [yen/kW]	Running cost c_j^R [yen/kWh]	Maintenance cost c_j^M [yen/year]
I	0.5	200,000	20	50,000
II	0.4	300,000	10	100,000
III	0.3	100,000	10	100,000
IV	0.4	400,000	20	50,000

	System IV		
Cases	Efficiency η_j	Initial cost c_j^I [yen/kW]	Characteristics
1	0.4	400,000	Standard
2	0.7	800,000	High efficiency and high cost
3	0.7	400,000	High efficiency and low cost

REFERENCES

Chang, S.E., Shinozuka, M. & Moore II, J.E. (2000) Probabilistic earthquake scenario: extending risk analysis methodologies to spatially distributed systems. *Earthquake Spectra*, 16 (3), 557–572.

Rose, A., Benavides, J., Chang, S.E., Szczesniak, P. & Lim, D. (1997) The regional economic impact of an earthquake: direct and indirect effects of electricity lifeline disruptions. *Journal of Regional Science*, 37 (3), 437–458.

Shinozuka, M., Cheng, T.C., Jin, X., Dong, X. & Penn, D. (2002) System performance analysis of power networks. In: *Procs. of the Seventh U.S. National Conference on Earthquake Engineering (7NCEE), Boston, MA, USA*.

Shinozuka, M., Chang, S.E. & Dong, X. (2004) Assessment of disaster resilience of lifeline systems. In: *Procs. of the 3rd Asian-Pacific Symposium on Structural Reliability and Its Applications, Seoul, Korea*.

Table 3. Optimal power backup systems for public facilities. I to IV denote the alternative systems in Table 1.

Target period [years]	Case 1				Case 2				Case 3			
	Seismic intensity			C [billions of yen]	Seismic intensity			C [billions yen]	Seismic intensity			C [billions yen]
	Low 5	High 5	Low 6		Low 5	High 5	Low 6		Low 5	High 5	Low 6	
5	I	I	–	63.4	IV	I	–	83.2	IV	IV	–	63.4
10	IV	I	–	103.0	III	IV	–	103.0	I	IV	–	103.0
15	IV	I	–	142.6	II	IV	–	142.6	I	IV	–	142.6
20	–	I	I	221.8	–	I	I	221.8	–	IV	IV	221.8
25	–	I	I	261.4	–	I	I	261.4	–	IV	IV	261.4

Life-Cycle Analysis and Assessment in Civil Engineering: Towards an Integrated Vision – Caspeele, Taerwe & Frangopol (Eds)
© 2019 Taylor & Francis Group, London, ISBN 978-1-138-62633-1

Non-Gaussian stochastic features hidden in earthquake motion phase

T. Sato
International Institute for Urban System Engineering, South-East University, Nanjing, China
The Faculty of Contemporary Social Studies, Kobe-Gakuen University, Kobe, Japan

ABSTRACT

Many natural phenomena have been modeled and analyzed by using the Wiener process (Karatzas & Shrves 1991). Especially methods based on the stochastic differential equation have achieved dazzling results (Mikosh 1998). But the probability distribution function of Wiener process is restricted to the normal distribution, sometimes, it cannot be applied to explain the physical process of which tail part distribution becomes very thick and the variance does not exist.

First we introduce a simple formula to proof the Levy deconvolution theorem by using the central limit theorem, then explain the mathematical feature of random process continuous with respect to (wrt) the parameter value. Because its probability distribution function must be expressed by the Gaussian distribution the modeling the physical process is inevitably dropped in the Gaussian regime.

However from the end of twentieth to this centuries we have observed many kinds of stochastic phenomena which do not obey to the Gaussian distribution. How to model these kind of phenomena is the main theme in this paper, especially what kinds of distribution characteristics are required to model them.

Searching this kind of phenomenon in the nearby research topics we have found that there exists a non-Gaussian stochastic phenomenon in the earthquake motion phase. Through researches on the fractal nature hidden in earthquake motions we investigate the stochastic characteristic of Fourier phase of earthquake acceleration time histories, and then analyze a stochastic characteristic of the mean phase gradient which is defined by the quotient of the phase difference wrt the concerned circular frequency interval.

Decomposing the earthquake motion phase into the linear delay and its fluctuation parts, we investigate the stochastic characteristics of the phase difference in the fluctuation part. In this discussion except clearly defining, when we mention the phase, it means this fluctuation part of phase. The probability density function of the mean phase gradient is expressed

by a unique stable distribution, named Levy-flight (Nolan 2015), for any arbitrary circular frequency intervals. Because the variance of the Levi-flight distribution cannot define it is analytically derived that the earthquake motion phase is the continuous but un-differentiable function wrt the circular frequency. We propose a new type of stochastic process which can present these stochastic characteristics of phase difference by the use of Lubesgue-Stieltjes type integral formula. In which the kernel plays a role to realize the self-affine (Falconer 2003) and auto correlation nature of phase difference and the integral function presents the main stochastic characteristics of earthquake motion phase. To compose the integral function based on Levy-flight distribution we developed a new type of stochastic process named the Levy-flight noise process.

The proposed stochastic process being able to use for generating earthquake motion phase is named as the modified fractional Levy-flight process. Comparison of several numerically simulated results with observed earthquake motion phase differences results in the efficiency of the newly proposed stochastic process to simulate a realistic earthquake motion phase.

REFERENCES

Falconer, K. (2003) *Fractal Geometry; Mathematical Foundation and Application,* 2nd ed. The Artium, Southern Gate, Chichester, John Wiley & Sons, Ltd.

Gnedenko, B.V. & Kolmogorov, A.N. (1968) *Limit Distribution for Sums of Independent Random Variables,* English translation by K. L. Chung, revised 1968, Addison-Wesley.

Karatzas, I. & Shreve, S.E. (1991) *Brownian Motion and Stochastic Calculus.* New York, Springer-Verlag, New York.

Mikosch, T. (1998) *Elementary Stochastic Calculus with Finance in View.* World Scientific Publishing Co. Pte. Ltd.

Nolan J.P. (2015/12/15) Reading: Stable Distribution: Models for Heavy Tailed Data, http://academic2.american.edu/~jpnolan/stable/chap1.pdf

Implications of performance-based seismic design of nonstructural building components on life cycle cost of buildings

G. Karaki
Department of Civil Engineering, Birzeit University, Palestine

ABSTRACT

Current trends in structural design emphasizes on robust, resilient and sustainable design of buildings, and this is achieved by multiple design objectives that support target performance levels for the intended design concept. Recent innovations towards a sustainable design are performance-based design (PBD) and building life-cycle assessment (LCA). Structural engineering and design against seismic actions mainly focuses on the integrity of the structures assigning a life safety performance level (code-based design) or multiple performance levels for the structural systems and elements (PBD). The structure has the critical role of protecting the nonstructural systems and components from damage, therefore, recent development of performance-based seismic design aims to combine the performance levels between structural and nonstructural components. This type of combined assessment can significantly affect the life cycle assessment results of the building. In this paper, performance-based seismic design is carried out for RC building considering different hazard levels. The target performance levels are considered in terms of inter-story drift and acceleration levels of building floors. The repair cost of structural and nonstructural components in the life span of the building under a seismic hazard is examined with respect to the type of the lateral structural system and consequently its ductility level. It is concluded that the increase of ductility of the structural system affects the nonstructural repair cost and consequently the life cycle cost. It is also shown

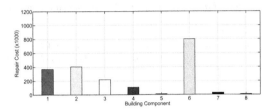

Figure 2. Break down of repair cost of EnMRS.

that the ratio of the repair cost for nonstructural components to the total repair cost depends on the type of the lateral structural system and the type of nonstructural components, i.e. displacement-sensitive or acceleration-sensitive. Therefore, an essential criterion for running a performance-based seismic design is not only to consider the performance of nonstructural components but balancing their controlling EDP values at the different floors for a better sustainable design and assessment of building's life cycle.

The following nonstructural elements are considered in the analysis: 1. Edge beams and columns, 2. Interior beams and columns, 3. RC slab, 4. Exterior window system, 5. Wall and floor finishing, 6. Wall partitions, 7. Suspended ceiling, and 8. Water piping.

REFERENCES

FEMA (2012) Next-Generation Methodology for Seismic Performance Assessment of Buildings, prepared by the Applied Technology Council for the Federal Emergency Management Agency, Report No. FEMA P-58, Washington, D.C.

Filiatrault, A. & Sullivan, T. (2014) Performance-based seismic design of nonstructural building components: the next frontier of earthquake engineering. *Earthquake Engineering and Engineering Vibration*, 13, 17–46.

Vamvatsikos, D. & Cornell, C.A. (2006) Direct estimation of the seismic demand and capacity of oscillators with multi-linear static pushovers through IDA. *Earthquake Engineering Structural Dynamics*, 35–39, 1097–1117.

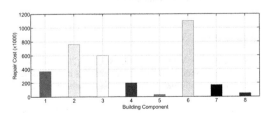

Figure 1. Break down of repair cost of MRS.

Reliability of base-isolated structures with sliding hydromagnetic bearings considering stochastic ground motions

L.C. Ding, R. Van Coile & R. Caspeele
Department of Structural Engineering, Ghent University, Ghent, Belgium

Y.B. Peng
State Key Laboratory of Disaster Reduction in Civil Engineering, Tongji University, Shanghai, P.R. China
Shanghai Institute of Disaster Prevention and Relief, Tongji University, Shanghai, P.R. China

J.B. Chen
State Key Laboratory of Disaster Reduction in Civil Engineering, Tongji University, Shanghai, P.R. China
College of Civil Engineering, Tongji University, Shanghai, P.R. China

ABSTRACT

As an efficient technique of structural control, base isolation is widely used in the response mitigation and performance enhancement of structural seismic hazard reduction. Considering there are some limitations of the traditional isolation technologies, a novel base isolation, the sliding hydromagnetic bearing, has been proposed. It offers several technical benefits (Villaverde 2017).

In this contribution, the dynamic reliability of a base-isolated structure with sliding hydromagnetic bearings is evaluated using the Probability Density Evolution Method (PDEM) (Chen et al. 2007). Meanwhile, three hundreds representative samples of artificial seismic ground motions are generated using the stochastic function model of seismic ground motions (Ding et al. 2018). The stochastic seismic response analyses are also implemented with a base-fixed structure to compare with the responses of the base-isolated structure. Finally, the reliabilities are evaluated based on the extreme value distribution of inter-story drifts of the based-isolated structure.

Compared to the response of inter-story of the base-fixed structure, the effect of base-isolation with sliding hydromagnetic bearings can reduce the responses by one degree of seismic intensity. In addition, the superstructure of the base-isolated structure vibrates like a rigid body under the seismic ground motions. This indicates the sliding hydromagnetic bearing is efficient in controlling the responses of the structure and can achieve the demanding of seismic protection.

According to reliability assessment of the extreme values of the inter-story drifts of the base-isolated structure, see Figure 1 and Table 1, the superstructure is safe enough. However, the reliability of the base story is 0.946, and thus there exists some risk of the inter-story drift or displacement limit being exceeded. The results provide quantified indices for decision making of the designer and the owner.

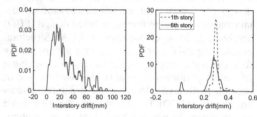

(a) The base story.   (b) The first and the sixth story.

Figure 1. The pdfs of the extreme values of the inter-story drift responses (0.2 g).

Table 1. Reliabilities of the inter-story drift responses of the base-isolated structure.

Number	Threshold*	Reliability
Base	63.5 mm	0.946
1	3.6 mm	1.000
2	3.6 mm	1.000
3	3.6 mm	1.000
4	3.6 mm	1.000
5	3.6 mm	1.000
6	3.6 mm	1.000

*Maximum displacement permitted.

REFERENCES

Chen, J.B., Liu, W.Q., Peng, Y.B. & Li, J. (2007) Stochastic seismic response and reliability analysis of base-isolated structures. Journal of *Earthquake Engineering*, 11 (6), 903–924.

Ding, Y.Q., Peng, Y.B. & Li, J. (2018) A stochastic semi-physical model of seismic ground motions in time domain. *Journal of Earthquake and Tsunami*, 1850006.

Villaverde, R. (2017) Base isolation with sliding hydromagnetic bearings: concept and feasibility study. *Structure and Infrastructure Engineering*, 13 (6), 709–721.

GS6: Traffic load modelling

Development and validation of a full probabilistic model for traffic load of bridges based on Weigh-In-Motion (WIM) data

J. Kim & J. Song
Seoul National University, Seoul, South Republic of Korea

ABSTRACT

It is important to accurately estimate the traffic load effects on bridges for the purpose of design or performance evaluation. Traffic load effects on a bridge form a stochastic process, which requires the development of a site-specific model representing random properties from traffic conditions. However, most design codes and many studies still do not fully consider the site-specific conditions of bridges and rely on conservative assumptions. Therefore, in this study, a full probabilistic traffic load model based on Weigh-In-Motion (WIM) data measured from three sites of highway system in South Korea is developed to consider the site-specific characteristics of vehicles and traffic flow.

To describe vehicle characteristics, four random variables representing the characteristics of each vehicle type, i.e., GVW, axle weights, total axle spacing and axle spacing are modeled by Gaussian mixture models fitted to the WIM data. Additionally, linear correlations between these random variables are fully considered using a Nataf model.

In the model for traffic flow characteristics, traffic volumes are fitted to empirical distributions in the unit of veh/5min to accurately describe the congestion state. The relationship between velocity and traffic volume in Greenshield's model is identified by the measured WIM data.

This relationship and the probability of congestion occurrence are incorporated into the model and

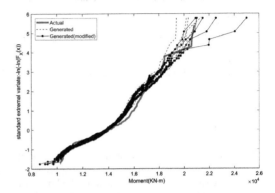

Figure 2. Gumbel probability paper of daily maximum moment at mid-span (span length: 100m) for Sunsan.

variability of heavy vehicle ratio over time is also incorporated. The headway distribution models are established by conditional distribution given free flow and congestion state separately. To consider the correlation between multiple lanes, some sections of WIM data on the lanes except the primary lane are regenerated appropriately.

After developing statistical models, WIM data are generated on all lanes at three sites and the corresponding traffic load effects are calculated using moment influence line of a simple beam bridge. The daily maximum moments by generated WIM data and actual WIM data are plotted in Gumbel probability papers to validate the developed model.

The results are similar between load effects by generated and actual WIM data, which validates the developed probabilistic model. In addition, the effect of modification of WIM data introduced to consider the correlation between multiple lanes is verified.

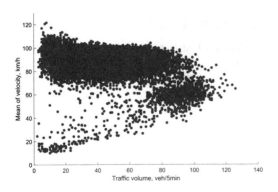

Figure 1. Traffic volume-velocity relationship on the primary lane of WIM data from Sunsan.

REFERENCES

Caprani, C.C., O'Brien, E.J. & McLachlan, G.J. (2008) Characteristic traffic load effects from a mixture of loading events on short to medium span bridges. *Structural Safety*, 30 (5), 394–404.

Enright, B. & O'Brien, E.J. (2013) Monte Carlo simulation of extreme traffic loading on short and medium span bridges. *Structure and Infrastructure Engineering*, 9 (12), 1267–1282.

The effect of traffic load model assumptions on the reliability of road bridges

M. Teichgräber
Engineering Risk Analysis Group, Technical University of Munich, Munich, Germany

M. Nowak
Chair of Concrete and Masonry Structures, Technical University of Munich, Munich, Germany

J. Köhler
Department of Structural Engineering, Norwegian University of Science and Technology, Trondheim, Norway

D. Straub
Engineering Risk Analysis Group, Technical University of Munich, Munich, Germany

ABSTRACT

Modern structural design standards, such as the Eurocode, follow a semi-probabilistic approach, in which the safety of a structure is controlled by partial safety factors. These standards are based on simplified models of load and structural performance, where approximations are made that typically lie on the safe side, leading to additional "hidden" safety. If these models are replaced by more sophisticated and potentially more accurate models, some of this hidden safety might disappear. Overall, increased accuracy is preferable, as it leads to a more economical and sustainable use of resources, but the question remains whether lost hidden safety leads to an overall reduction in structural reliability.

We established a purely contemplative definition of hidden safety. Moreover, we defined a measure of hidden safety in models within a dimensioning methodology in a relative sense. It is conditional on the dimensioning methodology, the failure definition and the applied reliability analysis.

In this case study, we applied the measure of hidden safeties to a three-field support reinforced concrete bridge under traffic load (see Figure 1). Firstly, the bridge is designed by means of a more sophisticated traffic load model. For this purpose, the traffic load is modeled with simulation using measured traffic data and the resulting load effects are modeled with extreme value statistics. Finally, the reliability of the different design variations of the bridge are calculated with a full probabilistic approach. Thereby, four different definitions of failure are compared.

The analysis results in a quantification of the hidden safeties associated with the traffic and the static model following Eurocode. High amounts of hidden safety in the traffic-load model LM1 of the Eurocode ere revealed. Elimination of this hidden safety reduces the amount of steel by 24 [%]. The evaluations of hidden safeties strongly depended on the type of reliability analysis, in particular on the definition of failure.

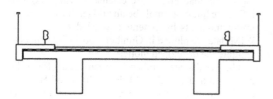

Figure 1. Cross-section of the three-field support reinforced concrete bridge.

REFERENCES

Bailer-Jones, D.M. (2009) *Scientific Models in Philosophy of Science.* University of Pittsburgh Pre.
DIN, E. (1991) 2 (2003) Eurocode 0: Grundlagen.

Extended extrapolation methods for robust estimates of extreme traffic load effects on bridges

M. Nowak
Chair of Concrete and Masonry Structures, Technical University of Munich, Germany

D. Straub
Engineering Risk Analysis Group, Technical University of Munich, Germany

O. Fischer
Chair of Concrete and Masonry Structures, Technical University of Munich, Germany

ABSTRACT

Bridge loading constitutes a complex random process, characterized by a mixture of different loading event types contributing to extreme load effects on bridges. In contrast, classic extreme value analysis applying the block maxima method generally assumes asymptotic models such as Gumbel or generalized extreme value distribution (Coles, 2011). To account for the complexity of the bridge loading process in statistical extrapolation of extreme traffic load effects on bridges, two different strategies are investigated. Tail fitting aims at data reduction to a subset significant for modeling the extreme value behavior of the random variable. Alternatively, fitting of composite distribution models attempts to directly consider the inhomogeneities arising from the underlying bridge loading process. Results from long term traffic simulations serve for benchmark testing and evaluating the effectivity of the proposed methods.

Tail fitting shows in general quite irregular performance compared to the benchmark results. The use of criteria with either fixed or sample size dependent ratios for upper tail definition provides only occasionally good results. Empiric evaluations of optimal tail ratios reveal significant variance, even when performing multiple simulation runs with the same analysis setting. This method appears unsuitable for general application. Other criteria for tail fitting were not investigated.

If more detailed data with separate block maxima series for different loading event types is available, the use of composite distribution models can be a good choice, to make use of the richer information at hand for assessing extreme load effects (Caprani et al., 2008). However, such models require more model parameters, making them in general more sensitive for model fit. In this context, a proper compromise regarding definition and classification of loading event types

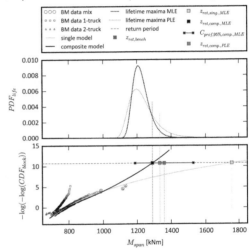

Figure 1. Comparison of model fit and resulting extreme value estimates for fit with single GEV distribution and composite distribution.

needs to be found, which should be part of future research efforts to improve the robustness of the method.

REFERENCES

Caprani, C.C., OBrien, E.J. & McLachlan, G.J. (2008) Characteristic traffic load effects from a mixture of loading events on short to medium span bridges. *Structural Safety*, 30 (5), 394–404.

Coles, S. (2001) *An Introduction to Statistical Modeling of Extreme Values*. London, Springer.

GS7: Bridge engineering

Life-Cycle Analysis and Assessment in Civil Engineering: Towards an Integrated Vision – Caspeele, Taerwe & Frangopol (Eds)
© 2019 Taylor & Francis Group, London, ISBN 978-1-138-62633-1

Steel bridge structural retrofit: Innovative and light-weight solutions

A. Pipinato & R. Pavan
AP&P srl, Italy

P. Collin & R. Hallmark
Lulea University of Technology, Sweden

S. Ivanov & R. Geier
Schimetta Consult, Austria

M. van der Burg
Movares, The Netherlands

ABSTRACT

Prolonging the life of existing steel trusses all over the world is a incoming necessity for all managing authorities wherever these structures have reached a particular state of use. The unsatisfactory behavior of these structures under the current type and increasing number of loads make clearer that innovative solutions are needed. For this purpose, the realization of composite decks is needed, in order to redistribute the common actual load onto a structure conceived with historical codes, characterized by lower design live loads. The possibilities of retrofitting bridges with this solution studied in a recent research program are presented and compared in this study. Increasing traffic density and vehicle weight have been increasing in recent decades. Many existing truss bridges, all over the world, were not designed for the high service loads and the increased number of load cycles that they are exposed to today. Some bridges will have to be either strengthened or replaced in the next decades. As late as the 1980s, steel truss bridges in Europe were often constructed with a non-composite concrete slab on top of steel girders. Different alternatives of interventions for truss bridges are presented in this paper, according to the different truss type to be reinforced. Sometimes the bridge should be retrofitted by deck replacing, other times a new structural configuration should be adopted, or in some cases only little interventions are needed in order to strengthen a truss. In this paper different strategies in order to prolong the lifetime of existing steel truss bridges have been presented. The first technique deals with the possibility to upgrade the load carrying capacity of a cultural heritage arched-truss bridge by deck replacing, maintaining the existent architectural profile, but increasing at the same time the structural strength and reliability; the second technique explored the possibility to build a new UHPC deck onto existing metro-railway trusses, calculating the beneficial reduction of stressed and deformations of

Figure 1. Lowering of the old Årsta midspan section.

the whole existing structure; the third solution provide refined calibrated FEM finalized to build a direct fastened rail incorporating an anti-derailment device, able to reduce the stress amplitude dealing with fatigue, and consequently prolonging the life of the whole truss; the fourth solution presented a parametric analysis of a road truss, to discover the most suitable retrofit solution, represented by the realization of a thin UHPC 50mm deck together with cove-plating applications; the last solution presented a parametric analysis to prolong the fatigue strength of existing bridges, to discover the most suitable retrofit solution, and to compare different retrofit techniques. As a matter of fact, the whole solutions presented are finalized to demonstrate that it is possible to prolong the life of existing trusses, with the innovative techniques illustrated herein, satisfying the EN codes actual loading scheme (EN 1991-2:2003).

REFERENCE

EN 1991-2 (2003) Eurocode 1: Actions on Structures – Part 2: Traffic Loads on Bridges. Brussels, Belgium, CEN-European Committee for Standardization.

Assessment procedures for existing bridges: Towards a new era of codes and standards

R. Pavan & E. Siviero
AP&P srl, Italy

ABSTRACT

The assessment of existing structures adopting a specific code or standard is a technical procedure that is pro-vided only in some nation. The possibility of harmonizing these type of procedure is needed in Europe, together with a completely new plan of infrastructural renewal. In this study new insight on this theme are pro-vided, together with some code proposal for the new Eurocode era. Many existing bridges all over the world were not designed for the high service loads and the increased number of load cycles that they are exposed to today. Some bridges will have to be either strengthened or replaced in the next decades. Advancing their age, bridges could be subjected to a large variety of not designed actions or could also be not accurately maintained, leading to a particular state in which they need particular care and attention, in order to remain in service. Moreover, existing structures were not exempted from serious failures and engineers had regularly to suffer setbacks.

These setbacks contributed in many cases, however, to research and advancements in better understanding structural behavior and developing new theories (Kuhn et al. 2010). In these cases, the assessment procedures are needed.

The assessment of an existing structure aims at producing evidence that it will function safely over a specified residual service life. It is mainly based on the results of assessing hazards and load effects to be anticipated in the future, and of assessing material properties and geometry taking into account the present state of the structure (JCSS 2001). Guidelines for existing structures exist in a large number of countries. Thereby many countries have presented documents for particular categories of structures. In Canada, Germany, the Netherlands, Switzerland, UK and USA such guidelines have been prepared at a detailed level. In any assessment, the problem of fixing risk acceptance criteria is difficult since it must be compatible to codes for new structures (limit state analysis, safety factor format, etc.), and with national

Figure 1. A multi span historical truss bridge.

determined parameters (generally partial safety factor values) (Kuhn et al. 2010).

As far as specific code and standards are not available, technical guide or pre-standard documents should be used. Moreover, procedures coming from evidenced based or field related experience, are quite extensively used also by bridge and infrastructures owner. Basing on past studies and research, a simplified approach has been developed by the author and applied in recent investigations of existing bridges. This approach is reported in the following paragraphs.

REFERENCE

Kuhn, B., Lukic', M., Nussbaumer, A., Gunther, H.-P., Helmerich, R., Herion, S., Kolstein, M.H., Walbridge, S., Androic, B., Dijkstra, O. & Bucak, Ö. (2008) Assessment of existing steel structures: Recommendations for estimation of remaining fatigue life. Joint report prepared under the JRC-ECCS cooperation agreement for the evolution of Eurocode 3 (programme of CEN/TC 250). In: Sedlacek, G., Bijlaard, F., Ge'radin, M., Pinto, A. & Dimora, S. (eds) *Background Documents in Support to the Implementation, Harmonization and Further Development of the Eurocodes* First Edition, February 2008 EUR 23252 EN-2008.

Life-Cycle Analysis and Assessment in Civil Engineering: Towards an Integrated Vision – Caspeele, Taerwe & Frangopol (Eds)
© 2019 Taylor & Francis Group, London, ISBN 978-1-138-62633-1

Stress concentration factor in concrete-filled steel tubular K-joints under balanced axial loading

I.A. Musa & F.R. Mashiri
School of Computing, Engineering and Mathematics, Western Sydney University, Australia

ABSTRACT

Uniplanar welded steel tubular K-joints are widely used in highway and railway bridges. In fatigue studies, focus has to be put on the stress concentration at the vicinity of the weld which is usually the location of fatigue crack initiation. The hot spot stress method is one of the widely used fatigue design methods and uses the stress concentrations at the vicinity of the weld in the fatigue design. The stress concentration factor (SCF) is the ratio between the hot spot stress at the joint and the nominal stress in the member due to a basic member load which causes this hot spot stress (Zhao et al., 2001). Few studies on CFST K-joints have been reported in the literature. Udomworarat et al., (2000) conducted experimental study on stress concentration in CFST K-joints under balanced axial loading. Tong et al., (2008) performed an extensive experimental study on stress concentration factor in CFST K-joints under balanced axial loading. Xu et al., (2015) conducted experimental study on CFST K-joints.

In this study, stress concentration has been studied experimentally and numerically in one CFST K-joint under balanced axial loading. Strain has been measured at 45° around the brace-chord intersection on both two braces and chord. The experimental stress concentration factor (SCF) has been compared with those predicted by formulae given in CIDECT (Zhao et al., 2001), and DNV.GL (2016) Manual. The non-dimensional geometrical parameters of the K-joint test specimen were chosen to be complimentary to existing knowledge.

The following observations were made:

- High stress concentration exists at the area between the two braces and specifically at crown toe positions at the chord and the two braces.
- The maximum SCF exists at the crown toe position.

- Minimum stress concentration exists at the crown heel position or close to crown heel position.
- Parametric equations for predicting SCFs in empty K-joints are not suitable for CFST K-joints.
- Parametric equations for predicting SCFs in CFST K-joint given in DNV.GL manual, ISO-19902:2007 and API (2014) provide unconservative SCF results and would result in unsafe design.

REFERENCES

API (2014) Recommended Practice 2A-WSD: 22nd edition. *Planning, Designing, and Constructing Fixed Offshore Platforms—Working Stress Design.* Washington, USA, API Publishing Services.

Dnvgl, D.N.V. (2016) *Fatigue Design of Offshore Steel Structures.* No. DNVGL-RP-C203.

ISO, E. (2007) *International Standard: Petroleum and Natural Gas Industries – Fixed Steel Offshore Structures.* Switzerland.

Tong, L.W., Sun, C.Q., Chen, Y.Y., Zhao, X.L., Liu, C.B. & Shen, B. (2008) Experimental comparison in hot spot stress between CFCHS and CHS K-joints with gap. *Tubular Structures XII.* Taylor & Francis.

Udomworarat, P., Miki, C., Ichikawa, A., Sasaki, E., Sakamoto, T., Mitsuki, K. & Hasaka, T. (2000) Fatigue and ultimate strengths of concrete filled tubular k-joints on truss girder. *Journal of Structural Engineering, 46,* 1627–1635.

Xu, F., Chen, J. & Jin, W.L. (2015) Experimental investigation of SCF distribution for thin-walled concrete-filled CHS joints under axial tension loading. *Thin-Walled Structures, 93,* 149–157.

Zhao, X.L., Herion, S., Packer, J.A., Puthli, R.S., Sedlacek, G., Wardenier, J., Weynand, K., Wingerde, A.M.V. & Yeomans, N.F. (2001) *Design Guide for Circular and Rectangular Hollow Section Welded Joints under Fatigue Loading,* Cologne-Germany. CIDECT and TÜV-Verlag.

Influence of fly-ash mixture to give to life cycle cost and constructability of composite girder bridge

K. Kubota
Kawada Industrial, Inc., Japan

H. Ito
Toyama Prefectural University, Japan

H. Kuriyama & T. Izumiya
Kawada Industrial, Inc., Japan

ABSTRACT

Composite girder bridges called "pre-flexed beam bridges" have been applied to each place in Japan. This type of bridge is made by the pre-stressing method, so the concrete needs to harden early and have high strength. The concrete needs high workability to be placed in the narrow space. When the bridge is subjected to the chloride attack near the sea shore, the steel parts of the bridge may be corroded by the chloride ion penetration into concrete.

The purpose of this study is to investigate the life cycle cost (LCC) of the bridge and the constructability of the FA concrete. For this purpose, the LCC estimation of the actual designed bridge was carried out. This LCC includes the initial construction cost and the maintenance cost. This maintenance cost was calculated based on the diffusion coefficient that was obtained by salinity penetration test. On the other hand, for the confirmation of the constructability of FA concrete, the L shape flow test, vibration slump test, V shape funnel test, and U shape filling test were carried out.

As result, the LCC of the composite girder bridge, the pre-flexed beam bridge, was improved by mixing FA. Based on the test result, FA concrete had the flow ability and the filling-ability in the same degree as the ordinary concrete.

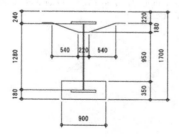

Figure 1. The pre-flexed beam girder cross-sectional view.

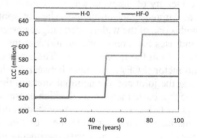

Figure 2. Life cycle cost.

As a result of the construction test, it was improved that the fluidity and filling property by securing the vibration can be satisfied. In addition, the compression strength test was proportional to the water binder materials ratio, regardless of the cement type. In the early-strength H case, the strength necessary for pre-stress introduction was satisfied by selecting the water binder materials ratio.

For calculation of LCC, assuming the coastal area susceptible to salt damage, the concrete cover was 70 mm. The lower flange concrete was 900 mm wide and 350 mm high. Another, the length of the bridge was 150 m and the height of the girder was 1.7 m. Out of 100 years, the repairing number was 4 times at H-0 and 2 times at HF-0. From this, it was found that by adding FA, the number of repair is reduced to about half.

As a result of LCC, by mixing FA, the initial cost increased due to the difference between cement and FA (2000 yen/m^3). On the other hand, by reduce these repair frequency, it was able to reduce maintenance cost and by about 66 million yen. (Assuming repair by surface coating) From these results, it was confirmed that by mixing the FA, it is possible to reduce about 65 million yen by comparing the life cycle including the initial cost and the maintenance cost.

Assessment of Barton High Level Bridge approach span superstructures

D.M. Day
Atkins SNC-Lavalin, Warrington, UK

ABSTRACT

Barton High Level Bridge (BHLB), one of the longest in the UK, carries the M60 motorway over the Manchester Ship Canal. It is of strategic importance in maintaining the smooth functioning of the North West of England's highway network. BHLB was opened in 1960 but, with traffic growth, the dual two-lane carriageway became a bottleneck on the network. Consequently, between 1986 and 1989 the bridge was widened to accommodate a dual three-lane carriageway. Prior to widening, the superstructure comprised eight longitudinal steel plate girders of riveted construction. The widening works added a new line of welded girders on each side of the bridge. Widening of the substructure did not prove to be a simple matter. Areas of map-cracking were discovered indicating the possibility of alkali aggregate reaction. Testing confirmed that the concrete was slightly reactive, raising concerns that expansion of the original pier concrete could occur in the future. The pier widenings were designed as independent structures, building-in the potential for vertical differential movement (VDM) between the bearings on the original piers and those on the new. Accordingly, the Bridge Maintenance Manual requires monitoring of bearing levels to be at intervals of not greater than 2 years and re-levelling of the bearings before the VDM exceeds 3mm.

In 2013, proposals for two privately funded schemes in the vicinity of the bridge warranted a structural assessment of BHLB to demonstrate that ground movement associated with the proposed construction would not affect its safe operation. The approach span structures were initially modelled using simple methods; grillage analysis for the deck structure and plane frames for the end restraints. The majority of elements within the superstructure passed the assessment with sufficient reserve to accommodate the historical maxima of VDM. The most critical elements were the end plate T-stub connections in the deck end frames. The initial rigid jointed stick models proved to be stiff, and when the maxima of seasonal VDM were introduced, usage factors in the end plate connections were generally greater than 1.0 (up to 3.8). With no signs of distress, it was thought that the real structure movements were being accommodated due to the inherent flexibility of the T-stub connections.

Re-analysis of the end bracing at the half joints was undertaken with a finite element (FE) model using LUSAS software. The stiffnesses of the connections were determined from the flexibilities of the basic joint components in accordance with BS EN 1993-1-8. Local FE models were produced for the joints when in tension in order to validate the use of the Eurocode method. For each connection a non-linear stiffness was determined to allow for the response of the tensile and compression regions of the joint. The local FE models were extended to investigate the effect that preload in each bolt row would have on joint stiffness. An initial elastic analysis of the end frame adopting the two-phase joint stiffness model showed that a number of connections still exceeded theoretical capacities. Further analysis was undertaken allowing for joint plasticity by introducing a third phase in the joint stiffness model.

The increased flexibility of the end plates improved the VDM that could be tolerated and changed the overall theoretical failure mechanism away from the connections to the other components in the frame. Ultimately the overall capacity was limited by buckling of the top brace in the lower K bracing frame and theoretical bolt failure in a deck end trimmer connection. The assessment result was improved from 'zero' allowable VDM, to values significantly exceeding the design maximum. The improvement in the assessment results allowed the two private schemes adjacent to BHLB to proceed.

Moving forward, the assessment result has provided Highways England with greater time and flexibility to manage relevelling of the bearings, as and when required by on-going monitoring.

Lifecycle performance of HSS bridges

M. Seyoum Lemma, C. Rigueiro, L. Simoes da Silva & H. Gervásio
ISISE, University of Coimbra, Portugal

J.O. Pedro
GRID, Lisbon, Portugal

ABSTRACT

Infrastructures have a primordial role for the society and consequently the lifetime performance of bridges is a subject that has received much attention from the scientific community in recent years. The lifecycle performance of bridges shall include all stages, from the production of all necessary materials and construction of the bridge, to ultimately its demolition and management of the resulting waste.

Previous researches have shown that the material production stage and the operational stage of bridges (when traffic is affected by maintenance activities) are the stages which have the major impact in the lifecycle performance of bridges.

In this respect, the use of high strength steel (HSS) in structures may represent a significant advantage since it usually enables to reduce the amount of material that is required for a structure to fulfil its function. In most cases, the reduction of material use and consequently the reduction of the environmental impacts due to the amount of steel produced, compensate the increase of the environmental impacts due to the production of high strength steels. However, increasing impacts must be considered, associated with larger alloy content and usually more complex process route than ordinary steel grades as well as with the reduction of the fatigue resistance of welded joints and additional stability problems.

Hence, in this paper a comparative life cycle analysis is performed: a bridge made of standard steel grade S355 (illustrated in Figure 1) is compared with an equivalent bridge made of steel grade S690. The analysis takes into account the complete life cycle of the bridge and includes safety, environmental and economic criteria.

The main goal of this comparative analysis is to show the advantages and drawbacks of the use of HSS.

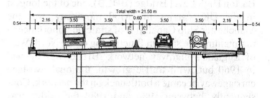

Figure 1. Cross-section of the deck.

Clear advantages are observed in the early stages of the design of the HSS bridge, as a significantly reduced amount of steel is required when using high-strength steel. Reduction of the steel volume is achieved by using HSS. Additional reduction is possible by using longitudinal stiffeners instead of vertical stiffeners.

In summary, the following advantages were identified for the high-strength steel bridge:

- LCA: Better environmental performance due to the reduction volume of steel required and subsequent reduction in weld volumes.
- LCC: Cheaper as a result of the lighter structures achieved and the use of improved knowledge on the structural behavior of HSS.
- LCS: reduced user costs as a result of reduced maintenance operations comes as a direct result of reduced surface are of the steel that requires corrosion protection.

ACKNOWLEDGEMENT

The work presented in this paper has been developed in the scope of the research project OPTIBRI "Optimal use of high strength steel grades within bridges" carried out with financial support from the European Union's Research Fund for Coal and Steel (RFCS).

Life-Cycle Analysis and Assessment in Civil Engineering: Towards an Integrated Vision – Caspeele, Taerwe & Frangopol (Eds)
© 2019 Taylor & Francis Group, London, ISBN 978-1-138-62633-1

Refurbishment of Swanswell Viaduct

C.G. West
Atkins, Member of the SNC-Lavalin Group, Birmingham, UK

ABSTRACT

Swanswell Viaduct forms part of the ring road around the city centre of Coventry, UK. The structure dates from 1970, and consists of 118 spans. Each span is simply-supported and is formed from precast, pretensioned beams.

Like many highway structures of its age, the viaduct has experienced a deterioration in its condition including leaking expansion joints and drainage, and reinforced concrete defects. The concrete fascia panels at the deck edges are in poor condition, resulting in parts of the panels spalling or even the risk of whole panels detaching. This presents a safety hazard for people and property below the viaduct and requires significant ongoing maintenance to remove loose material.

In addition, the containment provided by the parapets does not comply with current standards.

Coventry City Council wishes to address these problems by undertaking a refurbishment of the viaduct. The council commissioned SNC-Lavalin's Atkins business to:

– Carry out a high-level review of structural risks
– Specify, supervise and interpret the results of site investigations
– Undertake structural assessments
– Propose potential remedial options
– Design the maintenance and upgrade scheme

Resulting from the risk review, targeted investigations were undertaken which confirmed that the substructures were suffering extensive deterioration due to chloride-induced corrosion of the reinforcement. However, it did not reveal any results that were substantially worse than expected and which would impede the planned refurbishment of the deck. Furthermore, it confirmed that apart from the fascia panels, the deck itself was not suffering significant deterioration.

Initial assessments of the piers raised concerns about the capacity of the corbels that support the deck beams. Subsequent testing to identify the tensile strength of the reinforcement, and extensive scanning to determine the layout of the reinforcing bars allowed the assessment to be improved and alleviate the concerns.

Options for upgrading the edge protection were explored. To meet current standards, a proprietary metal parapet system will be provided. The existing deck-edge upstand will need to be replaced with a new edge of the appropriate shape and strength to support the parapet. It was decided to utilise precast units for the new deck edge construction as far as possible. The units will be anchored using steel bars hooked over the deck beams and fixed into vertical holes drilled in the infill between beams, to avoid drilling into the prestressed beams themselves.

The parapets will now be separated from the deck-edge construction. Hence if they are damaged by a vehicle impact they can be readily unbolted and replaced, whilst the deck edges remain unaffected.

Water management is vital to the ongoing serviceability of structures. Inevitably even the best drainage and expansion joint systems will leak at some time. In this project, the drainage pipes will be moved to the outside of the piers so that any future leaks will be obvious and in a position that cannot do significant damage before rectification. However, due to the form of construction, it will not be possible to introduce secondary measures to deal with leakage at the renewed expansion joints, and good quality, well maintained joints remain the only line of defence.

The refurbishment works will significantly extend the working life of the viaduct. They will replace deteriorated components of the structure with new parts that are detailed for greater durability, and they have adapted the structure to changed standards in respect of vehicle containment. The structural upgrades will be complemented by new fascia panels that will revitalise the appearance of the viaduct.

Life-Cycle Analysis and Assessment in Civil Engineering: Towards an Integrated Vision – Caspeele, Taerwe & Frangopol (Eds)
© 2019 Taylor & Francis Group, London, ISBN 978-1-138-62633-1

Resilience and economical sustainability of a FRP reinforced concrete bridge in Florida: LCC analysis at the design stage

T. Cadenazzi & M. Rossini
University of Miami, Coral Gables, Florida, USA

S. Nolan
Florida Department of Transportation, Tallahassee, Florida, USA

G. Dotelli & A. Arrigoni
Politecnico di Milano, Milano, Italy

A. Nanni
University of Miami, Coral Gables, Florida, USA

ABSTRACT

Fiber-reinforced polymer (FRP) technology is nowadays playing a major role in civil engineering solutions guaranteeing to concrete elements resilience and corrosion-resistance in aggressive environments. Glass Fiber Reinforced Polymer (GFRP) and Carbon Fiber Reinforced Polymer (CFRP) represent a proven non-metallic solution, able to ensure the required mechanical resistance of reinforced concrete (RC) and prestressed concrete (PC) structures. The research presented hereinafter addresses the LCC analysis of an FRP-RC/PC bridge in Florida, at the design stage. The analysis is performed in compliance with the standard ISO 15686-5:2008.

The bridge main structure consists of CFRP square PC bearing piles, CFRP-PC/GFRP-RC sheet piles, GFRP-RC girders, GFRP-RC bent caps, GFRP-RC bulkhead caps, traffic railings and approach slabs and a 20m long GFRP-RC gravity wall, resulting in the absence of any steel reinforcement for the entire design.

Focus of this paper is analyzing in detail the cost to face for constructing an FRP-RC/PC vehicular bridge; designed for a 1255-year service life, the initial cost of the composite materials is mainly recovered by their long-term durability. The majority of long-term maintenance costs of traditional bridges utilizing carbon steel are due to corrosion associated degradation of the structure. Generally, rehabilitation includes replacement, repair, externally bonded reinforcement or cathodic protection solutions applied to restore/strengthen the existing structure. All of these solutions are expensive and, in some cases, require challenging over-water or in-water working activities. In addition, transportation and environment play a determinant role: the material allows less haul costs, given its significantly lighter weight, and reduced environmental impact in terms of carbon emissions.

According to ISO 15686-5:2008, costs that should be included in life-cycle costing are those relative to the following bridge life stages: construction, operation, maintenance, and end-of-life. Construction costs represent the main focus of the present work. Indeed, an accurate and extensive amount of data is being collected directly during the ongoing bridge construction, while costs relative to other stages are estimated for the moment.

GS8: Life-cycle assessment

Life-Cycle Analysis and Assessment in Civil Engineering: Towards an Integrated Vision – Caspeele, Taerwe & Frangopol (Eds)
© 2019 Taylor & Francis Group, London, ISBN 978-1-138-62633-1

ProLCA—treatment of uncertainty in infrastructure LCA

O. Larsson Ivanov
Structural Engineering, Lund University, Sweden

D. Honfi & F. Santandrea
RISE Research Institutes of Sweden, Sweden

H. Stripple
IVL Swedish Environmental Institute, Sweden

ABSTRACT

The construction, operation and maintenance of transportation infrastructure require energy and materials which impact the environment. Large infrastructure projects use resources intensively and leave a significant environmental footprint. To demonstrate and support the sustainability of such large-scale projects, life cycle assessment (LCA) has become a common tool to evaluate environmental impacts in all stages of infra-structure life cycle, from raw material production through end-of-life management. However, the various phases of the assessment are all associated with uncertainties. If decisions are made without consideration of these uncertainties, they might be misleading and suboptimal.

The results from LCA are affected by uncertainties and variability at all stages. The natural variation between geographical locations, material properties and variations over time will cause an unavoidable variability in LCA results. Some uncertainties, however, arise from other problems such as lack of data, lack of knowledge and/or model differences. Such effects may be reduced by research, data collection, and by estimations and analysis of the validity and uncertainty of the results.

There are many possible methods for including uncertainties in a LCA-tool, the difficulty being to find a method that is not too complex but still provide reasonable results. The most common methods to use are Monte Carlo simulation techniques, but analytical methods may also be used especially for identifying the most important parameters.

In this paper, the effects of including various levels of uncertainty on an LCA result are investigated. Different types of uncertainties are categorized and described, followed by including uncertainty values in a LCA-case study to find how they propagate. The aim of the study is to analyze the effect of different choices that can be made for including uncertainties such as stochastic parameters, model uncertainties and distribution types. If uncertainties should be included in an LCA, it is important to establish the assumptions and boundaries on how to perform such an analysis. This study aims to show the importance of such boundary and parameter choices.

The most influential parameters can be identified with sensitivity analysis methods, since for LCA with a large number of parameters it may be unreasonable to incorporate all in a probabilistic simulation. For a limited amount of influential variables, Monte Carlo simulation has been used to assess the effects of uncertainties on the results. A bridge has been used as a case study to find important aspects in infrastructure LCA.

The results indicate that there are many different types of uncertainties related to an LCA and to environmental impacts in general. It is possible to categorize them to be able to have a better understanding of the uncertainties and to identify the sources. There is a lack of validated input data including uncertainty factors for LCA-applications, this is the major obstacle for implementing probabilistic methods in LCA.

The results also show that if the most influential parameters are considered as random variables, it is possible to estimate the uncertainty and increase the validity of the life cycle assessment.

The current practice of including uncertainties using variability for emission factors is not sufficient for estimating the total uncertainty. In this case study, it was shown that the effect of variations in material amounts and estimated service life for each input parameter will have a large influence on the results, which is often not included in today's methods that are more focused on data quality for the emission factors themselves.

The large number and magnitude of uncertainties related to LCA makes it necessary to incorporate such effects as a basis for decisions when comparing environmental impact of different solutions. If uncertainties are not considered, it is recommended that the LCA results are not used as a basis for procurement decisions.

Life-Cycle Analysis and Assessment in Civil Engineering: Towards an
Integrated Vision – Caspeele, Taerwe & Frangopol (Eds)
© 2019 Taylor & Francis Group, London, ISBN 978-1-138-62633-1

LCA of civil engineering infrastructures in composite materials—ACCIONA Construction's experience

M.M. Pintor-Escobar, E. Guedella-Bustamante & C. Paulotto
Technology & Innovation Division, ACCIONA Construction, Alcobendas, Madrid, Spain

ABSTRACT

Construction sector is aware of the need of measuring, avoiding and reducing environmental damage caused by construction materials, processes, services, construction methods and sites, under a life cycle perspective. ACCIONA Construcción has the strategy of considering LCA as a tool for both environmental and business management as decision making tool for their projects. ACCIONA has applied LCA methodology to innovative civil engineering infrastructures built with innovative materials and construction processes that help to reduce the environmental impacts and improve the sustainability of the infrastructures. Some LCA examples highlight encouraging results for innovative

materials like fiber reinforced polymers in comparison with traditional ones like concrete and steel. This paper shows the results obtained by the evaluation of the environmental impact analyzed under LCA methodology of the construction of a 30 m span and 12m width bridge, comparing the environmental behavior of a FRP bridge and a concrete bridge, and the comparison between steel and FRP external strengthening systems. Concrete bridge was made with prefabricated concrete troughs beams, precast concrete pre-slabs, steel bars and a concrete slab. FRP Bridge was made with FRP beams (infusion process), glass fiber pre-slabs (pultrusion method), a concrete slab, glass fibers bars (pultrusion method) and glass fiber upper flange (infusion process).

Method and assessment decisions in the evaluation of the LCA-results of timber construction components

S. Ebert & S. Ott
Chair of Timber Structures and Building Construction, Technical University Munich, Germany

ABSTRACT

With LCA-results as an important additional decision criteria for planers, the significance of the individual results shown gain more and more center stage. This paper discusses different aspects for LCA-planers regarding method and assessment decisions. Ecological studies and assessments can vary in many different assumptions, indicators and methods according to the intended goal and scope. This is why an integral approach, transparency and iteration are part of the main principles of life cycle assessment (CEN, 2009). This paper will outline proposals for the communication and presentation of the results of these indicators for better decision and optimization processes.

The LCA-results of the "dataholz.eu"-project provide the background data and argument stated in this paper. The 'dataholz.eu' database is an online-catalogue for building materials, components and details especially for timber structures (HFA Austria, 2018) with additional timber construction components for the German market as part of a cooperative project between the Technical University of Munich and 'Holzforschung' Austria funded by the 'Deutsche Bundesumweltstiftung' (DBU). The calculated components include 116 outer wall components consisting of 35 massive timber constructions and 81 timber frame components. The insulation material varies between mineral wool (43 components), wood fiber (41 components) and cellulose (32 components).

This contribution constitutes the relevant aspects regarding environmental impact, embedded biogenic carbon and use of resources, especially with the use of renewable resource based construction components. With the normative background (CEN, 2014) the necessity of a calculation of components containing renewable resources especially over the product and construction stage to the end of life stage becomes very clear. The following illustration demonstrates that a focus on single life stages may lead to incomplete and confusing conclusions.

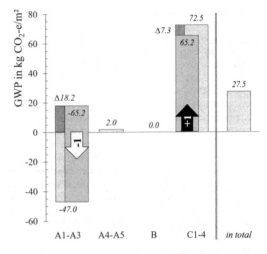

Figure 1. GWP results for one construction component over the whole LC with the illustration of the embedded carbon flow.

To achieve transparency and comprehension LCA-results should be presented according to clear recommendations. The results of this study show the ambivalence between the positive effects due to the use of wood products and the embedded carbon and the reduction of primary energy use, which cannot be played off against each other.

REFERENCES

CEN (2009–11) *Environmental Management – Life Cycle Assessment – Principles and Framework*. (EN ISO 14040). Brussel.
CEN (2014-03) *Round and Sawn Timber – Environmental Product Declarations – Product Cate-Gory Rules for Wood and Wood Based Products for Use in Construction*. (EN 16485). Brussel.
HFA Austria. (2018) dataholz.eu. Retrieved from: https://www.dataholz.eu/

Comparative evaluation of the ecological properties of timber construction components of the dataholz.eu platform

S. Ott & S. Ebert
Chair of Timber Structures and Building Construction, Technical University Munich, Germany

ABSTRACT

The comparison between different materials are quiet popular amongst construction products suppliers to promote the own product, if it succeeds in performance assessment against concurrent materials. Beyond this pure marketing related material comparison, in planning processes decision makers search trustful, transparent and independent assets of the ecologic impact of construction components. As surplus, they also request a bunch of facts to know about the component's performance, from structural load-bearing to heat and sound protection. The dataholz.eu web-platform supplies a multitude of such data for timber components either timber-framed as well as for mass timber, cp. (HFA Austria 2018). Based on a coherent calculation method according to EN standards LCA-results for components are calculated, followed by an in-depth comparative analysis between variations of the same component composition (CEN 2013).

The calculated components include 59 floor slabs consisting of the categories massive timber with 7 and timber frame components with 52 instances. Overall results from traditional, standard requested and important indicators like Global Warming Potential to address environmental impact or Primary Energy Use to address resource consumption are calculated for all floor slabs. For better analysis and interpretation, all results were clustered between timber frame components and mass timber components. Additionally further clustering on insulation material, underside lining and screed type are presented in detail.

The results reflect the improvement regarding the standard deviation of all accumulated results. In the further course of the research, the distinction between timber frame and massive timber construction leads to more differentiated results for the mean values and standard deviations within the respective group, especially to a decreased standard deviation for regrowing resources and therefore the embedded biogenic carbon. Another essential distinction deals with the different insulation materials. Again, more accurate

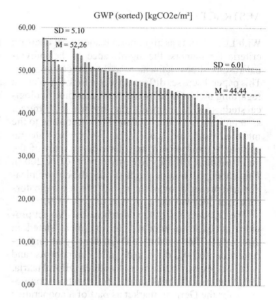

Figure 1. Distinction between structural material on GWP indicator.

characteristics can be determined and in the same move is also clearly identifiable for the biogenic insulation materials. In the same way the differentiation of characteristic properties can be shown in detail. Although, knowing the given data are only a first estimate for full building LCA, we have shown that it is close to a state-of-the-art full LCA of buildings, qed.

REFERENCES

CEN (2013) EN 15804:2013 -11 *Sustainability of Construction Works – Environmental Product Declarations – Core Rules for the Product Category of Construction Products.*
HFA Austria (2018) www.dataholz.eu – *Catalog of Building Physics and Ecologically Tested Wooden Components.*

The impact of structural system composition on reduced embodied carbon

M. Sarkisian, D. Shook, C. Horiuchi & N. Wang
Skidmore, Owings & Merrill LLP, San Francisco, California, USA

ABSTRACT

The mitigation of structural system environmental impacts during construction and in service is critical to the overall life-cycle sustainability goals set forth by jurisdictions and designers.

Using life-cycle analysis, hundreds of fully designed buildings have been evaluated for their embodied carbon impacts. Data suggests that traditional steel and concrete framed buildings have similar levels of embodied carbon when considering all structural elements in light of material procurement, on-site construction, and life-cycle seismic repair. In fact, higher diversity is observed when buildings are categorized by their gravity and lateral force resisting systems. These results challenge the traditional, material-based structural system selection process in favor of a systems-based structural system selection process. It also indicates that structural system selection including non-traditional materials has the greatest potential.

The database of buildings constructed around the world has allowed for a comprehensive investigation into the environmental impacts of structural systems. Analysis of this data allows for a calculation of an average embodied carbon value which can be used by engineers to make valuable comparisons when targeting improvements. Furthermore, investigations into the data can reveal important trends, relating embodied carbon averages to building height, lateral loading conditions, gravity system and lateral system.

The potential for structural system selection informed by databases of embodied carbon levels is possible. By observing data that is representative of actual embodied carbon levels in construction, designers can focus on the use of efficient structural systems and the development of new structural systems. Most importantly, designers now have an understanding of embodied carbon impacts in structural system selection.

Recently, designers have reconsidered traditional structural systems and focused on strategic structural

Figure 1. Rendering of Inlayed Slab Structural System.

system compositions that incorporate a variety of materials including mass timber and non-traditional applications of technologies such as post-tensioning informed by topology optimization. Synergies of significant embodied carbon mitigation, enhancement of seismic resiliency, improved serviceability through light-weight structural systems, and the creation of value-enhancing aesthetic options have been identified at a variety of scales and applications. One example of this is presented, in which inlayed timber panels are integrated into a post-tensioned concrete slab system and serve to reduce structural weight and also reduce embodied carbon. This interdisciplinary solution informs the architectural expression and reduces finishing costs while creating a unique residential experience.

The consideration of embodied carbon is critical due to the extensive impact of the built environment. The presented LCA dataset establishes important findings related to reducing life-cycle embodied carbon of structural systems. The described building case studies provides clear examples of a new language of structural systems which can dramatically improve life-cycle impacts of structural systems, reduce construction costs, and enhance value to developers.

Benchmarking embodied carbon in structural materials

C. De Wolf
Swiss Institute of Technology Lausanne, Lausanne, Switzerland

D. Davies
Magnusson Klemencic Associates, Seattle, USA

ABSTRACT

To meet the targets set in the Paris Climate Agreement, the Intergovernmental Panel on Climate Change states that the building sector should be "zero carbon" by 2050. One specific type of emissions, the operational carbon, has been the focus of sustainability in building design. Operational carbon is related to the use phase of the building: heating, cooling, ventilation, lighting, and hot water. However, the carbon related to the rest of the building's life cycle including material extraction, production, transport, construction and demolition or the embodied carbon, is often not reduced. Architects are involved in initiatives such as the American Institute of Architects (AIA) 2030 Commitment to lower the operational energy in their buildings. However, structural engineers play a major role in reducing carbon emissions.

In recognition of this, the Carbon Leadership Forum, an academic and construction industry collaboration, launched the Structural Engineers 2050 Commitment. The initiative aims to inspire structural engineers to contribute towards the global vision of net zero carbon buildings by 2050. To participate, structural engineers are encouraged to share their progress with data on material quantities and embodied carbon in their structures within a publicly available database of embodied Quantity outputs (deQo), which will be presented. This paper and presentation will provide an overview of the SE 2050 Commitment and its challenge, structural material tracking goals, and how it is proceeding in the early days of adoption.

The presentation will also provide case studies that consider this challenge and strategies for reductions in the embodied carbon of building structures: including structural system optimization and low carbon material choices that focus on the tracking and reporting of embodied carbon coefficients (kg_{CO2e}/kg) of the materials being used. Also presented, and in support of the SE 2050 challenge, will be a highlight of a recently released LCA specification guide, also published by

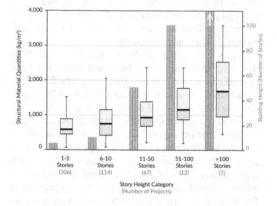

Figure 1. Output for deQo (Simonen et al., 2017).

the Carbon Leadership Forum, which allows a more consistent ask of LCA material information from suppliers.

In project case studies provided by Magnusson Klemencic Associates, variations of carbon footprints from conventional practices within the manufacturing of concrete, steel and timber are discussed, as well as new lower carbon construction materials under development. These newer materials will include carbon negative concrete aggregates, a byproduct of a CO_2 scrubber technology; new cement replacement materials, including ground recycled glass; and composite pressed bamboo dimensional lumber, which is more fire resistant, and stronger than conventional lumber.

This paper is complementary to a presentation being proposed by Mark Sarkisian and David Shook of SOM.

REFERENCE

Simonen, K., Rodriguez, B. & De Wolf, C. (2017) Benchmarking the embodied carbon of buildings. *TAD*, 1 (2), 88–98.

Life-Cycle Analysis and Assessment in Civil Engineering: Towards an Integrated Vision – Caspeele, Taerwe & Frangopol (Eds)
© 2019 Taylor & Francis Group, London, ISBN 978-1-138-62633-1

Sustainable model-based lifecycle cost analysis of real estate developments

M. Moesl & A. Tautschnig
Unit of Project and Construction Management, University of Innsbruck, Innsbruck, Austria

ABSTRACT

The development of real estates in relation to the design process is increasingly based on sustainable aspects in order to ensure the value of real properties at an early stage. Due to this aspect, an application of a holistic cost-analysis is important. Such an approach combines production costs and user costs for the life-cycle cost calculation on the one hand with the identification of the expected yield during the operating phase on the other hand. The main purpose of this article is to present this approach relating to the stage of development.

Comparing revenue and expenses, the predicted life-cycle ROI (Return on Investment) may be calculated and the results can be used as a prognostic factor for the profitability of an investment. Based on the analysis of all relevant costs and incomes for the life cycle it's possible to identify capabilities and variants of component combinations and building arrangements at an early stage. The added value to reflect all considerations in the life-cycle return enables the comparison of variations in the project as well as different types of buildings.

An automatic combination of the basic information included in a digital model (e.g. floor spaces, spatial volumes, levels) with other complementary indicators is required for an early stage cost-analysis. Based on the implication of these early perceptions an increase in productivity and efficiency is offered in the following design phases. As a result, specifications for the building construction and operational aspects may be planned in advance in order to create an optimally and economically built and operated property.

Subsequently these facts offer an economic advantage for all project participants ensuring a sustainable value of the property. How this is achieved using building information modeling techniques will be shown in this article.

EFIResources: A novel approach for resource efficiency in construction

H. Gervasio
EC-Joint Research Centre, Ispra, Italy
ISISE, Department of Civil Engineering, University of Coimbra, Portugal

S. Dimova
EC-Joint Research Centre, Ispra, Italy

ABSTRACT

The project *EFIResources* aims for the development of a performance based approach for sustainable design, which enables to assess resource efficiency of buildings. The proposed methodology is in line with the "limit state approach" of the Eurocodes and aims for the harmonization between environmental criteria and structural criteria in the design of buildings, leading to an enhanced design, coping with the required safety demands but with lower pressure on the environment and on the use of natural resources.

Hence, in this approach two variables are defined: (i) the environmental performance of the structure (E) and (ii) the reference value of the environmental performance of the structure (R), against which the performance of the building may be compared. Both variables are quantified based on a life cycle approach and therefore, they are subjected to a high degree of uncertainties and variabilities not only due to the long life span of buildings but also due to the inherent uncertainties of life cycle approaches. These uncertainties should be taken into account in the analysis and therefore, both variables are defined by vectors of basic random variables with respective probability density functions, as represented in Figure 1.

In this paper, the above approach is introduced but the paper focus on the development of the framework for the quantification of the benchmarks and on the adopted model for the environmental life cycle assessment of buildings, which is based on EN 15804:2012 and EN 15978:2011. The use of a standardized procedure ensures the use of a consistent approach that was developed specifically for the assessment of construction works (Gervasio and Dimova, 2018a).

In the end of the paper, a preliminary set of benchmarks, indicated in Table 1, is provided for 'conventional' and 'best' practices, based on the statistical analysis of a representative sample of buildings (Gervasio and Dimova, 2018b).

Table 1. Statistical analysis for each building type in GWP (kg CO_2 eq./m^2.yr) and PE (MJ/m^2.yr).

		Median	Standard deviation	**Quartile 25%**	Quartile 75%
SI	GWP1	7.22	3.80	**2.53**	8.71
	GWP2	8.94	4.20	**5.01**	11.27
	PE	139.25	54.17	**124.19**	186.18
MF	GWP1	6.30	3.76	**4.88**	9.94
	GWP2	7.32	3.89	**5.37**	10.75
	PE	105.60	48.80	**84.50**	159.77
HR	GWP1	5.53	2.05	**4.34**	6.91
	GWP2	6.57	1.81	**5.03**	6.94
	PE	88.89	24.78	**68.51**	94.93

REFERENCES

Gervasio, H. & Dimova, S. (2018a) *Model for Life Cycle Assessment (LCA) of Buildings*. EUR 29123 EN, Publications Office of the European Union, ISBN 978-92-79-79974-7, 10.2760/789069, JRC110082.

Gervasio, H. & Dimova, S. (2018b) *Environmental Benchmarks for Buildings*. Publications Office of the European Union, JRC110085.

Gervasio, H., Structural eco-efficiency: harmonizing structural and environmental assessments. *European Journal of Environmental and Civil Engineering*, Published online: 17 Jan 2017.

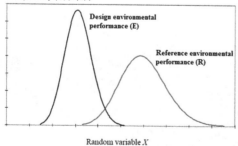

Figure 1. Probability density functions of E and R (Gervasio, 2017).

Life-Cycle Analysis and Assessment in Civil Engineering: Towards an Integrated Vision – Caspeele, Taerwe & Frangopol (Eds)
© 2019 Taylor & Francis Group, London, ISBN 978-1-138-62633-1

Sustainability rating of lightweight expanded clay aggregates using energy inputs and carbon dioxide emissions in life-cycle analysis

F.M. Tehrani & R. Farshidpour
California State University, Fresno, California, USA

M. Pouramini
University of Tehran, Tehran, Iran

M. Mousavi & A. Namadmalian Esfahani
LECA Company, Tehran, Iran

ABSTRACT

This paper highlights the result of a case study on energy inputs and carbon dioxide emissions of lightweight expanded clay aggregates. These aggregates have numerous applications in various areas of civil engineering and agriculture, including lightweight construction, geotechnical backfills, pavement sections, water treatment, and horticulture. Replacement of non-renewable natural aggregates with alternative lightweight aggregates in these applications offer new opportunities and challenges. The environmental consequence of such replacement is particularly significant for geotechnical fill applications involving large volume of materials, often in direct contact with natural ground and undisturbed soil. Such significance justifies implementation of sustainable means, methods, and materials of

construction. Evaluating the potential of these aggregates to enhance sustainability rating of infrastructures requires analyzing energy inputs and carbon dioxide emissions as major sustainability performance measures. These measures contribute to the production phase of any lifecycle cost analysis, where environmental footprints represent the cost. Similarly, physical and mechanical characteristics of materials, such as lightness, damping, insulation, and durability, will alter the consumption of energy and the release of greenhouse gas emissions during operation, maintenance, and decommissioning phases of the lifecycle analysis. The presented case study employs these techniques to reframe lifecycle analysis based on environmental measures and to compare the outcome with conventional cost analyses. Moreover, conclusions discuss the link between these comparative analyses and ENVISION sustainability rating measures.

Standardization of condition assessment methodologies for structures

J. Engelen
Chairman NEN Commission, The Netherlands

R. Kuijper
Program Manager Infrastructure Province Gelderland, The Netherlands

D. Bezemer & L. Leenders
Director & Consultant DON Bureau, The Netherlands

ABSTRACT

An increasingly number of asset owners are using asset management as a method for a cost-effective exploitation and management of their assets. For public space (built environment) in general and in particular for infrastructure and real estate. No matter where and how assets are managed, one of the basic requirements is the availability of complete information regarding the assets. It is vital to know your assets and in what state they are. Uniformity and objectivity in the way the technical state of assets is assessed and reported is essential.

It is therefore that about a decade ago the sector in the Netherlands developed a method for condition assessment that is applicable for assets in the public environment. This resulted in the Dutch standard for Condition assessment NEN2767, first for real estate and later for infrastructure.

The European Committee for Standardization (CEN) is planning to realise an European Technical Specification (TS) that describes an methodology to carry out a physical condition assessment for all types of non-movable constructed assets. The CEN/TC 319 'Maintenance' started a new workgroup, WG 11 'Condition assessment methodologies for structures' to draft a TS which describes a methodology for condition assessment. The TS will describe an objective inspection method, which is usable for the determination of the technical condition of assets in a uniform way. Starting point is the Dutch standard for Condition assessment of buildings and infrastructural assets, NEN 2767. This standard is now common practice in the building and infrastructural sector. Recently (January 2017) a new version has been released in named "Condition assessment of the built environment".

By applying NEN 2767 the condition can be expressed in a condition score rating. This condition

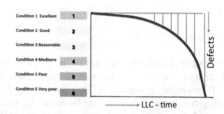

Figure 1. Condition rating based on defects assessment.

score is based on three parameters: the severity, extent and intensity of the defects. This will lead to:

- uniformity in both assessment and results;
- greater objectivity;
- similarity in definitions.

As a result, all parties involved in the process of condition assessments will benefit from:

- a uniform interpretation of condition scores on all possible aggregation levels of assets in the public environment;
- improved long-term asset planning and control;
- improved maintenance budget planning.?

REFERENCES

NEN 2767-1:2017 (2017) *Condition Assessment of Built Environment – Part 1: Methodology.* Delft, The Netherlands, NEN.

NEN 2767-2:2008 (2008) *Condition Assessment of Buildings and Installation Components – Part 2: List of Faults.* Delft, The Netherlands, NEN.

NEN 2767-4-2:2010 (2010) *Condition Assessment – Part 4: Infrastructure – Part 2: Web Application for the Decomposition and List of Defects.* Delft, The Netherlands, NEN.

Van der Velde, J. (2011) *Asset Management.* Utrecht, The Netherlands, Rijkswaterstaat.

Vrouwenvelder, A.C.W.M. (2011) *Onderhoud en beheer in de bouw.* Delft, The Netherlands, TNO/TU Delft.

Life-Cycle Analysis and Assessment in Civil Engineering: Towards an Integrated Vision – Caspeele, Taerwe & Frangopol (Eds)
© 2019 Taylor & Francis Group, London, ISBN 978-1-138-62633-1

Life cycle assessment of asphalt mixtures healed by induction heating

E. Lizasoain-Arteaga, I. Indacoechea-Vega & D. Castro-Fresno
GITECO Research Group, Universidad de Cantabria, Santander, Cantabria, Spain

ABSTRACT

Transport infrastructures have a positive impact on the economic growth by creating jobs and wealth, boosting trade between regions, and above all facilitating daily life for millions of citizens. However, the continuous increase of freight traffic on an already aged road network is imposing a heavy burden on road structures leading to increasingly frequent maintenance and rehabilitation interventions. Repairing pavements by traditional methods extends the lifetime of roads, but it is costly, requires a large amount of natural resources and causes unacceptable flow capacity losses, which incur in delay, reduced mobility and reduced reliability of the road network.

Induction-heated asphalt mixtures (HEALROAD) consist of replacing fractions of aggregates with magnetic particles. When micro cracks appear in the wearing course, an induction heating generator passes through the road surface, heating only the magnetic particles. Bitumen melts, flowing through the micro cracks and closing them, extending more than 90% the lifetime of the road when only one healing treatment is applied (Ajam et al., 2017) (Gómez-Meijide et al., 2016). This preventive maintenance, that postpones the replacement of the asphalt surface for several years, is nearly zero-intrusive and has a minimum impact on the road network capacity.

To establish the environmental characteristics of the HEALROAD technology, a preliminary life-cycle analysis (LCA) has been carried out following the standards ISO 14040:2006 (ISO, 2006a) and ISO 14044:2002 (ISO, 2006b). To this aim, a porous induction-healing asphalt surface has been compared with a conventional one traditionally maintained. Results, in terms of Global Warming Potential (GWP) and Energy consumption, have been referred to the impacts produced by 1 km of a 3.5 m width lane during a 30-years analysis period.

After evaluating the material production, road construction, road maintenance, congestion and end-of-life of both technologies, it has been concluded that the HEALROAD technology generates less GWP and consumes less energy than the conventional case, mainly due to the consideration of the traffic affection during maintenance activities. However, a more detailed study is being carried out.

ACKNOWLEDGEMENTS

This paper is based on the results obtained in the HEALROAD project (grant number 31109806.0003), financed by the ERA-NET Plus Infravation 2014 Call. The HEALROAD project has been carried out by the University of Cantabria, University of Nottingham, German Federal Highways Research Institute (BASt), European Union Road Federation (ERF), Heijmans Integrale projecten B.V. and SGS IN-TRON B.V.

REFERENCES

Ajam, H., Lastra-González, P., Gómez-Meijide, B., Airey, G. & Garcia, A. (2017) Self-healing of dense asphalt concrete by two different approaches: Electromagnetic induction and infrared radiation. *Journal of Testing and Evaluation*, 45. https://doi.org/10.1520/JTE20160612

Gómez-Meijide, B., Ajam, H., Lastra-González, P. & Garcia, A. (2016) Effect of air voids content on asphalt self-healing via induction and infrared heating. *Construction and Building Materials*, 126, 957–966. https://doi.org/https://doi.org/10.1016/j.conbuildmat.2016.09.115

ISO (2006a) *ISO 14040: Environmental Managemen – Life Cycle Assessment – Principles and Framework, 2 end. International Organization for Standarization.*

ISO (2006b) *ISO 14044: Environmental Management – Life Cycle Assessment – Requirements and Guidelines, 1 edn. International Organization for Standarization.*

Service life prediction of pitched roofs clad with ceramic tiles

R. Ramos
Portuguese Air Force, Pêro Pinheiro, Portugal

A. Silva & J. de Brito
IST, University of Lisbon, Lisbon, Portugal

P.L. Gaspar
Faculty of Architecture, University of Lisbon, Lisbon, Portugal

ABSTRACT

Buildings and their components are subjected to different degradation processes during their life cycle, thus suffering from various types of deterioration. The degradation of elements of the buildings' envelope, such as pitched roofs, occurs essentially due to the aging process associated with the absence of appropriate maintenance policies (Adeniyi & Farayola, 2012). The degradation of the roof system increases the probability of failure of the building envelope and of the entire building, having a significant impact on the annual expenditures on maintenance.

In this sense, this study proposes a systemic methodology to predict the service life of ceramic claddings on pitched roofs, considering various steps: i) inspection and diagnosis of pitched roofs' ceramic claddings, in service conditions, through visual inspections; ii) mathematical modelling of the sample obtained during fieldwork; iii) identification of the degradation factors that influence the deterioration of pitched roofs' tiles; and iv) estimation of their service life. For that purpose, a sample of 85 buildings and 146 claddings, with a total area of 43.991,6 m² were analysed.

A numerical index, called severity of degradation (Gaspar & de Brito, 2011), is applied to evaluate the overall degradation condition of the pitched roofs' ceramic claddings analysed. Based on this index, a degradation curve or pattern is defined, correlating the degradation of the claddings and their age. In this study, the service life of ceramic claddings on pitched roofs' is calculated through the intersection between the degradation curve and the maximum acceptable degradation level, which corresponds to a severity of degradation of 20% (Figure 1).

According to the model proposed in this study, the pitched roofs clad with ceramic tiles present an estimated service life of 60 years. This value seems coherent with the empirical knowledge and the related studies addressing the durability and service life of ceramic claddings and roofing systems.

The loss of performance of ceramic claddings in pitched roofs during their life cycle is strongly

Figure 1. Illustrative example of a pitched roof ceramic cladding with a severity of degradation of 20%.

influenced by the claddings' characteristics. Thus, in this study, the influence of the roofs' design (roofs' slope and geometry), the characteristics of the ceramic claddings (type of ceramic tile) and the environmental exposure conditions (building's location and exposure, height of the building, climatic zone and distance from the sea) on the ceramic claddings' service life are evaluated. Different degradation curves are defined for each characteristic considered, which allows predicting the estimated service life for the pitched roofs' ceramic claddings according to the characteristic analysed.

The results obtained show logical trends, in which claddings subjected to more unfavourable conditions are more prone to degradation, with a high incidence of defects, reaching the end of service life sooner than claddings in buildings subjected to more favourable conditions. The results of this study can be extremely useful for the adoption of strategic proactive maintenance policies, optimizing funds and resources, as well as improving the performance of pitched roofs' ceramic claddings during the buildings' life cycle.

REFERENCES

Adeniyi, Y.B. & Farayola, O.H. (2012) Financial planning model for sustainable building maintenance. In: *International Conference on Sustainable Development & Environmental Protection 2012*, Edited by Adedeji Daramola, Bells University of Technology, Published by Institute for Environment Research and Development, Nigeria. pp. 198–200.

Gaspar, P. & de Brito, J. (2011) Limit states and service life of cement renders on façades. *Journal of Materials in Civil Engineering*, 23 (10), 1396–1404.

Energy consumption evaluation of a passive house through numerical simulations and monitoring data

C. Tanasa, V. Stoian, D. Stoian & D. Dan
Politehnica University Timisoara, Timisoara, Romania

ABSTRACT

This paper presents the investigations on an existing passive house using numerical simulations and monitoring data. The work involved developing the energy model of the building using the EnergyPlus dynamic simulation tool and calibrating the building energy using measured data. A set of data measured in 2012 was used: hourly interior temperature for heating and cooling temperature set points schedules, hourly exterior temperature, hourly lighting and electric equipment energy consumption for internal loads. The accuracy of the building energy model was assessed through the statistical indicators Normalized Mean Bias Error (NMBE) and Coefficient of Variation Root Mean Square Error (CVRMSE). The simulation is considered calibrated if CVRSME is less than 15% and NMBE is in the range ±5% (ASHRAE 14 2002, FEMP 2015). A value of 5.25% was obtained for the CVRMSE and −2.67% for NMBE. The graph in Figure 1 shows the monthly measured and simulated energy consumption. The house's 2012 annual total electricity consumption was 6590.16 kWh.

The energy model simulation resulted in an annual electricity consumption of 6831.05 kWh. The annual simulated energy consumption is 3.65% higher than the annual measured energy consumption. The major

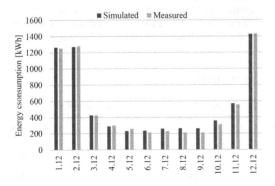

Figure 1. Simulated – Measured total energy consumption.

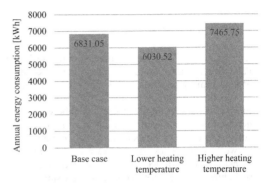

Figure 2. Annual energy consumption for the three scenarios.

discrepancy between measured and simulated lies in the heating, cooling and ventilation energy consumption. This suggests that a more accurate input data is necessary related to the building systems characteristics and operation schedules in order to increase the performance of the calibration. This paper also investigated the effect of the temperature set-point for heating on the total energy consumption. Two scenarios were simulated in order to assess how the increase or decrease of the interior temperature contributes to the energy consumption change. The graph in Figure 2 shows the results for the two scenarios compared to the base case scenario. It was concluded that the total energy consumption can vary with approximately ±10% at a temperature variation of ±2°C.

REFERENCES

ASHRAE Guideline 14 (2002) *Measurement of Energy Demand and Savings*. Atlanta, GA, American Society of Heating, Refrigerating and Air-Conditioning Engineers.

Federal Energy Management Program (FEMP), M&V Guidelines (2015) *Measurement and Verification for Federal Energy Projects Version 4.0*. Washington, DC, U.S. Department of Energy Federal Energy Management Program.

Life-Cycle Analysis and Assessment in Civil Engineering: Towards an Integrated Vision – Caspeele, Taerwe & Frangopol (Eds)
© 2019 Taylor & Francis Group, London, ISBN 978-1-138-62633-1

A review of retrofit strategies for Large Panel System buildings

E. Romano, O. Iuorio & N. Nikitas
School of Civil Engineering, University of Leeds, Leeds, UK

P. Negro
European Commission, Joint Research Centre (JRC), Ispra (VA), Italy

1 INTRODUCTION

A large stock of the 50s-70s precast concrete Large Panel System (LPS) buildings, based on the assembly of load-bearing panels for walls and slabs connected by on-site wet or dry joints, is still present in all Europe. This structural typology is inherently vulnerable to extreme loadings and prone to progressive collapse because of the lack of integrity coming from its weak joints. Thus, safety and reliability requirements are not satisfied, as demonstrated by the rate of failure and damages due to gas explosions above all in UK and in Romania in the last decade. The interest in progressive collapse started after the 1968 Ronan Point partial collapse in London resulting in the first robustness provisions, even if the lack of specific codes against extreme events is still evident. Moreover the majority of LPS buildings approach the end or exceeded their expected life leading to higher probability of serviceability deficiencies. LPS structural retrofit to resist normal and accidental loads becomes a high-priority research challenge. In that line, this study is a literature review based paper on the potential LPS retrofit techniques as a first step of an on-going research project focused at investigating the behavior of local strengthened LPS residential buildings with the purpose to provide guidelines for the best practice. This should consider also environmental and economic issues beyond the structural performance in the perspective of integrated life-cycle oriented retrofits.

2 CONCRETE STRUCTURES RETROFIT

Retrofit needs to change from a reactive approach into a predictive and performance based life-cycle approach. Thus, service life profiles under ordinary and exceptional conditions should be considered. Attention is gained to concrete structure retrofit approaches, also focusing on strengthening measures to mitigate progressive collapse.

3 RETROFIT OF LPS UNDER ACCIDENTAL LOADS

An overview on retrofit options for LPS buildings is presented with attention on local and global scale. The former include strengthening solutions for load-bearing components and joints. The latter refers to vertical and horizontal tying or installation of structural members (Matthews & Reeves 2012). A consequent focus on structural retrofit of the flank wall/floor horizontal joint is devoted, underlining that structural deficiencies in LPS dwelling blocks are still evident in all Europe.

4 A MULTI-PERFORMANCE RETROFIT

The importance of defining an integrated retrofit methodology to assess the most appropriate solution in terms of structural, environmental and economic performance is briefly presented. The Sustainable Structural Design (SSD) methodology (Romano et al. 2014), which consists in three main steps to combine environmental aspects and structural design in economic terms throughout the entire life-cycle of a structure, is suggested in order to extend integrated approaches to retrofit under accidental loads. A specific methodology will be developed in the future steps of the research and applied to LPS buildings.

REFERENCES

Matthews, S. & Reeves, B. (2012) *Handbook for the Structural Assessment of Large Panel System (LPS) Dwelling Blocks for Accidental Loading*. Garston, Watford, IHS BRE Press.

Romano, E., Negro, P. & Taucer, F. (2014) *Seismic Performance Assessment Addressing Sustainability and Energy Efficiency*. Joint Research Centre, Report EUR 26432 EN, Luxembourg, Publications Office of the European Union.

Life-Cycle Analysis and Assessment in Civil Engineering: Towards an Integrated Vision – Caspeele, Taerwe & Frangopol (Eds)
© 2019 Taylor & Francis Group, London, ISBN 978-1-138-62633-1

Investigation, assessment and suggestion to the existing traditional house of low-income family in Cambodia based on the principles of passive house design

A. Vann & G.Q. He
College of Civil Engineering and Architecture, Zhejiang University, Hangzhou, China

ABSTRACT

A Cambodian traditional house is a building that comprises various sizes and heights of the area, that is specifically built in the purpose of airflow and light. Although the traditional building styles are followed, due to the small size opening window, the indoor environment is not so good. It does not have enough sunlight and weak airflow; so there should be some particular changes of the design characteristics to enhance air movement and to provide enough sunlight. Consequently, this paper is to study the existing traditional houses in Cambodia through the investigation and the energy consumption in the houses is analyzed using the principle of passive house design. Improvements in energy saving of residential traditional houses detailing were put forward. Furthermore, the economic advantage of the proposed measurements was validated by comparison between the proposed passive house and traditional house through the analysis using simulation software. Therefore, an analysis through the simulation software, DesignBuilder, and case studies are used for the study on lighting and energy consumption on cooling load. According to the simulation result, the proposed house with bigger area needs less energy consumption than the case study; moreover, the proposed house also has better sunlight than the case study according to the GreenStar Credit IEQ4 Report after the simulation. As data collection through the Temperature Data Logger (TDL) in Cambodia is time-consuming, so the study is shortened to one month study which might be insufficient for a whole year result. Moreover, since the available time of the study is in rainy season, future research should study dry season. Nevertheless, the proposed house design shows a good result in DesignBuilder simulation software. This study can provide reference on energy saving of traditional houses on lighting with renewable materials in urban low-income families in Cambodia for further research.

REFERENCES

Aflaki, A., Mahyuddin, N., Awad, Z.A.-C.M. & Baharum, M.R. (2015) A review on natural ventilation applications through building facade components and ventilation opening in tropical climates. *Energy and Buildings*, 101, 153–162.

Al-Obaidi, K.M., Ismail, M. & Abdul Rahman, A.M. (2014) Passive cooling techniques through reflective and radiative roofs in tropical houses in Southeast Asia: A literature review. *Frontiers of Architectural Research*, 3, 283–297.

Bhikhoo, N., Hashemi, A. & Cruickshank, H. (2017) Improving thermal comfort of low-income housing in Thailand through passive design strategies. *MDPI Sustainability*, 9 (8), 1440.

Chalermwat, T., Torwong, C. & Maniporn, P. (2006) Integrative passive design for climate change: a new approach for tropical house design in the 21st century. *International Journal of Ventilation*, 30 (3), 259–282.

Haslam, M.P.G. & Farrell, A. (2014) Natural ventilation strategies in near-zero-energy building. In: *48th International Conference of the Architectural Science Association*. pp. 619–630. Institute, P.H. (24 Nov. 2015) *House in Mediterranean Climate*.

Institute, P.H. (2011) *Quality Approved Passive House Certification-Criteria for Residential Passive House*.

Komitu, A. (2015) *Eco-Friendly Youth Center*.

Puruj, A., Elena, D. & Kevin. M. (2006) *Lessons from Traditional Architecture. Design for Climatic Responsive Contemporary House in Thailand*.

Truong, N.-H.-L. (2014) Lessons from climatic response in Vietnamese Vernacular House. *Journal of Civil Engineering and Environment Technology*, 1, 41–46.

Trust, B. (2015) *Framework Project*.

Wimmer, R. (2013) Adapting zero carbon houses for tropical climates-passive cooling design in the Philippines. *SB13 Sustainable Buildings, Infrastruture and Communities in Emerging Economies*.

Wimmer, R., Hohensinner, H. & Drack, M. *S-House-Sustainable Building Demonstrated by a Passive House Made of Renewable Resources*.

Interchange of economic data throughout the life cycle of building facilities in public procurement environments

F. Salvado
LNEC, Lisbon, Portugal

N. Almeida
IST-UL, Lisbon, Portugal

Á. Vale e Azevedo
LNEC, Lisbon, Portugal

ABSTRACT

The Life Cycle Cost (LCC) concept enables costs prediction throughout the life cycle of building projects. The scientific community and the practitioners of the Architecture, Engineering and Construction (AEC) sector have been developing this concept for the past decades. Presently, it is encouraged by several international and regional standards, procurement guidelines and regulations applicable to this same sector. However, public procurers of building projects still face difficulties regarding the costs estimation and time ratios of systems and components over the life-cycle of building facilities. It has been recognized that dedicated public databases with the adequate quantity and quality of economic data are needed in this regard, and that the existing ones often present problems such as those of inadequate data granularity, incompleteness and inaccuracy or data structures with formats that make comparison and extrapolation difficult. The present document discusses a framework for economic data collection throughout the whole life-cycle of building facilities (LCC-EDC), which may contribute to address these problems. This proposal seeks to harmonize and empower previous efforts in the realm of LCC applications in the AEC sector, enabling information exchanges according to a standardized taxonomy that can fulfill gaps and limitations of current practice. A case study, related to buildings, is also described

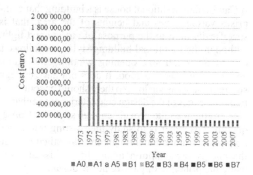

Figure 2. Example of building LCC nominal costs.

and discussed as to show the potential of the proposed framework for widespread LCC application.

LCC-informed decisions in BM depend on the widespread and consistent application of the LCC concept to the AEC sector, namely by generating and making available the adequate quantity and quality of economic data.

This framework is applicable to any type of building and intends to assist decision making processes at any stage of the building life cycle. The main goal of is to improve the economic performance of buildings while addressing the recent challenges imposed by regulations such as the European Directive 2014/24/EU.

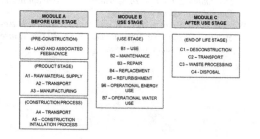

Figure 1. Framework for economic data collection (LCC-EDC).

REFERENCES

Directive 2014/24/EU of the European Parliament and of the Council of 26 February 2014 on public procurement. *Official Journal of the European Union.*
EN 16627:2015 – *Sustainability of Construction Works. Assessment of Economic Performance of Buildings – Calculation Method* European Committee for Standardization.
Goh, B. & Sun, Y. (2016) The development of life cycle costing for buildings. *Building Research & Information*, 44 (3), 319–333.

Life-Cycle Analysis and Assessment in Civil Engineering: Towards an
Integrated Vision – Caspeele, Taerwe & Frangopol (Eds)
© 2019 Taylor & Francis Group, London, ISBN 978-1-138-62633-1

Smart grid integration towards sustainable retrofitting of large prefabricated concrete panels collective housing built in the 1970s

D.M. Muntean
Department of Steel Structures and Structural Mechanics, Politehnica University Timisoara, Timisoara, Romania

V. Ungureanu
Department of Steel Structures and Structural Mechanics, Politehnica University Timisoara, Timisoara, Romania
Laboratory of Steel Structures, Romanian Academy - Timisoara Branch, Timisoara, Romania

ABSTRACT

Buildings are intensive energy consumers, using up to 40% of the energy worldwide during their lifespan. Changes and measures in buildings modus operandi can lead to significant energy savings and reduced carbon footprint. Moreover, in the near future, building should be able to generate the amount of energy required for daily consumption by becoming zero or nearly zero energy buildings.

Built in the early 1970s due the rapid and vast growth of the urban population, buildings from large prefabricated concrete panels (LPCP) represent around 2% of the entire building stock. More than half of the urban population lives in them. Currently these buildings consume huge amounts of energy and are outdated by failing to meet modern living standards. Because they were built based on project types and configured in various ways for a better use of land, these buildings are suitable for an extensive retrofitting strategy to bring them up to date while improving not only the quality of life within but also the surrounding communities.

With the advances in design, operation optimization and control of energy-influencing building components (HVAC, solar, CHP, natural ventilation, shading, fuel cells etc.) the potential for realizing significant energy savings and efficiencies in a buildings operation is unleashed.

Today, governments and power companies across the world have recognized that the traditional grid, which has not significantly changed over the last decade, must be replaced by more efficient, flexible and intelligent energy-distribution networks, called "smart grids". The building connected to the new power network becomes both receiver as well as energy distributor. The grid optimizes power delivery and

Figure 1. Smart grid concept diagram.

facilitate two-way communication across the grid, enabling end-user energy management, minimizing power disruptions and transporting only the required amount of power. The result is a lower cost to the utility and the customer, more reliable power, and reduced carbon emissions.

Furthermore, by considering incorporation of smart metering, demand response, distributed systems and interoperability, the now called "smart buildings" can create a suitable environment for urban populations, creating a place of advanced social progress and environmental regeneration as well as a point of economic growth (Figure 1).

GS9: Assessment of existing structures

Condition assessment based on results of qualitative risk analyses

A. Panenka & F. Nyobeu
Federal Waterways Engineering and Research Institute, Karlsruhe, Germany

ABSTRACT

Large parts of the German transport infrastructure were built in the middle of the 20th century. Nowadays, since an increasing number of structures is in serious condition, there is a need for differentiated but still comprehensive key figures for the establishment of a meaningful prioritization of urgently necessary maintenance measures. Supported by the "BMVI Network of Experts" the Federal Waterways Engineering and Research Institute (BAW) developed a procedure using a *failure mode and effect analysis* (FMEA) with modified risk criteria and extended risk assessment, to provide responsible authorities with such key figures.

The FMEA is a systematic and inductive method. Its idea is the assessment of every possible failure modes for any system or system member. At the same time, possible failure consequences and failure causes are identified. Finally, the procedure leads to a risk assessment and the determination of optimization measures. The method aims at identifying risks and weak spots as early as possible in order to implement improvements early enough. The risk assessment based on the three criteria *occurrence*, *detectability* and *severity* results in a ranking of the identified risks supporting the prioritization of optimization measures for the most effective improvement in comparison to the actual situation (c.f. Bowles 2004). In context with civil engineering structures the criteria detectability was extended to *maintainability* in order to cover all aspects regarding effective maintenance measures.

The FMEA establishes a link between the data used for analysis and the functional requirements of the assessed object by means of the cause-and-effect chains. The link allows creating *specific condition grades* only considering certain functional requirements. The determination and calibration of the risk criteria makes use of a large variety of all forms of available data. The customized risk factors are flexible enough to allow the consideration of expert knowledge or detailed information derived from thorough structural analyses. The resulting key figures like *risk*

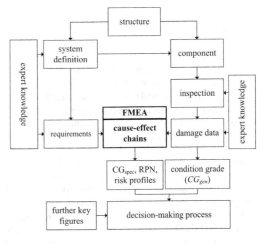

Figure 1. FMEA of civil engineering structures.

priority numbers or *risk profiles* are an expressive representation of influencing factors regarding the condition of a structure.

First trial applications show promising results. Current developments concern the definition of an indicator and its measure for the risk factor *severity* based on advanced structural analyses. Further, the generated key figures shall be implemented in a risk classification methodology for waterway infrastructures (c.f. Schmidt-Bäumler 2017).

REFERENCES

Bowles, J.B. (2004) An assessment of RPN prioritization in a failure modes effects and criticality analysis. *Journal of the IEST*, 47 (1), 51–56.

Schmidt-Bäumler, H. (2017) Risk-based maintenance management system for waterway infrastructures in Germany. In: Bakker, J., et al. (ed.) *Life-Cycle of Engineering Systems: Emphasis on Sustainable Civil Infrastructure*. London: Taylor & Francis Group. pp. 559–566.

Life-Cycle Analysis and Assessment in Civil Engineering: Towards an Integrated Vision – Caspeele, Taerwe & Frangopol (Eds)
© 2019 Taylor & Francis Group, London, ISBN 978-1-138-62633-1

Development and operation of non-destructive inspection device for stay cables

H. Sakai
Engineering/Construction Headquarters, Central Nippon Expressway Company Limited, Japan

ABSTRACT

Stay cables of cable-stayed bridges and exstradosed bridges are very difficult to inspect by visual inspection, because there are in high places. The high strength steel wire or strand used for stay cables may be broken due to corrosion due to penetration of deterioration factors, or fatigue due to stress fluctuation.

The stay cables consist of cables (steel wires or strands), protective tubes, and filling materials. The steel material used in a stay cable must be inspected routinely to confirm that its function has been maintained, because it is protected from corrosion with a protective tube and filling material. However, a inspecting of the cables, protective tubes, and filling materials is difficult to achieve owing to the height and difficulty in approaching the site. In order to solve these problems, the automatic stay cable inspection device was developed and operated on the prestressed concrete cable-stayed bridge. This inspection device can visually inspect the protective tube with image sensors (cameras) and can detect non-destructively the breakage of the steel wire or strand in the protective tube with the eddy current flaw detector.

The inspection device equipped with image sensors and eddy current flaw detection devices can be raised and lowered and stopped by wireless operation from a personal computer on the bridge deck. The mechanical specifications of the inspection device are as shown in Table 1.

Table 1. Mechanical specifications of the inspection device.

Item	Specification
Inspectable diameter	100 to 200 mm
Inspectable angle	65 degrees or less
Inspectable length	200 m or less
Device dimensions	$70 \times 60 \times 120$ cm
Body weight	48.0 kg
Eddy current sensor weight	7.2 kg
Inspection speed	7.0 to 10.0 m/min
Video camera pixels	2 million pixels
Power supply	Lithium ion battery 24 V

The inspection device was used at the Tomei Ashigara Bridge (prestressed concrete 3 span continuous cable-stayed bridge, pylon height 81.75 m, bridge length 370 m) under service as a bridge of the Tomei expressway. This bridge was completed in 1991 as the first prestressed concrete cable-stayed bridge on the Japanese expressway.

The protective tubes of 48 stay cables are inspected by the inspection device, 53 degradations were confirmed by the inspection device. Degradations were indentation, minute abrasions, and etc. In this inspection, the result of the inspection of the degradation of the protective tube by the inspection device and the result of the close visual inspection by the inspector were in agreement. From this fact, it was verified that the inspection device can be fully utilized as a substitute for the close visual inspection by the inspector. The occurrence of breakage of steel wire could not be inspected from the eddy current flaw detection result carried out at the degradation part of the protective tube. This is presumed that the corrosion of the steel wire did not occur because the protective tube was relatively slightly degraded and the deteriorating factor did not penetrate from the degradation part. However, since it has been verified that the breakage of the steel wire can be confirmed from the test result of the specimen conducted beforehand, it is inferred that it is possible to inspect the breakage of the steel wire even in the stay cable of the actual bridge. From these facts, it is concluded that the inspection device is sufficiently effective for the inspection for conservation of the stay cables.

REFERENCES

Sakai, H., Shirahama, S. & Hosoi, K. (2017) Maintenance method for stay-cable system and anchor zone. *Journal of Prestressed Concrete*, 59 (5), 49–56. Japan Prestressed Concrete Institute. *In Japanese.*

Sakai, H. & Ohashi, G. (2017) Development of non-destructive inspection device for inspection of stay-cables. *Concrete Journal*, 55 (8), 651–656. Japan Concrete Institute. *In Japanese.*

Life-Cycle Analysis and Assessment in Civil Engineering: Towards an Integrated Vision – Caspeele, Taerwe & Frangopol (Eds)
© 2019 Taylor & Francis Group, London, ISBN 978-1-138-62633-1

Point-based POMDP risk based inspection of offshore wind substructures

P.G. Morato, Q.A. Mai & P. Rigo
Department of ArGEnCo, University of Liege, Belgium

J.S. Nielsen
Department of Civil Engineering, Aalborg University, Denmark

ABSTRACT

The harsh marine environment becomes an obstacle as the structures are located further from shore and it is vital to perform an operation and maintenance optimization including information gathered from inspection and monitoring data to significantly reduce O&M costs. The traditional maintenance optimization methods impose stationary decision rules by including simplifications such as arranging the inspections when a damage threshold is reached or with constant intervals to make the analysis computationally tractable.

This article presents a flexible and reliable support to decision-making seeking to balance inspection, repair and failure costs leading to select the optimal maintenance strategy of one structural component by means of an innovative methodology which is capable to incorporate time-variant policies while remaining computationally tractable. The developed methodology firstly updates an accumulated fatigue deteriorating model with quantitative evidences from inspections and measurement data about mean wind speed and year-to-year uncertainty of the fatigue damage leading to a reduction of the existing epistemic uncertainties.

Afterwards, the minimum life-cycle cost policy is identified using a Partial Observable Markov Decision Process (POMDP), which is a generalization of a Markov Decision Process where the current belief of the state is employed to yield the optimal solution. The difficulties of solving a POMDP within a reasonable computational time are overcome by using a point-based algorithm where the backup of the value function is only performed to a certain number of reachable belief points at each time step. After comparing the policies obtained by the proposed method and traditional RBI procedures, it is demonstrated that the proposed methodology attains a lower maintenance cost, especially, when seasonal variations are taking into consideration. Ultimately, the proposed maintenance optimization methodology can be also used for other wind turbine components or civil structures by only incorporating the corresponding deterioration, repair and cost models.

Life-Cycle Analysis and Assessment in Civil Engineering: Towards an Integrated Vision – Caspeele, Taerwe & Frangopol (Eds)
© 2019 Taylor & Francis Group, London, ISBN 978-1-138-62633-1

Acoustic emission based fracture analysis in masonry under cyclic loading

N. Shetty & E. Verstrynge
Department of Civil Engineering, Katholieke Universiteit Leuven, Leuven, Belgium

M. Wevers
Department of Materials Engineering, Katholieke Universiteit Leuven, Leuven, Belgium

G. Livitsanos, D. Aggelis & D. Van Hemelrijck
Department of Mechanics of Materials and Constructions, Vrije Universiteit Brussel, Brussels, Belgium

ABSTRACT

Several collapses of historical masonry structures under static loading have indicated the need to understand their mechanical behavior under compression (Binda, 2008; Verstrynge & Van Gemert, 2018). The damage progression in masonry structures should be evaluated for a safe and long service life. This has led to a growing interest in laboratory experiments to evaluate the performance of masonry and its constituent materials under axial compression tests (Hilsdorf, 1969).

Using non-destructive testing (NDT) methods, the damage assessment of masonry structures under compression is still very limited. Among the available NDT techniques, Acoustic Emission (AE) sensing is one of the techniques which detects the release of high frequency waves from a localized source generated by a crack within a material (Grosse & Ohtsu, 2008).

A damage scheme was proposed in concrete by combining two AE parameters i.e., Load ratio and Calm ratio (Ohtsu, 2006). However, the application of damage assessment in masonry under cyclic compressive load is yet to be investigated. The aim of the paper is to assess the stress-induced progressive damage in masonry under cyclic compressive loads using AE analysis.

Historical masonry is usually built with soft bricks and lime-based mortar. By taking that into consideration, similar materials were chosen for this study. The masonry walls were built with clay bricks in combination with four different mortar types; cement, hybrid lime-cement, hydraulic lime and lime hydrate mortar. Two samples for each type of masonry constructed with four types of mortar were tested at the age of 28 days.

The masonry walls were subjected to cyclic uni-axial loading and unloading, where the cycle's peak stress increases from cycle to cycle. To investigate the Kaiser and Felicity effect, AE events generated in each cycle were analysed. As the stress increases from cycle to cycle, the Felicity effect was witnessed as an indication of micro fracture initiation and growth. AE events registered during the unloading phase of subsequent loading cycle justified the presence of secondary AE activity.

AE parameters, being Load and Calm ratios, were calculated for each loading cycle. The decrease in Load ratio and increase in Calm ratio indicates the damage growth. The trend of the Load-Calm ratios shows a good correspondence with the stress-strain curves. A transition towards a state of micro-structural failure ("heavy damage") was observed at normalised load levels of 35–45% of masonry with cement and hydraulic lime mortars, and at 75–80% for masonry with lime hydrate mortar. This method shows that it is a promising approach to quantify the progressive damage in masonry.

In the next step, Digital Image Correlation (DIC) will be applied to have an in-depth visual investigation into damage progression. Further research will focus on analysing additional AE parameters such as average frequency, RA value and localizing AE sources to characterize the fracture modes.

REFERENCES

Binda, L. (ed.) (2008) Learning from failure – Long-term behavior of heavy masonry structures. In: *Advances in Architecture*. Vol. 23. WIT Press, Southampton.

Grosse, C.U., Ohtsu, M. (eds.) (2008) *Acoustic emission testing – Basics for research – Applications in civil engineering*. Berlin, Springer.

Hilsdorf, H.K. (1969) An investigation into the failure mechanisms of brick masonry loaded in axial compression. In: *Designing, Engineering and Construction with Masonry Productions*, pp. 34–41.

Ohtsu, M. (2006) Quantitative AE techniques standardized for concrete structures. *Advanced Materials Research*, 13–14, 183–192.

Verstrynge, E. & Van Gemert, D. (2018) Creep failure of two historical masonry towers: analysis from material to structure. *International Journal of Masonry Research and Innovation*, 3 (1), 50–71.

A comparative study on load response of long-span bridges derived by the macro and micro scale methods

Z.R. Jin, X. Ruan & Y. Li
Department of Bridge Engineering, Tongji University, Shanghai, China

ABSTRACT

Traffic load is an essential component in the design and maintenance of bridges. Especially for those with long spans, the variability of traffic load throughout the life-cycle is too salient to be addressed by simple calculation models. Some efforts have been taken by the research community to develop new traffic models appropriate for the estimation of load response of long span bridges (Soriano et al. 2016). Simulation-based ones are promising since it can reproduce the traffic flow reliably and provide large amounts of data for the extrapolation of load response. Nevertheless, the current simulation models, i.e., the macro-scale method and the micro-scale method, cannot meet the requirement of computational efficiency and calculation accuracy at the same time. From this point of view, a study on the performance of these two methods in terms of accuracy needs to be conducted.

This paper compares the features of load response derived by simulation methods at micro scale and macro scale, respectively. The micro-scale method is developed based on stochastic traffic cellular automata, which is commonly used in transportation engineering (Ruan et al. 2017, OBrien et al. 2012). While the macro-scale method is formed by the vehicle bootstrapping, only considering the total vehicle weight exerted on the bridge at a certain time (Crespo-Minguillon & Casas 1997). These two methods are applied to the estimation of two types of load response, i.e., the mid-span displacement and the mid-span moment of the same cable-stayed bridge. The equivalent load intensity and the value of load response are analyzed statistically.

It is found that the probability distributions of equivalent load intensity of the two methods are comparable, indicating the loads applied on the bridge are at a similar level. However, when the results of the micro-scale method are taken as the bench-mark, the macro-scale method shows an obvious deviation. Therefore, it is not appropriate to use the macro-scale method alone for the estimation of load response. Moreover, the deviation varies drastically between instances of the mid-span displacement and the mid-span moment. It is

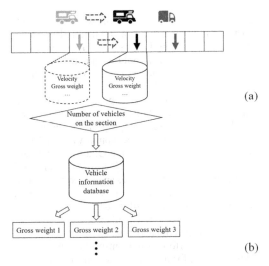

Figure 1. (a) The micro-scale method and (b) The macro-scale method.

more likely to impacted by the geometric variation pattern of the influence line. Further work still needs to be performed in this aspect.

REFERENCES

Crespo-Minguillon, C. & Casas, J.R. (1997) A comprehensive traffic load model for bridge safety checking. *Structural Safety*, 19, 339–359.
Obrien, E.J., Hayrapetova, A. & Walsh, C. (2012) The use of micro-simulation for congested traffic load modeling of medium- and long-span bridges. *Structure and Infrastructure Engineering*, 8, 269–276.
Ruan, X., Zhou, J., Tu, H., Jin, Z. & Shi, X. (2017) An improved cellular automaton with axis information for microscopic traffic simulation. *Transportation Research Part C: Emerging Technologies*, 78, 63–77.
Soriano, M., Casas, J.R. & Ghosn, M. (2016) Simplified probabilistic model for maximum traffic load from weigh-in-motion data. *Structure and Infrastructure Engineering*, 13, 454–467.

Reschedule or not? Use of benefit-cost indicator for railway track inspection

M.H. Osman & S. Kaewunruen
University of Birmingham, Birmingham, UK

ABSTRACT

A statistical approach has been extensively adopted to model railway track degradation path. Upon arrival of recent inspection data, the prior knowledge about parametric uncertainty could be updated, resulting in model calibration, and might involve condition prediction (Andrade & Teixeira 2011). A new inspection time and/or with additional inspection(s) in some parts of the published schedule might be proposed (Osman et al. 2016). However, the proposal of rescheduling action, should exhibit an economic value added to a planned maintenance program. Otherwise, it receives very little to no attention (Stenström et al. 2013). Therefore, this study proposes an easy-to-read benefit-cost indicator. This involves deciding whether an option to reschedule the affected part of the inspection schedule should be carried out or not.

Given an operational period, for instance $\ell : t_o \rightarrow t_n$, a decision to perform m track inspections under a periodic interval, for instance τ, is said to be rational if the following condition which is derived from benefit-to-cost ratio of preventive maintenance (PM) can be satisfied.

$$I(m,\tau)/C_{CM} < \alpha\varepsilon < 1 \qquad (1)$$

where $I(m, \tau)$ and C_{CM} represent an inspection cost and (estimated) cost of corrective maintenance (CM), respectively. Parameter α is probability of defect detection. Where the likelihood of track condition experiencing a sudden shift in a degradation path β is established, an absolute value of difference between β and the ratio of C_{PM}/C_{CM} defines ε.

For various selections of α and ε, a mesh plot associated with Eqn. 1 is presented in Fig. 1(a). In general, an (initial) inspection schedule with $I(m, \tau)$ positioned at higher coordinate from (α, ε)-axis has more opportunity to adjust its inspection interval. The possibility of violating the condition due to periodicity change is substantially increased for the front edge area (blue area) in Fig. 1(a).

Any suggestions for rescheduling the remaining number of inspections $\widehat{m}$ with new interval $\widehat{\tau}$ are most

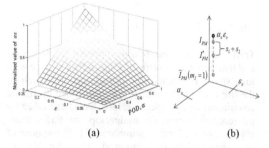

Figure 1. Indicators of feasible reschedule solutions. a) Upper bound of inspection cost, b) Permitted additional investment.

probably economically feasible as long the position of $I(\widehat{m}, \widehat{\tau})$ is below $I(\widehat{m}, \tau)$ at least by $s = s_1 + s_2$, as illustrated in Fig. 1(b). Note that, inspection reschedule requests would incur both track possession cancellation charge s_1 and reschedule fee s_2. The latter is definitely the cost of 'late' track possession order. A corresponding cost function is used to determine a refund from the cancelled possession order.

Overall, either using Eqn. 1, Fig. 1 or both tools, the proposed indicator offers an insight into the logic of rescheduling a disrupted track inspection. The indicator, however, should be used with caution because its parameters require industry values in order to clearly exhibit potential benefits of changing inspection interval and or frequency changes.

REFERENCES

Andrade, A.R. & Teixeira, P.F. (2011) Uncertainty in rail-track geometry degradation: Lisbon-Oporto line case study. *Journal of Transportation Engineering*, 137 (3), 193–200.

Osman, M.H. Bin et al. (2016) Need and opportunities for a "Plan B" in rail track inspection schedules. *Procedia Engineering*, 161 (Suppl. C), 264–268.

Stenström, C. et al. (2013) Performance indicators and terminology for value driven maintenance. *Journal of Quality in Maintenance Engineering*, 19 (3), 222–232.

Life-Cycle Analysis and Assessment in Civil Engineering: Towards an Integrated Vision – Caspeele, Taerwe & Frangopol (Eds)
© 2019 Taylor & Francis Group, London, ISBN 978-1-138-62633-1

Instrumentation, truck, track and bridge on operation railway

R. Montoya
Railway Bridge Consultant, Montoya Alves Engenharia and Pontifical Catholic University of Rio de Janeiro, Rio de Janeiro, Brazil

L. Fernando Martha
Associate Professor in Civil Engineering. Pontifical Catholic University of Rio de Janeiro, Rio de Janeiro, Brazil

J. Fernando Rodriguez, A. Merheb, A. Sisdelli & F. Masini
LSE and MRS Logistic, Juiz de Fora, Brazil

ABSTRACT

The uncertainties of the design, execution and maintenance participate in the increase of the dynamic impact on the structures. In the maintenance of the bridges we can observe that each factor is relevant on the total value of design dynamic impact. Knowing each percentage of importance makes the process of maintenance of each responsible area easier, which was initially done intuitively to be able to locate where the trigger of the problem was. Theoretically we lack parameters to obtain the variables that allow us to use an adequate analysis of reliability. However, it was necessary to approach the problem experimentally, by assembling a known vehicle, bridge and track, isolating known defect patterns, to be able to measure the relevance of each variable within the dynamic amplification. This global instrumentation project is certainly the first instrumentation on the world, isolating defects of the vehicle, track and bridge, more interestingly, the whole company managed to work together with a common purpose which was and efficient maintenance of bridges.

The understanding of dynamic impact and all uncertain seem up to now, allows us to minimize the effect of the dynamic impact, also, we can have an adequate control of all the operation and maintenance of the bridge.

We can observe that all these factors, when not adequately maintained, can overcome the standards factors.

Finally every part of the system has an interface and responsibility with the dynamic impact. Damage

Vehicle (spring, wheel and load) is most important, because they affect all the system.

Defects on the zone transition amplify the dynamics response vehicle, when the vehicle on the bridge, it request the bridge with greater energy and amplification of force. For speeds of 30 km/hr the structure, the bridge begins to feel the changes of stiffness in the zone transition, then vehicles whit defect springs, requesting the bridge with dynamic load superiors to 30% static load.

The behavior of damaged bridge is similar to spring, in this moment the spring absorbs less energy, then not produce a good interaction between bridge and vehicle.

Dynamic effects are limited for speeds less, for instance, when we have damage on bridge, we can use speed less than 15 km/hr to limit the dynamic impact.

REFERENCES

Afonso, D. (2007) *Check the Bridges Fatigue Steel Railway.*
AREMA (2013) Manual for Railway Engineering.
Esveld, C. (2001) *Modern Railway Track.* 2nd edition.
Merheb, A. (2017) *Dynamic Loads Evaluation in Existing Railroad Infrastructure Under Increasing Axle Loads and Speed.*
Montoya, R. (2009). *Cartool Fatigue.*
Montoya, R., et al. (2011) *Vale Infrastructure Guide.*
Montoya, R. (2017) Prioritization of the time rehabilitation in railway bridges based on risk and fatigue index.

Life-Cycle Analysis and Assessment in Civil Engineering: Towards an Integrated Vision – Caspeele, Taerwe & Frangopol (Eds)
© 2019 Taylor & Francis Group, London, ISBN 978-1-138-62633-1

Adaptive direct policy search for inspection and maintenance planning in structural systems

E. Bismut & D. Straub

Engineering Risk Analysis Group, Technische Universität München, Germany

ABSTRACT

Inspection and maintenance (I&M) contribute significantly to the total life-cycle cost of structures and infrastructures. Inspections reduce the uncertainty associated with deterioration processes, such as fatigue or corrosion, that might lead to structural failure. Stochastic deterioration models can be used to inform I&M strategies through risk-based inspection (RBI) planning.

The optimal decision strategy S_{opt} for I&M actions should balance the expected cost of failure with the expected I&M cost of the structure over its lifetime, so that

$$S_{opt} = \text{argmin}_S(\mathbf{E}[C_{\text{tot}}|S]), \qquad (1)$$

where $E[C_{\text{tot}}|S]$ is the expected discounted total life-cycle cost, including inspections, repairs, and possible failures of the system, for a given strategy S (Bismut et al. 2017).

A methodology for optimal I&M planning for structural systems was developed in (Luque & Straub 2017, Bismut et al. 2017), which combines a hierarchical dynamic Bayesian network (DBN) model of the deteriorating system with a direct policy search with planning heuristics to perform the I&M planning optimization, as shown in Figure 1.

The strategies are described by $\boldsymbol{\theta} = \{W_1, W_2, \dots W_m\}$, a small set of heuristic parameters. The resulting strategy $S_\theta = f(\boldsymbol{\theta})$ deterministically assigns an action for each state of the system conditioned on past observations. The I&M planning optimization in Equation 1 is thus performed on the space of strategies described by the parameters $\boldsymbol{\theta}$.

This paper extends this methodology to enable adaptive planning, whereby the initial optimal I&M plan is adapted through learning from past observations and deterioration history of the structure. This adaptive planning and allows the operator to improve the I&M strategy at any point during the life of the structure.

The benefit of adaptive planning is evaluated for a structural frame, by comparing the expected total life-cycle costs of the initial optimal strategy, defined by parameters $\boldsymbol{\theta}_0^*$, and the adapted optimal strategy, defined by parameters $\boldsymbol{\theta}_{1|z}^*$, as per Equation 2:

$$\mathbf{E}[C_{\text{tot}}|\boldsymbol{\theta}_0^*] - \left(\tfrac{1}{1+r}\right)^{t_1} \cdot \left(\mathbf{E}\left[C_{\text{tot}}|\boldsymbol{\theta}_{1|z}^*\right] + c_{ini}\right), \qquad (2)$$

where t_1 is the time of first inspection, $\mathbf{z}$ contains the inspection outcomes, c_{ini} is a fixed initial cost depending on $\boldsymbol{\theta}_0^*$, and r is the discount rate. The optimization $\boldsymbol{\theta}_{1|z}^*$ characterizes the optimal strategy for the rest of the service life, with the deterioration model updated to include the information $\mathbf{z}$.

The gain calculated in Equation 2 comes from a reduction in the uncertainty in the posterior model on the initial system state of deterioration and on the deterioration parameters.

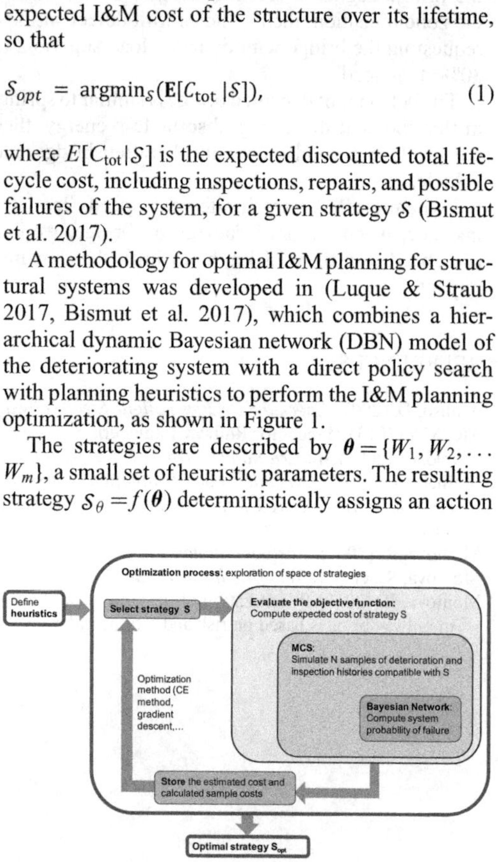

REFERENCES

Bismut, E., Luque, J. & Straub, D. (2017) Optimal prioritization of inspections in structural systems considering component interactions and interdependence. In: *12th International Conference on Structural Safety & Reliability*, Vienna, Austria.

Luque, J. & Straub, D. (2017) Risk-based optimal inspection strategies for structural systems using dynamic Bayesian networks. Submitted to *Structural Safety*.

Figure 1. Flowchart for the proposed strategy optimization method.

Life-Cycle Analysis and Assessment in Civil Engineering: Towards an Integrated Vision – Caspeele, Taerwe & Frangopol (Eds)
© 2019 Taylor & Francis Group, London, ISBN 978-1-138-62633-1

Author index

Aba, M. 472
Abdulkadeer, A.A. 125
Abi-Rached, H. 232
Abu, A.K. 359
Abualdenien, J. 113
Adey, B.T. 247
Aggelis, D. 546
Aggelis, D.G. 428
Agusta, A. 378
Agustí-Juan, I. 105, 107
Ahamed, A. 463
Ahrens, M.A. 199
Ai, Q. 281
Akiyama, M. 196
Akın, Y. 397
Al-Sabah, S. 342
Alalade, M. 217
Alexander, N.A. 497
Allacker, K. 159, 163, 166
Allah Bukhsh, Z. 266
Allaix, D.L. 80, 101
Almansour, H. 175
Almeida, N. 538
Alsem, D. 274
Amado-Mendes, P. 221
Amditis, A. 371
Anastasopoulos, D. 71
Andrews, J. 350
Andrews, J.D. 257
Ang, A.H.-S. 11
Angst, U.M. 185, 186, 241
Antonopoulou, S. 341
Araki, S. 84
Arangjelovski, T. 120
Arrigoni, A. 402, 520
Ashley, R. 137
Atanasiu, G.M. 207
Auping, W.L. 188
Aurtenetxe, J. 262
Aviso, K.B. 406

Baji, H. 328, 407
Bakker, J. 258, 267, 268, 269, 272
Baldassino, N. 388
Ballio, F. 302
Balouktsi, M. 156
Banerjee, S. 177
Banisoleiman, K. 345, 346
Barontini, A. 221
Barrias, A. 353
Basso, N. 88, 355

Bastidas-Arteaga, E. 309, 311
Bayane, I. 450
Beaudouin, F. 210
Bektas, K.E. 268, 269
Belletti, B. 197, 491
Benbow, E. 261
Benham, N. 85
Benitez, K. 59
Bergmeister, K. 234, 316
Berntatz, B. 411
Bertacca, E. 95
Bezemer, D. 532
Bianchi, S. 211
Bigaj van Vliet, A. 80
Bilotta, E. 284
Binder, F. 331
Bindu Inti, V. 162
Biondini, F. 209, 211
Birgisdóttir, H. 160
Bismut, E. 203, 550
Björnsson, I. 310
Blin, R. 303
Blomfors, M. 87
Bödefeld, J. 189
Böhms, M. 274
Bonifaz, J. 57
Borrmann, A. 113
Boschmann Käthler, C. 185
Bosschieter, C. 146
Botte, W. 467
Boumakis, I. 325
Bouwman, E.C.J. 139
Bowman, M.D. 270
Braml, T. 490
Breitung, K. 466
Bressi, S. 296
Broos, E.J. 146
Brownjohn, J. 347
Brühwiler, E. 68, 450, 451
Buckland, T. 261
Bunea, G. 207
Burggraaf, H.G. 80
Bursa, J. 64
Burtscher, S.L. 331
Byrne, G. 341

Cadenazzi, T. 520
Cai, X.S. 286
Callens, R. 403
Camarinopoulos, S. 371
Capacci, L. 209
Caprani, C. 383

Carmichael, D.G. 259
Casas, J.R. 227, 315, 316, 351, 353
Caspeele, R. 237, 467, 468, 504
Castaldo, P. 213
Castro-Fresno, D. 533
Causse, G. 349
Caverzan, A. 212
Ceprián-Pamplona, J.J. 293
Cerezo, V. 264, 296
Chacha Costa, L.L. 124
Chang, K.C. 228
Chang, Z.-M. 489
Chatzi, E. 68, 70
Chatziandreoglou, C. 371
Chatzis, M.N. 67
Chen, A. 16, 200, 248, 252
Chen, F. 328, 407
Chen, H.P. 220
Chen, J.B. 504
Chen, S. 348
Chen, X. 53
Chen, Y. 462
Chiaia, B. 134
Chijiwa, N. 479
Chlayon, T. 479
Chryssanthopoulos, M.K. 75
Churchill, C.J. 119
Cigada, A. 302
Collette, M. 368
Collin, P. 513
Connolly, L. 179
Contento, A. 308
Cooper, R. 398
Corti, D. 446
Cosma, M.P. 197
Costas de la Peña, L. 343
Courage, W.M.G. 80
Couto, R.A. 238
Craeye, B. 237, 421
Creach, A. 311
Criel, P. 399
Croce, P. 189, 312
Crotti, G. 302
Cuenca, E. 402
Cui, C. 252
Cui, F. 195
Cunha, A. 218
Curt, C. 138
Czarnecki, B. 260
Czernuschka, L. 325

D'Antimo, M. 387, 390
Daduna, H. 189
Dahmen, J. 170
Damigos, Y. 371
Damjanovic, D. 68
Dan, D. 535
Darsono, D. 74
Davies, D. 528
Dawy, Z. 232
Day, D.M. 517
De Backer, H. 200, 394
De Belie, N. 399, 401, 403
de Brito, J. 534
de Gijt, J.G. 146
de Jong, J.-J. 198
de la Fuente, A. 327
De Leon, D. 11
De León, D. 461
de Paor, C. 179
De Roeck, G. 71
De Schryver, T. 399
De Schutter, G. 237
De Silva, S. 335
De Smedt, E. 468
De Smedt, M. 423
De Troyer, F. 159
de Vos-Effting, S. 291
De Wilder, K. 423
De Winter, L. 237, 421
De Wolf, C. 528
Dean, R. 257
Debacker, W. 159
Debruyne, D. 487
Del Rosario, A. 406
Delem, L. 159, 166
Delgado, D. 461
Demirel, H.Ç. 275
Demonceau, J.-F. 387, 390
den Heijer, F. 137
Deng, P. 464
Dhakal, R.P. 359
Diamantidis, D. 95
Dimova, S. 530
Ding, L. 254
Ding, L.C. 504
Ding, Y.L. 231
Diniz, S.M.C. 238
Dinmohammadi, F. 265
Dixit, M.K. 169
Dizon, A.M. 406
Döhler, M. 379
Donev, V. 183
Dong, Y. 173
Dotelli, G. 402, 520
Dotzler, C. 488
Du, C. 222
Du, G. 376

Ebert, S. 525, 526
Edler, P. 204
Eguchi, K. 123

Elsener, B. 185
Engelen, J. 532
Erdenebat, D. 63
Erşan, Y.Ç. 397
Esteva, L. 461
Esu, O.E. 75
Evangeliou, P. 53
Evette, A. 417

Faleschini, F. 499
Fan, W. 464
Fan, W.L. 11
Fantilli, A.P. 134
Farshidpour, R. 531
Fathi, A. 380
Faulkner, K. 347
Favre, D. 168
Fernandes, J. 263
Fernando Martha, L. 549
Fernando Rodriguez, J. 549
Fernández Torres, M.A. 97
Ferrara, L. 402
Ferreira, A. 264
Fischer, K. 101
Fischer, O. 509
Fisher, C.R. 365
Flintsch, G. 264
Foestl, F. 118
Formichi, P. 312
Frangopol, D.M. 173, 193, 196, 307, 366, 367, 443
Frankenstein, B. 371
Freitag, S. 204
Friedl, N. 444
Frischknecht, R. 157, 158
Fröhle, P. 148
Frøseth, G.T. 72
Frossard, P.-A. 417
Fu, G. 335
Fülöp, L. 434
Furuta, H. 455

Galbusera, L. 212
Galvão, N. 319
Gao, X. 122
Garavaglia, E. 88, 355
Garcia de Soto, B. 247
García-Martínez, A. 161
García-Segura, T. 492
García-Sánchez, D. 262
Gardoni, P. 23, 299, 308
Gaspar, P.L. 534
Ge, X. 497
Gehlen, C. 48, 118
Geier, R. 513
Geiker, M.R. 184, 187
Gerhard, R. 371
Gersonius, B. 137
Gervasio, H. 530
Gervásio, H. 518
Geyer, P. 112

Geyer, S. 90
Ghosh, B. 301
Ghosh, J. 176, 383
Ghosh, S. 383
Ghosn, M. 195
Giannopoulos, G. 212
Gijbels, W. 421
Gino, D. 213
Giorgi, M. 168
Goi, Y. 224, 226
Gomes, J. 218
Gómez de Cózar, J.C. 161
Gonzalez Merino, A. 343
Gonzva, M. 311
González, A. 343, 345, 346
González, I. 375
Gordt, A. 371
Gose, J. 368
Graubner, C.-A. 91
Gruyaert, E. 237
Gründer, K.-P. 426
Guan, Z. 333
Guedella-Bustamante, E. 293, 524
Guerrier, S. 308
Guo, H. 464
Guo, S. 391

Haberl, A. 142
Habert, G. 107, 134, 155, 168, 401
Hafner, A. 292
Haitsma, R. 271
Hájek, P. 153
Hallmark, R. 513
Hamdan, A. 456
Hanley, C.M. 318
Hanák, M. 153
Harik, I. 61
Harter, H. 109
Hartl, T. 142
Hartmann, A. 266
Hasegawa, S. 228
Hasegawa, T. 404
Hashimoto, K. 427
Hashimoto, N. 473
Hassoun, K. 232
Hauck, M. 291
Havelaar, M. 188
Hayashizaki, N. 126
He, C. 196
He, G.Q. 537
He, Y. 122
He, Y.T. 254
He, Z. 196
Heath, A. 398
Heitner, B. 349
Henderson, A. 361
Hendy, C.R. 85
Henriques, A.A. 263
Hernandez, E.S. 62

Hertogh, M. 258, 275
Hester, D. 347
Hingorani, R. 99, 100
Ho, S.H. 121
Hoder, G. 131
Hoepfner, M. 330
Hoerbinger, S. 412
Hofer, L. 499
Hoffmann, M. 142, 183
Hoffmann, N. 389
Hofman, R. 291
Hofmann, M. 204
Holicky, M. 82
Hollberg, A. 105, 107, 155, 168
Honda, R. 208
Honfi, D. 310, 523
Honfi, D. 87
Hopkin, D. 361
Hopkin, D.J. 362
Hordijk, D.A. 54, 222, 427
Horiuchi, C. 527
Horne, I. 398
Hradil, P. 433, 434
Huang, F.K. 73
Huang, J. 283
Huang, R. 463
Hudson, J. 61
Huibregtse, J.N. 139
Huseynov, F. 347
Hüsken, G. 426

Iannacone, L. 299
Ignjatović, I. 327
Im, D. 130
Inaudi, D. 262, 303
Indacoechea-Vega, I. 533
Ishikawa, Y. 121
Ito, H. 471, 516
Ito, S. 129
Iuorio, O. 536
Ivanov, S. 513
Iwanami, M. 479
Iyoda, T. 124, 125, 128, 129, 404
Izumiya, T. 471, 516

Jacques, D. 131
Janczyk, K. 436
Jansen, B. 291
Jaspart, J.-P. 387, 390
Jaspers, W. 188, 198
Jaymond, D. 417
Jeremiah, M. 98
Jiang, X. 281
Jin, Z.R. 332, 547
Johansson, J. 310
John, V. 168
Jonkman, S.N. 101
Jordan, P. 148
Julnipitawong, P. 130
Jung, K. 86

Kácsor, E. 164
Käding, M. 58
Kadoke, D. 426
Kaewunruen, S. 445, 548
Kaito, K. 84
Kalidromitis, V. 371
Kalny, G. 411
Kamrath, P. 435, 436
Kanraj, D. 119
Kapoor, M. 382
Kapteina, G. 240
Karabulut, B. 487
Karaki, G. 503
Karaki, J. 232
Karlsson, F. 69
Karoumi, R. 375, 377
Kashani, M.M. 497
Kassir, A. 232
Kato, E. 141, 145, 475
Kato, Y. 123, 473
Katpady, D.N. 123
Katsanos, E. 382
Kawabata, Y. 141, 145, 475
Kayser, J. 150
Kazato, T. 84
Kedar, A. 320
Keijzer, E. 291
Kerkhofs, S. 294
Kerr, A. 398
Kerwin, S. 247
Kesikidou, F. 400
Keßler, S. 242
Keuser, M. 490
Khan, M.S. 383
Kibe, J.W.B. 458
Kihl, D. 365
Kikuchi, S. 121
Kim, C.W. 224, 226, 228
Kim, J. 170, 507
Kim, S. 367
Kimura, N. 225
Kioumarsi, M. 132, 133
Kiss, B. 164, 165
Kitahara, T. 219
Kjellström, E. 310
Klanker, G. 266
Klatter, H.E. 300
Klerk, W.J. 137, 140, 143
Klüber, N. 105
Kobayashi, K. 84
Köhler, J. 89, 96, 508
König, M. 108
Konuma, K. 455
Konzel, C. 142
Kotrotsiou, V. 400
Kourti, N. 212
Krause, K. 292
Krausmann, E. 212
Kravchuk, R. 429
Kreiner, H. 111
Kremer, K. 204

Krug, B. 481
Kubo, Y. 121, 126
Kubota, K. 471, 516
Kuhlmann, U. 389
Kuhnhenne, M. 436
Kuijper, R. 532
Kunz, C. 90, 144, 147
Kuriyama, H. 471, 516
Kurumatani, M. 337
Kušar, M. 319
Kušter Marić, M. 304
Kwiecień, A. 405

Laefer, D.F. 348
Lahmer, T. 217
Lam, W.C. 159
Lampalzer, T. 412, 414
Landi, F. 312
Landis, E.N. 429
Lang, W. 109, 488
Lantsoght, E.O.L. 54, 55, 59, 60
Larocque, I. 416
Larsson Ivanov, O. 87, 310, 523
Lassing, B. 149
Lasvaux, S. 168
Lau, I. 335
Lay, S. 361
Le Gat, Y. 138
Leander, J. 375, 377
Leenders, L. 532
Leendertse, W. 275
Leendertse, W.L. 294
Lehký, D. 482
Leiva, S.L. 270
Lelon, J. 428
Lemmens, L. 131
Lemmens, R.L.G. 295
Lemos, J.V. 218
Lenas, S. 371
Lener, G. 441
Leon, F. 207
Lepech, M. 184
Lepech, M.D. 187
Li, A.Q. 231
Li, C.-Q. 29, 279, 328, 335
Li, C.Q. 407
Li, F. 287, 474
Li, J. 122
Li, L. 248, 252
Li, Q.T. 360
Li, Q.Y. 167
Li, R.H. 285
Li, Y. 309, 332, 547
Li, Y.R. 73
Li, Z. 464
Liao, H. 58
Lichtenheld, T. 105
Limbeck, A. 331
Limongelli, M.P. 317, 380
Lin, P. 249
Liskounig, R. 442

Liu, B. 253
Liu, S. 131
Liu, Y. 366
Livitsanos, G. 546
Lizasoain-Arteaga, E. 533
Llorens, M. 233
López López, A. 97
Lo Presti, D. 296
Loebjinski, M. 89
Lombaert, G. 69, 487
Long, L. 379
Loshkov, D. 429
Lounis, Z. 175, 336
Loupos, K. 371
Lourenço, P.B. 221
Lu, D.-G. 194, 489
Lu, F.Y. 462
Lu, L.M. 360
Lu, P. 495
Luiten, B. 274
Lundgren, K. 87
Lupíšek, A. 153
Lützkendorf, T. 156

Ma, R. 16, 252
Maes, K. 69
Maes, M. 421
Maes, N. 131
Magalhães, F. 218
Mahnoodian, M. 279
Mai, Q.A. 545
Maier, S. 371
Malekjafarian, A. 352
Malerba, P.G. 446
Malliou, C. 371
Manchenko, M.M. 326
Mancini, G. 213
Mandic, A. 68, 316
Mandić Ivanković, A. 304
Mangina, E. 348
Mankar, A. 449
Manojlovic, N. 148
Marinković, S. 327
Mark, P. 199
Markovski, G. 120
Marková, J. 83, 86, 243
Marques, H.R. 220
Marsili, F. 189, 312
Martin, T. 437
Martí, J.V. 492
Martín-Castellote, M.S. 293
Martín-Sanz, H. 68
Martínez Otero, A.D. 352
Marx, S. 56, 58
Masciotta, M.-G. 221
Mashiri, F.R. 515
Masini, F. 549
Matis, P. 437
Matos, J.C. 263, 315, 316, 318
Matsuda, N. 125
Matsushima, I. 500

Mattern, H. 108
Matzenberger, C. 234
McGetrick, P.J. 437
McNally, C. 341, 342
Meier, T. 56
Mercier, D. 311
Merheb, A. 549
Meschke, G. 204
Mesgarpour, M. 354
Meson, V.M. 184
Mewisse, C. 210
Michel, A. 184, 186, 187
Mignon, A. 401
Milana, G. 346
Mimasu, T. 226
Minne, P. 237
Minnenbo, P. 428
Miraglia, S. 380
Miyamoto, A. 117
Mizuno, H. 128
Mizuta, M. 126
Moesl, M. 529
Mohammed, A. 175
Mold, L. 316
Molkens, T. 480
Mollaert, M. 468
Mondoro, A. 307
Montoya, R. 549
Morato, P.G. 545
Moreno, B. 233
Moss, P.J. 359
Moughty, J.J. 227, 351
Mousavi, M. 531
Müller, D. 91
Müller, H.S. 334
Mundell, C. 85
Muñoz Black, C.J. 97
Muntean, D.M. 539
Musa, I.A. 515
Mustapha, S. 232
Muttoni, A. 491
Myers, J.J. 62

Nagao, T. 495
Nagata, K. 219
Nahshon, K. 365
Nakagawa, M. 424
Nakajima, H. 208
Nakajima, S. 228
Nakanishi, Y. 129
Nakov, D. 120
Nalpantidis, L. 382
Namadmalian Esfahani, A. 531
Nanni, A. 520
Nasr, A. 310
Navarova, V. 243
Negro, P. 536
Nehasilová, M. 153
Nesterova, M. 451
Neves, A.C. 375
Neves, L.C. 257, 261, 354

Nicolai, R.P. 143
Nielsen, J.S. 381, 545
Nikitas, N. 536
Nimura, K. 404
Nishida, H. 223
Niu, X. 220
Nogal, M. 344
Nolan, S. 520
Novák, D. 481, 482, 483
Novák, L. 482, 483
Nowak, M. 508, 509
Núñez, A. 273
Nyobeu, F. 81, 543

O'Brien, E. 347
O'Brien, E.J. 349, 352
O'Byrne, M. 301
O'Connor, A. 179, 344
O'Keeffe, A. 274
O'Keeffe, E. 348
Okazaki, S. 337, 425
Okosun, F. 372
Okuma, C. 337, 425
Okutsu, M. 424
Oliveira, D.V. 263
Ongpeng, J.M.C. 406
Onishi, H. 225
Oosterveld, E. 271
Orcesi, A. 317
Osello, A. 110
Osman, M.H. 445, 548
Otake, Y. 126
Ott, S. 525, 526
Ouchi, K. 225
Ouellet-Plamondon, C.M. 170
Ožbolt, J. 304

Pacejka, H.E. 146
Padgett, J.E. 178
Paine, K. 398
Pakrashi, V. 301, 318, 321, 372
Palaniappan, S. 162
Palmer, D. 398
Pan, Z. 16
Panenka, A. 81, 543
Panesar, D.K. 119
Panetsos, P. 371
Papaioannou, I. 90
Paratscha, R. 412, 414
Pardo, S. 311
Paredes, J.E. 55
Parry, T. 261, 354
Passer, A. 111, 163
Pasternak, H. 89
Patel, R.A. 131
Paulissen, J.H. 300
Paulotto, C. 524
Pauwels, P. 295
Pavan, R. 513, 514
Pavón, S. 233
Pech Lugo, L.E. 97

Pedro, J.O. 518
Peelen, W.H.A. 300
Peiris, A. 61
Pellegrino, C. 499
Penadés-Plà, V. 492
Peng, J. 250, 498
Peng, Y.B. 504
Pereira, S. 218
Pérez Rocha, L.E. 97
Pérez-Hernando, A. 262
Perko, J. 131
Perrotta, F. 261, 354
Petraschek, T. 444
Petry, S. 490
Petzold, F. 106
Peuportier, B. 154
Phung, Q.T. 131
Pilien, V. 406
Pintor Escobar, M.M. 293
Pintor-Escobar, M.M. 524
Pipinato, A. 513
Pirskawetz, S. 426
Piton, G. 417
Pittau, F. 401
Pospíšilová, B. 153
Pot, R. 140
Pouramini, M. 531
Praxedes, C. 392
Pulakka, S. 433
Pyl, L. 468
Pyschny, D. 436

Qin, J. 376
Quirk, L. 318

Rabe, R. 81
Rados, K. 411
Rahimi, A. 127, 239
Rainieri, C. 76
Rajasingam, G.J. 444
Rama, D. 257
Ramon, D. 163
Ramos, L.F. 221
Ramos, R. 534
Rapetti, N. 110
Rashedi, R. 249
Rasmussen, F.N. 160
Rasoli, Z. 219
Rastayesh, S. 449
Rauch, H.P. 411, 412, 413, 414, 417
Ravasini, S. 491
Raymond, P. 416
Recking, A. 417
Reinhardt, M. 81
Remenyte-Prescott, R. 350
Ren, J. 167
Reynders, E.P.B. 71, 76
Rigamonti, S. 402
Rigo, P. 545
Rigueiro, C. 518

Rijke, J. 137
Robalino, A. 57
Robert, D. 279
Robinson, D. 437
Röck, M. 163
Rodriguez Burneo, A. 60
Roebers, H. 272
Röger, C. 488
Rogge, A. 203
Romano, E. 536
Rønnquist, A. 72
Roscoe, K.L. 143
Rossbacher, L. 444
Rossi, B. 487
Rossi, E. 159
Rossini, M. 520
Roth, M. 148
Roubos, A.A. 101
Roverso, G. 388
Rózsás, Á. 80
Ruan, X. 251, 254, 332, 547
Rug, W. 89
Ruiz Alfonsea, M. 161
Rychkov, D. 371
Ryjáček, P. 83

Saito, C. 424
Sakai, H. 544
Sakoi, Y. 472
Salman, A.M. 309
Salvado, F. 538
Sancharoen, P. 130
Sánchez, T. 57
Sánchez-Silva, M. 174
Sangalli, P. 415
Sanio, D. 199
Sanja, A. 68
Sanjarovskiy, R.S. 326
Sanna, C. 371
Sansom, M. 433
Santandrea, F. 523
Santos, J. 264, 296
Sap, J. 143
Sarkisian, M. 527
Sarkisian, M.P. 37
Sasaki, E. 223
Sato, T. 463, 502
Sato, Y. 329
Sauter, E.M. 295
Sawicki, B. 451
Sayers, P. 137
Schalbart, P. 154
Schalk, R. 273
Schaper, M. 148
Schelland, M. 139
Scherer, R.J. 456
Scherz, M. 111
Schins, F. 143
Schirmann, M. 368
Schmid, J. 441
Schmidt, J.-C. 148

Schmidt-Bäumler, H. 81
Schmidt-Döhl, F. 127
Schmiedel, S. 334
Schneider, R. 203
Schneider-Marin, P. 109, 488
Schnellenbach-Held, M. 114
Schoefs, F. 301, 349
Schoenmaker, R. 267
Schotte, K. 394
Schraven, D. 258
Schäffner, M.E.M. 294
Schüttrumpf, H. 150
Seeman, T. 131
Seetharam, S. 131
Sein, S. 319, 320
Semeraro, F. 110
Señís, R. 233
Serban, O. 393
Servaes, R. 159
Seumenicht, L. 148
Seyoum Lemma, M. 518
Sgambi, L. 88, 355
Shaikh, S. 148
Sharanbaswa, B. 177
Sharma, T. 398
Shekhar, S. 176
Shetty, N. 546
Shi, X. 67
Shibuya, T. 124
Shiotani, T. 427
Shoji, G. 500
Shook, D. 527
Shrivastava, M. 359
Shu, Q.J. 360
Shuku, T. 201
Sihombing, F. 496
Silva, A. 534
Simoes da Silva, L. 518
Simon, S. 142
Simoner, M. 142
Sinfield, M.F. 365
Singaravel, S. 112
Singh, M.M. 112
Sinsamutpadung, N. 223
Sisdelli, A. 549
Siviero, E. 514
Skarmoutsos, G. 389
Slobbe, A. 80
Slowik, O. 481, 483
Smutny, R. 412, 414
Snoeck, D. 399
Socorro, R. 262
Soetens, T. 421
Solorio, E. 461
Somodikova, M. 234
Šomodíková, M. 482
Song, J. 507
Sørensen, J.D. 38, 381, 449
Sorgatz, J. 81, 150
Soriano, J. 100
Sourav, M.S.N.A. 342

Sperstad, I.B. 381
Spyridis, P. 330
Staindl, U. 444
Stang, H. 184, 187
Steenbergen, R.D.J.M. 101
Stefanidou, M. 400
Steiner, D. 114
Stipanovic, I. 68, 266
Stoian, D. 535
Stoian, V. 535
Straub, D. 45, 90, 203, 508, 509, 550
Strauss, A. 197, 234, 316, 412, 414, 443, 481, 483
Stripple, H. 523
Sun, Y. 259
Sunaga, H. 126
Svendsen, B.T. 72
Sýkora, M. 83, 86, 95
Synek, J. 64
Szalay, Zs. 164, 165

Tahir, A. 144
Takahashi, H. 424
Takase, K. 223
Takeda, K. 329
Tan, L. 398
Tan, R.R. 406
Tanaka, A. 84
Tanaka, Y. 145, 475
Tanasa, C. 535
Taneja, P. 146
Tang, M.C. 3
Tangtermsirikul, S. 130
Tanner, P. 99, 100
Tao, L.J. 285
Tao, Y. 282
Tardío, G. 415
Tasaki, Y. 202
Tatsis, K. 68, 70
Tautschnig, A. 529
Tedstone, N. 257
Teferle, F.N. 63
Tehrani, F.M. 531
Teichgräber, M. 508
Teixeira, R. 344
Tekieli, M. 405
Ter-Emmanuilyan, T.N. 326
Thiel, C. 48, 118
Thöns, S. 203, 378, 379, 382
Tian, H. 248, 252
Tijssen, A. 143
Tirca, L. 393
Tiso, P. 70
Tondolo, F. 134
Torbol, M. 74, 98, 496
Torres, S. 174
Toulemonde, C. 210
Tourment, R. 138
Tošić, N. 327
Tran, N.C. 210

Treiture, R. 267
Trigaux, D. 159
Tron, S. 416
Truong-Hong, L.C. 348
Tsangouri, E. 428
Tsaoussidis, V. 371
Tschümperlin, L. 158
Tsertou, A. 371
Tsionis, G. 212
Tsukinaga, Y. 472
Tsuruta, H. 455
Tu, X. 287
Tuttipongsawat, P. 223

Ueda, N. 455
Ungureanu, V. 433, 434, 539
Uno, K. 141

Vafaeinejad, H. 132, 133
Vagnoli, M. 350
Valcke, E. 131
Vale e Azevedo, Á. 538
Valenzuela, M.A. 322
Van Belleghem, B. 403
Van Bogaert, Ph. 394
Van Coile, R. 362, 504
Van den Boomen, M. 267
Van den Heede, P. 401, 403
van der Burg, M. 513
van der Ham, H. 53
van der Hammen, J.M. 140
van der Klashorst, E. 465
van der Mark, P.J. 268
van der Meer, L. 267
van der Veen, C. 54
van der Vlist, M.J. 271
van der Voort, M. 268
van Duffelen, L.S. 198
van Es, S.H.J. 300
Van Hemelrijck, D. 428, 546
van Herk, S. 149
van Meerveld, H. 269, 291
van Nederveen, G.A. 188
van Poorten, P. 149
Van Steen, C. 422, 423
Van Tittelboom, K. 403
Vandewalle, L. 423
Vanermen, D. 421
Vann, A. 537
Vares, S. 433
Varzina, A. 131
Vecchi, F. 197, 491
Veit-Egerer, R. 64, 444
Venkatraj, V. 169
Vereecken, E. 467
Verstrynge, E. 422, 423, 546
Vidovic, A. 317, 443
Vikan, H. 187
Viljoen, C. 465
Villalba, S. 353
Viner, H. 261

Virgoe, J. 398
Vishnu, N. 178
Vogel, M. 334
Vogel, P. 155
Volker, L. 275
von der Thannen, M. 412, 413, 414
Vonk, B. 149
Vorel, J. 325

Wagemann, F. 127
Waldmann, D. 63
Walraven, M. 139, 149
Wan, C. 463
Wan-Wendner, R. 325
Wang, B. 200
Wang, D. 16, 248
Wang, G.S. 73
Wang, H. 283
Wang, J. 283
Wang, J.-S. 489
Wang, K. 280
Wang, N. 527
Wang, P. 333
Wang, W. 194
Wang, X. 282
Wang, X.J. 251
Wang, Y. 75
Wasim, M. 279
Wastiels, L. 159, 166
Wedel, F. 56
Wei, J. 464
Wei, Y. 254
Weissteiner, C. 413
Welte, T.M. 381
Wenner, M. 56, 58
Wenzel, H. 321
Werey, C. 138
Wessels, J.F.M. 268, 269
West, C.G. 519
West-Russell, M.A. 465
Wevers, M. 422, 546
Williams, M.S. 67
Windsor, R. 149
Winkler, B. 411
Winkler, J. 382
Wirges, W. 371
Wojciechowska, K. 140
Wolfert, A.R.M. 188
Wolfert, R. 267
Wu, L. 70
Wu, Y. 250, 498
Wuttke, F. 217

Xiang, D. 122
Xiang, Y. 391
Xiao, N. 462
Xie, Y. 258
Xu, H. 308
Xu, S. 16
Xu, Y. 347

Yaegashi, D. 225
Yaguchi, T. 455
Yalamas, T. 349
Yamamoto, T. 455
Yan, H. 253
Yanagi, S. 424
Yang, D.Y. 193
Yang, L. 167
Yang, W. 167, 407
Yang, X.Q. 167
Yang, Y. 53, 59, 222, 427
Yao, X. 280
Ye, Z. 286
Yepes, V. 492
Yianni, P.C. 257
Yin, Z. 250, 498
Yokota, H. 141, 145
Yoshida, H. 337, 425
Yoshida, I. 201, 202
Yoshimura, Y. 126
Yu, H.T. 284

Yuan, G.L. 360
Yuan, X.-X. 392
Yuan, X.X. 249
Yuan, Y. 281, 282, 284, 286, 287

Zahedi, A. 106
Zaimes, G. 415
Zając, B. 405
Zalbide, M. 262
Zambon, I. 443
Zandi, K. 87
Zandonini, R. 388
Zandvoort, M. 271
Zanini, M.A. 499
Zanuy, C. 99
Zaruma, S. 57
Zdanowicz, Ł. 405
Železná, J. 153
Zhang, F. 427
Zhang, J. 287, 333

Zhang, W.-H. 194
Zhang, X. 287
Zhang, Y. 248
Zhang, Z. 186
Zhang, Z.M. 284
Zhao, H.L. 284
Zhao, H.W. 231
Zhao, J. 474
Zhao, L. 463
Zhao, M. 285
Zhao, W. 253
Zhao, X. 285, 457
Zhelyazov, T. 484
Zhou, K.P. 251
Zhou, L. 250, 498
Zimmermann, T. 234
Žitný, J. 83
Zoeteman, A. 273
Zou, G. 345
Zwicky, D. 79

Life-Cycle of Civil Engineering Systems Series

ISSN Print: 2161-3907

Series editor
Dan M. Frangopol
Lehigh University, Bethlehem, PA, USA

Life-Cycle Civil Engineering: Proceedings of the International Symposium on Life-Cycle Civil Engineering (IALCCE'08), Varenna, Lake Como, Italy, 11–14 June, 2008 (2008)
Edited by: Fabio Biondini & Dan M. Frangopol
ISBN: 978-0-415-46857-2

Life-Cycle and Sustainability of Civil Infrastructure Systems: Proceedings of the Third International Symposium on Life-Cycle Civil Engineering (IALCCE'12), Vienna, Austria, October 3–6, 2012 (2012)
Edited by: Alfred Strauss, Dan M. Frangopol & Konrad Bergmeister
ISBN: 978-0-415-62126-7

Life-Cycle of Structural Systems: Design, Assessment, Maintenance and Management: Proceedings of the Fourth International Symposium on Life-Cycle Civil Engineering (IALCCE 2014), Tokyo, Japan, 16-19 November, 2014 (2014)
Edited by: Hitoshi Furuta, Dan M. Frangopol & Mitsuyoshi Akiyama
ISBN: 978-1-138-00120-6

Life-Cycle of Engineering Systems: Emphasis on Sustainable Civil Infrastructure: Proceedings of the Fifth International Symposium on Life-Cycle Civil Engineering (IALCCE 2016), Delft, The Netherlands, 16–19 October 2016 (2016)
Edited by: Jaap Bakker, Dan M. Frangopol & Klaas van Breugel
ISBN: 978-1-138-02847-0

Life-Cycle Analysis and Assessment in Civil Engineering: Towards an Integrated Vision: Proceedings of the Sixth International Symposium on Life-Cycle Civil Engineering (IALCCE 2018), Ghent, Belgium, 28–31 October, 2018 (2018)
Edited by: Robby Caspeele, Luc Taerwe & Dan M. Frangopol
ISBN: 978-1-138-62633-1